An Electrifying Introduction to ELECTROMAGNETICS

Krishna Srinivasan

First Edition

With the deepest respect to
Srinivasa Ramanujan.

Cartoons by Krishna Srinivasan. Adapted from cartoons in a handout.

Contents

Acknowledgements

I am often in disbelief over the beautiful thoughts in this book given to me. In an unusual experience, the book came to me in a flash when I was studying at the library. Unexpectedly, in a moment, the contents started falling into place. I was filled with an enormous desire to start writing this book, which took 7 years to finish. I hit upon a plan: I will write science books, and try and make science education a better learning experience.

Often times, I feel that it was my aunt, who had then passed away a short time earlier, felt my deep love of science, and illumined me with these thoughts. A distinguished researcher, an educator, and a brilliant physician, she authored many books and innumerable publications. But regardless, I owe my aunt and my uncle, Shanthi and Suresh, and graduate advisor Prof. Swaminathan, my heartfelt gratitude for fulfilling my dream of pursuing a higher education.

I am deeply indebted to all my teachers, family, relatives, and colleagues for helping me in achieving my dream of wanting to be learned, and to be well versed in arts. Above all, it has been my greatest honor to learn from all my exceptionally talented violin instructors over the past 12 years. It has been an effective way to relax.

I have compiled this book from course textbooks, and various sources that are available on the omniscient, omnipresent, and omnipotent internet. This work is possible only because of organizations, such as Internet Archive *http:\\archive.org*, who have done an enormous help to the scientific community by archiving textbooks written in the "1800s" and earlier, online. The authors who have taken the time and the effort to post tutorials and experiments on the internet, are the pillars of science and the quintessential scientist. These resources helped me feel the Spirit of science. I could tell myself, "This is what science is all about!"

This book is *not* a history textbook. There are 3 objectives to this book: (1) change the opinion of students who feel that electromagnetics is not so "cool" and boring, (2) present the artistic experiments that developed the science of electromagnetics, and (3) derive Maxwell's equations in a logical manner from these experiments, integrated with our modern understanding, making them intelligible. Newton's laws and calculus are the prerequisites for this book.

Krishna Srinivasan
2018

On the Cognition of Science
and the Science of Cognition

I always had a sense of dissatisfaction, no matter how much time I devoted to learning science. But learning how electromagnetic theory developed, documented in this book, provided me a very deep level of fulfillment. I found the proper way to learning science. My feeling of dissatisfaction has since been replaced by a feeling of euphoria.

Hardy wrote of Ramanujan: "The limitations of his knowledge were as startling as its profundity." It is my belief that it is not vast knowledge that would enable one to feel the Spirit of science, rather it is the depth in which even "little" is known. "Anyone who would track language to its lair must indeed end as omniscient.", wrote Yogananda.

Advancement of scientific knowledge is not possible without looking at the past to understand how we got to the present. Until then, one may only *know*, but not *understand*. One may know, for example, Hydrogen exists as H_2. However, one may not understand why that is the case, without learning the theory and the experiments leading to that conclusion. Science will remain cryptic until we document the relevant and necessary information from the past, and integrate them with our present understanding. The greatest tragedy that can befall upon us is experiments, and the roots of ideas, slowly forgotten.

Everyone would likely agree that science is not about memorizing equations and knowing facts. My vision for science education is to present the theory, the mathematical constructs, the experiments, and their conclusions in a logical and gradual manner, enabling one to feel the Spirit of science, rather than learn only the facts. Together we can make this a reality. If you are not already involved, you can help by picking a topic of interest that you could not follow along from your past education experience, researching and understanding it well, and publishing your findings for yours and others' benefit. Please join me in my quest on making science education a more fulfilling learning experience. Thank you.

Krishna Srinivasan
2018

1

Experiments With Charged Objects

In Richard P. Feynman's response to the question "Why do magnets attract or repel each other?", he points out, the answers to the *why* questions are never ending [1]. Modifying Feynman's response in the form of a conversation between two people A and B,

> A: I slipped on the ice and fell down.
> B: Why?
> A: Because the ice was slippery.
> B: *Why* is ice slippery?
> A: ....

There is no end to asking why, and one can never get a fulfilling answer. Lets keep this in mind, while seeking answers to the questions that we pose.

Let us begin our study of electromagnetics with electrostatics, the study of stationary charges. A few experiments with charged objects will be presented first. These experiments show that two types of charges exist in nature: positive charges and negative charges. The two mechanisms by which objects can be charged, and the apparatus to detect charged objects, will be outlined in this chapter.

1.1   Positive and Negative Charges

The experiments from Reference [2] are presented here. Two objects charged alike, repel each other. Two uncharged balloons are shown in Figure 1.1(a). When the two balloons are charged alike by rubbing each of them against hair, and when brought close together, they repel each other, as illustrated in Figure 1.1(b). The balloons in the figure have been shown as charged negative. The method to detect if an object is charged positive or negative, will be discussed later in Section 19.5. Since the balloons have been charged in the same manner, the charge that they acquire must also be of the same polarity. This experiment shows that like charges repel each other.

In the second experiment shown in Figure 1.2, a balloon has been charged by rubbing against hair. When the balloon is brought near a metal can without touching the can, Figure 1.2(a), the balloon attracts the can, and the can starts rolling down the table, Figure 1.2(b).

(a)   (b)

Figure 1.1: (a) Two uncharged balloons. (b) The balloons with like charges repel each other.

The experiment result can be explained by analyzing the charge distribution. From the first experiment, it was concluded that like charges repel each other. Since the can is attracted to the balloon, a charge of opposite polarity must have been induced in the can. This shows that there must exist charges of two polarities: positive and negative.

If the balloon is uncharged, there is no non-uniform charge distribution in the can, shown in Figure 1.3(a). When a charged balloon is brought near the can, however, a non-uniform charge distribution is created in the can, as shown in Figure 1.3(b). If the balloon has a negative charge, this repels the like charges in the can, leaving a net positive charge distribution in the can, as shown in Figure 1.3(b).

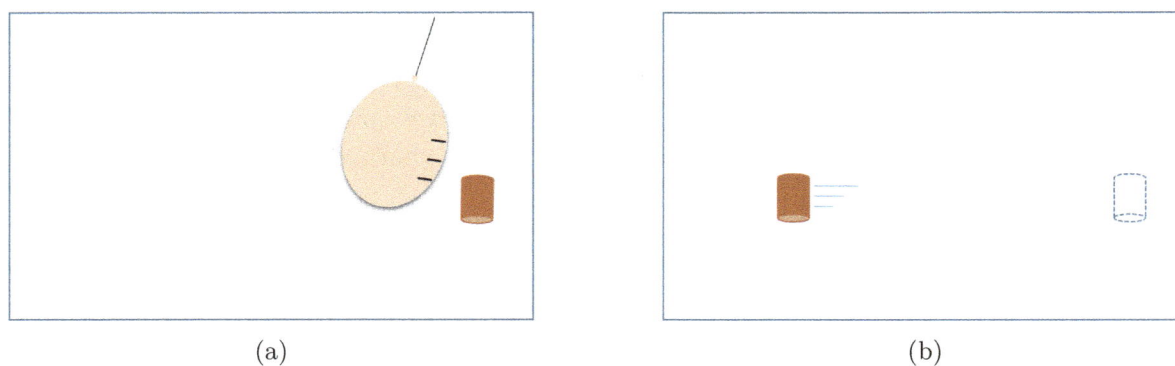

(a)   (b)

Figure 1.2: (a) A charged balloon brought near a metal can, and without touching the can. (b) The metal can is attracted to the charged balloon, and starts rolling down the table.

The positive charge distribution in the can is closer to the negative charges of the balloon, than the negative charge distribution in the can. As a consequence, overall, the force of attraction dominates between the unlike charges in the balloon and the can, compared to the force of repulsion between the negative charges in the balloon and the can. The net force is the attraction of the can to the balloon, causing it to roll down the table.

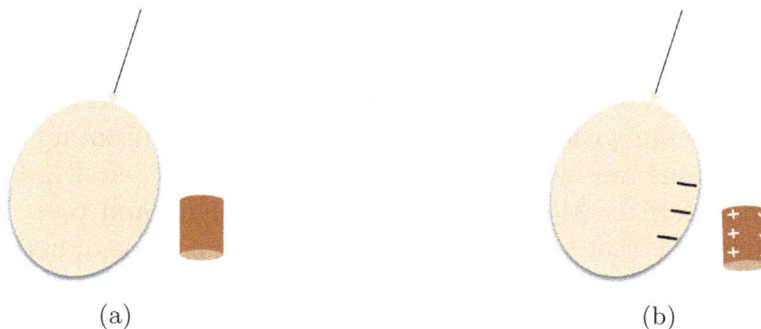

Figure 1.3: (a) An uncharged balloon near a metal can. (b) A charged balloon near the metal can, inducing a non-uniform charge distribution in the can.

A similar scenario occurs in the experiment, where a charged and an uncharged balloon are brought near a thin water stream. The water stream, shown by the dotted line in Figure 1.4, is not attracted to an uncharged balloon, shown in Figure 1.4(a). If the balloon is charged, however, the stream is attracted to the charged balloon, shown in Figure 1.4(b).

Similar to how the can was attracted to the charged balloon in the previous experiment, the thin water stream is also attracted to the charged balloon. The charged balloon repels the like charges in the water, and attracts the unlike charges. Overall, the force of attraction dominates, causing the water stream to be attracted, and bending its path.

This experiment, and numerous other experiments similar to the ones presented, show that there are two types of charge polarities: positive and negative. Like charges repel, and unlike charges attract each other.

1.2 Identity of Electricity

One can think of the identity of electricity, the flow of charges, as the flow of positive charges, or negative charges, or both. We can think of positive charges as the ones that flow, but in the opposite direction to the negative charges, without any contradiction. For example, a positively charged object can be viewed as negative charges having flowed out of the object, or positive charges having flowed in the opposite direction to the negative charges, into the object, making it positively charged, or both.

The flow of positive charges is still the convention that is followed today. The bias held towards the positive charge will become evident after the definitions of current and voltage are presented in later chapters.

The identity of electricity was established as the electron, from the work done by Heinrich Geissler, Sir William Crookes, Jean Perrin, and Sir J. J. Thomson, in 1897. The reader is referred to other resources for more details on J. J. Thomson's experiment with cathode ray tubes, and the discovery of the electron. The electromagnetic theory was fully developed by 1864, well before the discovery of the electron! From the equations and the conventions followed in electromagnetic theory, which will be presented with great clarity in this book, and from how the electrons behave in experiments, the electrons are assigned as negatively charged particles. Negatively charged objects acquire their charge from electrons transferring to them, and positively charged objects have electrons removed from them. However, since there is no contradiction in assuming that positive charges are the ones that flow, the convention has not been changed.

(a)   (b)

Figure 1.4: (a) A water stream near an uncharged balloon. (b) The water stream is attracted to the charged balloon.

1.3   Electroscope

An electroscope is an instrument to determine if an object is charged or not. Different types of electroscopes exist. Two of the common types are shown in Figure 1.5 and Figure 1.6. The electroscope in Figure 1.5 will be referred to as the needle electroscope, and the one in Figure 1.6 as the leaf electroscope.

Both these electroscopes work on the same principle. To start with, the electroscope is uncharged, and the needle is not deflected, as shown in Figure 1.5(a). If a charged object, such as an ebonite rod rubbed with wool, is brought near the needle electroscope, the like charges are repelled to the needle and the metal structure near the pivoting needle. The like charges repel each other, causing the needle to pivot about its center, as shown in Figure 1.5(b). The more charged an object is, the stronger the needle gets repelled.

The markers on the electroscope allows a way to make relative comparisons of the charge contained in different charged objects. If the charged object is moved away, the charges are no longer repelled to the needle, and the needle returns to its original unrepelled state.

If the charged ebonite rod is put in contact with the metal sphere on top of the electroscope,

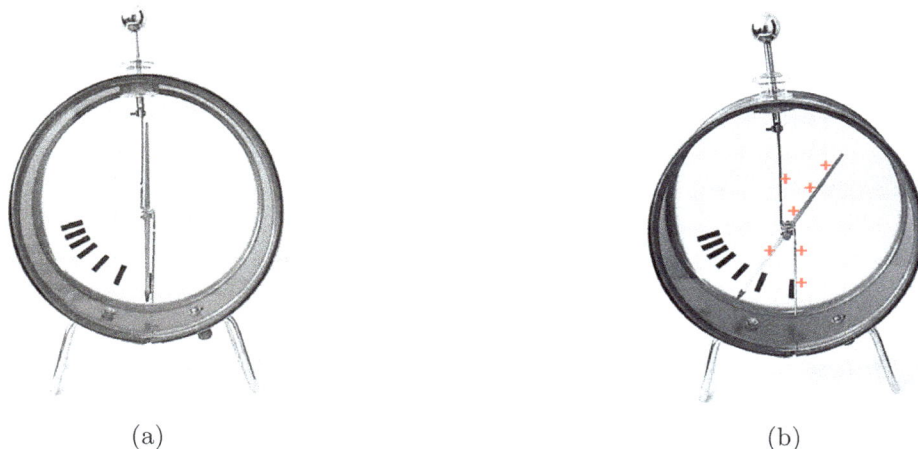

(a)   (b)

Figure 1.5: (a) A needle electroscope. (b) The needle pivots about its center, when a charged object is near, or is put in contact with the metal sphere at the top.

part of the charge in the ebonite rod gets transferred to the needle and the metal structure of the electroscope, and the needle gets repelled.

The leaf electroscope works in the similar manner. The leaf is a small thin metal sheet glued to the fixed electroscope rod. In the absence of a charged object near the electroscope, the leaf is undiverged, as shown in Figure 1.6(a). However, if a charged object is brought near the metal cap at the top of the electroscope, the like charges are repelled to the end of the electroscope, and the leaf becomes diverged, as shown in Figure 1.6(b). If the charged object is put in contact with the metal cap at the top of the electroscope, the electroscope acquires a net charge, and the leaf becomes diverged. Touching the electroscope discharges the electroscope. The excess charges flow to the ground, and the leaf becomes undiverged.

(a)   (b)

Figure 1.6: (a) A leaf electroscope. (b) The leaf diverges, when a charged object is near, or is put in contact with the metal at the top.

1.4   Charge Flow in a Metal vs a Non-Metal

From the experiments presented, it can observed that charges can move about freely in a metal. In the experiment in Figure 1.3(b), the charges redistribute in the can, causing it to be attracted to the charged object. This does not happen if the metal can is made of wood, for example. A charged object brought near an electroscope, even without any contact, repels the charges in the electroscope, causing the leaf or needle to diverge, as seen earlier. This also shows that an uncharged object contains equal quantities of positive and negative charges. Metals are also called as conductors. In insulators, also called as dielectrics, on the other hand, the charges in the material do not move freely.

1.5   Determining the Relative Polarity of Two Charged Objects

The method to determine the absolute polarity of a charged object, if it is positively or negatively charged, is presented later in Section 19.5. The relative polarity of two charged objects, Object 1 and Object 2, if they are charged alike or oppositely charged, can be determined using an electroscope and a charge scoop [3].

Figure 1.7: A charge scoop.

A charge scoop (C), shown in Figure 1.7, which is simply a piece of metal attached to an insulated handle, can be used to scoop charge by swiping its metal tip against a metal surface. Using a charge scoop, the charge on Object 1 is sampled. The charge scoop is put in contact with an electroscope, charging the electroscope with the same polarity as Object 1. If Object 1 is positively charged, for example, this case is illustrated in Figure 1.8(a).

The charge on Object 2 is then sampled using a different charge scoop. When the charge scoop is brought near the electroscope, if the leaf of the electroscope becomes more diverged, then Object 1 and Object 2 have the same charge polarity. If both Object 1 and Object 2 are positively charged, for example, as shown in Figure 1.8(b), more positive charges are repelled to the leaf of the electroscope, making the leaf more diverged. If the leaf of the electroscope becomes less diverged, then Object 1 and Object 2 have different charge polarities. If Object 1 is positively charged, while Object 2 is negatively charged, for example, the negative charges sampled from Object 2 would attract the positive charges from the electroscope leaf, making the leaf less diverged.

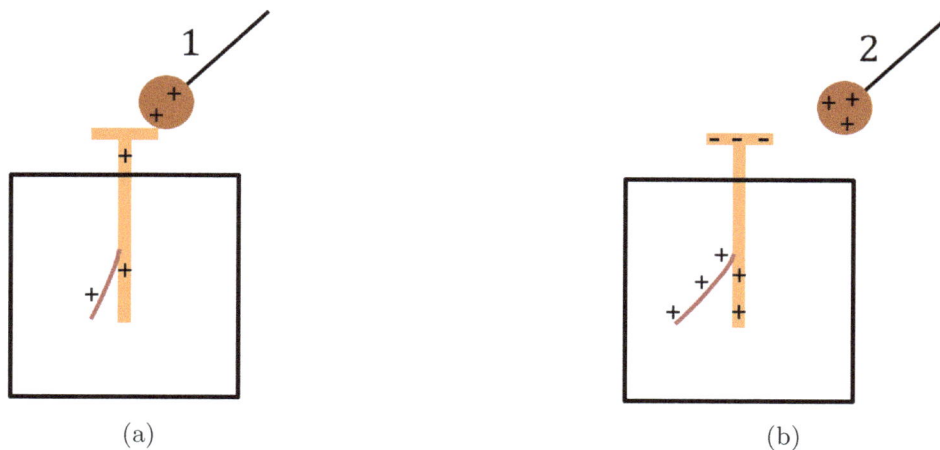

(a)   (b)

Figure 1.8: (a) Charging an electroscope with the charge sampled from Object 1. (b) The charge sampled from Object 2 is brought near the electroscope to determine the relative charge polarity of Object 1 and Object 2.

1.6 Charging by Conduction vs Charging by Induction

Charging an object by conduction requires a contact between two objects, such as rubbing an ebonite rod with wool or fur. If the charged ebonite rod touches a neutral metal sphere, the metal sphere also becomes charged. These are examples of charging by conduction.

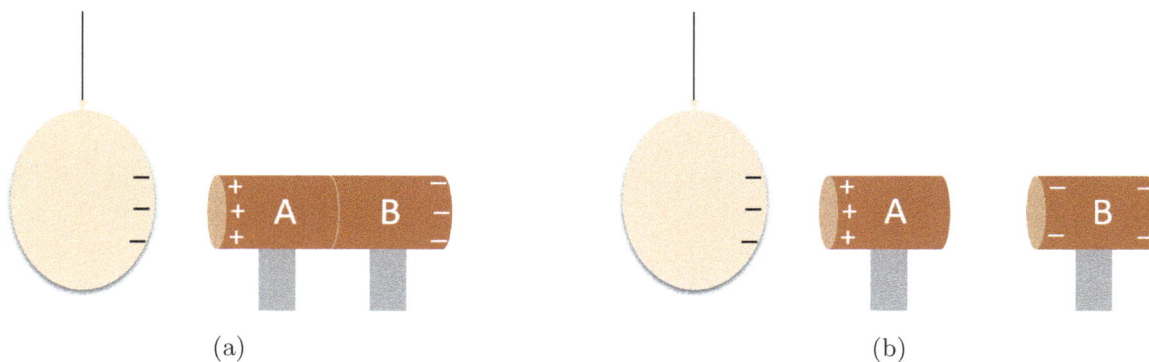

(a)   (b)

Figure 1.9: (a) A charged balloon near two metal cans joined together. (b) The cans are separated while keeping the charged balloon near the cans, and the cans are charged by induction.

Charging by induction, however, does not require a contact between two objects. An example of charging by induction is shown in Figure 1.9 [4]. A negatively charged balloon, for example, is brought near two metal cans that are in contact, as shown in Figure 1.9(a). The charged balloon repels the negative charges (or attracts the positive charges) in metal can A, making A positively charged, and B negatively charged. Keeping the balloon in its place, while moving can B away, as illustrated in Figure 1.9(b), leaves A positively charged and B negatively charged. The relative charge polarities can be determined using the technique presented in Section 1.5.

2

Systems of Units in Electromagnetics

Any physical quantity, such as length, velocity, time, etc. must be specified by a magnitude or the value of the quantity, a direction if its a vector, and the units associated with it. Without all of these properties, the value of the physical quantity is incomplete. For example, a stick of some length can serve as the definition of the unit length, with the new unit, say *stk*. The definition of the unit length as the length of the stick or its unit *stk*, both mean the same. The definition of the unit length, allows one to make measurement of length. A measured value of length will be a multiple, and/or a fraction of this stick length, say 2 *stk* or 3.5 *stk*.

In the past, the CGS system of units was widely used. CGS stands for the initials of the 3 base units: centimeters (*cm*) for length, grams (*gm*) for mass, and seconds (*s*) for time. The units of other variables are made up of the base units, and are called derived units. Several examples will be presented. The reader is referred to other books for more details on how the unit values of these base units are defined.

The *Système International d'Unités* (International System of Units), or the SI system, is the system of units used today. In the SI system of units, there are 7 base units: meter (*m*), kilogram (*kg*), second (*s*), ampere (*A*), kelvin, mole, candela. Meters, kilograms, and seconds are the counterparts of centimeters, grams, and seconds in the CGS system. In this book, the only base units used are meters, kilograms, seconds, and amperes. The definition of the ampere will be discussed in detail later in the book. The reader is referred to other books for more details on the definitions of kilograms, meters, and seconds.

By the fundamental rule of dimensional analysis, two quantities can be added or subtracted only if they have the same dimensions. It wouldn't be meaningful to add length and time, for example. However, quantities with different dimensions can be multiplied or divided. For example, if an object travels distance $\Delta \vec{x}$ *cm* in time Δt, its average velocity $\vec{v}$ is

$$\vec{v} = \frac{\Delta \vec{x}}{\Delta t},$$

(2.1)

and the unit of $\vec{v}$, denoted using square brackets $[\vec{v}]$, is

$$[\vec{v}] = \frac{cm}{s}. \tag{2.2}$$

Two quantities with different dimensions are divided to calculate the velocity.

Likewise, the average acceleration $\vec{a}$ is the change in velocity $\Delta\vec{v}$ in time Δt,

$$[\vec{a}] = \frac{cm}{s^2}. \tag{2.3}$$

From the above examples, $[\vec{v}]$ and $[\vec{a}]$ are called derived units that are made up of the base units of length and time.

In any equation, the dimensions of both sides of an equation must match. For example, the relation between meters m and centimeters cm is

$$1\,m = 100\,cm, \tag{2.4}$$

where both sides of the equation are the dimension of length.

Physical quantities are related to each other through mathematical equations. From the relations specified by mathematical equations, by defining the units of a few variables, the units of other variables get automatically defined. For example, in the made-up equation,

$$y = x^2, \tag{2.5}$$

if $[x]$ is cm, $[y]$ is the unit cm^2.

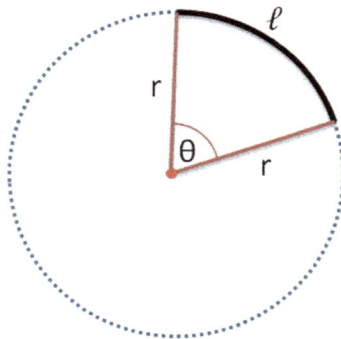

Figure 2.1: The definition of an angle in radians.

Some derived units have aliases. For example, Newton's 2nd law states that the acceleration $\vec{a}$ of an object is proportional to the force $\vec{F}$ acting on it, and inversely proportional to the object's mass m, with the proportionality constant set to the dimensionless value of 1,

$$\vec{a} = \frac{\vec{F}}{m}. \tag{2.6}$$

24

From the above derived units,

$$[\vec{F}] = \frac{gm\ cm}{s^2}. \tag{2.7}$$

The alias of the unit of force is the *dyne*, and

$$1\ dyne = 1\ \frac{gm\ cm}{s^2}. \tag{2.8}$$

Angles are always dimensionless, degrees or radians. By definition of radians, angles in radians have no units. The definition of an angle in radians θ is shown pictorially in Figure 2.1, as the ratio of the length of the arc ℓ to the radius r,

$$\theta = \frac{\ell}{r}. \tag{2.9}$$

Since the radian is the ratio of lengths, the units cancel, and is unitless, although the angle is written with the keyword *rad*. The angle of a full circle is 2π radians, which is defined as $360°$ on the degrees scale.

$$360° = 2\pi\ rad. \tag{2.10}$$

From the above equation, dividing both sides by 360, the unit degree is defined as

$$1° = \frac{2\pi}{360}\ rad. \tag{2.11}$$

Since radians is unitless, and degrees is a unitless fraction multiplier of radians, degrees are also unitless, although written with the keyword *deg* or the symbol $°$.

The argument x in functions such as $sin\,x$, e^x, etc. must be dimensionless. An easy way to understand this is that such functions can be written as an infinite series, for example,

$$e^x = 1 + x + x^2 + ... \tag{2.12}$$

In the above equation, if $[x] = cm$, then for the different powers of x, the values added are length, area, volume, etc., which has no physical meaning [5]. Therefore, the arguments of trigonometric, exponential functions, etc. must be dimensionless.

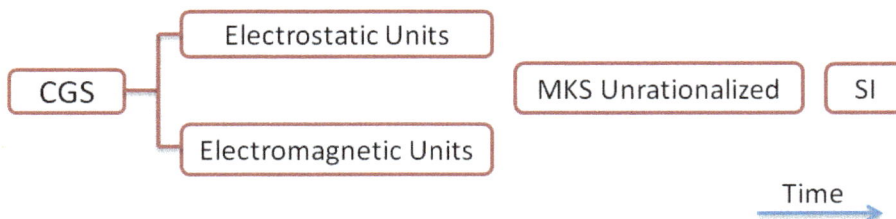

Figure 2.2: The evolution of electromagnetic units.

In electromagnetics, many different systems of units were used before the present SI units [6]. Some of these systems of units will be introduced in chronological order to describe how electromagnetic units evolved over time. This is absolutely essential to understanding electromagnetics. As shown in Figure 2.2, the electrostatic units (ESU) and the electromagnetic units (EMU) were widely used in the beginning. ESU and EMU use the CGS units for the base units of length, mass, and time. The next major system used was the MKS unrationalized units, and finally our modern SI units.

3

Coulomb's Work on the Torsion of Wires

Coulomb's work on the torsion of wires led to the development of Coulomb torsion balance, using which, he proved the force law governing electric charges. In this chapter, Coulomb's work on the torsion of wires, and its application in the construction of a torsion balance will be presented. A review of kinematics, and Newton's second law of rotation, which will help understand the equations of Coulomb's torsion balance, will be presented first. Coulomb's torsion-balance experiment will be presented in detail in Chapter 4.

3.1 A Review of Kinematics

A 1D x-axis is shown in Figure 3.1. The position of an object on the axis at $x = -2$, for example, can be described by the vector

$$\vec{s} = -2\,\hat{i}, \tag{3.1}$$

where $\hat{i}$ is the unit vector of the x-axis. In general, the position vector $\vec{s}$ can be written as

$$\vec{s} = x\,\hat{i}, \tag{3.2}$$

where x is a scalar value.

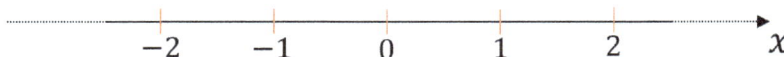

Figure 3.1: 1D x-axis.

The velocity vector $\vec{v}$ is the derivative of the position vector,

$$\vec{v} = \frac{d\vec{s}}{dt} \tag{3.3}$$

$$= \dot{x}\,\hat{i}, \tag{3.4}$$

where the first time derivative of x is denoted as $\dot{x}$. The above equation is equivalent to

$$\vec{v} = (-\dot{x})\left(-\hat{i}\right), \tag{3.5}$$

or velocity of $\dot{x}$ in the $\hat{i}$ direction, is the same as velocity $-\dot{x}$ in the $-\hat{i}$ direction.

The acceleration vector $\vec{a}$ is the derivative of the velocity vector,

$$\vec{a} = \ddot{x}\,\hat{i}. \tag{3.6}$$

In 1D problems, the unit vector $\hat{i}$ can be excluded in the position, velocity, and acceleration equations. The vector information is captured in the sign of the value. The scalar form of the equations are

$$s = x \tag{3.7}$$
$$v = \dot{x} \tag{3.8}$$
$$a = \ddot{x}. \tag{3.9}$$

In the above position equation, if x is a positive value, the direction of the position vector is to the right, and to the left, if negative. Likewise, from the above equations, this is also true for v and a.

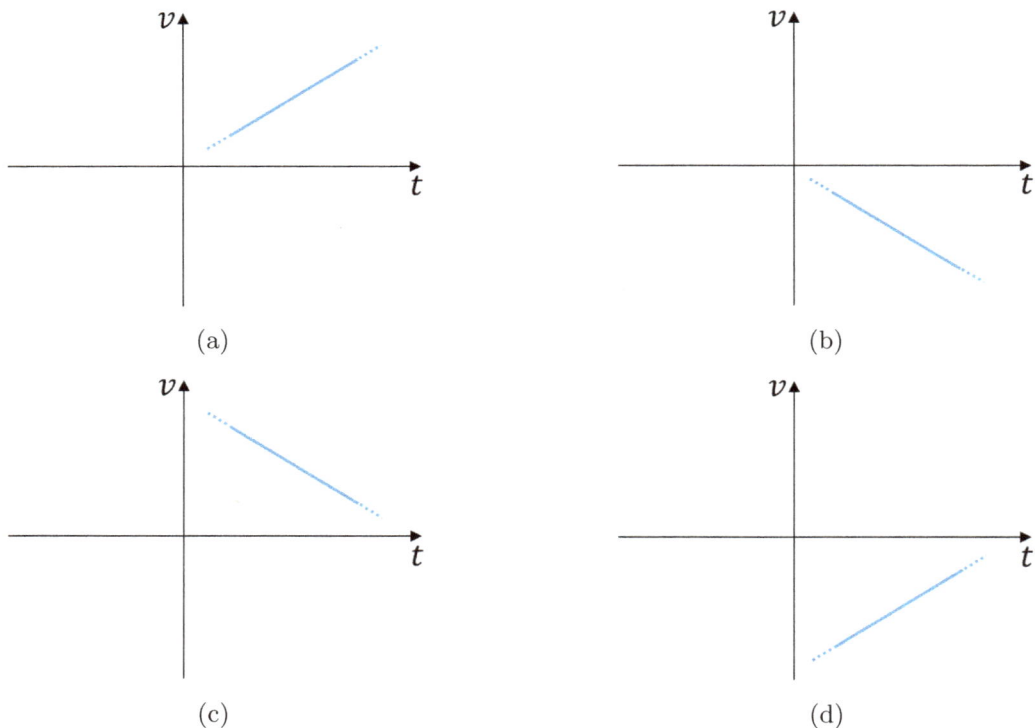

Figure 3.2: The velocity of an object, (a) positive and increasing, (b) negative and decreasing, (c) positive and decreasing, (d) negative and increasing.

If the position x increases as a function of time, the object is moving to the right. Its slope, or velocity, is a positive number. Therefore, a positive value of $v = \dot{x}$, means that the object is moving to the right. Similarly, a negative value of velocity means that the object is moving to the left.

If the velocity of the object is positive and increasing, as shown in Figure 3.2(a), the object is speeding up, and moving to the right. The slope of the velocity waveform, acceleration, is a positive number. Therefore, if the velocity and acceleration are both positive, the object is speeding up, and moving to the right. Likewise, if the velocity and acceleration are both negative, as shown in Figure 3.2(b), the object is speeding up, but moving to the left. If the signs of velocity and acceleration are opposite, as shown in Figure 3.2(c)-Figure 3.2(d), the object is slowing down in its direction of motion. The direction of motion can be obtained from the sign of velocity, as explained in the previous paragraph.

3.2   Period of Oscillation of a Torsion Pendulum

Coulomb's work on electrostatics originated from his study on the torsion of wires, or the twisting of wires. Examples of torsion pendulums are shown in Figure 3.3 [7]. If the wire is twisted an angle θ, by turning the mass attached to the wire, and releasing it, the mass keeps twisting and untwisting in an oscillating motion.

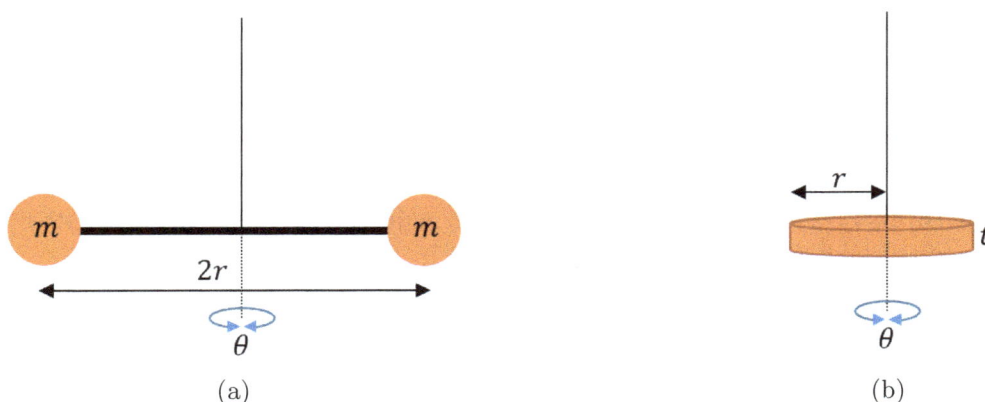

Figure 3.3: Two examples of torsion pendulums.

In differential equations, it is more convenient to work with θ in radians, rather than in degrees. The derivative of trigonometric functions,

$$\frac{d}{d\theta} \sin \theta = \cos \theta, \tag{3.10}$$

for example, is correct only if θ is in radians [8]. If θ is in degrees, there is an additional scaling factor of $\frac{\pi}{180}$ that will have to be included. To simplify the equations, θ will be in radians.

The top view of a torsion pendulum in torsional oscillation is shown in Figure 3.4. The untwisted pendulum is shown by the dotted circles. Although the figure shows acute angles of θ and

$-\theta$, it can be any value, including $\theta > 2\pi$ radians, which would be the case if the wire is twisted by more than one full rotation. In a lossless system, also referred to as the undamped system, which will be the assumption made here, the pendulum oscillates forever.

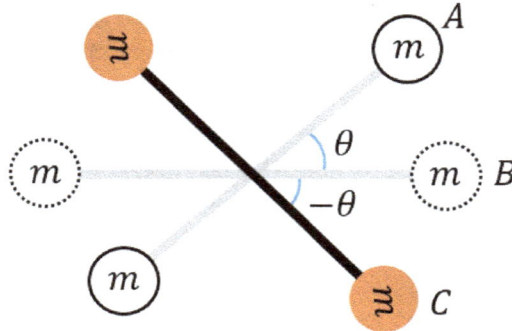

Figure 3.4: The torsional oscillation of a wire-mass system.

The counter-clockwise angles of rotation θ are chosen as positive values, and the clockwise angles are negative. This is arbitrarily chosen. The position vector of the mass is

$$\vec{s} = \theta \, \hat{u}, \tag{3.11}$$

where $\hat{u}$ represents the counter-clockwise direction unit vector. $-\hat{u}$ is the unit vector in the opposite direction to $\hat{u}$, which is the clockwise direction.

One could also think of $\vec{s}$ as the position vector on a linear θ axis, as shown in Figure 3.5. $\hat{u}$ is the unit vector in the positive θ direction, and $-\hat{u}$ in the negative θ direction.

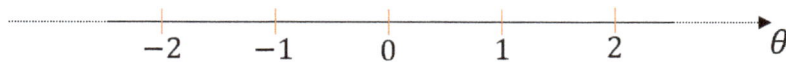

Figure 3.5: 1D θ-axis.

Angular velocity $\vec{\omega}$ is the time derivative of the position vector,

$$\begin{aligned} \vec{\omega} &= \frac{d\vec{s}}{dt} \\ &= \dot{\theta} \, \hat{u}. \end{aligned} \tag{3.12}$$

Angular acceleration $\vec{\alpha}$ is

$$\begin{aligned} \vec{\alpha} &= \frac{d\vec{\omega}}{dt} \\ &= \ddot{\theta} \, \hat{u}. \end{aligned} \tag{3.13}$$

30

Newton's 2nd law of rotational motion in vector form is

$$\vec{\tau} = I\vec{\alpha}, \tag{3.14}$$

where $\vec{\tau}$ is the torque acting on a body with mass moment of inertia I, and $\vec{\alpha}$ is the angular acceleration. The reader is referred to other resources for more details on this equation. I is always a positive value, and $\vec{\alpha}$ is in the same direction as $\vec{\tau}$.

In the 1D case, substituting Equation 3.13 in the above equation,

$$\vec{\tau} = I\ddot{\theta}\,\hat{u} \tag{3.15}$$
$$= \tau\,\hat{u}. \tag{3.16}$$

In the 1D case, as explained in the previous example, the unit vector can be excluded from Equation 3.11, and the equation can be treated as a scalar expression. The vector information is captured in the sign of θ. The positive values of θ are in the counter-clockwise direction, and negative values in the clockwise direction. From the above equations, if $\dot{\theta}$, $\ddot{\theta}$, and τ are positive values, $\vec{\omega}$, $\vec{\alpha}$, and $\vec{\tau}$ are vectors in the counter-clockwise direction, and clockwise if negative. The scalar versions of the above equations are

$$s = \theta, \tag{3.17}$$
$$\omega = \dot{\theta}, \tag{3.18}$$
$$\alpha = \ddot{\theta}, \tag{3.19}$$

and the scalar version of Newton's 2nd law of rotational motion is

$$\tau = I\ddot{\theta}. \tag{3.20}$$

Assuming that superposition of torques holds true, in the above equation, τ is the sum of all the torques acting on the system. I is always a positive value, and the direction of $\ddot{\theta}$ is the same as τ. $\dot{\theta}$ and $\ddot{\theta}$ are the equations of angular velocity and angular acceleration, regardless of whether the clockwise or counter-clockwise angle has been assigned as the positive direction.

Similar to the example in the kinematics review, if ω and α are both positive (negative), then the mass is accelerating in the counter-clockwise (clockwise) direction. If the angular velocity and angular acceleration are opposite signs, the mass is decelerating in the direction of angular velocity, where the mass is heading towards.

It can expected that the oscillation of the pendulum follows a periodic motion of twisting and untwisting of the wire. $\theta(t)$ would be a "sinusoidal" waveform. This will be confirmed in a moment. To get an intuitive understanding of the harmonic motion, an example of a sinusoidal motion will be presented first.

θ, $\dot{\theta}$, and $\ddot{\theta}$ as a function of time are plotted in the three subfigures in Figure 3.6. $\theta(t)$ is a cosine waveform, and the pendulum is twisted by an angle θ at time $t = 0$, and allowed to oscillate. $\dot{\theta}$ and $\ddot{\theta}$ can be easily calculated as the two time derivatives of the cosine waveform. The peak amplitudes of the twists in Figure 3.4 have been marked A and C.

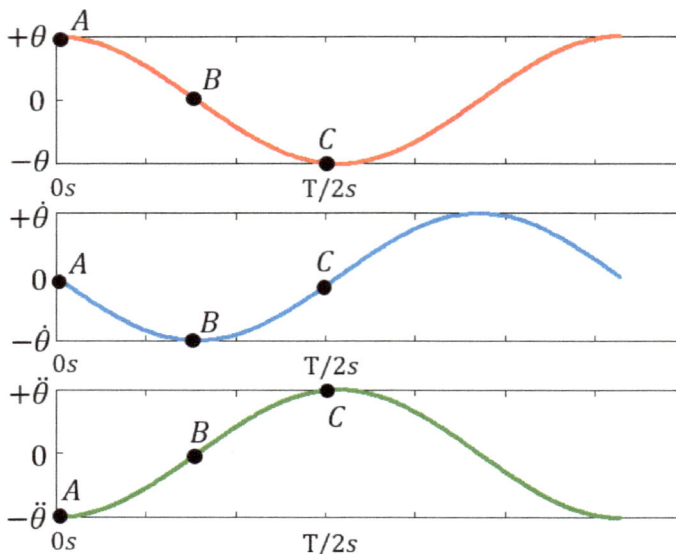

Figure 3.6: An example of the angular position, angular velocity, and the angular acceleration of a torsion pendulum.

The half period of oscillation $T/2$ for the mass to move from A to C will be analyzed. θ is positive from A to B, and negative from B to C, consistent with the convention chosen for the sign of θ. $\dot{\theta}$ is negative from A to C, which shows that the mass is moving in the clockwise direction. From A to B, the increasing negative value of $\dot{\theta}$ shows that the mass is speeding up, and slowing down from B to C. From A to B, $\ddot{\theta}$ is negative, having the same sign as $\dot{\theta}$. This means that the object is speeding up, but opposite signs between angular velocity and angular acceleration from B to C, means the object is slowing down.

When the wire is twisted by angle θ, either a positive or a negative value, and released, the wire untwists itself. This behavior is captured by Hooke's law for rotational motion,

$$\tau_r = -\kappa\theta, \tag{3.21}$$

where τ_r is the restoring torque, or the torque resulting from the wire untwisting itself when the wire has a twist of θ radians, and κ is a constant, the torsion constant, which depends upon the material and the physical properties of the wire. The above equation will be verified in a moment, from the solution of the differential equation. κ is always a positive value, and the negative sign in the above equation results in opposite signs of τ_r and θ. At any point in time, this means that the direction of τ_r is always opposite to that of the direction in which the wire has been twisted. The equation captures the behavior of the wire untwisting itself.

The solution to Equation 3.20 is $\theta(t)$, where the variation of θ as a function of time is to be determined. At any time t, when the torsion pendulum is twisted by an angle θ, there is a restoring torque acting on the system, captured in Equation 3.21. Substituting Equation 3.21 in Equation 3.20,

$$-\kappa\theta = I\ddot{\theta}. \tag{3.22}$$

If the mass is twisted by an angle θ in the counter-clockwise direction, where θ is a positive value, and released, the mass begins to accelerate in the clockwise direction. Solving for the angular acceleration $\ddot{\theta}$ in the above equation for a positive value of θ, $\ddot{\theta}$ is a negative value, or an acceleration in the clockwise direction. This is consistent with the expected behavior, and is a checkpoint to verify the setup of the above equation.

The solution to the above equation is the torsional oscillation of the pendulum back and forth,

$$\theta(t) = a\cos(\omega_o t + \phi) \tag{3.23}$$

$$\omega_o = \sqrt{\frac{\kappa}{I}}, \tag{3.24}$$

where ϕ is phase offset that can be used to set the value of $\theta(t)$ at $t = 0$, a is the amplitude of the oscillation, ω_o is the undamped, or lossless, angular frequency of oscillation. The unit of the argument of the cosine function is radians,

$$[\omega_o t] = [\phi] = rad, \tag{3.25}$$

resulting in

$$[\omega_o] = \frac{rad}{s}. \tag{3.26}$$

Since radians has no units, as explained in Chapter 2, $[\omega_o]$ can be simplified as

$$[\omega_o] = s^{-1}. \tag{3.27}$$

In Equation 3.23, if
$$\omega_o t = 2n\pi \; rad, \; n = 0, 1, 2, 3, \; ... \tag{3.28}$$

where $\omega_o t$ is an integer multiple of $2\pi \; rad$, then $\theta(t)$ would be the same value as $\theta(t = 0)$. When $t = 0$, $\omega_o t = 0$, and

$$\theta(t = 0) = a\cos\phi. \tag{3.29}$$

The next occurrence when $\theta(t) = \theta(0)$ is when $\omega_o t = 2\pi \; rad$, where

$$\theta\left(t = \frac{2\pi}{\omega_o}\right) = a\cos(2\pi + \phi)$$

$$= a\cos\phi. \tag{3.30}$$

33

This means that the period of oscillation T, the time taken for a full oscillation is

$$T = \frac{2\pi}{\omega_o}. \tag{3.31}$$

From dimensional analysis applied to the above equation, $[T]$ is seconds, since 2π has no units. The frequency of oscillation, or the number of oscillations in a second, is $f = \frac{1}{T}$ with units s^{-1}. Rearranging the above equation,

$$\omega_o = 2\pi f. \tag{3.32}$$

Substituting Equation 3.24 in Equation 3.31,

$$T = 2\pi \sqrt{\frac{I}{\kappa}}. \tag{3.33}$$

The period of oscillation can be used to empirically verify the rotational equivalent of Hooke's law that was assumed at the beginning of the derivation. Using the above equation, the torsion constant κ can be determined for a load of known moment of inertia I. Fixing κ by using the same wire, and varying the moment of inertia, Equation 3.33 can be empirically verified. This exercise can be repeated using wires of different torsion constants, implicitly proving the rotational equivalent of Hooke's law.

3.3 Torsion of Wires

Coulomb showed that κ for a wire of length l and diameter D is given by

$$\kappa = w \frac{D^4}{l}, \tag{3.34}$$

where w is a constant that depends on the material of the wire. By measuring the period of oscillation with a load of known moment of inertia, κ can be calculated from Equation 3.33. This exercise can be repeated on wires of different diameters and lengths to derive the above relation.

Note that the diameter of a wire, assuming a cylinder for the wire shape, can be calculated from its mass, volume, and the density of the wire material. Since, $\kappa \propto \frac{1}{l}$, longer length wires will result in smaller values of κ. This has the advantage of creating a greater angle of twist in a torsion pendulum for smaller applied torques, as explained next.

If an applied torque τ_a is balanced by the restoring torque τ_r of a twisted wire, from the rotational equivalent of Newton's 2nd law in Equation 3.20,

$$\tau_a + \tau_r = 0, \tag{3.35}$$

since the torsion pendulum remains stationary. Substituting Equation 3.21 in the above expression, and rearranging the equation,

$$\theta = \frac{\tau_a}{\kappa}. \tag{3.36}$$

From the above equation, smaller values of κ will result in larger values of θ for the same values of τ_a. Coulomb realized that by using a long wire, small forces, such as the force of repulsion between charged objects of the same polarity, higher turn angles θ can be obtained that would be easier to measure. This is reflected in his apparatus, which he used to demonstrate the inverse square behavior of the force between charged objects, and described in Chapter 4.

4

Coulomb's Law:
"The Fundamental Law of Electricity"

In 1785, Charles Augustin Coulomb presented to the Paris Academy of Sciences, a torsion balance using which he demonstrated what he called the "fundamental law of electricity", now known as Coulomb's law in his honor. His experiment results confirmed the force relation between electric charges.

Coulomb was a retired military engineer at the time. Between $1785 - 1791$ Coulomb read seven memoirs to the Academy: the first two memoirs are on torsion balance, the third on the leakage of electric charge, fourth-sixth on the charge distribution of conducting bodies, and the seventh memoir on magnetism. Although Coulomb is most famous for his work on electrostatics, his research work spanned a wider domain, such as mechanics, and plant physiology [9].

The motivation to surmise the force relation between electric charges will be explained first. Coulomb's experiment, and the replication of Coulomb's experiment in Reference [10], will be presented next. The principle of superposition applied to the force between electric charges will also be discussed.

4.1   Coulomb's Law in Electrostatic Units

From Newton's law of gravitation, there was a belief that a similar relationship should also hold true for the force of repulsion and attraction between electric charges. Drawing similarities to Newton's law of gravitation, Coulomb's law in ESU can be surmised to be

$$\vec{F} = \frac{q_1 \, q_2}{r^2} \, \hat{r}, \tag{4.1, ESU}$$

where $\vec{F}$ is the force between two point charges of magnitudes q_1 and q_2, separated by a distance r, and $\hat{r}$ is the unit vector in the direction of the force. q_1 and q_2 are called point charges, where charge is concentrated at a point. In a practical sense, point charges can be viewed as tiny charged metal spheres centered at the point. Although the concept of a point charge may seem abstract, surface charge density or volume charge density, are often used to represent charge distribution over an area or a volume. If charges are distributed over an area (volume), the area (volume) can be divided into smaller area (volume) elements, and the charge in each of the tiny elements can be approximated by a point charge at the center of the element. Even in such a case, where charges are distributed, point charges are helpful in analysis.

Given two tiny charged spheres, their force of repulsion or attraction is along the line joining the two spheres, which can observed from experiments. As seen in Chapter 1, in the case of like charges, the force of repulsion is shown in Figure 4.1(a), and the force of attraction between unlike charges in Figure 4.1(b).

(a)   (b)

Figure 4.1: (a) The force of repulsion between charges with the same polarity. (b) The force of attraction between charges with different polarities.

From Equation 4.1, the unit charge in electrostatic units (ESU) is defined such that two unit charges placed $1\,cm$ apart, will repel each other with a force of $1\,dyne$. The dyne was defined in Chapter 2, and its units are given in Equation 2.8, as the force needed to accelerate a $1\,gm$ mass by $1\,\frac{cm}{s^2}$. Applying dimensional analysis on Equation 4.1, the unit of charge in ESU is

$$[q] = \frac{gm^{\frac{1}{2}}cm^{\frac{3}{2}}}{s}. \tag{4.2, ESU}$$

There is no base unit for charge in ESU, rather it is a derived unit, written in terms of the base units of mass, length, and time.

To distinguish between electrostatic and electromagnetic units, the convention used is that the electrostatic units is preceded by the word *stat*, and the electromagnetic units by the prefix *ab* that stands for *absolute*. The unit charge in ESU is the statcoulomb, as opposed to the abcoulomb in electromagnetic units,

$$1\,statcoulomb = 1\,\frac{gm^{\frac{1}{2}}cm^{\frac{3}{2}}}{s}. \tag{4.3}$$

It will be explained in Chapter 57, the unit charge has different definitions in different systems of units. The unit charge in ESU is not equal to the unit charge in EMU. In general, therefore, Coulomb's law is written as

$$\vec{F} = k_e\frac{q_1\,q_2}{r^2}\hat{r}, \tag{4.4}$$

where k_e is a proportionality constant that depends on the system of units being used. The proportionality constant k_e in Equation 4.4, is equal to 1 in ESU. Coulomb's law assumes air, or strictly speaking vacuum, as the medium in which the charges are present.

Coulomb's apparatus to show the inverse-square relation in Equation 4.4 is shown in Figure 4.2. The sphere in the glass enclosure connected to Rod Q is charged, for example, by an ebonite rod that has been rubbed vigorously with fur. The charged sphere divides its charge, upon contact with a movable sphere that is attached to the edge of a torsion bar. In Coulomb's time, the small spheres were made of pith, a light and spongy tissue in the stems of large plants [11]. But in recent times, these have been replaced by spherical objects spray coated with a conductive material.

The force between the two charged spheres causes them to repel each other, and the torsion bar to rotate about its center, on the horizontal plane. The restoring force of the torsion of the silver wire is balanced by the force of repulsion of the two pith balls. The micrometer attached to the apparatus can be turned to twist the wire more, and increase the restoring force. This causes the repelled sphere on the torsion bar to move closer to the fixed sphere on Rod Q. The micrometer and the glass enclosure around the torsion bar are marked with angle between 0° to 360°. This is used to determine the angle that the micrometer has been turned, and the angle between the spheres.

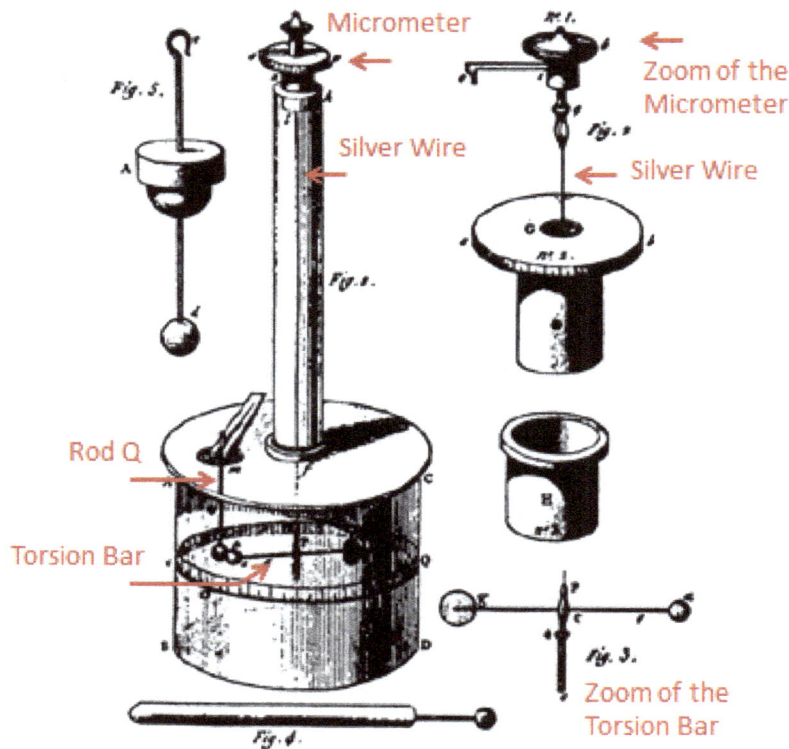

Figure 4.2: Coulomb's torsion balance of 1785 [10].

Coulomb's memoir has three measurement data points [10]. The analysis of these data points in Reference [10] will be presented next that shows that the force between electric charges must obey the inverse-square law. From his memoir,

"First Trial. Having electrified the two balls with the pinhead, the index of the micrometer pointing to 0, the *ball a* on the needle is displaced from the *ball t* by 36°.

Second Trial. Having twisted the suspended filament, by means of the *knob o* of the micrometer to 126°, the two balls approached one another and stopped at 18° of distance the one from the other.

Third Trial. Having twisted the suspended filament by 567°, the two balls approached one another to 8° and a half."

The angle of separation between the pith balls will be denoted as α, and the angle the micrometer is turned, contributing to the additional twist in the wire α_m. The total twist on the wire α_{total} is

$$\alpha_{total} = \alpha + \alpha_m. \tag{4.5}$$

From these results, Table 4.1 summarizes the results of the 3 data points.

Lets assume that the inverse-square relation in Coulomb's law is unknown. Coulomb's law is written as

$$F = k_e \frac{q_1 q_2}{r^n}, \tag{4.6}$$

where n is an unknown that is to be measured from experiment results. The force of repulsion between the charged spheres is balanced by the restoring force of the twisted wire. This can be used to derive the expression for the angle the micrometer needs to be turned, as a function of the angle of separation between the pith balls, for a specific value of n. It will be shown that for $n = 2$, the measurement results match very closely to the theoretical calculation, thereby proving the inverse-square law.

Data	α	α_m	$\alpha + \alpha_m$
1	36°	0°	36°
2	18°	126°	144°
3	8.5°	567°	575.5°

Table 4.1: Coulomb's experiment results.

A cartoon of the top view of the torsion balance is shown in Figure 4.3. The stationary pith ball is marked A and the pith ball attached to the torsion bar that rotates about O is marked B. The center-center distance between A and B is r. $\angle AOB$ is α and the length OA and OB is ℓ. The force of repulsion between A and B is F, and the normal and the parallel components of F to the torsion bar, are marked F_N and F_P.

From trigonometry, the linear center-center distance between A and B is

$$r = 2\ell \sin\left(\frac{\alpha}{2}\right),$$ (4.7)

and the force of repulsion between A and B is

$$F = \frac{k_e q_A q_B}{2^n \ell^n \left(\sin\frac{\alpha}{2}\right)^n},$$ (4.8)

where q_A and q_B are the charge in spheres A and B. Let

$$K_E = \frac{k_e q_A q_B}{2^n \ell^n},$$ (4.9)

which is a term independent of α and α_m, and is a constant value for a given value of n. From the above equations, Equation 4.8 can be rewritten as

$$F = \frac{K_E}{\left(\sin\frac{\alpha}{2}\right)^n}.$$ (4.10)

As shown in Figure 4.3, the force perpendicular to the arm of the torsion bar F_N is

$$F_N = \frac{K_E \cos\frac{\alpha}{2}}{\left(\sin\frac{\alpha}{2}\right)^n}.$$ (4.11)

The torque created by F_N on the torsion bar is balanced by the restoring torque of the twisted wire in Equation 3.21. From Newton's 2nd law of rotational motion, the torques must balance in equilibrium,

$$\ell F_N = \kappa(\alpha_m + \alpha).$$ (4.12)

Substituting Equation 4.11 in Equation 4.12, and rearranging the terms,

$$\alpha_m = \frac{\cos\frac{\alpha}{2}}{h\left(\sin\frac{\alpha}{2}\right)^n} - \alpha,$$ (4.13)

where

$$h = \frac{\kappa}{\ell K_E}.$$ (4.14)

h is independent of α or α_m, and is a constant value for a given n. Solving for h from Equation 4.13,

$$h = \frac{\cos\frac{\alpha}{2}}{(\alpha_m + \alpha)\left(\sin\frac{\alpha}{2}\right)^n}.$$ (4.15)

From the above equation, h can be calculated from one of the measured (α_m, α) data point, and a given value of n. Equation 4.13 can then be used to plot α_m as a function of α for different values of n.

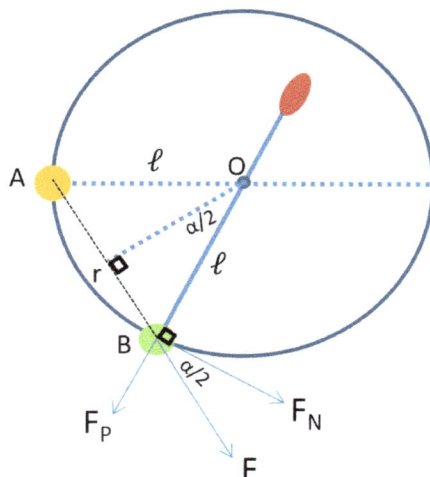

Figure 4.3: The top view of the torsion balance.

The values of h for $n = 1$, $n = 2$, and $n = 3$ have been calculated using the first data pair $(\alpha_m, \alpha) = (0°, 36°)$. The plots of α_m as a function of α for $n = 1$ (A), $n = 2$ (B), and $n = 3$ (C) are shown in Figure 4.4. The filled circles are markers representing Coulomb's data, and the diamonds are the measurement data from Reference [10] who replicated Coulomb's experiment.

The measurement results show good agreement with $n = 2$, and the experiment data from Reference [10] is in very close agreement with Coulomb's results. From the above trials, Coulomb established the inverse-square law of the force between electric charges.

Coulomb's law gives rise to additional questions. What if n is not exactly 2? Could it be that $n = 2.0001$ or $n = 1.9999$? The torsion balance most certainly is subject to experiment error and uncertainty. What if the electric charges exhibit a behavior

$$F \propto q_1^p \, q_2^p, p \neq 1? \tag{4.16}$$

The results from the torsion balance only shows the inverse-square law behavior, but does not address the above question. The above questions will be answered in Chapter 9, where Coulomb's law will be proved analytically, and established beyond any uncertainty.

4.3   Superposition of Forces Caused by Electric Charges

The principle of superposition will be assumed to be true, and states that if charges $\{q_1, q_2, ..., q_N\}$ exert forces $\{\vec{F}_1, \vec{F}_2, ..., \vec{F}_N\}$, respectively, on a charge q_0 individually, then the charges together exert a net force of $\vec{F} = \vec{F}_1 + \vec{F}_2 + ... + \vec{F}_N$ on q_0. This is one of the fundamental axioms of physics.

A force is a push or a pull, and in this regard, there is no difference between an electrical force and a mechanical one. The principle of superposition is similar to the idea in mechanics, where the mechanical forces acting on an object is equivalent to the vector sum of all the forces acting on the object. Many of the equations formulated, such as Gauss's law that will be derived later, depends on the validity of the principle of superposition. As electromagnetic theory is developed, many checkpoints are presented along the way, making sure that the formulations are correct. These checkpoints validate the assumptions made, including the principle of superposition.

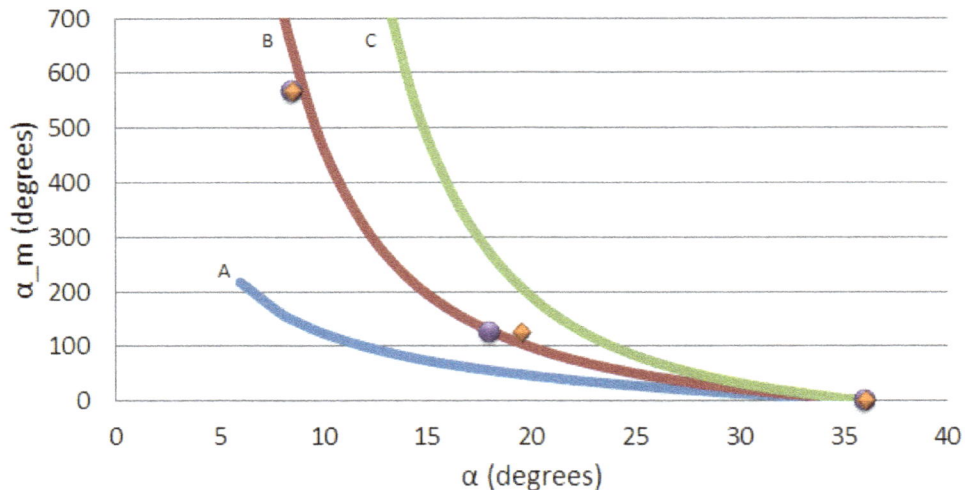

Figure 4.4: α_m vs α plots for $n = 1$ (A), $n = 2$ (B), $n = 3$ (C). The circle markers are Coulomb's original measurement data, the diamond markers are the measurement data from the reproduction of Coulomb's experiment in Reference [10].

5

Electric Field

A field can be either a vector field or a scalar field. A vector field has both a magnitude and a direction at a point. A scalar field, however, has only a magnitude defined. For example, the measurement of temperature at every point in a room will be a temperature field, which is a scalar field. A classic example of a vector field is the gravitational field $\vec{g}$,

$$\vec{g} = \frac{\vec{F}}{m},\tag{5.1}$$

which is the force acting on a unit mass at different points in space. From Newton's 2nd law, the right-hand side of the above equation is the same as the acceleration $\vec{a}$ of a free falling unit mass at different points in space, and the resulting field is the gravitational field [12].

The gravitational field around Earth is plotted in Figure 5.1. The convention used in vector-field plots is densely spaced lines indicate a greater magnitude of the field, and sparsely spaced lines indicate a weaker field in the region. The gravitational field near the surface of the Earth is higher compared to farther away, as shown by the densely spaced lines around the surface of the Earth.

The electric field is a vector field. The definition of the electric field serves the following purposes:

☐ Mathematically describe the space around charged objects.

☐ Calculate the force experienced by charges at any point, which propel them from one point to another. Unlike Coulomb's law, however, the details about the charges exerting the force is abstracted out.

If a charge q experiences the force $\vec{F}$ at a point, the electric field at the point is defined as

$$\vec{E} = \frac{\vec{F}}{q}.\tag{5.2}$$

From the above equation, the electric field at a point is the force experienced by a unit positive charge at the point. Note the similarity to the definition of the gravitational field in Equation 5.1.

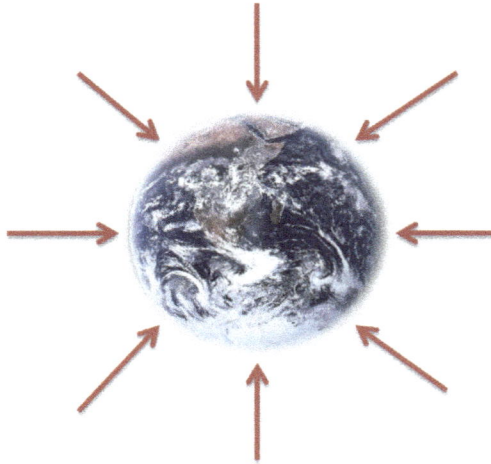

Figure 5.1: The gravitational field around Earth. [13]

An electric-field pattern $\vec{E}$ is shown in Figure 5.2. The direction of the field is shown by the arrows. Using Equation 5.2, a point charge q, shown by the solid black circle, experiences the force

$$\vec{F} = q\vec{E}, \tag{5.3}$$

in the direction shown by the dashed arrow in the figure, if q is positive.

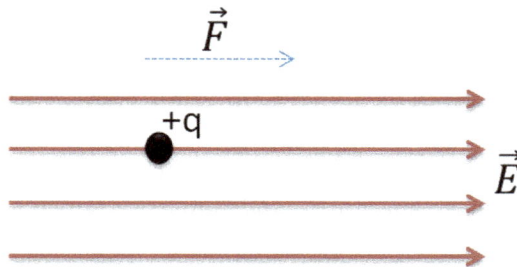

Figure 5.2: The force experienced by a charge in an electric field.

In ESU, the unit of force is the dyne, and the unit of charge is the statcoulomb. Applying dimensional analysis on Equation 5.2, the unit of electric field, denoted by $[\vec{E}]$, is

$$[\vec{E}] = \frac{gm^{1/2}}{cm^{1/2}\,s}. \tag{5.4, ESU}$$

5.1   Examples of Electric-Field Patterns

The electric field at distance r from a point charge q can be calculated from Coulomb's law, and is the same as the force exerted on a unit positive charge placed at the point,

$$\vec{E} = \frac{q}{r^2}\,\hat{r}, \tag{5.5, ESU}$$

44

where $\hat{r}$ is the unit vector from q to the point at which the electric field is calculated. Assuming that the point charge is positive, the electric-field pattern around the point charge is shown in Figure 5.3(a). The cross section of a sphere with radius r, is shown by the dotted circle. The electric field exhibits spherical symmetry, since the field has the same magnitude at all points on a sphere centered at q.

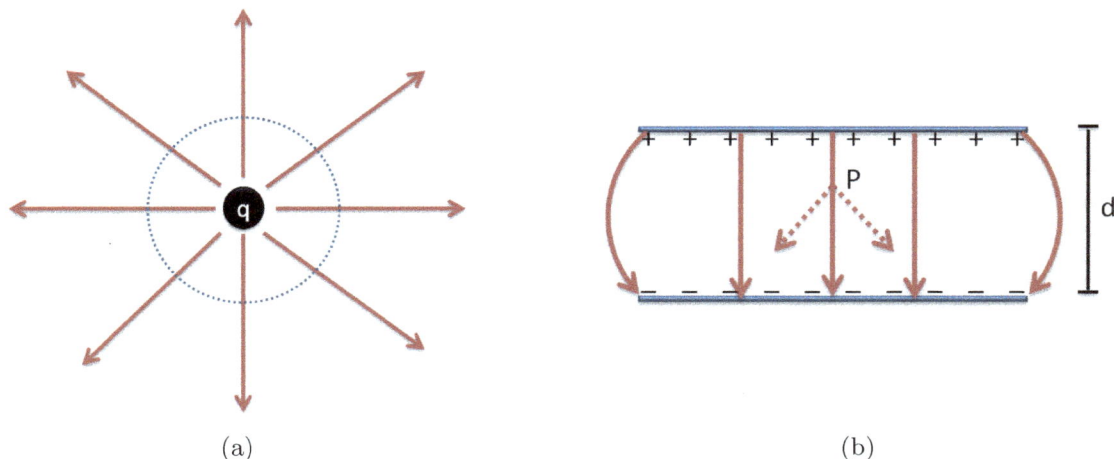

Figure 5.3: (a) The electric-field pattern around a point charge. (b) The electric-field pattern of a parallel-plate capacitor.

The second example is the electric-field pattern between two parallel plates, charged with opposite polarities, shown in Figure 5.3(b). The top plate is charged positive, and the bottom negative. It will be shown by experiments in Chapter 8, the excess charges in a conductor lie on the outer surface of the plates, and none within the metal. Since unlike charges attract, assume that all the charges lie on the inner faces of the metal plates, as shown. A uniform surface charge density on both the plates will be assumed for simplicity.

The electric-field pattern can be plotted by noting the direction of the force exerted on a unit positive charge. If the separation of the plates d is very small relative to the plate dimensions, the plates can be viewed as infinite planes. If a unit positive charge is placed at Point P, away from the edges of the plates, the forces on the charge by the ones on the plates is illustrated by the dotted vectors. By symmetry, the horizontal components of the forces cancel, and the resulting field is vertical at the center region of the plates, as shown by the solid lines. The uniform electric field assumption at the center region of the plates, away from the edges, will be verified mathematically in Section 34.1. On the edges of the plates, the field lines are not vertical, but form fringing fields, which can be noted from the direction of the force on a unit positive charge.

Alternately, the electric-field pattern can be obtained by numerically solving for the electric field generated by a positive and a negative sheet of charge of uniform charge density, representing the plates. The plates are discretized into patches of areas, small enough that the force on the unit charge at P, by the charges on each of the patches, can be approximated to be uniform. Using superposition, see Section 4.3, the field generated by each of the patches are added together, to calculate the electric field at various points, and to construct the field pattern.

The Conservative Property of the Electrostatic Field

A vector field $\vec{E}$ is a conservative field, if the path integral of a closed loop C is always 0,

$$\oint_C \vec{E} \cdot \vec{dl} = 0. \tag{6.1}$$

It will be proven in this chapter, using the law of conservation of energy, the electric field in electrostatics must be conservative.

6.1 Review of Work and Energy

By definition, the work W done by a force $\vec{F}$, in moving an object from Point A to Point B is

$$W = \int_A^B \vec{F} \cdot \vec{d\ell}, \tag{6.2}$$

where the integral is evaluated over the path in which the object is moved.

In any action or reaction, there is an entity called energy that is always conserved. This is known as the law of conservation of energy, a fundamental axiom of physics: energy is neither created, nor destroyed, but only converted between different energy forms. The reader is referred to other books for more details on work and energy. A brief review is presented here.

An object of mass m, shown in Figure 6.1, is moved in the vertical direction. $\vec{F}_a$ is the applied force, $\vec{F}_f$ is the force of friction acting on the mass during its motion, and $\vec{F}_g$ is the force of gravity. The object is moved up from A to B in Figure 6.1(a), and down from B to A in Figure 6.1(b), at a constant velocity. The object is at rest, at its initial and final locations. Lets calculate the work done by the applied force $\vec{F}_a$ in both these cases.

In Figure 6.1(a), the work W_{AB} done by $\vec{F}_a$ in moving the object from A to B, is the sum of the work done to move the object along each of the N tiny segments Δy making up the path,

$$W_{AB} = \sum_{i=1}^N \vec{F}_a \cdot \vec{\Delta y_i}, \tag{6.3}$$

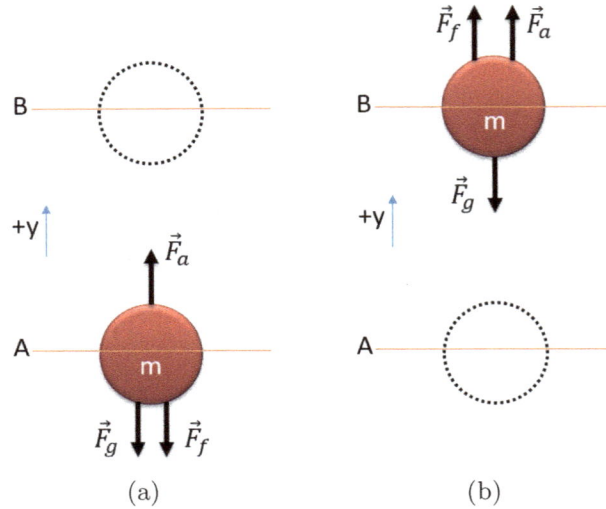

Figure 6.1: An object is moved in the vertical direction from (a) A to B, and (b) B to A.

where $\vec{\Delta y}_i$ is the ith segment, which points in the direction in which the object is moved along the segment. In the limit as $N \to \infty$, the above Riemann sum becomes the integral,

$$W_{AB} = \int_A^B \vec{F}_a \cdot \vec{dy}.$$ (6.4)

Using Newton's 2nd law of motion, if the object is moving at a constant velocity, the sum of the forces acting on the object must be 0,

$$\vec{F}_a + \vec{F}_g + \vec{F}_f = 0,$$ (6.5)

Rearranging the above equation, the applied force needed to move the object up is

$$\vec{F}_a = -\vec{F}_g - \vec{F}_f,$$ (6.6)

which is the sum of the forces to overcome gravity and friction. In the formulation of Equation 6.6, the positive work done to accelerate the mass from rest, at the starting point A to the steady velocity during transit, cancels the negative work done to decelerate the mass to rest, at the ending point B. This will be excluded from the work equation, since both the values cancel each other. Alternately, it can be assumed that the mass is moved very slowly, and the work to accelerate (decelerate) the mass from (to) rest can be assumed to be negligible.

From the above equations,

$$\begin{aligned} W_{AB} &= -\int_A^B \vec{F}_g \cdot \vec{dy} - \int_A^B \vec{F}_f \cdot \vec{dy} \\ &= \int_A^B F_g\, dy + \int_A^B F_f\, dy, \end{aligned}$$ (6.7)

48

where $\vec{F}_g = -F_g\,\hat{y}$, $\vec{F}_f = -F_f\,\hat{y}$, and F_g and F_f are positive values. The terms in the above equation are positive, and there is positive work done by $\vec{F}_a$ in moving the object from A to B.

Rewriting the above expression,

$$W_{AB} = \int_A^B F_g\,dy + E_{loss}^{AB}, \tag{6.8}$$

where W_{AB} is the sum of the energy spent to overcome the force of gravity to move the object up, and any loss incurred E_{loss}^{AB} during the transit, such as friction. From this example, it can be noted that positive work by an applied force is energy spent or transferred to an object, increasing its potential energy.

The work W_{BA} done in moving the object from B to A is

$$W_{BA} = \int_B^A \vec{F}_a \cdot \vec{dy}. \tag{6.9}$$

The applied force $\vec{F}_a$ to balance the other forces, so that the object is moved at a constant velocity is

$$\vec{F}_a = -\vec{F}_g - \vec{F}_f. \tag{6.10}$$

From the above two equations,

$$\begin{aligned} W_{BA} &= -\int_B^A \vec{F}_g \cdot \vec{dy} - \int_B^A \vec{F}_f \cdot \vec{dy} \\ &= -\int_A^B F_g\,dy + \int_A^B F_f\,dy, \end{aligned} \tag{6.11}$$

where $\vec{F}_g = -F_g\,\hat{y}$, $\vec{F}_f = F_f\,\hat{y}$, and F_g and F_f are positive values. As before, the above relation can be written as

$$W_{BA} = -\int_A^B F_g\,dy + E_{loss}^{BA}. \tag{6.12}$$

Without $\vec{F}_a$, the object would gain kinetic energy as it accelerates from B to A. $\vec{F}_a$ keeps the object moving at a constant velocity.

The first term in Equation 6.12 is a negative number. From the standpoint of the person doing the work, negative work is a gain in energy. If no work is done, and if the object is allowed to free fall, energy is gained as kinetic energy of the accelerating object. As before, from the above equation, the work done is the sum of the energy spent to overcome the force of gravity, and any loss in energy incurred E_{loss}^{BA}, such as friction.

The total work done in moving the object up from A to B, and down from B to A is

$$W_{AB} = \int_A^B F_g \, dy + E_{loss}^{AB} - \int_A^B F_g \, dy + E_{loss}^{BA}$$
$$= E_{loss}^{AB} + E_{loss}^{BA}. \tag{6.13}$$

From this example, it can be seen that the net work done to overcome the force of gravity cancels. The energy spent to increase the gravitational potential energy of the object, when it is raised, is equal to the energy gained when the object is brought down. The net positive work done is the energy spent on the loss incurred during the transit, which is converted to other forms such as heat due to friction. The energy accounting is properly balanced, and there is no unexplained gain or loss of energy.

6.2   Work Done by an Applied Force to Move a Charge in an Electric Field

The electric field $\vec{E}$ in Figure 6.2(a), is generated by some charge distribution, but satisfying the electrostatic condition that the charges are stationary. Lets do a thought experiment to calculate the work done by an applied force in moving a charge q along path C, from Point A to Point B, shown by the dotted line [14]. The point charge q will be viewed as a very tiny charged metal sphere. The path C is divided into N small segments, each of which is small enough to be

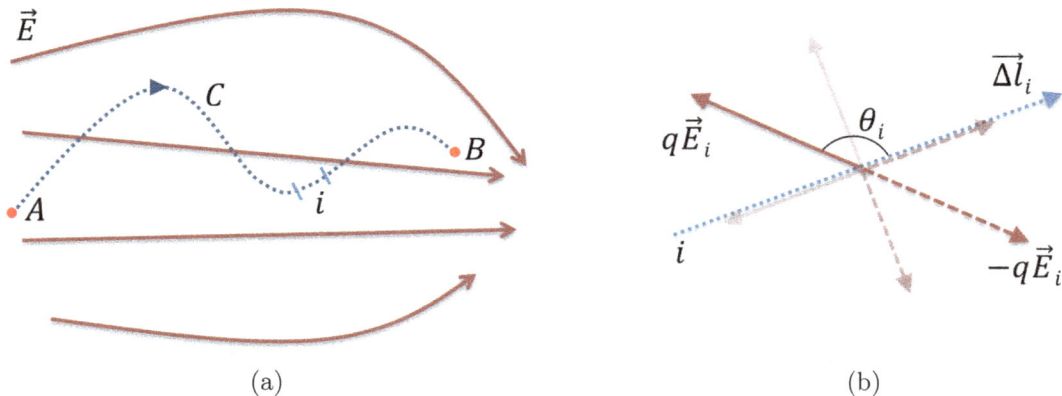

Figure 6.2: (a) An electric-field pattern, and a contour C to calculate the path integral. (b) The electric field at the center of the ith segment of the contour C.

approximated as a linear segment, and the electric field at the midpoint of the segment can be assumed to be uniform in the segment. The charge is moved at a constant velocity in each of the segments. The charge is at rest at the starting point of each of the segments, is accelerated to reach the uniform velocity, and then decelerated, before coming to rest at the end of the segment. This is repeated in each of the segments to traverse the path.

Similar to the example in Section 6.1, the positive work done to accelerate the charge from its rest state to uniform velocity at the beginning of the segment, is cancelled by the negative work to decelerate the charge at the end of the segment, before coming to rest. The two terms cancel each other, and will not be included in the work equation.

The ith segment in Figure 6.2(a) is drawn in Figure 6.2(b), and represented by the vector $\vec{\Delta\ell_i}$, whose magnitude is the length of the segment, and its direction is along the path traversed in the segment. $\vec{E}_i$ is the electric field at the ith segment. $\vec{E}_i$ forms angle θ_i with $\vec{\Delta\ell_i}$.

Using Equation 5.2, the force $\vec{F}$ exerted on the charge q in the ith segment is

$$\vec{F} = q\vec{E}_i. \tag{6.14}$$

The force is separated into two components, one parallel to $\vec{\Delta\ell_i}$, and the second perpendicular to it, as shown in the figure. By Newton's 2nd law of motion, the applied force to move the charge at a constant velocity must balance these force components, shown by the dashed line. Therefore, the applied force $\vec{F}_a$ must be

$$\vec{F}_a = -\vec{F} \tag{6.15}$$
$$= -q\vec{E}_i, \tag{6.16}$$

to move the charge at a constant velocity along the segment.

The work done by the applied force to move the charge along the path, is the sum of the work done to move the charge along each of the segments making up the path,

$$W_C = \sum_{i=1}^{N} -q\vec{E}_i \cdot \vec{\Delta\ell_i}$$
$$= \sum_{i=1}^{N} -qE_i\Delta\ell_i \cos\theta_i. \tag{6.17}$$

In the limit as $N \to \infty$, the above Riemann sum becomes the integral,

$$W_C = \int_C -q\vec{E} \cdot \vec{dl}. \tag{6.18}$$

The applied force to overcome the force of gravity of the charged sphere, would be a second term added to the right-hand side of the above equation. However, for simplicity, this is not taken into account. The gravitational field and the electric field are two independent fields, and can be analyzed separately. Although not presented in this chapter, a proof similar to the one presented here, can be used to show that the gravitational field is a conservative field.

6.3   Electric Potential Energy

A uniform electric-field pattern $\vec{E}$ of strength E in the horizontal direction, marked by the arrows, is shown in Figure 6.3. The force exerted by $\vec{E}$ on the charge q is

$$\vec{F} = q\vec{E}, \tag{6.19}$$

using the definition of the electric field in Equation 5.2. If a charge q is moved from Point B to Point A in the horizontal direction at a constant velocity, antiparallel to $\vec{E}$, using Newton's 2nd law of motion, the applied force is

$$\vec{F}_a = -q\vec{E}, \tag{6.20}$$

which is the force needed to counter balance the force on the charge by the electric field, as explained earlier.

Figure 6.3: A uniform electric field $\vec{E}$.

Using Equation 6.18, the work done to move the charge is

$$W_{BA} = \int_B^A \vec{F}_a \cdot \vec{d\ell},$$
$$= qEd, \tag{6.21}$$

where d is the distance between B and A. For example, say if q is positive, the work done in moving the charge is positive. Positive work is the energy spent by the applied force. The energy spent is converted to electric potential energy. This is similar to the work done to raise an object to a higher level in the previous example, where energy spent is converted to gravitational potential energy. Similar to the gravitational potential energy, the electric potential energy is a function of the location of the point charge in space.

If the charge is moved from Point A to Point B, the work done W_{AB} is

$$W_{AB} = -qEd. \tag{6.22}$$

The work done is negative, and this means that energy is gained by the applied force. Similar to the example in the previous section, when the object is brought down from B to A, if no work is done, then the charge accelerates due to the force of the electric field, and energy is gained in the form of kinetic energy.

The net work W done by an applied force in moving the charge from B to A, and then A to B is

$$W = W_{BA} + W_{AB}$$
$$= 0. \tag{6.23}$$

This shows that there is no net gain or loss of energy, when the charge is moved in a closed loop. The energy spent by the applied force in moving the charge from B to A is gained when the object is moved back from A to B.

6.4 Path Independence in a Conservative Field

If a vector field is conservative, it will be shown that the path integral of the field, or the work done by an applied force, depends only on the start and end points, and does not depend on the path taken. This is first shown by an example [15], followed by a simple proof.

The work W done by an applied force $\vec{F}_a$, in moving a charge q', between Point A and Point B, in the electric field $\vec{E}$ generated by a point charge q, is

$$W = \int_A^B \vec{F}_a \cdot \vec{dl}. \tag{6.24}$$

Substituting Equation 6.20 in the above expression,

$$W = -q' \int_A^B \vec{E} \cdot \vec{dl}. \tag{6.25}$$

It will be shown that the work done in moving the charge q' from A to B, in the two different paths shown in Figure 6.4, yield the same result.

The work done by the applied force on Path 1 in Figure 6.4(a), can be written as the sum of the work done along segments AC and CB,

$$W_{Path1} = -q' \int_A^C \vec{E} \cdot \vec{dl} - q' \int_C^B \vec{E} \cdot \vec{dl}. \tag{6.26}$$

Since $\vec{E}$ is always perpendicular to $\vec{dl}$ along path AC, the first integral in the above equation reduces to 0, and simplifying the above integral to

$$W_{Path1} = -q' \int_C^B \vec{E} \cdot \vec{dl}. \tag{6.27}$$

Similarly, Path 2 can be separated into the work done along segments AD, DE, EF, and FB, as shown in Figure 6.4(b).

$$W_{Path2} = -q' \int_A^D \vec{E} \cdot \vec{dl} - q' \int_D^E \vec{E} \cdot \vec{dl} - q' \int_E^F \vec{E} \cdot \vec{dl} - q' \int_F^B \vec{E} \cdot \vec{dl}. \tag{6.28}$$

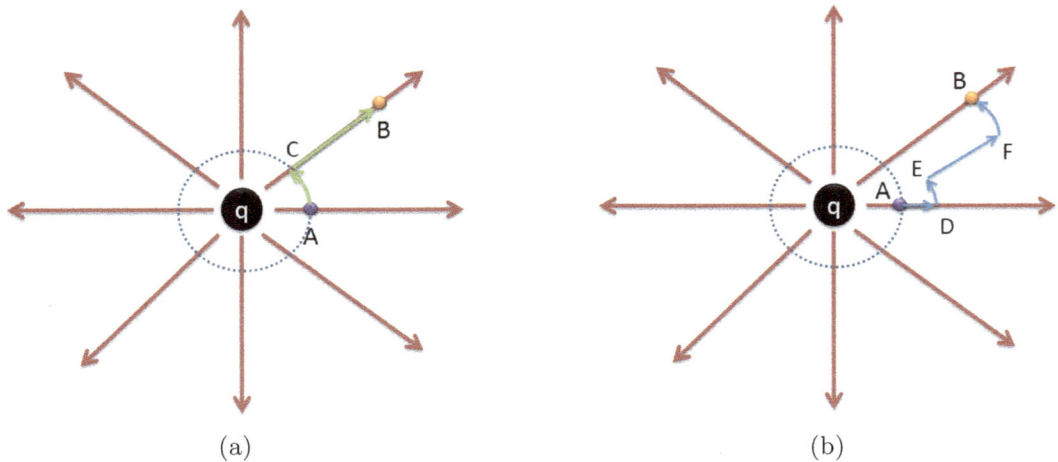

Figure 6.4: Two of the many possible paths from Point A to Point B: (a) Path 1, (b) Path 2.

Along Path 2, segments DE and FB are arcs on a circle centered on charge q. Since $\vec{E}$ is directed radially outward, $\vec{E} \cdot \vec{dl}$ is 0 along paths DE and FB. The work done along Path 2 can be simplified to

$$W_{Path2} = -q' \int_A^D \vec{E} \cdot \vec{dl} - q' \int_E^F \vec{E} \cdot \vec{dl}. \tag{6.29}$$

By radial symmetry, the sum of the line integrals along paths AD and EF in Figure 6.4(b), is the same as the path integral on CB in Figure 6.4(a). Therefore,

$$W_{Path1} = W_{Path2}. \tag{6.30}$$

From this example, visually, it can be noted that a vector field can be identified as conservative, if it is "radial", as shown in Figure 6.5(a), and non-conservative if the field is swirling around, as shown in Figure 6.5(b). If the field is "swirling", it can be expected that

$$\oint \vec{E} \cdot \vec{dl} \neq 0. \tag{6.31}$$

Intuitively, it can be seen that a closed path integral, shown by the dotted circle in Figure 6.5(b) for example, is clearly non-zero, since $\vec{E}$ and $\vec{dl}$ are always parallel (anti-parallel) in the clockwise (counter clockwise) path of integration on the dotted circle. In Figure 6.5(a), however, the circular dotted path is orthogonal to the field, and the path integral is zero.

It will be formally shown by a simple proof, if a field is conservative,

$$\oint \vec{E} \cdot \vec{dl} = 0, \tag{6.32}$$

54

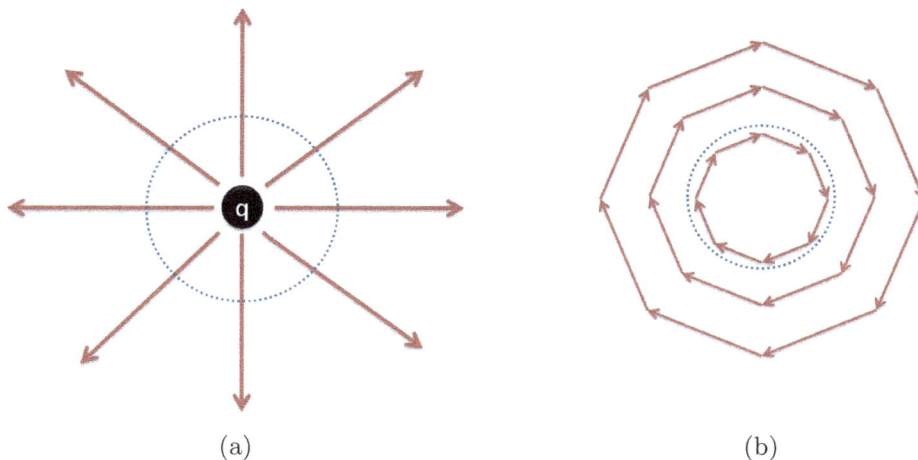

(a) (b)

Figure 6.5: Examples of vector fields in electromagnetics: (a) a conservative field, (b) a non-conservative field.

then this implies path independence. The above integral can be divided into two parts,

$$\oint \vec{E} \cdot \vec{dl} = \int_{Path\ 1} \vec{E} \cdot \vec{dl} - \int_{Path\ 2} \vec{E} \cdot \vec{dl}$$
$$= 0, \tag{6.33}$$

where Path 1 and Path 2 are shown in Figure 6.6, two different paths from A to B. From calculus, the negative sign is added to the Path 2 integral, to reverse the direction of integration. Rearranging the above equation,

$$\int_{Path\ 1} \vec{E} \cdot \vec{dl} = \int_{Path\ 2} \vec{E} \cdot \vec{dl}, \tag{6.34}$$

which means that the path integrals along Path 1 and Path 2 are equal to each other, proving that Equation 6.32 implies path independence. Multiplying Equation 6.34 by $-q'$, the work done by an applied force in moving q' between points A and B is also path independent.

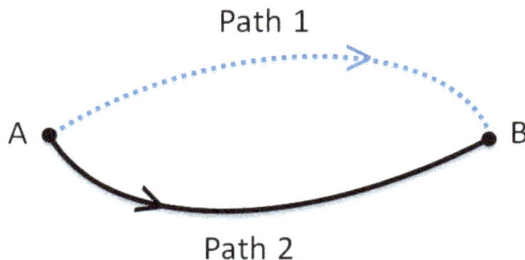

Figure 6.6: Two different paths from A to B.

55

6.5   Proof that the Electrostatic Field is Conservative

The law of conservation of energy will be used to show that the electrostatic field is conservative. The proof presented here from Reference [14] is a general proof, and is valid for any material medium, uniform or non-uniform, that the field may be present in. The effect of electric field in a material medium other than air, will be studied in a later chapter, but is not a prerequisite for the proof presented in this section.

Using Equation 6.18, the work done by an applied force W_C, to move a charge q in a closed loop C, in an electrostatic field $\vec{E}$, is

$$W_C = \oint_C -q\vec{E} \cdot \vec{dl}. \tag{6.35}$$

It will be proven that W_C must be 0, thereby proving that the electrostatic field must be conservative,

$$\oint_C \vec{E} \cdot \vec{dl} = 0. \tag{6.36}$$

Following a proof by contradiction, assume that the work done to move a charge in a closed loop W_C is not zero. Lets say that if

$$W_C > 0, \tag{6.37}$$

energy is spent by the person moving the charge. By the property of path integrals of vector fields, reversing the direction of integration C, written as C', would make the result negative,

$$W_{C'} = \oint_{C'} -q\vec{E} \cdot \vec{dl}$$
$$< 0, \tag{6.38}$$

or energy is gained. As seen in the example in Section 6.1, if the charge is allowed to accelerate or decelerate from the force of the electric field, the above equation means that the charge has more kinetic energy when it returns back to the starting point.

Where does this gain in energy come from? By the law of conservation of energy, something must lose energy for the energy gain to occur, and it is unaccounted for. May be the energy spent to charge the object, or the energy spent to generate the static electric field, is the source of the energy gain. However, in any given time interval, the change in energy during the time interval must be properly accounted for. If the beginning of the time interval is when the charge starts making the loop, and if the end of the time interval is when the charge returns to its starting point, there is an unexplained gain in energy in this time interval.

Some energy is spent traversing the loop, which are the segments where positive work is done, but from the above equation, overall more energy is gained by the person moving the charge than spent. As seen in the example in Section 6.1, negative work is a gain in the kinetic energy of the charge, resulting from the force exerted on the charge by the electric field. The charge excursion around the loop can be continued indefinitely, thereby gaining infinite kinetic energy, and attaining infinite velocity. However, this is clearly non-physical. This is similar to the case where, if a ball was to be dropped on the floor from a height h, after it bounces back, and when it returns back to the height h, it still has kinetic energy and travels higher than h. The ball bounces higher and higher, which is clearly non-physical.

Perhaps, there is a loss in the electric potential energy that balances the energy gain. However, the start and end points are the same, and therefore, there can be no change in the electric potential energy, unlike the result in Equation 6.22.

The only result that makes physical sense is if the closed path integral in Equation 6.35 must always be 0, or

$$\oint \vec{E} \cdot \vec{dl} = 0, \tag{6.39}$$

thereby proving that the electrostatic field is conservative.

In Chapter 42, from Faraday's experiments, in time-varying fields, the conservative property of an electric field is not satisfied. However, it will be shown that the gain in energy in Equation 6.38, comes at the expense of the energy spent in generating the electric field, during any time interval.

7

Electrostatic Generators

A conductor can be charged by conduction or induction using an "electric machine", also known as an electrostatic generator. There are two types of electrostatic generators: friction machine and influence machine. A friction machine uses friction to charge a metal by conduction. Friction machines, such as the ones in References [16][17], are examples of early electrostatic generators used to charge a conductor. A popular friction machine, found in science museums and classrooms, is the Van de Graaff generator. The reader is referred to other resources for more details on friction machines.

An influence machine, on the other hand, works on the principle of induction. There is no friction involved. It is called an influence machine, as its operation is based on "influence", or induction. A commonly used influence machine, even today, is the Wimshurst machine. The operation of the Wimshurst machine is presented in great detail in this chapter. Lord Kelvin's water dropper, which is an influence machine, helps understand the operation of the Wimshurst machine, and is presented first.

7.1 Lord Kelvin's Water-Dropper Experiment

Lord Kelvin is one of the most distinguished scientists who contributed to the development of electromagnetic theory. Outlining all his contributions is beyond the scope of this book. In addition to the experiment presented in this section, his technique of solving electrostatic problems using image theory will be presented in a later chapter.

Two metal objects may appear to be completely discharged. However, one of them always has a small net charge more than the other. This difference is not observable by an electroscope. One of the conductors may have more net positive or negative charge than the other. Believe it or not, this charge difference can be amplified, which can then be detected. Lord Kelvin's water-dropper experiment can be used to demonstrate this in a very dramatic way. This result is important to understand the working of a Wimshurst machine.

The setup of the water-dropper experiment is shown in Figure 7.1. Four metal cans A, B, C, and D are connected by wires as shown. A and D are connected together, as well as, B and C. W is the source of two thin streams of water. This can be created by a container with water, and two holes for the water to fall as thin streams into the cans. A and C have an open bottom, and the water that falls through these cans is collected in B and D. Two metal spheres S in close proximity, are attached to the cans B and D. Periodically, every 5 to 10 seconds, a spark is observed between the spheres [18]–[20].

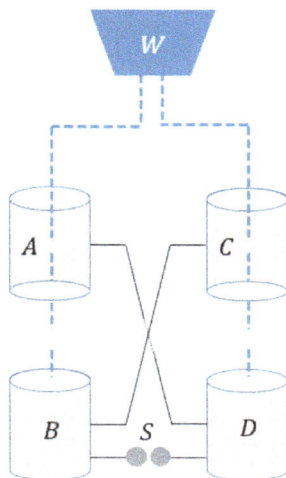

Figure 7.1: The setup of Lord Kelvin's water-dropper experiment.

The outcome of the experiment can be explained by starting with the assumption that A has a small net charge more than C. For example, lets assume that A has a small net positive charge more than C, shown in Figure 7.2(a). Tap water does mildly conduct electricity [21], and in this regard, it can be viewed as a metal, exhibiting properties of conduction and induction. The positive charges in A repel the positive charges in the stream, leaving the water falling through A negatively charged. This charges B negative.

A water falling as a thin stream, after a certain distance, becomes drops of water, and not a continuous stream. This is captured in the clip in Reference [18]. A and C are placed such that the charges are repelled in the water stream, leaving the water droplets oppositely charged. In the figures, the water droplets is illustrated by a discontinuous dashed line. This is very important, since the negatively charged B cannot repel the negative charges in the falling drops away from it. This forces the negatively charged falling droplets to charge B negative.

Since B is connected to C, C is also negatively charged, shown in Figure 7.2(b). Similar to A, the negative charges in C, repel the negative charges in the water stream, so that the droplets have a positive charge, shown in Figure 7.2(c). As a result, D is charged positive. Since A is connected to D, this is a positive feedback action, that increases the positive charge in A. The positive feedback, results in a quick build up of charge in the cans, and beyond a certain threshold, seen as a spark between the spheres S.

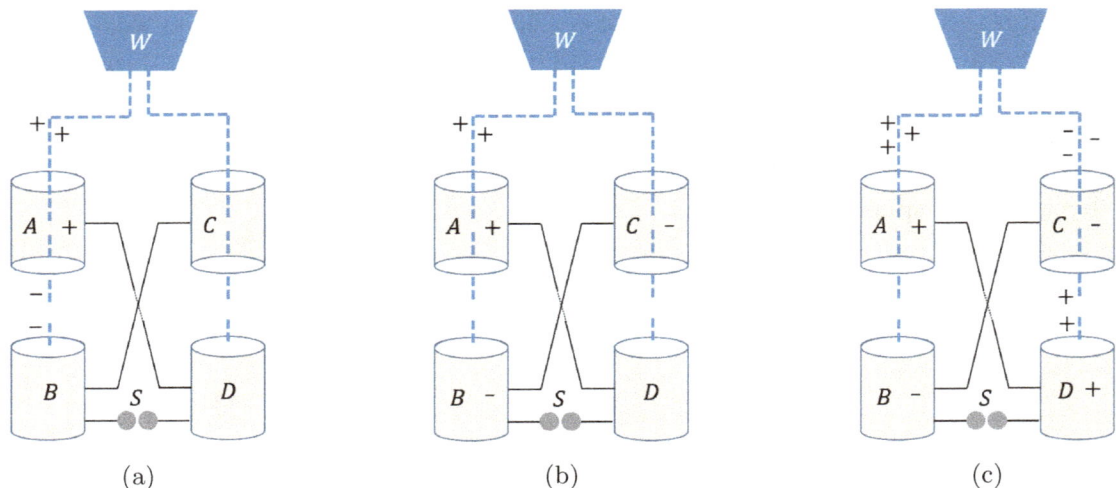

Figure 7.2: The positive feedback creating a charge buildup in the metal cans.

Optionally, an electroscope can be attached to B or D. It can be observed that each time a spark occurs, the electroscope needle no longer deflects [19]. This shows that the spark between the spheres is just a stream of electric charges discharging the cans.

This experiment serves as an empirical evidence that there is always a small net charge difference that exists between two metal objects that are not electrically connected together. This property will be used to explain the operation of the Wimshurst machine.

The Kelvin water dropper is an electrostatic generator. A metal object connected to B or D will charge the object. However, it is not as effective, and less convenient to use than the Wimshurst machine, as explained next.

7.2   The Wimshurst Machine

A Wimshurst machine is an influence machine, named after its creator James Wimshurst, and is widely used even today [22]. It is an improvisation of earlier influence machines of Holtz, Voss, and others [23]. The front and the back sides of the Wimshurst machine are shown in Figure 7.3(a) and Figure 7.3(b). The schematic of the Wimshurst machine is shown in Figure 7.4. The labels in the figures are kept consistent. The structure of the Wimshurst machine will be described first, followed by its operation [23][24].

There are two circular insulator discs D_1 and D_2 that rotate in opposite directions, as shown in Figure 7.5(a). To simplify the illustration, the two rotating discs in the schematic are drawn as two concentric circles. In reality, however, the two discs are parallel, as shown in Figure 7.5(a).

Etched on the discs are metal sectors M. As the lever L is turned manually, the discs rotate

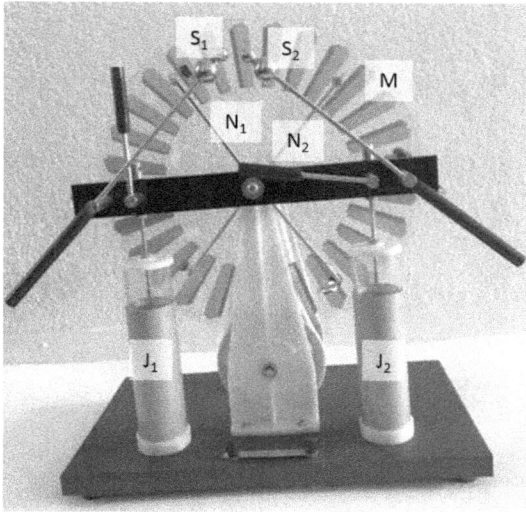
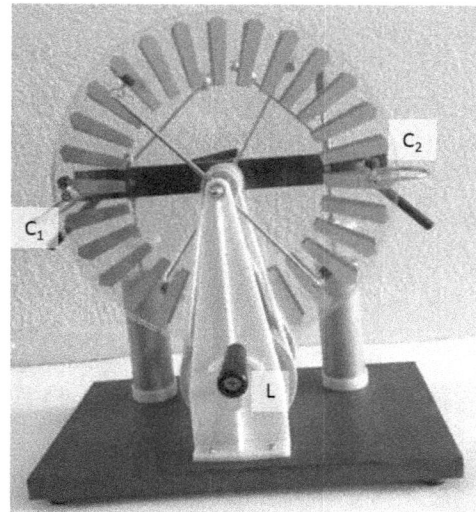

(a)   (b)

Figure 7.3: The Wimshurst machine.

in opposite directions. This is achieved by making one of the belts B_1 that turns one of the discs, crossed like an '8', compared to B_2 that turns the other disc, shown in Figure 7.5(b). The belts are connected to the pulleys P_1 and P_2.

There are two "neutralizing bars" N_1 and N_2, one for each disc, which are approximately orthogonal to each other. Each of the bars has metal brushes on its two ends. The metal brushes at one of the ends of, one of the neutralizing bars, is shown in Figure 7.6(a). As the discs rotate, the brushes of N_1 (N_2) come in contact with the metal sectors of D_1 (D_2). The neutralizing bars remain stationary when the discs rotate.

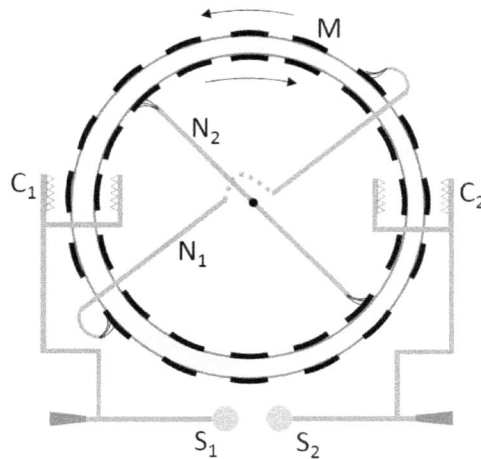

Figure 7.4: A schematic of the Wimshurst machine.

(a)

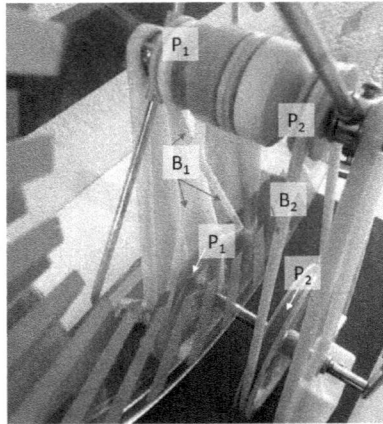
(b)

Figure 7.5: (a) The discs of the Wimshurst machine turning in opposite directions. (b) The belts and pulleys that enable the discs to turn in opposite directions.

There are two collector combs C_1 and C_2. The collector combs remain stationary, and do not rotate with the discs. Each of the collector combs is made of a metal with a sharp corner, in the shape of U, that is in close proximity to the two metal sectors facing each other, shown in Figure 7.6(b). C_1 (C_2) is connected to the metal sphere S_1 (S_2). This is illustrated in the schematic in Figure 7.4.

As the discs rotate, charge quickly builds up on the metal sectors M. When the spheres are placed close to each other, a spark is observed between the spheres S_1 and S_2. A conductor can be charged by connecting it to S_1 or S_2.

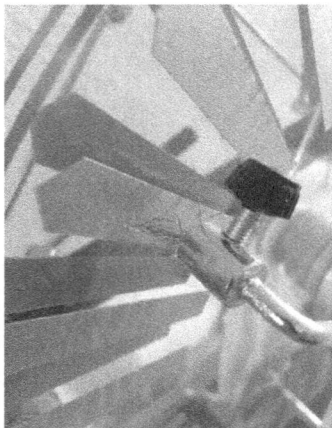
(a)

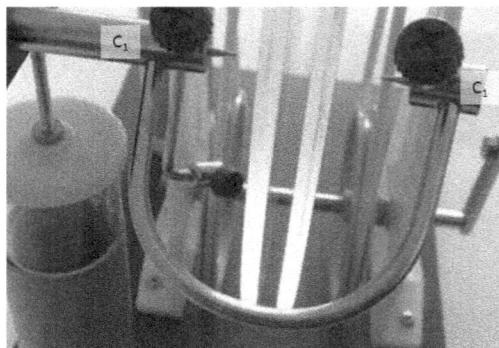
(b)

Figure 7.6: (a) The metal brushes of a neutralizing bar. (b) The collector combs.

There are two Leyden jars J_1 and J_2 to store charge. Leyden jars will be covered in Section 17.2, but is not required to understand the operation of the Wimshurst machine.

7.3 Operation of the Wimshurst Machine

The operation of the Wimshurst machine, although its construction is entirely different from the Kelvin water dropper, works on the same principle: there is a positive feedback mechanism that amplifies the charge build up. The schematic in Figure 7.4 will be used to explain the operation of the Wimshurst machine. The labels in the figure are not marked again in the future diagrams to avoid repetition.

Lets assume that the metal sector marked a has a small net positive charge more than b, shown in Figure 7.7(a). How does charge spread to the other sectors? How does the charge on the sectors get amplified? These questions will be answered next. Note the similarity to the water-dropper experiment, where A was assumed to have a small net positive charge more than C.

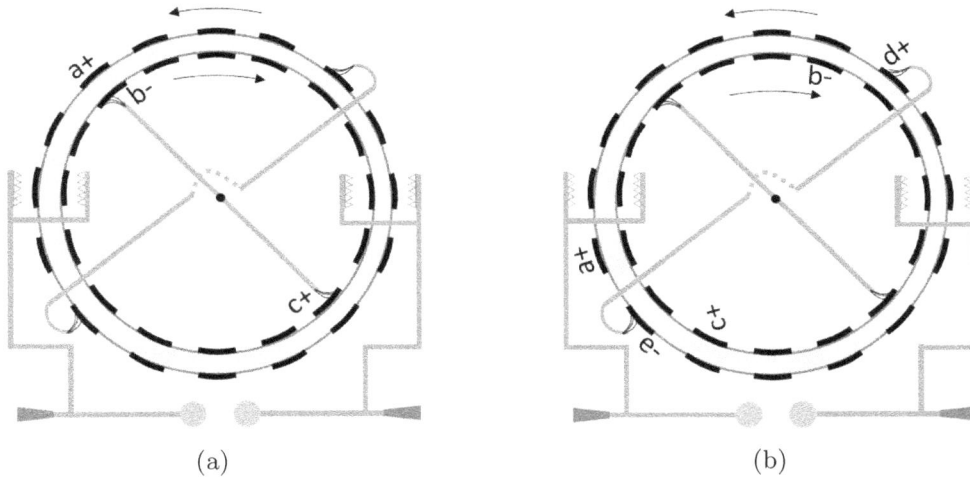

(a)   (b)

Figure 7.7: A Wimshurst machine in operation.

The neutralizing bar N_2 joins sectors b and c. The positive charge in a induces a negative charge in b, leaving c positively charged. The discs are assumed to rotate in the opposite directions shown by the arrows.

b and c reach the sectors adjacent to the neutralizing bar N_1, shown in Figure 7.7(b). The proximity of negatively charged b to d, where d is the metal sector that is in contact with N_1, induces a positive charge on d. This charges e negative, which is the other end of N_1. The negative charge on e is also induced by the positive charge on c.

As the discs rotate by 1 metal sector, the snapshot of the discs is shown in Figure 7.8(a). Positively charged c induces a negative charge on a, making f positive. Negatively charged b induces a positive charge on f as well.

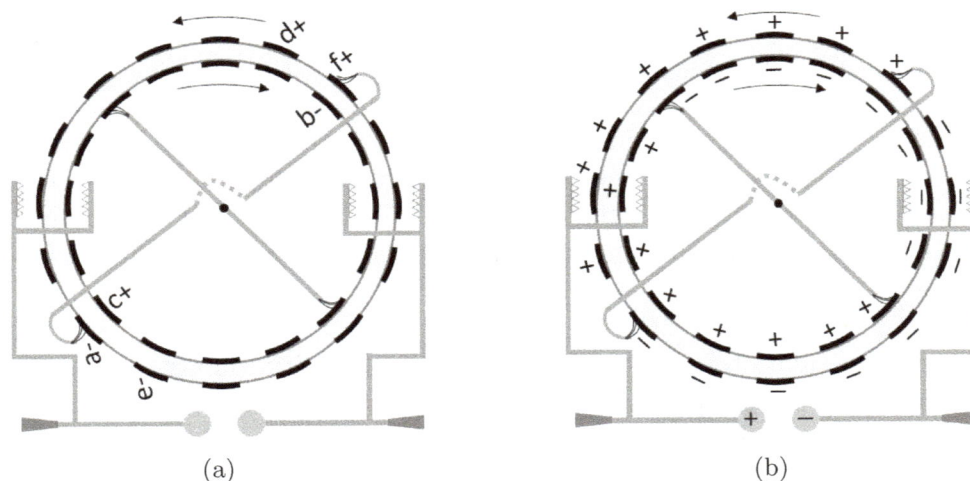

Figure 7.8: A Wimshurst machine in operation.

Starting with a single sector that has a small net charge, charges are induced on other sectors. In this manner, all the sectors get charged, shown in Figure 7.8(b). This is the mechanism by which the charge spreads to all the sectors.

As more sectors get charged, the inductive action becomes greater. This is how charge gets amplified. For example, in Figure 7.9, four metal sectors p through s are shown. The brushes of a neutralizing bar N is in contact with p. q, r, and s are the metal sectors that are in proximity to p, which generate an induced charge on p. If q and s are uncharged, then less charge is induced on p. However, as more sectors get charged, more charge is induced on p. More charge on the sectors in one of the discs, induces a larger charge on the other disc. This positive feedback between the two discs, creates a charge amplification on the metal sectors.

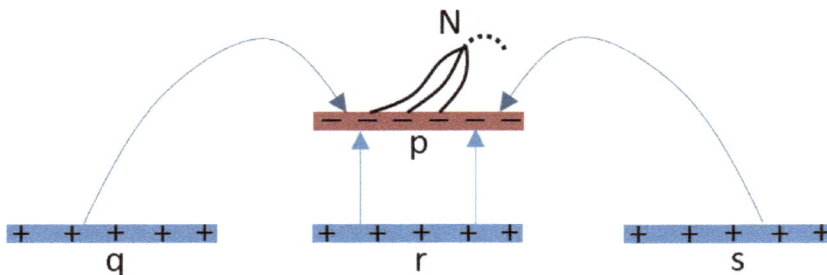

Figure 7.9: An illustration of the mechanism by which the charge amplification occurs.

In Figure 7.8(b), note that the charges on the metal sectors where the collector combs are present, are of the same polarity on both the discs. The collector combs are not in physical contact with the metal sectors. It will be shown later in Section 8.7 that charges are more concentrated in sharp corners than on flat surfaces. Charge is induced on the sharp ends of the collector combs by the metal sectors, leaving the spheres connected to the metal combs charged opposite. As the charge becomes greater, if the spheres are sufficiently close to each other, beyond a certain threshold, an electric spark is observed between the metal spheres. It was observed from Kelvin's water-dropper experiment that an electric spark is a stream of electric charges.

If the Wimshurst machine is operated in a fully dark environment, electric discharges between the sharp corners of the collector combs and the metal sectors, as well as between the metal sectors and the metal brushes of the neutralizing bars will be observed as sparks. If electric discharges occur as the metal sector moves away from the metal brush, then this reduces the charge on the metal sector. These discharges prevent the metal sectors from acquiring a charge beyond a certain maximum value.

8

Properties of Conductors in Electrostatics

A conductor, repeating the definition in Section 1.4, is an object, such as metal, where charges flow freely within them. The distribution of the excess charges in a charged conductor is not something that is arbitrary. It will shown in this chapter by experiments, the net charges distribute themselves in a manner that the electric field within a conductor in the electrostatic condition is 0. Other properties of charge distribution in a conductor, such as charge being more concentrated in sharp corners compared to flat surfaces, will also be empirically proven.

8.1 Distribution of the Net Charge in a Conductor in Electrostatics: Experiment 1

An experiment is demonstrated in this section, which shows that in a charged conductor, the net charge lies entirely on its outer surface [26]. As shown in Figure 8.1, a hollow metal sphere (B) is charged using a Wimshurst machine (A). A charge scoop (C), which is simply a piece of metal

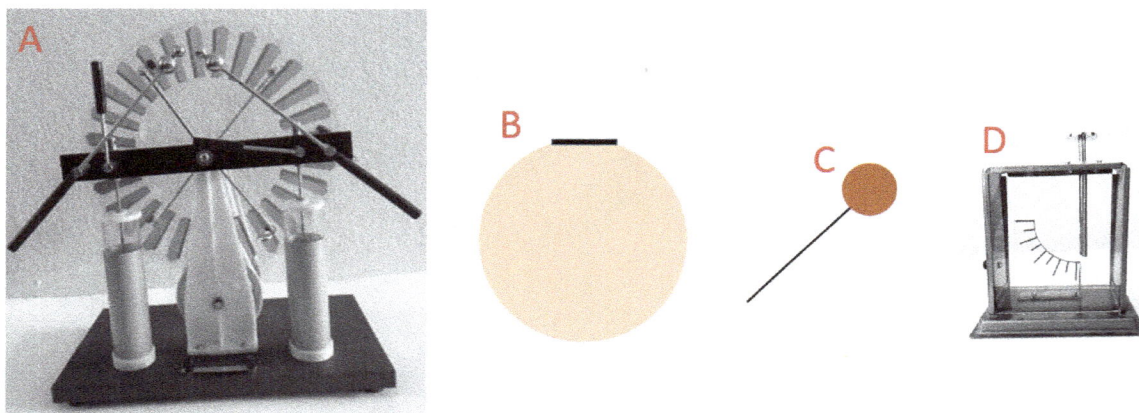

Figure 8.1: An experiment setup to show that the net charge of a charged metal object lies entirely on its outer surface. (A) a Wimshurst machine, (B) a hollow metal sphere, (C) a charge scoop, (D) an electroscope.

attached to an insulated handle, can be used to scoop charge by swiping its metal tip against a metal surface. The hollow metal sphere has a small opening at its "north pole", which can be used for testing if any excess charge resides on its *inner* surface. The charge on the scoop can be detected using an electroscope (D).

The outer surface of the metal sphere is connected to one of the terminals of the Wimshurst machine, to charge the sphere. The charge scoop is swiped against the outer sphere, as shown in Figure 8.2(a). The charge collected by the charge scoop, shows that excess charge does exist on the outer surface, as seen by the diverged leaf in the electroscope in Figure 8.2(b). The same steps are repeated to test for charge on the inner surface of the sphere, as shown in Figure 8.3(a) and Figure 8.3(b). However, the electroscope detects no charge present on the inner surface, as seen by the undiverged leaf in Figure 8.3(b). This shows that the net positive or negative charge, resides only on the outer surface of a metal object, either hollow or filled.

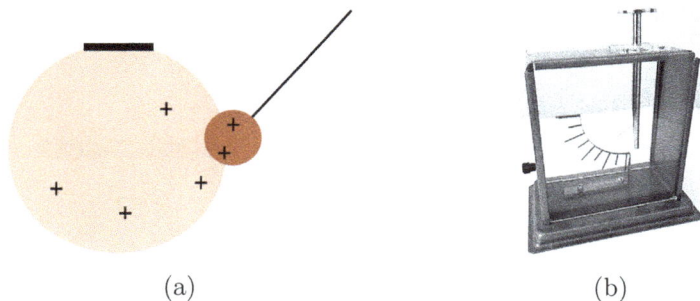

(a)   (b)

Figure 8.2: (a) A charge scoop is used to check if charge is present on the outside surface of the conductor. (b) An electroscope to detect the charge on the charge scoop.

(a)   (b)

Figure 8.3: (a) A charge scoop is used to check if charge is present on the inside surface of the hollow conductor. (b) An electroscope detects no charge on the inner surface.

8.2   Distribution of the Net Charge in a Conductor in Electrostatics: Experiment 2

Other experiments also arrive at the same conclusion. A hollow spherical shell is separated into two hemispheres, as shown in Figure 8.4, and both the hemispheres are charged with the same polarity using a Wimshurst machine. The two hemispheres are joined together, and in whose cavity is placed another uncharged conductor. The outer sphere is then made to contact the inner sphere, while keeping the hemispheres still together forming a closed metal surface. The hemispheres are then removed, and the inner sphere is tested to see if it is charged. The outcome of this experiment is that no charge is detected on the inner sphere. This experiment shows that the net charge *within* the surface of a conductor is zero, and the net charge resides only on the *outer* surface of a conductor.

A variation of this experiment is to charge the inner sphere using a Wimshurst machine, instead of the hemispheres. The two hemispheres are joined together forming a closed surface, and then put in contact with the enclosed inner sphere. The hemispheres are then removed, and the conductors are tested for their net charge. It will be observed that the inner sphere is now uncharged, and the hemispheres are now charged. The net charge on the inner sphere, upon contact with the enclosing hemispheres, flow to the outer surface of the hemispheres.

Figure 8.4: An experiment to show that the net charge on a conductor resides on its outer surface [27].

These experiments demonstrate Coulomb's law in an implicit way. The excess charges in a conductor, since they are all of the same type, repel each other, as described by Coulomb's law, and trying to "get away" from each other. The surface of the conductor is the maximum distance that they can separate from each other, and therefore, no charge is present within the metal surface.

Here's a thought experiment to show the existence of the Coulomb force between charges, although they are in a metallic medium. A charged metal sphere A, upon contact with the inner surface of a hollow metal shell B, and before the charges begin redistributing themselves, is shown in Figure 8.5(a). As seen from the experiments presented earlier, the charges repel each other, and flow to the outer surface of B, and A is uncharged, shown in Figure 8.5(b)–8.5(c). This result shows that there exists a Coulomb force between charges, although they may be "submersed" in a metallic medium. There exists a force of repulsion between the charges joined by the dotted line, for example, in Figure 8.5(c), although they are separated by air and metal.

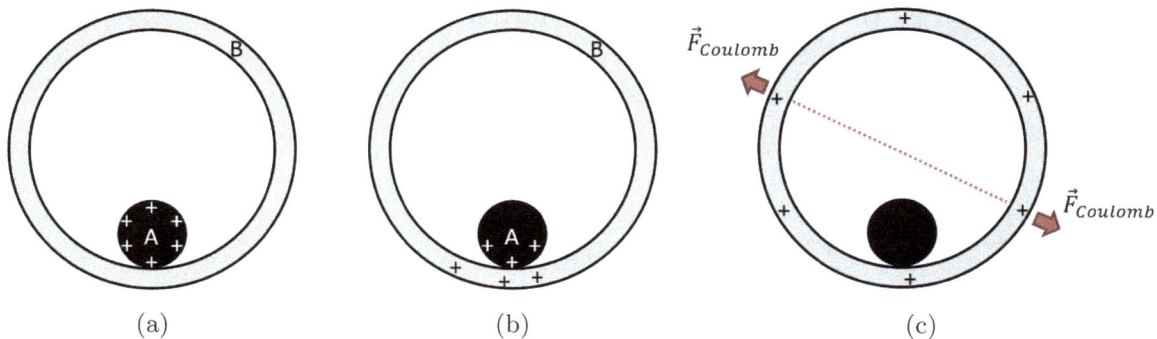

Figure 8.5: (a) A charged sphere is put in contact with the inner surface of another conductor. (b) The charges repel each other, and (c) the charges are present only on the outer surface in the steady state.

8.3 Electric Field in a Metallic Region in Electrostatics

In the steady state, after the charges redistribute themselves, the electric field in a metallic region must be 0. The charges are free to move in a metal, as explained in Section 1.4. A non-zero electric field in a metal would exert a force on the charges, resulting in an endless motion of the charges. It will be proven later in Chapter 19, charges flow in a wire connected to a battery. A thin wire glows red hot when a large quantity of charge flows in the wire [28]. This shows that flow of charges in a conductor generates heat. If the charges are in an endless motion, there is a continuous dissipation of heat, which is clearly a violation of the law of conservation of energy.

In the experiment shown in Figure 8.5, there is a transient non-zero electric field that exists in the metal, propelling the charges. The charges eventually come to rest on the surface. Although a transient electric field momentarily exists in the metal, the electric field in the metal must become 0 in the steady state, to not violate the law of conservation of energy.

8.4 Electric Field Within a Closed Metal Surface in Electrostatics

Experiments show that the electric field within a metal surface must be 0 in the steady state, regardless of whether it is filled, or a cavity is present within the metal surface. A variation of the experiment from Reference [29] will be used to prove this.

When a charged object C, positively charged for example, is brought near an electroscope in a glass casing G, the electroscope leaf diverges, as shown in Figure 8.6(a). The charged object attracts the opposite charges in the electroscope, creating a non-uniform charge distribution, and causing the leaf to diverge, as shown in Figure 8.6(b).

In the presence of the charged object C, a transient electric field exists in the metal A that exerts a force on the charges, creating the charge distribution shown in the figure. Since the electric field in A must be 0 in the steady state, the charges distribute themselves such that the electric field within the metal structure A is 0 in the steady state. The electric-field lines in the steady state are shown in the figure, which originate on the positive charges and end on the negative charges.

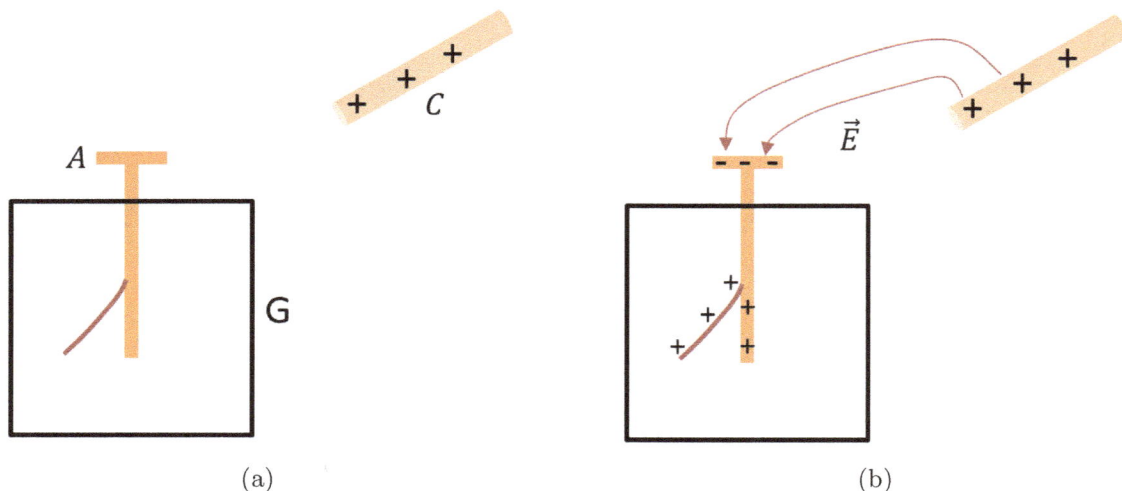

Figure 8.6: (a) A charged object near an electroscope. (b) The electric-field lines in the region around the electroscope.

If the electroscope is enclosed by a metal surface M, the electroscope leaf remains undiverged, as shown in Figure 8.7(a). This is observed, although M does not completely enclose the electroscope, to keep the electroscope leaf visible during the experiment. If the electric field within M is non-zero, then this would exert a force on the charges in A, and redistribute them in a way that the electric field within A is 0. This will result in a diverged leaf, as seen in the example in Figure 8.6. Since the leaf remains undiverged, this shows that the electric field within M is 0 in the steady state. The electric-field diagram is shown in Figure 8.7(b).

It was proven in Section 8.6 that induced charges lie on the outer surface of a metal. The induced charges, in the presence of C, distribute themselves in a manner that the steady-state electric-field within the surface of M is 0. Since M is uncharged to begin with, the net charge in M is 0. This is shown by the presence of positive and negative induced charges on the surface of M.

A similar example is shown in Figure 8.8, which confirms again that the excess charges distribute themselves on the outer surface of a metal, and in a manner that the electric field within the enclosed surface is 0.

A meshed metal sphere, instead of solid filled, through which it can be observed if the electroscope is diverged or at rest, is shown. The meshed metal sphere is made up of two hemispheres that can be opened or closed, as shown in the figure. An electroscope is attached to one of the

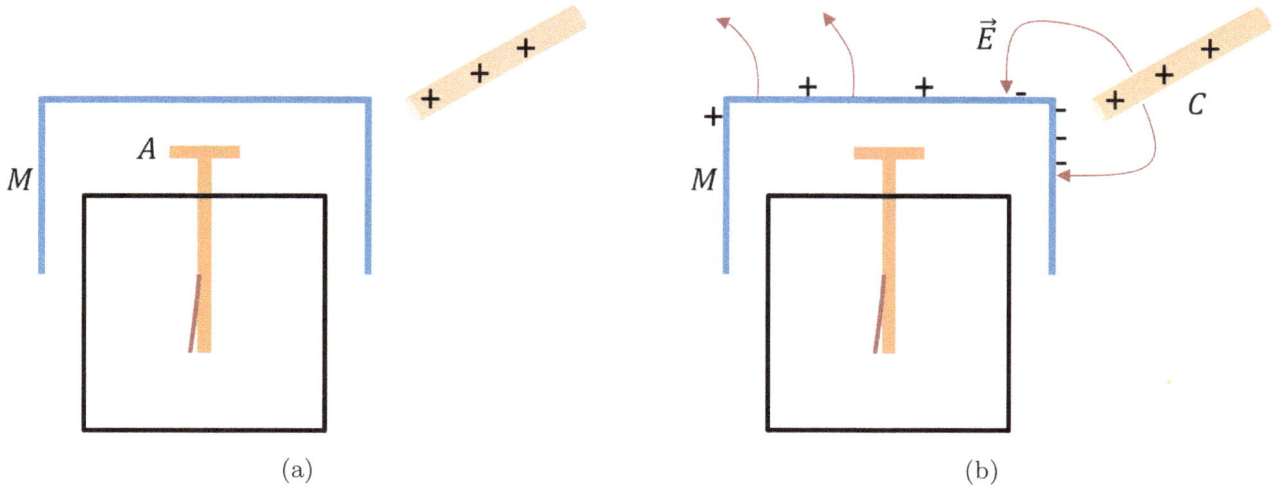

Figure 8.7: (a) An electroscope with a metal shield. (b) The electric-field pattern around the metal shield.

hemispheres, such that when the hemispheres are joined together, the electroscope lies within the enclosed sphere. A is fixed, but the needle N, can pivot about its center that is attached to A, when N and A repel each other.

The two hemispheres are connected together by a wire, and charged using a Wimshurst machine. The two hemispheres will then have the same charge polarity. In Figure 8.8(a), when the hemispheres are left open, N is repelled. However, when the hemispheres are joined together, resulting in the electroscope enclosed within the closed sphere, the needle is no longer repelled, as shown in Figure 8.8(b).

The observed results can be explained by analyzing the charge distribution. When the hemispheres are separated, the charges are present on the outer surface, and since it is not a closed surface, the charges are also present on the surfaces of A and N, as shown in Figure 8.9(a). This repels the needle N.

When the hemispheres are joined together forming a sphere, the charges present on the electroscope that are lying within the closed surface, move to the outer surface of the sphere. In this case, the charges are no longer present on A and N, and N is no longer repelled, as shown in Figure 8.9(b). This shows that charges don't reside within a closed metal surface, as concluded earlier in the chapter.

If there exists an electric field within the closed surface in the steady state, this exerts a force on the charges in A and N, creating a non-uniform charge distribution in the electroscope, as observed in the earlier experiment in Figure 8.6. This charge distribution in the electroscope causes the needle to be repelled, but this is not the outcome of the experiment. Therefore, this result

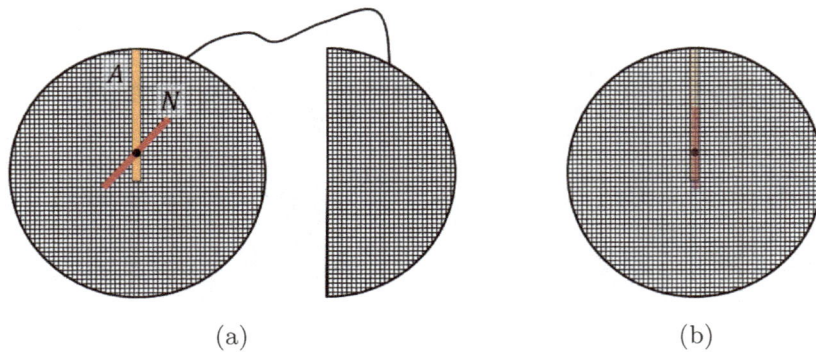

Figure 8.8: (a) An electroscope detects a charge when the hemispheres are left open. (b) No charge is detected when the hemispheres are closed.

also shows that the electric field within the surface of the metal is 0. A closed metal structure which shields the electric field is known as a Faraday cage.

These examples show that charges distribute themselves on a metal surface, such that the electric field within the metal surface is 0, independent of whether it is completely filled with metal or there exists a cavity within the metal surface. This important result will be used to analytically prove Coulomb's law, discussed in Chapter 9.

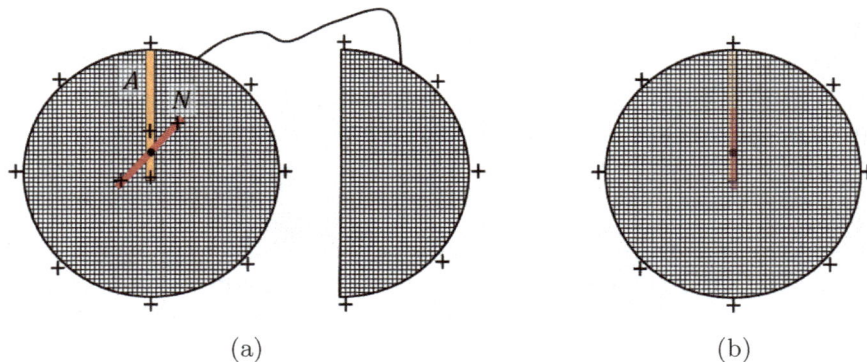

Figure 8.9: (a) The charge distribution when the hemispheres are left open. (b) The charge distribution when the hemispheres are closed.

8.5 Boundary Condition

In this section, it will be proven that the electric field just outside a metal surface is perpendicular to the metal surface. This is equivalent to stating that the tangential component of $\vec{E}$, just outside a metal surface is 0. This is shown in Figure 8.10(a), where the field lines along a section of the metal surface are drawn perpendicular to the surface, assuming a positively charged object. Two proofs are presented.

The metal surface is partitioned into tiny surface elements, each of which has a uniform charge density. As shown in Figure 8.10(b), very close to the metal surface, the surface looks like an infinite uniformly charged plane. The electric field at a point is the force experienced by a unit positive charge at the point. Just outside the metal surface, the force experienced by a unit positive charge would be in various directions, shown by the arrows. However, the tangential components cancel each other, leaving only the normal component, shown by the solid arrow.

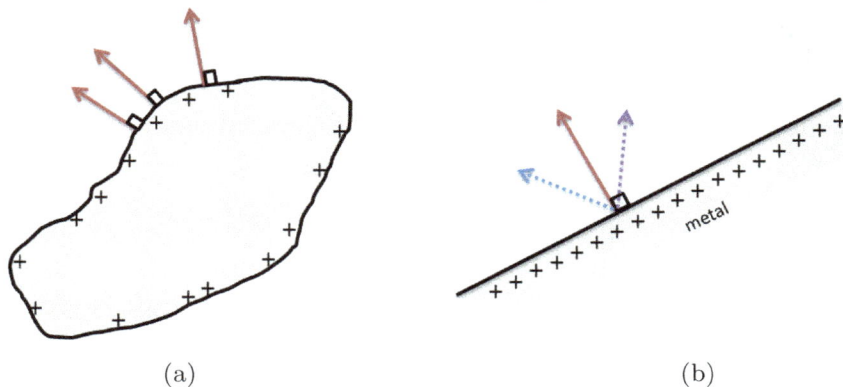

(a)   (b)

Figure 8.10: (a) $\vec{E} = 0$ within the metal surface. (b) $\vec{E}$ is perpendicular to the metal surface.

The second way to show that the electric field is normal to a metal surface is using the conservative property of the electrostatic electric field. In a conservative field, repeating Equation 6.31,

$$\oint \vec{E} \cdot d\vec{l} = 0, \tag{8.1}$$

for any closed path integral of the electric field $\vec{E}$.

A metal and a non-metal region boundary is shown in Figure 8.11. The above path integral is evaluated on a rectangle, in the direction shown by the arrows. The tangential electric field on either side of the boundary are marked $\vec{E}_{1t}$ and $\vec{E}_{2t}$. Assuming that $\Delta y \to 0$, the two shorter sides can be ignored in the integration, resulting in

$$\oint \vec{E} \cdot d\vec{l} = 0$$
$$= E_{1t}\Delta x - E_{2t}\Delta x. \tag{8.2}$$

It was shown earlier that the electric field within the metal surface in the steady state must be 0,

$$E_{2t} = 0. \tag{8.3}$$

From the above equations,

$$E_{1t} = 0. \tag{8.4}$$

Therefore, the tangential electric field just outside a metal surface is 0.

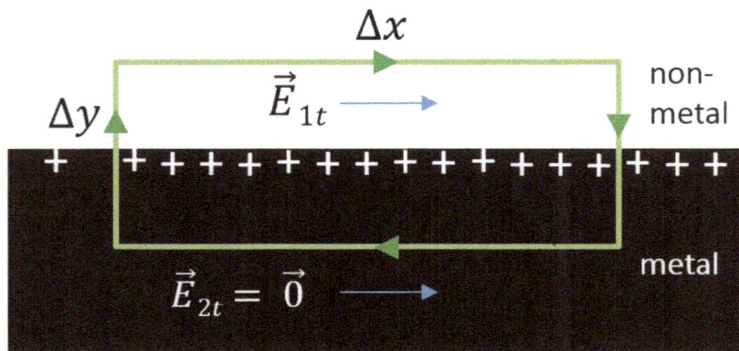

Figure 8.11: The tangential component of the electric field just outside a metal boundary must be 0.

8.6 Distribution of the Induced Charge in a Metal

It was shown earlier in the chapter, in a charged conductor, the net charges lie on the outer surface of the conductor. It will be shown empirically in this section, this is also true for an induced charge distribution. The induced charges in a metal also lie on the outer surface.

A narrow, deep, and hollow uncharged conductor M is shown in Figure 8.12(a). Its a closed surface except for the small opening at the top. A small spherical conductor S, attached to an insulator string, is placed in contact with the outer surface of the conductor M. A charged object R, positively charged for example, is brought near S. This attracts the negative charges to S, leaving S negatively charged. S can be shown that it is charged using an electroscope. This is known as charging by induction, as explained in Section 1.6.

The same experiment is repeated, however, now placing S inside the cavity of M, and touching its inner surface, shown in Figure 8.12(b). It will be observed that S cannot be charged by induction. This shows that the net induced charge distribution on M lies on its outer surface, thereby preventing S from being charged by induction in the second case.

8.7 Uniformity of Charge Distribution

It can be shown from electrostatic experiments that charge is not uniformly distributed in a metal. A variation of the experiment in Reference [30] has been used in this study. A conductor M with flat and sharp corners, on an insulated handle H, is shown in Figure 8.13. A small sphere p attached to an insulator string i can be used as a charge scoop to compare the relative charge densities on the conductor surface.

M is charged using a Wimshurst machine. p is put in contact with M at Point A that is on the flat region of the conductor. The angle that the leaf diverges in a leaf electroscope, is used as an indicator of the charge density at Point A. p is discharged and the experiment is repeated on

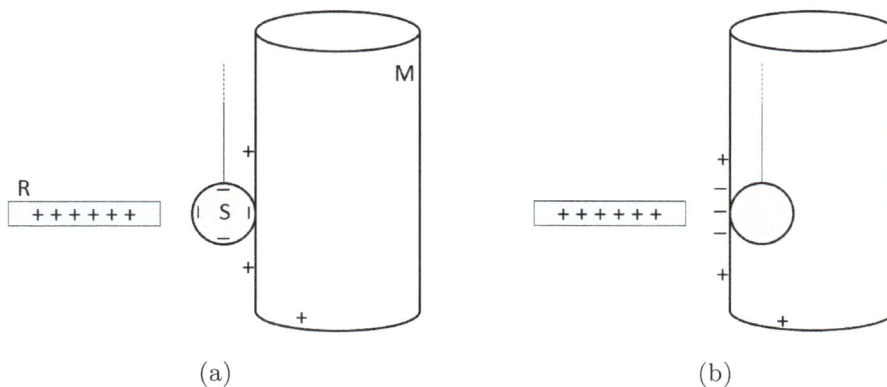

Figure 8.12: The charging of a sphere S by induction. (a) S is charged when placed in contact with the outside surface of the conductor M. (b) S is not charged when placed in contact with the inner surface of the conductor M.

Point B, which is at the sharp corner of M.

It will be observed that the leaf of the electroscope diverges more in the case of Point B than Point A. More charge is present at the sharp corner, although the charge at Point A on the flat surface was sampled first. This shows that more charge is concentrated in the region near a sharp corner than a flat surface. The charge density is higher near sharp corners than flat surfaces.

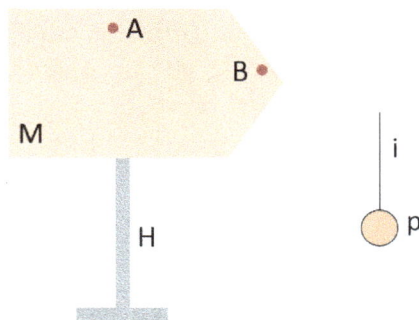

Figure 8.13: An experiment to show that charge is more concentrated at sharp corners than flat surfaces.

8.8 Faraday's Ice-Pail Experiment

Faraday's ice-pail experiment describes how charges distribute themselves when a charged object exists within a conductor cavity. The apparatus in Faraday's experiment consists of a metal container P, a charged metal sphere S, and an electroscope E, shown in Figure 8.14. Faraday used an ice pail in the experiment for P, and hence the name of the experiment. However, any metal container may be used. The electroscope is connected to the outer surface of the ice pail. When the charged metal sphere is lowered into the ice pail without touching, the leaves on the electroscope diverge.

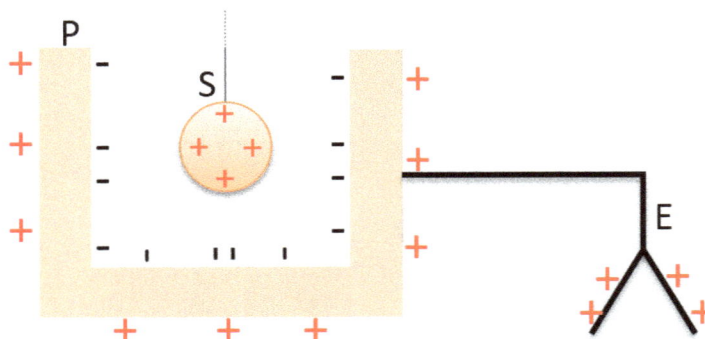

Figure 8.14: Faraday's ice-pail experiment.

The technique to determine the relative polarity of two charged objects, if they are charged alike or oppositely charged, was presented in Section 1.5. The outcome of the experiment is that the charge on the electroscope, as well as the outer surface of P, are found to be of the same polarity as the charged metal sphere. The inner metal surface of P has the opposite polarity as the metal sphere. The positively charged sphere S, for example, lowered into P, repels the positive charges to its outer surface, and to the electroscope, leaving the inner surface negatively charged, as shown in the figure.

The results from Faraday's experiment can be applied to the case shown in Figure 8.15(a). A positively charged object A, for example, lies within the cavity of B. The Coulomb force from A momentarily exists in the metal B, attracting the negative charges to the inner surface of B, leaving the outer surface positively charged. In the steady state, the charges distribute themselves in a way that the electric field in the metallic regions of A and B are 0. It will become clear after

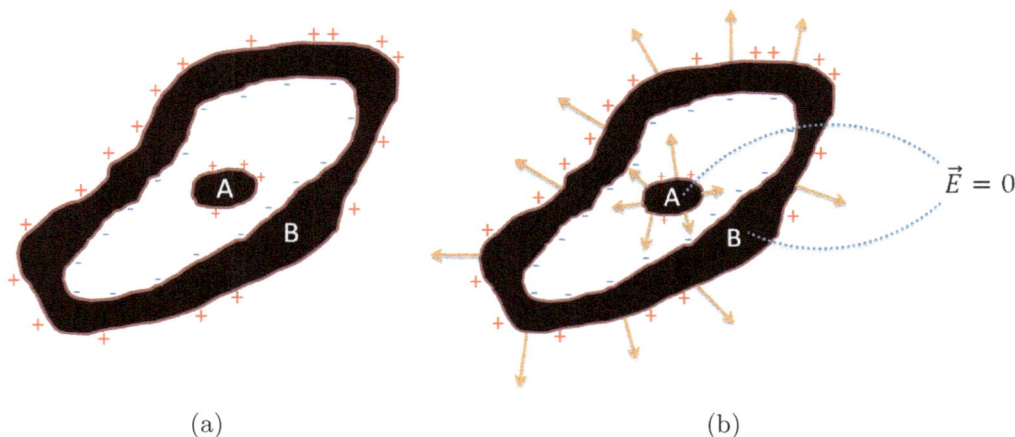

(a)

(b)

Figure 8.15: (a) A charged object inside a metal cavity. (b) The electric-field pattern of a charged object inside a metal cavity.

76

Gauss' law in Chapter 36, the presence of a charged object in a cavity, creates a non-zero electric field in the cavity, as shown in Figure 8.15(b), where the arrows represent the electric-field lines.

From this example, it can be noted that the caveat to the rule in Section 8.4, the electric field is 0 within a closed metal surface, is true, only if no charged objects exist in the cavity of the closed metal surface. By the law of conservation of energy, however, as explained in Section 8.3, the electric field in a metallic region is always 0, marked $\vec{E} = 0$, regardless of whether or not charged objects are present in the cavity.

9

An Analytical Proof of
Coulomb's Law [Optional]

If you are not yet convinced in the validity of Coulomb's law, read on! Is the exponent in the inverse-square distance of Coulomb's law, exactly equal to 2, or could it be a value very close to 2, such as 2.001? This small difference would be impossible to detect from the torsion-balance experiment. Also, could it be that the force between charges q_1 and q_2, varies as

$$F \propto q_1^p \, q_2^p, \tag{9.1}$$

where p is a value different from 1.0? Fortunately, there is an analytical proof of Coulomb's law [15], and in this chapter, Coulomb's law will be verified beyond a shadow of uncertainty.

It was shown by experiments in Chapter 8, charges reside on the surface of a conductor, and the electric field is 0 within the conductor. This property will be used to prove that Coulomb's law must be of the form

$$F \propto \frac{q_1 \, q_2}{r^2}. \tag{9.2}$$

The proof presented in this chapter uses solid angles. A tutorial on solid angles is presented in Appendix $\mathcal{A}$.

A metal shell that is positively charged with a uniform charge density, is shown in Figure 9.1. By spherical symmetry, the charges are uniformly distributed on its surface. As explained earlier in Section 8.2, there is a Coulomb force between charges in a metallic medium. It will be assumed that although the medium between the charges and a point P within the shell are both metal and air, not air alone, Coulomb's law is still applicable.

A 2D cross section of two identical cones with their apexes at Point P are drawn in Figure 9.2. The circle in the figure represents the outer surface of the charged shell, and scaled for clarity. The cones intersect the outer surface of the shell, subtending areas QS and VT at Point P. The solid angles are small enough that QS and VT can be assumed to be planar areas on the sphere. The cones and the subtended areas are redrawn in Figure 9.3(a) and Figure 9.3(b) for more clarity.

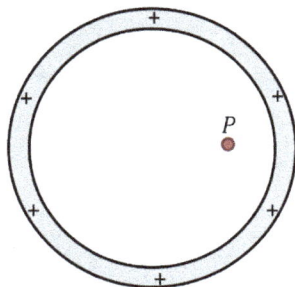

Figure 9.1: A point P within a uniformly charged metal shell.

The center of the areas have been marked R and U. The lengths $\overline{RP}$ is r_1 and $\overline{PU}$ is r_2. It is assumed that the solid angles are small enough that the lengths from P to any point on the areas QS and VT are r_1 and r_2.

As the solid angles become smaller and smaller, the planar areas QS and VT can be regarded as tangential planes to the sphere, at their center points R and U, illustrated in Figure 9.4. A property of the sphere is that the tangential plane is perpendicular to the radius of the sphere s. Therefore, the normal vectors to the areas QS and VT, $\hat{n}_1$ and $\hat{n}_2$, pass through the center of the sphere O, as drawn in Figure 9.2 and Figure 9.4. In Figure 9.2, $\overline{OR} = \overline{OU} = s$ and $\triangle ORU$ forms an isosceles triangle. Therefore, $\angle ORU = \angle OUR = \theta$. The angles between $\hat{r}_1$ and $\hat{n}_1$, as well as $\hat{r}_2$ and $\hat{n}_2$, are vertical angles to $\angle ORU$ and $\angle OUR$, and are also equal to θ. As a result, their dot products must also be equal,

$$\hat{r}_1 \cdot \hat{n}_1 = \hat{r}_2 \cdot \hat{n}_2 = \cos\theta. \tag{9.3}$$

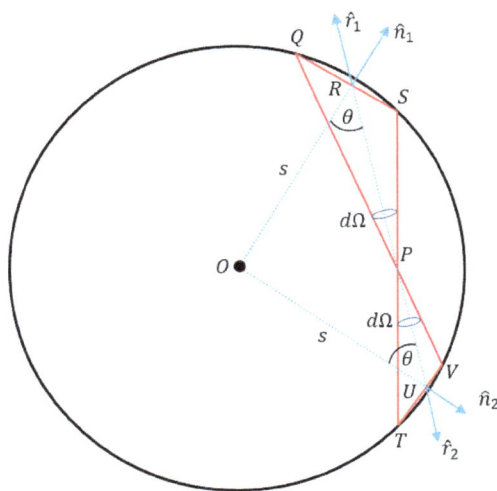

Figure 9.2: Equal pairs of solid angles at Point P.

(a)

(b)

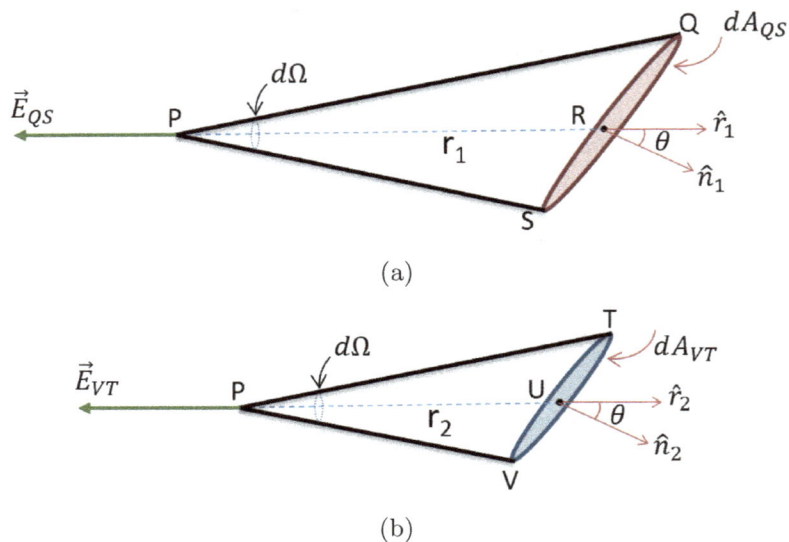

Figure 9.3: (a) Solid angle subtended by dA_{QS} at Point P. (b) Solid angle subtended by dA_{VT} at Point P.

The solid angles of the cones $d\Omega$, subtended by the areas QS and VT are equal to each other, since the cones are identical. By definition of solid angles in Equation $\mathcal{A}.19$,

$$d\Omega = \frac{dA_{QS}\,\hat{r}_1 \cdot \hat{n}_1}{r_1^2} \tag{9.4}$$

$$= \frac{dA_{VT}\,\hat{r}_2 \cdot \hat{n}_2}{r_2^2}. \tag{9.5}$$

From the above equations, the relation

$$\frac{dA_{QS}}{r_1^2} = \frac{dA_{VT}}{r_2^2}, \tag{9.6}$$

can be derived.

By spherical symmetry, the sphere is uniformly charged, and let its surface charge density be σ. The charge present in the areas dA_{QS} and dA_{VT} are $\sigma\,dA_{QS}$ and $\sigma\,dA_{VT}$.

Coulomb's law in Equation 4.1 will be written as

$$\vec{F} = \frac{q_1^p\,q_2^p}{r^n}\,\hat{r}, \tag{9.7, ESU}$$

with unknown exponents, and it will be shown next that $p = 1$ and $n = 2$ must be satisfied for the electric field within the sphere to be 0. In general, Coulomb's law in Equation 4.4, is written using the proportionality constant k_e, to account for other systems of units. Although the derivation

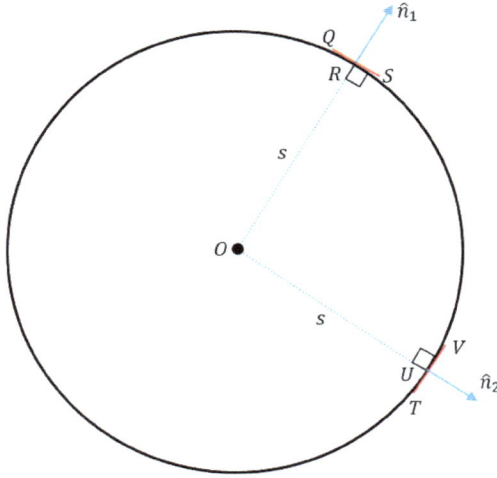

Figure 9.4: Areas dA_{QS} and dA_{VT}, subtending the solid angles at Point P, can be viewed as tangential planes to the surface of the sphere centered at Point O.

presented assumes $k_e = 1$ in ESU, similar steps can be followed to show that it is applicable for any value of k_e.

From the above Coulomb's relation, the definition of the electric field surrounding a point charge q in Equation 5.5 is rewritten as

$$\vec{E} = \frac{q^p}{r^n}\,\hat{r}. \tag{9.8, ESU}$$

The electric field $\vec{E}_{QS}$ at Point P, generated by the charge $\sigma\,dA_{QS}$ is

$$\vec{E}_{QS} = -\frac{\sigma^p\,dA_{QS}^p}{r_1^n}\,\hat{r}_1, \tag{9.9, ESU}$$

and $\sigma\,dA_{VT}$ is

$$\vec{E}_{VT} = -\frac{\sigma^p\,dA_{VT}^p}{r_2^n}\,\hat{r}_2, \tag{9.10, ESU}$$

which are illustrated in Figure 9.3. Since $\hat{r}_2 = -\hat{r}_1$, the net electric-field strength E_P at Point P is

$$E_P = \left|\frac{\sigma^p\,dA_{QS}^p}{r_1^n} - \frac{\sigma^p\,dA_{VT}^p}{r_2^n}\right|. \tag{9.11, ESU}$$

If $p = 1$ and $n = 2$, and substituting Equation 9.6, the above result reduces to

$$E_P = 0. \tag{9.12}$$

A pair of cones with a circular cross section is used only as an example. Any set of pairs of "cones", with a cross section of any shape, can be used to divide the sphere into pairs of areas of equal solid angles. Solid angles can be defined for a cone with a cross-section area of any shape, as shown in the example in Figure $\mathcal{A}.3$(b). Since the electric field cancels for each pair of the cones, the net electric field is 0, considering the total charge on the surface of the sphere. This analysis shows that the force F between charges q_1 and q_2, separated by distance r, must be written as the inverse square law,

$$F \propto \frac{1}{r^n}, \ n = 2,$$
(9.13)

and

$$F \propto q_1^p q_2^p, \ p = 1.$$
(9.14)

10

Voltage

Unlike electric field, which is a vector field, voltage is a scalar field. The motivation to introduce a new electrical quantity will become clear in the future sections. Since voltage is a scalar field, rather than a vector field, it is easier to use. For example, the electric field of a charged metal object is perpendicular to its surface, and are in different directions. It will be shown in this chapter, the charged metal object can be described by one scalar value, its voltage.

The term *potential* is sometimes used instead of voltage. The conservative property of the electric field in electrostatics, proved in Chapter 5, leads naturally to the definition of voltage. This property of the electric field in electrostatics was proved in Chapter 5.

By the fundamental theorem of calculus for line integrals [32], a conservative field can be written as the gradient of a scalar field. The reader is referred to a calculus textbook for more details. Likewise, if a vector field is written as the gradient of a scalar field, then the vector field is conservative. Since the electric field in electrostatics is conservative, it can be written as the gradient of a scalar field,

$$\vec{E} = -\nabla V, \qquad (10.1)$$

where V is voltage, and $\vec{E}$ is electric field.

From calculus, the gradient of the scalar field voltage is a vector, which points in the direction of the steepest ascent, from a lower to a higher voltage, at every point. The negative sign is included in Equation 10.1, so that the electric field is in the direction of the steepest descent, from a higher to a lower voltage. Therefore, the force on a positive charge, whose direction is the same as the direction of the electric field, is in the direction from a higher to a lower voltage. It was thought that positive charges flowed from a higher voltage to a lower voltage, similar to heat flowing from a higher to a lower temperature, and hence, the negative sign.

From the above equation and the definition of a gradient, if there is a spatial variation of voltage, there exists an electric field. If the voltage field is a constant in space, then the electric field is 0.

It is helpful to visualize Equation 10.1 in the discrete domain, as a discretization of space into tiny cells, shown in Figure 10.1. The center of the cells is marked by a small sphere. The distance between the center of a cell to the center of its nearest neighbor is Δx, Δy, and Δz, along the x, y, and z axes, for all the cells. Each cell is identified by indices (i,j,k) of its center point, and marked as shown in the figure. Each of the indices i,j, and k, corresponding to the x, y and z directions, takes on discrete values 1,2,3, ... From the definition of voltage, in the discrete domain, the electric field at the center of a cell $\vec{E}(i,j,k)$ is

$$\vec{E}(i,j,k) = \frac{V(i+1,j,k)-V(i,j,k)}{\Delta x}\hat{x} + \frac{V(i,j+1,k)-V(i,j,k)}{\Delta y}\hat{y} + \frac{V(i,j,k+1)-V(i,j,k)}{\Delta z}\hat{z}. \tag{10.2}$$

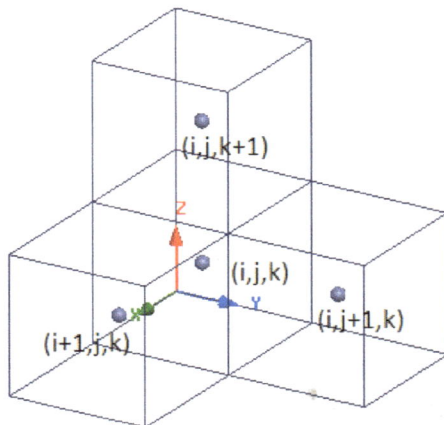

Figure 10.1: Discretization of space into tiny cells.

Integrating Equation 10.1 to calculate the path integral between any two points A and B,

$$\int_A^B \vec{E} \cdot \vec{dl} = \int_A^B -\nabla V \cdot \vec{dl}. \tag{10.3}$$

Since $\vec{E}$ is a conservative field in electrostatics, any path from A to B may be chosen for the path integral, as proved in Section 6.4. The negative sign in the right-hand side of the above equation, can be eliminated by reversing the limits of the integral,

$$\int_A^B -\nabla V \cdot \vec{dl} = \int_B^A \nabla V \cdot \vec{dl}. \tag{10.4}$$

By the fundamental theorem of calculus for line integrals,

$$\int_B^A \nabla V \cdot \vec{dl} = V(A) - V(B)$$

$$= V_{AB}, \tag{10.5}$$

where V_{AB} denotes the potential difference between points A and B. If the above result is positive, then Point A is said to be at a higher potential than Point B. From the above equations, the voltage difference between two points A and B is the path integral of the electric field between the two points,

$$V_{AB} = \int_A^B \vec{E} \cdot \vec{dl}. \tag{10.6}$$

From the above equation, note that voltage is always calculated between two points.

$$V_{AB} = V_A - V_B, \tag{10.7}$$

is the relative voltage at Point A with respect to Point B. Since only relative voltages are calculated, any point can be designated as a reference point. The potential at all other points are calculated with respect to the reference point. Although the reference point can be assigned any reference voltage, it is assigned 0 for convenience. If Point Z is the reference point with reference voltage V_Z, adding and subtracting V_Z to the above equation,

$$\begin{aligned} V_{AB} &= (V_A - V_Z) - (V_B - V_Z) \\ &= V_{AZ} - V_{BZ}. \end{aligned} \tag{10.8}$$

The above equation verifies that the voltage between two points A and B, is the difference in voltages with respect to the reference point Z.

The unit of voltage in ESU is the statvolt, and applying dimensional analysis on the above equation,

$$\begin{aligned} 1\ statvolt &= \frac{dyne}{statcoulomb} \cdot cm \\ &= \frac{gm^{1/2}cm^{1/2}}{s}. \end{aligned} \tag{10.9}$$

10.1   Voltage Near a Point Charge

A point charge, and the electric field around the point charge is shown in Figure 10.2. The electric field around the point charge exhibits spherical symmetry. From Coulomb's law in Equation 4.1, and the electric field definition in Equation 5.2, the electric field at distance r from q in spherical coordinates is

$$\vec{E} = \frac{q}{r^2}\,\hat{r}, \tag{10.10, ESU}$$

where $\hat{r}$ is the unit vector in the radial direction from the point charge to the point at which the electric field is calculated.

The path integral from A to B in Figure 10.2 can be divided into the path from A to C, and from C to B,

$$V_{AB} = \int_A^C \vec{E} \cdot \vec{dl} + \int_C^B \vec{E} \cdot \vec{dl}. \tag{10.11}$$

Since $\vec{E}$ is always perpendicular to $\vec{dl}$ along path AC, the first term in the above equation vanishes, and therefore,

$$V_{AB} = \int_C^B \vec{E} \cdot \vec{dl}. \tag{10.12}$$

In Figure 10.2, let $B \to \infty$, and the reference point for voltage calculations is assigned at $r = \infty$, and set to a reference voltage of 0. The origin is located at the point charge. Substituting Equation 10.10 in Equation 10.12,

$$V_A = V_B + \int_C^B \frac{q}{r^2} \, dr. \tag{10.13, ESU}$$

Since $V(B = \infty) = 0$ is the reference potential, the above equation simplifies to

$$V_A = \int_C^\infty \frac{q}{r^2} \, dr. \tag{10.14, ESU}$$

Evaluating the integral,

$$V_A = \frac{q}{r}, \tag{10.15, ESU}$$

where r is radial distance between the point charge and A (or C). Note that V_A has the same value, no matter where A is located on the dotted circle. The dotted circle is called an *equipotential* line, or the line at the same potential. In 3D, the dotted circle will be the cross section of a sphere, and points on the sphere of any radius, from the above example, will be at the same potential. Such a surface is called an equipotential surface.

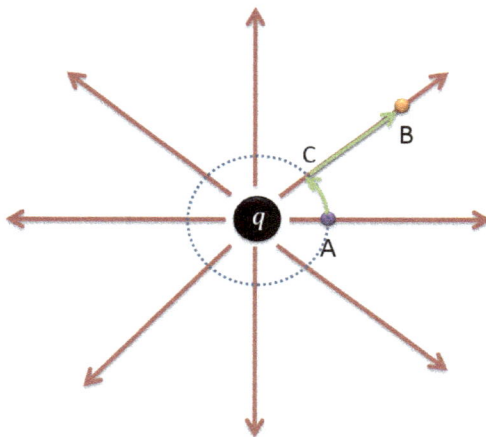

Figure 10.2: The electric field around a point charge.

The electric field from a charge distribution decays to 0 at ∞. If the electric field is 0 at ∞, then the path integral of the electric field between any two points at ∞ is 0. Therefore all points at ∞ are at the same potential. If the reference potential of 0 is assigned to any point at ∞, then all points at ∞ are at 0 potential.

10.2   Voltage Near Many Point Charges

Superposition of electric fields can be used to calculate the voltage at a point, when many point charges are present. N point charges $\{q_1, q_2, ..., q_N\}$ are shown in Figure 10.3, at distances $\{r_1, r_2, ..., r_N\}$ from Point P. Applying the definition of voltage, the voltage $V(P)$ at Point P is

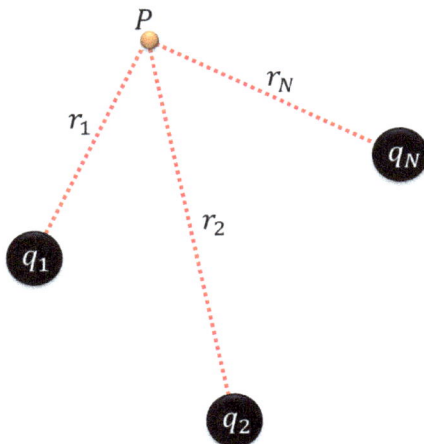

Figure 10.3: The voltage at Point P generated by point charges $\{q_1, q_2, ..., q_N\}$.

$$V(P) = \int_P^\infty \vec{E} \cdot \vec{dl}, \tag{10.16}$$

where the reference potential at ∞ is set to 0. Using superposition, $\vec{E}$ is the total electric field generated by all the point charges,

$$\vec{E} = \vec{E}_1 + \vec{E}_2 + ... + \vec{E}_N, \tag{10.17}$$

where $\{\vec{E}_1, \vec{E}_2, ... \vec{E}_N\}$ are the individual electric fields of the N point charges. From the above two equations,

$$V(P) = \int_P^\infty \vec{E}_1 \cdot \vec{dl} + \int_P^\infty \vec{E}_2 \cdot \vec{dl} + ... + \int_P^\infty \vec{E}_N \cdot \vec{dl}$$
$$= V_1 + V_2 + ... + V_N, \tag{10.18}$$

where $\{V_1, V_2, ..., V_N\}$ are the individual voltages of the N point charges. Using the result from Equation 10.15, the voltage at Point P is

$$V(P) = \frac{q_1}{r_1} + \frac{q_2}{r_2} + ... + \frac{q_N}{r_N}. \tag{10.19}$$

10.3   Voltage of a Parallel-Plate Capacitor

Two oppositely charged metal plates, separated by a small distance d, is shown in Figure 10.4. The top plate is charged positive, and the bottom plate negative. Such a configuration is known as a capacitor, and will be discussed in detail in future chapters. As explained in Section 5.1, assuming that the separation d is small relative to the plate dimensions, a uniform electric field $\vec{E}$ of magnitude E can be assumed at the center region of the plates, far away from the edges. The voltage between the two parallel plates is

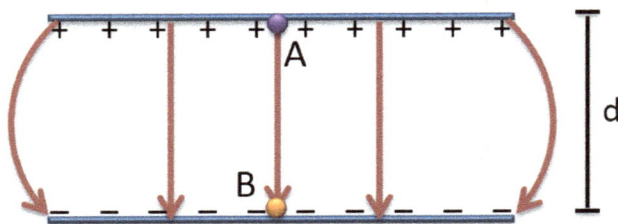

Figure 10.4: The voltage between two oppositely charged parallel plates, separated by a small distance d, with respect to the width of the plates.

$$V_{AB} = \int_A^B \vec{E} \cdot d\vec{l}$$
$$= Ed, \tag{10.20}$$

since $\vec{E}$ is assumed to be uniform along the path.

Substituting Equation 5.2,

$$V_{AB} = \frac{1}{q} \int_A^B \vec{F} \cdot d\vec{l}, \tag{10.21}$$

where $\vec{F}$ is the force experienced by the charge q due to the electric field $\vec{E}$. The path integral

$$W_{AB} = \int_A^B \vec{F} \cdot d\vec{l}, \tag{10.22}$$

is the work done by the electric field in moving the charge q from A to B. From the above equations,

$$V_{AB} = \frac{W_{AB}}{q}. \tag{10.23}$$

Using the above equation, the voltage between points A and B, can also be defined as the work done by the electric field in moving a unit positive charge from Point A to Point B. Although the definition of voltage may look very abstract, it is possible to measure voltage, as demonstrated in Chapter 48.

10.4 A Conductor is an Equipotential Volume

Summarizing Chapter 8, the following statements hold true in electrostatics: given a closed metal surface, filled or with a cavity, but no charged objects in the cavity,

☐ The electric field is 0 within the closed outer metal surface.

☐ The net charges, or the induced charges, lie on the outer metal surface.

A metal object is shown in Figure 10.5. Since the electric field is 0 within the surface, the path

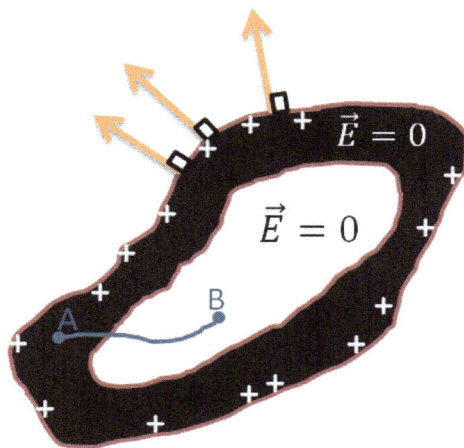

Figure 10.5: A metal in electrostatics is an equipotential volume.

integral of the electric field, or voltage, between two points A and B is 0,

$$V_{AB} = \int_A^B \vec{E} \cdot \vec{dl}$$
$$= 0. \tag{10.24}$$

Therefore, a metal is an equipotential volume, where all the points within the metal surface are at the same potential. By the same reasoning, conductors in contact, or connected by a wire, are at the same potential.

10.5 Charge Transfer Between Conductors

The charge transfer between conductors will be described using voltage and electric field. The voltage of a conductor V_{cond} is

$$V_{cond} = \int_P^{ref} \vec{E} \cdot \vec{dl}, \tag{10.25}$$

where ref is the reference point that is at 0 potential, and P is any point on the surface or within the conductor. Since a conductor in electrostatics is an equipotential volume, the location of P

within the conductor does not change the value of V_{cond}.

Conductors not physically connected may be at different potentials. A positively charged and an uncharged metal spheres, A and B, are shown in Figure 10.6. By spherical symmetry, the electric field of the positively charged sphere A is in the radial direction, shown by the arrows. B is uncharged and has no electric field.

If B is brought close to A, but without touching A, the positively charged A will attract the negative charges in B closer to it, creating a non-uniform charge distribution on the surface of B. The induced charges on B also lie on its outer surface, as shown by the experiment in Section 8.6. The charge distribution of the spheres influence each other. For simplicity, an approximation will be made that the charges on the spheres don't influence each other.

Applying the above definition of V_{cond} on Sphere A, integrating along the dotted line from a on its surface to $-\infty$, $\vec{E} \cdot \vec{dl}$ is a positive value along the path, and V_{cond} is a positive number. Making the approximation B is not influenced by A, and the electric field due to A, along the path from b to ∞ will be assumed to be negligible. Applying the definition of V_{cond}, Sphere B is at 0 potential. A is at a higher potential than B.

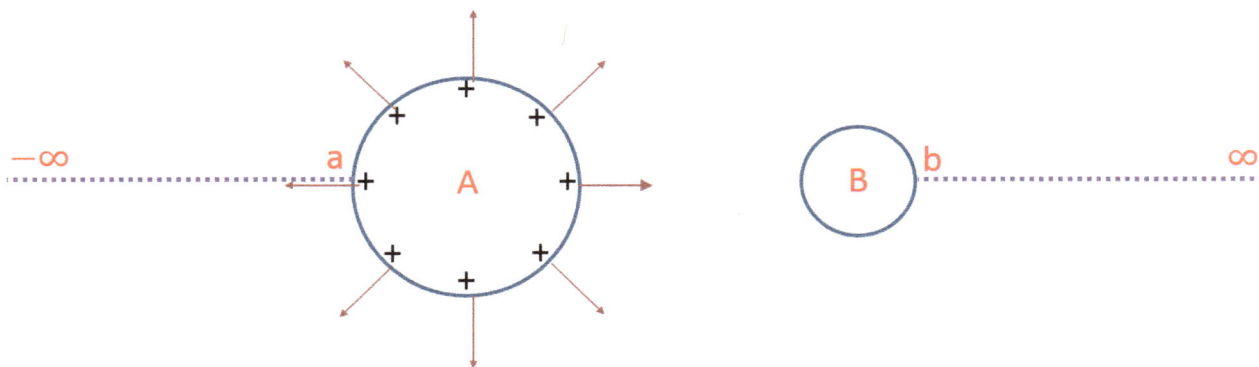

Figure 10.6: A charged metal sphere A, and an uncharged metal sphere B.

The two spheres are connected by a wire in Figure 10.7(a). It can be observed using an electroscope that B gets charged. When the spheres are connected, there must exist a transient electric field in the spheres, and in the wire connecting them, exerting a force on the charges in A, and moving some of them to Sphere B. The final charge distribution after equilibrium is shown in Figure 10.7(b), and the electric field is 0 in the steady state.

An understanding of the electric field in the spheres and the wire, resulting in the rearrangement of the charges, will be explained using Figure 10.8. The wire dimensions are greatly exaggerated to show the behavior of the field. A part of the wire connected to A is marked w. The electric-field lines at the instant the wire is connected, and before any charge transfer occurs, are shown by the arrows in Figure 10.8(a), and its corresponding voltage map in Figure 10.8(b). Both Figure 10.8(a)

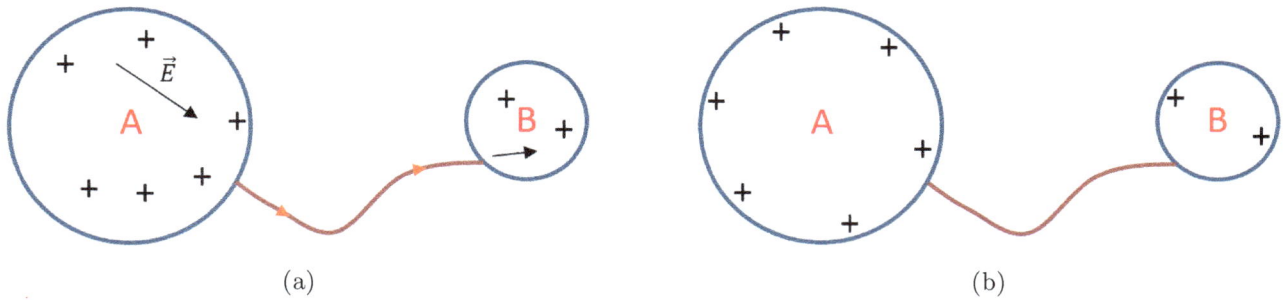

Figure 10.7: (a) A transient electric field in the conductors, forces the charges to flow from the charged sphere A to the uncharged sphere B, until steady state is reached. (b) The electric field is 0 in the conductors in the steady state.

and Figure 10.8(b) convey the same information. By spherical symmetry, the electric-field lines are radially outward of the sphere. This field is also present in the wire connection, as shown. The Coulomb force between the charges redistributes them.

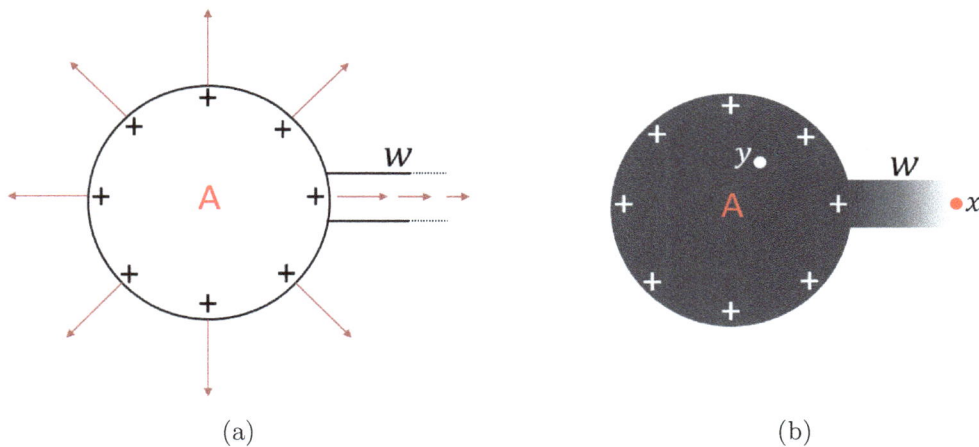

Figure 10.8: (a) An illustration of the electric field in the wire w joining two conductors A and B, before any charge transfer occurs. (b) The voltage gradient in the wire w, before any charge transfer occurs, shown by a color map.

By definition of voltage in Equation 10.1, the existence of an electric field in the spheres implies a spatial variation of voltage, or a voltage difference between the spheres. Since voltage is a scalar field, the spatial variation of voltage within the metal is shown by the color map in Figure 10.8(b). The point marked x serves as the reference point for the voltage map. A higher voltage difference between any point and x is indicated by a darker shade.

Using the electric-field lines in Figure 10.8(a), the voltage map can be constructed from Equation 10.6, between x and any other point y. The electric-field lines, and the voltage map vary over time, until equilibrium of the charges is attained. In the steady state, the conductors that are connected together form an equipotential volume, or zero electric field within the surface of the conductors.

10.6   The Law of Conservation of Charge

Experiments with charged objects, such as the ones presented until now, led to the belief that charges are not created or destroyed, but transferred from one object to another object. This is known as the law of conservation of charge, a fundamental axiom of physics.

11

A Brief History of Magnetism

Naturally occurring magnets, called lodestone, have been known to exist since antiquity, and it was common knowledge that they attracted pieces of iron. There is a story that a shepherd in Magnesia, Greece, accidentally discovered lodestone when his metal stick got stuck to it. The origin of the word magnet, comes from the place where it was first discovered.

A compass, invented in China over 1000 years ago, is a small magnet, when suspended freely, points towards the north [33]. By convention, the end of a freely suspended magnet pointing north is called the north pole of the magnet, and the end pointing south is the south pole of the magnet, and both poles are always present in any magnet. Similar to electric charges, like poles repel each other, while unlike poles attract.

A homemade magnetic compass is shown in Figure 11.1(a). A magnet is rubbed against an iron needle about 10 times and the needle becomes magnetized. The needle inserted into a wooden cork, and floated in water, as shown in the figure, always comes to rest pointing north [34].

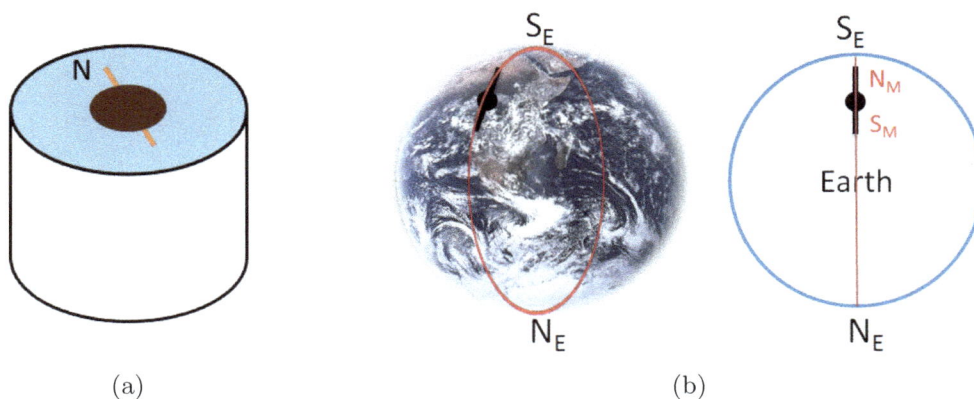

Figure 11.1: (a) A homemade magnetic compass pointing north. (b) The Earth is a giant spherical magnet [13].

William Gilbert, who was also the physician to Queen Elizabeth I, recognized that the Earth is a giant magnet. Using a spherical magnet as a miniature model for Earth, he was able to explain that a magnetic compass always pointed north because the north pole of the magnetic compass is attracted to the south pole of the terrestrial magnet.

In Figure 11.1(b), N_E, S_E are the north and south poles of the terrestrial magnet, and N_M, S_M are the north and south poles of the magnetic compass. A magnetic compass located at some point on Earth is shown in Figure 11.1(b). A circle through the compass and the terrestrial magnetic poles of Earth, is shown in the first picture of the figure. The view looking directly in front of the circle, is shown in the second picture of Figure 11.1(b). N_M is attracted to S_E, and S_M attracted to N_E, causing the magnet to point north. Note that the south pole of the terrestrial magnet is at the geographic north pole, and the north pole of the terrestrial magnet at the geographic south pole.

12

Magnetic Charge and its Importance in the Development of Electromagnetic Theory

It was believed that magnetic charges exist, similar to electric charges. Some of the ways by which magnetic charges can be detected, if they do exist, will be presented in Section 17.7. Although naturally occurring magnetic charges have not been detected until now, the hypothesis was a useful one: it allowed electromagnetic theory to develop.

The irony is that once the electromagnetic equations are formulated assuming that magnetic charges exist, magnetic charges are no longer required, and their existence can be treated as agnostic. Alternate definitions will be formulated, without the need of magnetic charges. To get to that point, however, magnetic charges are essential. Therefore, without magnetic charges, it is not possible to build electromagnetic theory. Unfortunately, the topic of magnetic charges in electromagnetic theory is often avoided. It is considered "out of fashion", and sometimes even treated as a taboo!

Some of the significant developments, made possible by the hypothesis of the existence of magnetic charges will be highlighted. These topics will be discussed in great detail in the future chapters. Magnetic field will first be defined using magnetic charges. Without this definition, it is not possible to formulate Faraday's law, Ampere's law, or Biot-Savart's law. Early instruments used for the measurement of current and magnetic field, directly or indirectly, depended on magnetic charges. Magnetic charges cannot be avoided, and its importance cannot be overemphasized!

13

Coulomb's Law of Magnetic Charges

A hypothesis will be made that the force governing magnetic charges is similar to electric charges. Similar to positive and negative electric charges, there are positive and negative magnetic charges. Charges of the same polarity repel each other, while opposite polarities attract. The equivalent Coulomb's law of magnetic charges, assuming symmetry with electric charges, is

$$\vec{F} = \frac{m_1 m_2}{r^2}\,\hat{r}, \tag{13.1, EMU}$$

where $\vec{F}$ is the force between two point magnetic charges of magnitudes m_1 and m_2, separated by a distance r, and $\hat{r}$ is the unit vector that points in the direction of the force. The force of repulsion between two point magnetic charges is illustrated in Figure 13.1(a), and the force of attraction in Figure 13.1(b). Gauss's empirical evidence to verify the inverse-square law of magnetic charges will be presented in Chapter 15.

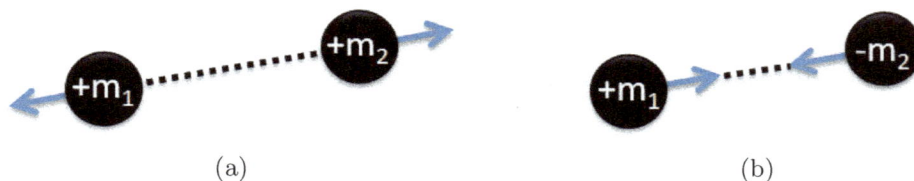

(a)   (b)

Figure 13.1: (a) The force of repulsion between two like magnetic charges. (b) The force of attraction between two unlike magnetic charges.

Equation 13.1 is valid only in EMU. This will be explained in Chapter 17.6, when the flowchart of definitions of electrical quantities in ESU and EMU will be presented.

The unit of force $[\vec{F}]$ in CGS units is the dyne, and repeating Equation 2.8,

$$1\,dyne = \frac{gm\,cm}{s^2}. \tag{13.2}$$

The unit of distance $[r]$ is cm. Applying dimensional analysis on Equation 13.1, the unit of magnetic charge $[m]$ is

$$[m] = \frac{gm^{1/2}cm^{3/2}}{s}$$
$$= maxwell \tag{13.3, EMU}$$

The unit of magnetic charge in EMU is known as maxwell. The ab/stat convention mentioned in Section 4.1 does not apply to the unit maxwell, or any other proper noun used as a unit.

14

Early Definition of Magnetic Field

A magnetic compass experiences a torque that aligns the needle to point towards the north. Similar to electric charges experiencing a force in an electric field, it was thought that magnetic charges in a magnetic needle, experience a force in the terrestrial magnetic field. In Figure 14.1(a), the north and the south poles of a magnetic compass are modeled with positive and negative magnetic charges $\pm m$. By convention, the north pole is modeled by a positive magnetic charge, and the south pole by a negative magnetic charge.

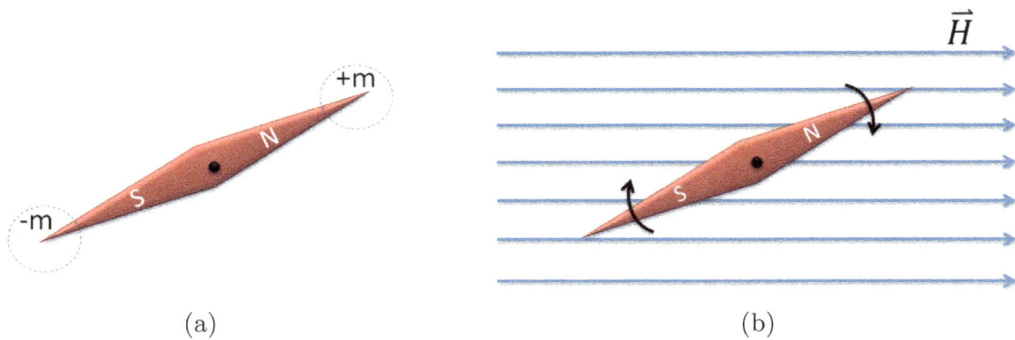

(a)   (b)

Figure 14.1: (a) The model of a magnetic compass with magnetic charges at its end points. (b) The torque on a magnetic compass causing it to align to the magnetic field $\vec{H}$.

The force from the terrestrial magnetic field on the magnetic charges generates a torque, and causes the magnetic needle to rotate around its center pivot and align itself to the field, as shown in Figure 14.1(b). From this example, note that the direction of the magnetic field at a point is the direction in which the north pole of a compass is pointing towards.

Similar to the definition of the electric field in Equation 5.2, the early definition of magnetic field is

$$\vec{H} = \frac{\vec{F}}{m},$$

(14.1)

where the magnetic field $\vec{H}$ is the force $\vec{F}$ exerted on a unit positive magnetic charge. From $[m]$

in Equation 13.3, and applying dimensional analysis on the above equation, the unit of magnetic field $[\vec{H}]$ in EMU is

$$[\vec{H}] = \frac{g m^{1/2}}{cm^{1/2}\, s}$$
$$= oersted. \tag{14.2, EMU}$$

The Earth is a giant magnet whose south/north poles are at the geographic north/south poles, as explained in Chapter 11. The terrestrial magnet is modeled by the magnetic charge $\pm m_E$, shown in Figure 14.2. Using the definition of the magnetic field in Equation 14.1, and the Coulomb's law of magnetic charges in Equation 13.1, the terrestrial magnetic field is plotted in the figure. The direction of the magnetic field of a magnet is from its north pole to the south pole.

The magnetic field of Earth can be resolved into a horizontal component that is tangential to the Earth's surface, and a normal component that is perpendicular to the surface. Near the equator, marked by the dotted line in the figure, the terrestrial magnetic field is mostly tangential to the Earth's surface. At the geographic poles, the terrestrial magnetic field is mostly normal to the surface. The compass is free to rotate on the plane that is parallel to the Earth's surface. The models of a compass and the terrestrial magnet with magnetic charges, well captures the behavior of a compass that points towards the north. This is a mild verification of the definitions, the models, and the assumptions made.

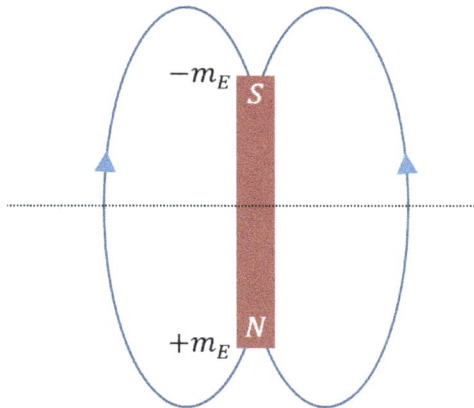

Figure 14.2: The terrestrial magnetic field.

14.1   Magnetic and Electric Dipole Moment

The top view of a magnetic needle is shown in Figure 14.3(a), and is free to rotate on the plane of the page about its center pivot. The needle is modeled with magnetic charges $\pm m$ at its end points. The direction of the terrestrial magnetic field is shown by the arrow marked $\vec{H}$. The magnetic charges experience a force

$$\vec{F} = \pm m \vec{H}, \tag{14.3}$$

which generates a torque on the needle, rotating the needle about its center pivot point, and aligning it to the magnetic field $\vec{H}$. If ℓ is the length of the needle, the magnitude of the torque on the needle is

$$\tau = 2 \times m \left(\frac{\ell}{2}\right) H \sin\theta. \tag{14.4}$$

The factor $2\times$ arises from the force acting on both the magnetic charges $\pm m$.

The above equation is written as a cross product,

$$\vec{\tau} = m\vec{\ell} \times \vec{H}, \tag{14.5}$$

where $\vec{\ell}$ is the length vector in the direction from $-m$ to $+m$, as shown in the figure. This is the convention followed for the direction of $\vec{\ell}$, and the order of the operands in the cross product. $m\vec{\ell}$ is known as the magnetic dipole moment.

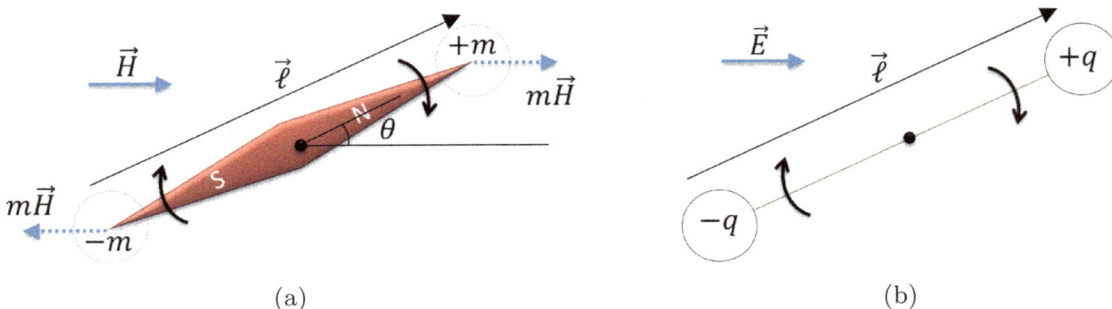

Figure 14.3: (a) A magnetic dipole. (b) An electric dipole.

The same exercise can be repeated on an electric dipole shown in Figure 14.3(b). The electric dipole is made up of two tiny charged metal spheres of electric charges $\pm q$, joined together by a dielectric rod, and is free to rotate on the plane of the page about its center pivot. $\vec{\ell}$ is the length vector from $-q$ to $+q$, similar to the magnetic dipole. In an electric field $\vec{E}$, the force on the charges creates a torque, similar to Equation 14.5,

$$\vec{\tau} = q\vec{\ell} \times \vec{E}. \tag{14.6}$$

$q\vec{\ell}$ is known as the electric dipole moment. This definition will be used in the optional Section 37.3 on bound charge.

15

Gauss's Experiment to Prove the Inverse-Square Law of Magnetic Charges

Lets assume that the inverse-square relation in Coulomb's law of magnetic charges is not known. The exponent of r in Equation 13.1 is replaced by p, whose value is to be verified by Gauss's experiment to be the numerical value 2. The modified equation is

$$\vec{F} = \frac{m_1 m_2}{r^p}\,\hat{r}, \qquad\qquad (15.1,\ \text{EMU})$$

where m_1 and m_2 are the magnitudes of two point magnetic charges separated by a distance r, and $\hat{r}$ is the unit vector that points in the direction of the force $\vec{F}$.

Gauss's experiment requires only two small magnets to show that indeed p does equal 2. One of the magnets stays fixed through out the experiment, and the second magnet is a small compass that is free to rotate about its center, on a plane parallel to the Earth's surface. He positioned the magnets in two different configurations, which will be referred to as broadside and inline, shown in Figure 15.1(a) and Figure 15.1(b), respectively. Point Q is "broadside" to CD in Figure 15.1(a), and is "inline" to CD in Figure 15.1(b).

Magnet 1 stays fixed in the experiment, and Magnet 2 is a magnetic needle from a compass that is free to rotate about its center, as marked in Figure 15.1. The shaded triangle of Magnet 1, and the shaded tip of Magnet 2, are the north poles of the magnets. The magnetic north-south (N-S) direction is the direction that Magnet 2 points in the absence of Magnet 1. This is marked by the dotted line, and is also called the magnetic meridian. The distance between the centers of the magnets is r. Using the angle of rotation of Magnet 2 from the magnetic meridian in the two configurations, as a function of r, Gauss was able to prove the inverse-square relationship of the force between magnetic charges. This derivation will be presented in detail first, followed by empirical results.

Subscripts b and i will be used to differentiate between broadside and inline configurations. Subscripts M and E differentiate between the source of the magnetic field at the location of Magnet 2, as that due to Magnet 1 or Earth, respectively. Subscripts 1 and 2 differentiate between Mag-

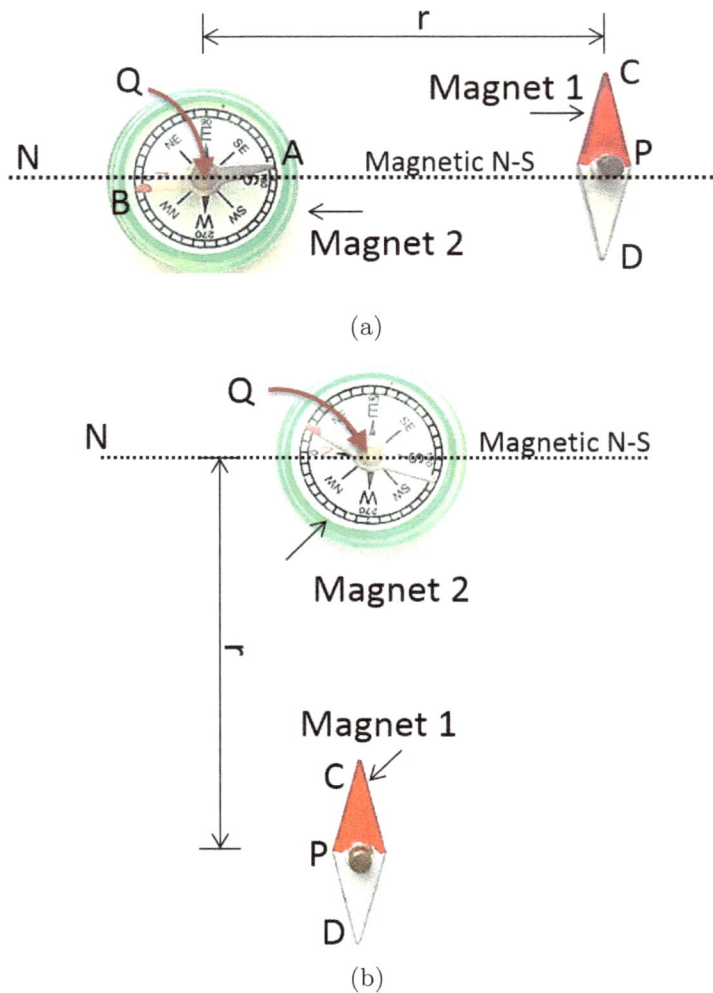

Figure 15.1: Magnets placed in two different configurations: (a) broadside configuration, (b) inline configuration, to prove the inverse-square relation of magnetic charges.

net 1 and Magnet 2, respectively. Only the component of the terrestrial magnetic field that is tangential to the Earth's surface is of interest. This is the component that causes a compass to rotate horizontally on a plane parallel to the Earth's surface, and the normal component will be ignored.

15.1   Broadside Configuration

Magnet 1 creates a magnetic field at Point Q, the center of Magnet 2, as shown in Figure 15.2(a) that illustrates the broadside configuration. Using the model in Chapter 14, Magnet 1 is modeled with magnetic charges $+m_1$ and $-m_1$ at its north (Point C) and south poles (Point D). The force acting on a unit positive magnetic charge located at Q, due to $+m_1$ and $-m_1$, by definition from Equation 14.1, is the magnetic field caused by Magnet 1 $\vec{H}_{Mb}$ at Q.

By symmetry, the horizontal components of the force acting on a unit positive magnetic charge at Point Q, due to $\pm m_1$, cancel each other, resulting in only the vertical component of the field, shown by the solid line $\vec{H}_{Mb}$ in the figure. As mentioned earlier, subscript M refers to the field generated by Magnet 1, and the subscript b stands for broadside configuration.

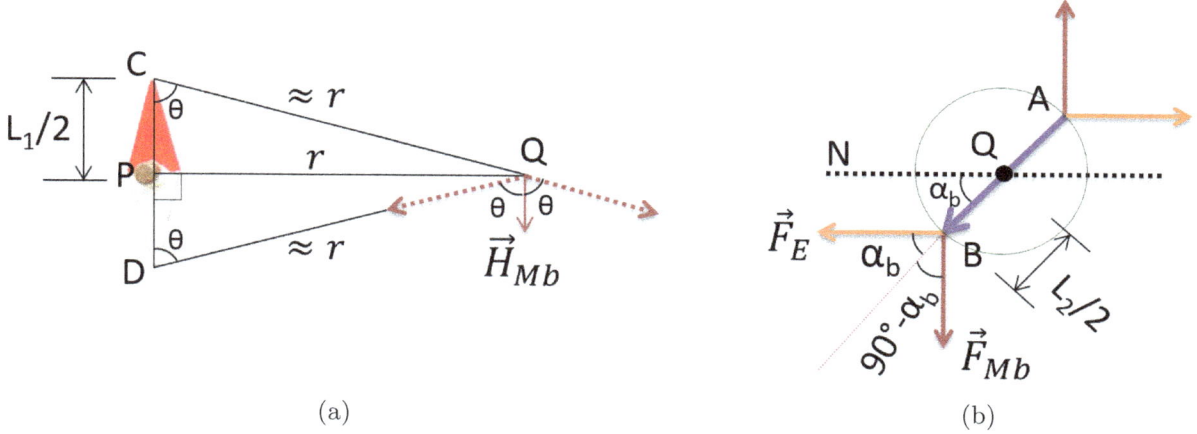

Figure 15.2: (a) Magnet 1 and (b) Magnet 2 in the broadside configuration.

Using Equation 15.1, and from trigonometric relations, the force on a unit positive magnetic charge at Point Q is

$$H_{Mb} = 2\frac{m_1}{r^p} \cos\theta, \qquad (15.2, \text{EMU})$$

where H_{Mb} is the magnitude of the magnetic field at Point Q due to Magnet 1, θ is angle $\angle DCQ$, $\pm m_1$ are the magnetic charges at Point C and Point D. To simplify the experiment, the length of the segments QC and QD will be approximated to be r. This approximation becomes more accurate for a larger separation between Point Q and Points C and D, or using a smaller Magnet 1. From the right-angled triangle formed by Points Q, C, and P,

$$\cos\theta = \frac{L_1}{2r}, \qquad (15.3)$$

where L_1 is the length of Magnet 1, or Segment CD. Substituting the above expression into Equation 15.2,

$$H_{Mb} = \frac{m_1 L_1}{r^{p+1}}. \qquad (15.4, \text{EMU})$$

The schematic of Magnet 2 is drawn in Figure 15.2(b). The magnetic needle of the compass (Magnet 2) is shown as an arrow, with its center marked Q in the figure. In the absence of Magnet 1, Magnet 2 points towards the north marked N, and the magnetic needle lies along the magnetic meridian shown by the dotted line.

Similar to Magnet 1 modeled with magnetic charges $\pm m_1$ at its north/south poles, Magnet 2 is modeled with magnetic charges $\pm m_2$. The magnetic field $\vec{H}_{Mb}$, created by Magnet 1 at the location of Magnet 2, exerts a force on the magnetic charges of Magnet 2, indicated as $\vec{F}_{Mb}$ in Figure 15.2(b). Let α_b be the angle the needle rotates from the magnetic meridian.

The magnetic field of Earth exerts a force $\vec{F}_E$ on the poles of Magnet 2, wanting to align the magnetic needle towards the magnetic meridian. Since the compass is in rotational equilibrium, the torque caused by $\vec{F}_{Mb}$ must balance the torque caused by $\vec{F}_E$. The resulting equation is

$$2 \left[\frac{L_2}{2} F_E \sin \alpha_b \right] = 2 \left[\frac{L_2}{2} F_{Mb} \sin \left(90° - \alpha_b \right) \right], \tag{15.5}$$

where L_2 is the length of Magnet 2. The multiplication factor of 2 on both sides of the above equation arises from the forces acting on the two poles $\pm m_2$ of the magnet. The reader is referred to a mechanics textbook for more details on the equations related to rotational equilibrium.

By definition of the magnetic field in Equation 14.1, the magnitude of the force due to Earth's magnetic field strength H on Magnet 2 is

$$F_E = m_2 H. \tag{15.6, EMU}$$

Likewise, the magnitude of the force on Magnet 2, caused by the magnetic-field strength generated by Magnet 1 H_{Mb} is

$$F_{Mb} = m_2 H_{Mb}. \tag{15.7, EMU}$$

Substituting Equation 15.4 in the above equation,

$$F_{Mb} = m_2 \frac{m_1 L_1}{r^{p+1}}. \tag{15.8, EMU}$$

Using the trigonometric identity

$$\sin(90° - \alpha_b) = \cos(\alpha_b), \tag{15.9}$$

and substituting Equation 15.6 and Equation 15.8 in Equation 15.5,

$$\tan \alpha_b = \frac{m_1 L_1}{H r^{p+1}}, \tag{15.10, EMU}$$

where α_b is the angle of rotation of Magnet 2 in the broadside configuration.

15.2   Inline Configuration

Magnet 1 and Magnet 2 in the inline configuration are shown in Figure 15.3. The magnets are modeled with magnetic charges $\pm m_1$ at Point C/Point D, and $\pm m_2$ at Point A/Point B, as before. The magnetic field generated at the center of Magnet 2, Point Q, by Magnet 1, in the inline configuration, will be derived next.

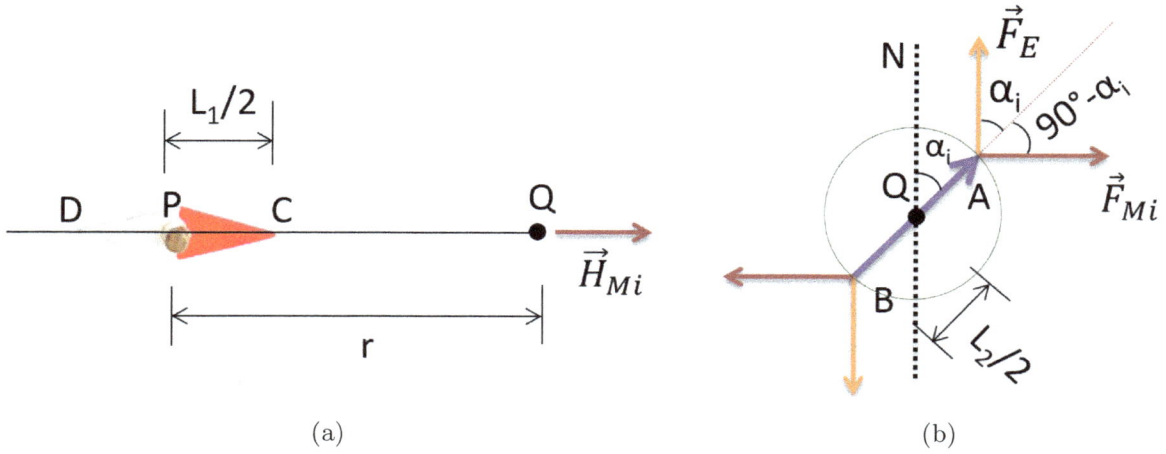

Figure 15.3: (a) Magnet 1 and (b) Magnet 2 in the inline configuration.

Using Equation 15.1, the force on a unit positive magnetic point charge at Point Q, which by definition is the magnetic field H_{Mi}, is

$$H_{Mi} = \frac{m_1}{\left(r - \frac{L_1}{2}\right)^p} - \frac{m_1}{\left(r + \frac{L_1}{2}\right)^p}$$
$$= \frac{m_1}{r^p \left(1 - \frac{L_1}{2r}\right)^p} - \frac{m_1}{r^p \left(1 + \frac{L_1}{2r}\right)^p}, \qquad \text{(15.11, EMU)}$$

where L_1 is the length of Magnet 1, and r is the length of segment PQ. $+m_1$ at Point C of Magnet 1 is closer to Point Q than $-m_1$ at Point D. As a result, the net force on the unit positive magnetic charge at Point Q, or the magnetic field at Point Q, is in the direction shown by the arrow at Point Q in Figure 15.3(a).

Assuming that $\frac{L_1}{2r} \ll 1$, and using the approximation $(1 \pm x)^p \approx 1 \pm px$ for $x \ll 1$,

$$H_{Mi} \approx \frac{m_1}{r^p \left(1 - \frac{pL_1}{2r}\right)} - \frac{m_1}{r^p \left(1 + \frac{pL_1}{2r}\right)}$$
$$= \frac{m_1 p L_1}{r^{p+1} \left(1 - \frac{pL_1}{2r}\right)\left(1 + \frac{pL_1}{2r}\right)}. \qquad \text{(15.12, EMU)}$$

Further simplifying, assuming $\frac{pL_1}{2r} \ll 1$,

$$H_{Mi} \approx \frac{m_1 p L_1}{r^{p+1}}. \qquad \text{(15.13, EMU)}$$

Magnet 2 in the inline configuration is shown in Figure 15.3(b). The north pole of the magnetic needle is marked the arrow. In the absence of Magnet 1, Magnet 2 points in the direction of the magnetic meridian shown by the dotted line, towards N marked in the figure. By definition of the magnetic field in Equation 14.1, the force F_{Mi} exerted on each of the magnetic compass poles due to Magnet 1 is

$$F_{Mi} = m_2 H_{Mi}. \tag{15.14, EMU}$$

Substituting Equation 15.13 in the above expression,

$$F_{Mi} = m_2 \frac{pm_1 L_1}{r^{p+1}}. \tag{15.15, EMU}$$

The force $\vec{F}_{Mi}$ on $\pm m_2$ causes Magnet 2 to rotate, and is balanced by the force caused by the magnetic field of Earth $\vec{F}_E$, as shown in Figure 15.3(b). At equilibrium, the magnetic needle has rotated α_i degrees from the magnetic meridian. Note that the direction that Magnet 2 turns is as drawn in the figure.

Magnet 2 is in rotational equilibrium, and by Newton's 2nd law of rotational motion, the torque caused by $\vec{F}_{Mi}$ and $\vec{F}_E$ must balance each other. The reader is referred to a mechanics textbook for more details on the equations related to rotational equilibrium. Therefore,

$$2 \left[\frac{L_2}{2} F_E \sin \alpha_i \right] = 2 \left[\frac{L_2}{2} F_{Mi} \sin (90° - \alpha_i) \right], \tag{15.16}$$

The two sides of the above equation are multiplied by 2, to account for the force acting on the north and the south poles, resulting in doubling of the torques.

Using the trigonometric identity

$$\sin(90° - \alpha_i) = \cos(\alpha_i), \tag{15.17}$$

and substituting Equation 15.6 and Equation 15.15 in Equation 15.16,

$$\tan \alpha_i = \frac{pm_1 L_1}{Hr^{p+1}}. \tag{15.18, EMU}$$

Note that $\tan \alpha_i$ is p times greater than $\tan \alpha_b$. Therefore,

$$p = \frac{\tan \alpha_i}{\tan \alpha_b}. \tag{15.19, EMU}$$

By doing two experiments, one on broadside configuration, and the other on inline configuration, and calculating the ratio of $\tan \alpha_i$ and $\tan \alpha_b$ in the two experiments, p can be determined, and the inverse-square law of magnetic charges can be proven. Gauss's experiment and the results are presented next.

15.3 Empirical Results of Gauss's Experiment

The experiment setup of the broadside and inline configurations are shown in Figure 15.1(a) and Figure 15.1(b). The values of p calculated from Equation 15.19, have been empirically determined

Smaller Compass Results		
r(cm)	α_i	α_b
4	70°	32.5°
5	40°	20°
6	20°	10°
8	10°	5°

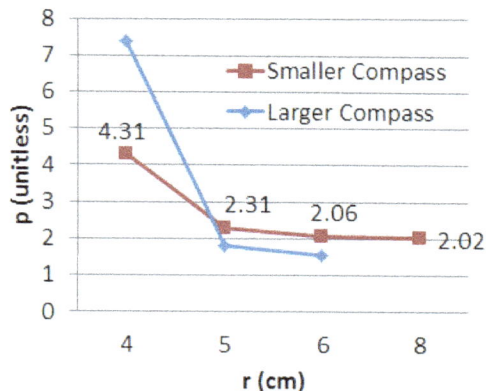

Figure 15.4: The values of p in Equation 15.19 for different values of separation between Magnet 1 and Magnet 2.

for $r = 4, 5, 6, 8$ cm.

Two different compasses were used in the experiment for Magnet 2, of which, only the smaller compass shown in Figure 15.1 produced the expected result. The values of α_i and α_b in the smaller compass case are tabulated in Figure 15.4. The values of p as a function of r, have been plotted in the figure. Results from the smaller compass are plotted using square data markers, and the diamonds are the markers from the larger compass. The value of p is almost equal to 2 for larger values of r, in the smaller compass case.

The results show the inverse-square law behavior only for larger values of r. The approximations made to simplify the equations depend on $r \gg L_1$, justifying the observed inverse-square law behavior only for larger values of r.

A smaller compass results in a smaller error for the approximations made in the derivation. For example, the magnetic field generated by Magnet 1, was calculated at the center of Magnet 2. This value was used to calculate the force on the magnetic charges of Magnet 2 at its endpoints. The spatial variation of the magnetic field over Magnet 2 due to Magnet 1, would be smaller across a smaller compass than a larger compass.

From this experiment, Gauss was able to provide experimental evidence for the inverse-square law of magnetic charges. Extending this work, Gauss was the first to develop a technique to measure the strength of a magnetic field, presented next.

16

Measurement of the Strength of a Magnetic Field Using Gauss's Magnetometer

A magnetometer is a device to measure the strength of a magnetic field, and Gauss invented the first magnetometer. It will be explained in Chapter 27, early instruments, such as the galvanometer to measure current, require the terrestrial magnetic-field strength at the place of measurement to be known. This technique plays an important role in the development of electromagnetic theory.

Gauss's powerful method requires only two small magnets to make the measurement. Gauss's method can be used to measure terrestrial magnetism, magnetic field near magnets, or the magnetic field generated by a current, which will be discussed later. Measurement of the magnetic-field strength of Earth will be discussed in this chapter [37].

16.1 Gauss's Method

Gauss's method will be used to measure the magnetic-field strength of Earth in Chandler, Arizona, USA. Unless otherwise specified, terrestrial magnetic field will always refer to the component of Earth's field that is tangential to the Earth's surface, and the field component normal to the Earth's surface will be ignored.

Two small magnets, which will be referred to as Magnet 1 and Magnet 2, will be required to make the measurement. The two data points needed for the calculation are (1) rotation angle of Magnet 2 when placed at some distance from Magnet 1, and (2) the period of oscillation of Magnet 1 after it has been displaced from the magnetic meridian.

16.1.1 Data Point 1

Two small magnets, Magnet 1 and Magnet 2, are placed in the inline or the broadside configuration, as shown in Figure 16.1, similar to how it was done in Chapter 15. In the example presented in this chapter, the broadside configuration is used for the magnetic-field calculation. The inline configuration could have also been used. Rearranging Equation 15.10, using the value $p = 2$ for the inverse square law that was shown empirically to be true, and using the same variable naming

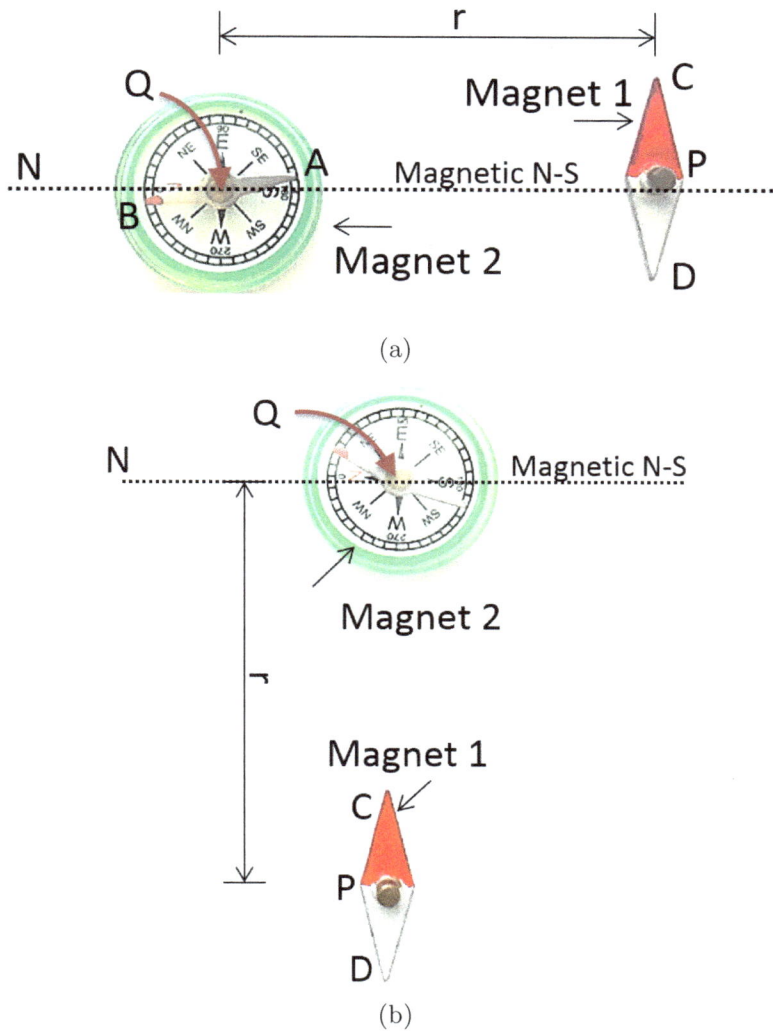

Figure 16.1: For the measurement of the terrestrial magnetic-field strength, the magnets can be placed either in (a) broadside configuration, or (b) inline configuration.

convention,

$$\frac{m_1 L_1}{H} = r^3 \tan \alpha_b, \qquad \text{(16.1, EMU)}$$

where $\pm m_1$ are the magnetic charges at the end points of Magnet 1, H is the magnetic-field strength of Earth that is to be determined, r is the distance between Magnet 1 and Magnet 2, α_b is the angle of Magnet 2 from the magnetic meridian in the broadside configuration, and L_1 is the length of Magnet 1. There are two unknowns in the above equation, m_1 and H. Using the second data point presented next, H can be solved.

16.1.2   Data Point 2

For the second data point, only Magnet 1 is required. Magnet 1 is placed on a needle, as shown in Figure 16.2(b). The period of oscillation is measured when the magnet is displaced by some angle and allowed to oscillate. The equation for the period of oscillation is solved next.

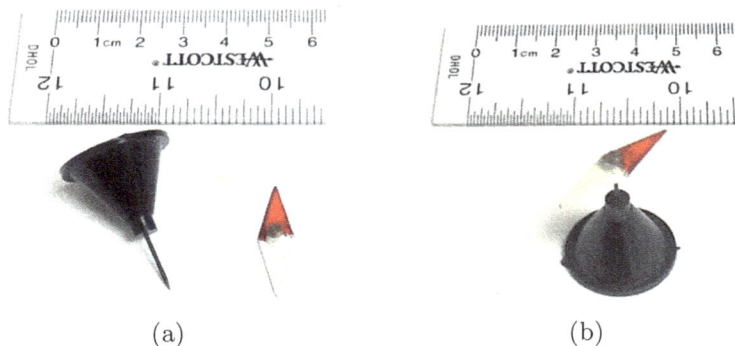

(a)   (b)

Figure 16.2: (a) Magnet 1 and a needle. (b) Magnet 1 placed on the needle to enable torsional oscillation around the pivot.

Repeating Equation 3.20, the rotational equivalent of Newton's 2nd law is

$$\tau = I\ddot{\theta}, \tag{16.2}$$

where τ is the torque applied on the rotating object, I is the mass moment of inertia, and $\ddot{\theta}$ is the angular acceleration. As explained in Section 3.3, angles will be measured in radians, rather than in degrees, for simplified expressions of integrals and derivatives.

Similar to the exercise in Section 3.3, positive values of θ are set as angles in the counter-clockwise direction. As explained before, positive values of $\dot{\theta}$, $\ddot{\theta}$ and τ are also in the counter-clockwise direction, and clockwise if negative. Magnet 1 is released after displacement by an angle θ in the counter-clockwise direction, and allowed to settle down, as illustrated in Figure 16.3. Magnet 1 is modeled with $\pm m_1$ at its north and south poles. The magnetic field of Earth exerts a force on the magnetic charges of the magnet,

$$F_E = \pm m_1 H. \tag{16.3}$$

Similar to the restoring torque in a twisted wire in Equation 3.21, F_E exerts a restoring torque on the magnet,

$$\tau_r = -2\left(\frac{L_1}{2}\right) F_E \sin\theta, \tag{16.4}$$

to restore the needle to the magnetic N-S direction, always acting in the opposite direction to the displaced angle of the magnet θ. This is the source of the negative sign in the above equation. The multiplying factor 2 arises from the force acting on the two magnetic charges $\pm m_1$, at the two endpoints of the magnet.

Substituting Equation 16.3 into the above equation and simplifying,

$$\tau_r = -L_1 m_1 H \sin\theta. \tag{16.5}$$

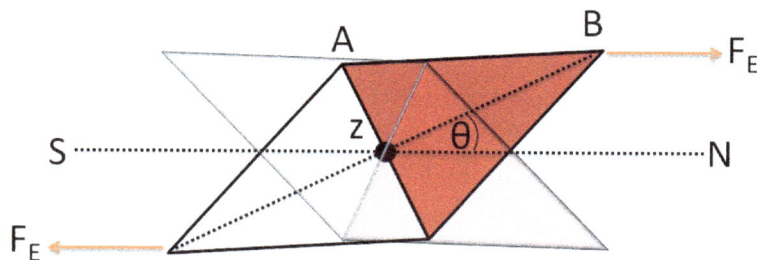

Figure 16.3: The top view of Magnet 1, oscillating after released by some angle, and allowed to settle down. The magnetic field of Earth exerting a force $\vec{F}_E$ on the magnetic compass.

Substituting the above expression in Equation 16.2,

$$-L_1 m_1 H \sin\theta = I\,\ddot{\theta}. \tag{16.6}$$

For small angles of θ, using the approximation $\sin\theta \approx \theta$,

$$-L_1 m_1 H\theta = I\,\ddot{\theta}. \tag{16.7}$$

The solution to the differential equation is the harmonic motion of the magnet,

$$\theta = a\cos\left(\omega_o t + \phi\right), \tag{16.8}$$

where ϕ is a phase offset that can be used to set the value of θ at time $t = 0$, a is the amplitude of the periodic motion, and ω_o is the undamped angular frequency of oscillation. Repeating Equation 3.31, using radians for angles,

$$\omega_o = \frac{2\pi}{T} = \sqrt{\frac{L_1 m_1 H}{I}}. \tag{16.9}$$

Solving for the period of oscillation T,

$$T = 2\pi\sqrt{\frac{I}{L_1 m_1 H}}. \tag{16.10}$$

Solving for H from Equation 16.10 and Equation 16.1,

$$H = \frac{2\pi}{T}\sqrt{\frac{I}{r^3 \tan\alpha_b}}. \tag{16.11, EMU}$$

The terms on the right-hand side of the above equation can be measured from the experiment, and the magnetic-field strength can be calculated. From H, $M = m_1 L_1$ can also be calculated, if needed, using Equation 16.1,

$$\begin{aligned} M &= m_1 L_1 \\ &= H\,r^3 \tan\alpha_b. \end{aligned} \tag{16.12}$$

The mass moment of inertia of Magnet 1 can be calculated from the dimensions of the magnet and the mass of the magnet. The top view of the magnet is shown in Figure 16.4. The magnet lies on the xy-plane, and the z-axis passing through Point O in Figure 16.4(a) is the axis of rotation at the center of the magnet.

The mass moment of inertia is

$$I = \int_A r^2 \rho \, dA, \tag{16.13}$$

where ρ is the area density or mass per unit area, r is the distance on the xy-plane from the axis of rotation to a differential area element dA. By symmetry, the mass moment of inertia can be calculated for a quarter of the area of the magnet, shown in Figure 16.4(b), and the value can be scaled $4\times$ to account for the complete area of the magnet. The dimensions of the magnet are shown in the figure.

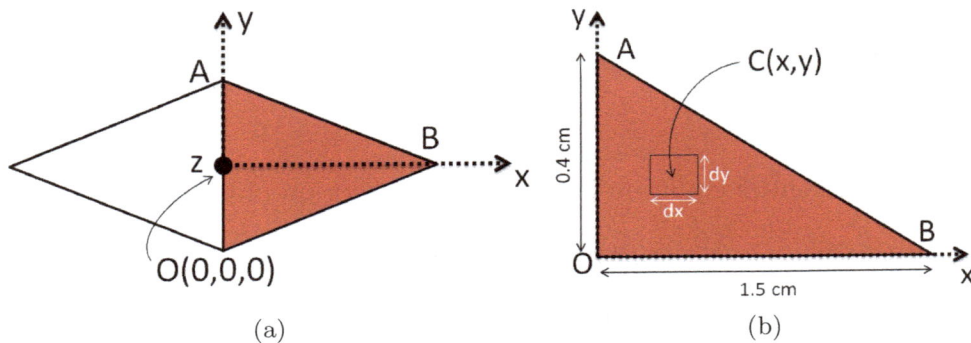

Figure 16.4: (a) The top view of Magnet 1. (b) The upper right triangle of Magnet 1.

The mass of the magnet used in the experiment was measured to be $0.7\,gm$ using a balance, and the area of the magnet is $1.2\,cm^2$, resulting in an area density of $\rho = 0.7833\,\frac{gm}{cm^2}$.

The equation of the line segment AB is

$$y = -0.2667x + 0.4. \tag{16.14}$$

Substituting

$$r^2 = x^2 + y^2 \tag{16.15}$$

for the square of the length of the segment OC in Figure 16.4(b), in Equation 16.13, the mass moment of inertia is

$$I = 4 \int_0^{1.5} \int_0^{-0.2667x+0.4} \left(x^2 + y^2\right)(0.7833)\, dy\, dx$$
$$= 0.2812\,gm\,cm^2. \tag{16.16}$$

Placing Magnet 1 over the magnetic needle, as shown in Figure 16.2(b), the magnet is twisted by some angle and allowed to oscillate. Allowing the oscillations to settle down, and keeping time for 5 oscillations, the average period of the oscillation T was determined to be $2.2\,s$.

From the table in Figure 15.4, $\alpha_b = 10°$ at $r = 6\ cm$. Substituting the measured values in Equation 16.11, the value of the terrestrial magnetic-field strength, at the location where the measurement was made in Chandler, Arizona, USA is

$$H = 0.245\,\frac{gm^{1/2}}{cm^{1/2}s}$$
$$= 0.245\ Oe. \tag{16.17, EMU}$$

17

Oersted's Experiment and the Birth of Electromagnetism

Until Oersted's famous experiment, it was thought that electricity and magnetism were independent of each other. Oersted's celebrated experiment marked the beginning of electromagnetism in the year 1820.

Volta invented the battery in 1800. The battery will be discussed in more detail in Chapter 19. For now, it is sufficient to know that it is a two terminal device, and charges flow in a wire connected across its terminals. The flow of charges in a wire is known as current. As explained in Section 1.2, although current can be viewed as the flow of positive charges or negative charges, by convention, current is the flow of positive charges because of the bias held towards the positive charge. This will be explained in more detail in Chapter 18.

How do we know that charges flow in a wire connected to a battery? This will be proven later in the chapter. Oersted made his discovery 20 years after the invention of the battery, which shows that nobody would have imagined that electricity and magnetism would be related.

17.1 Oersted's Experiment

Oersted discovered that a magnetic needle, similar to the one shown in Figure 11.1(a), deflects when charges flow in a wire near the needle. The side view of the experiment setup is shown in Figure 17.1(a) [35]. A wire w runs over a magnetic compass C. The battery connected to the wire is not shown for simplicity. The top view of the setup is shown in Figure 17.1(b)–(c). The direction of the compass when the battery is not connected to the wire is shown in Figure 17.1(b). When the battery is connected to the wire, however, as shown in Figure 17.1(c), the needle deflects. Oersted's experiment amazed the scientific community.

A magnet brought near a compass changes the direction of the compass. A similar effect is observed when a wire is connected to a battery. A Leyden jar can be used to show that charges flow in a wire connected to a battery, explained next. The logical conclusion is that the flow of

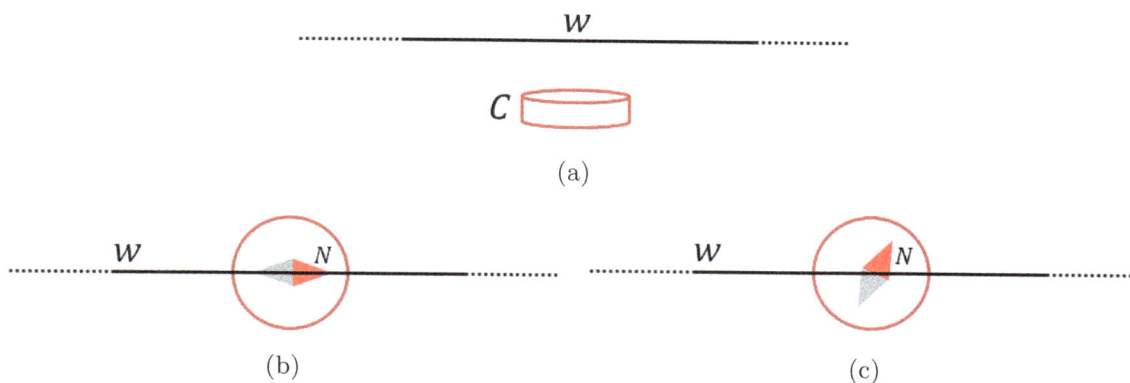

Figure 17.1: (a) The side view of Oersted's experiment setup. A wire w is routed over a magnetic compass. (b) The top view of the wire over a magnetic compass, showing the direction of the compass when no current flows in the wire. (c) The compass deflects when current flows in the wire.

charges in a wire generates a magnetic field, and changes the direction of the compass.

17.2   Charge Storage in a Leyden Jar

A Leyden jar is a charge-storage device. It is named after the city Leiden (Leyden), Netherlands, where the jar was invented by Pieter van Musschenbroek in 1745. In the present nomenclature, the Leyden jar is called a capacitor.

A Leyden jar can be constructed using a glass bottle covered with metal foil on the inner surface and the outer surface of the glass, as shown in Figure 17.2. It will be shown later in Section 38.2, the reason the Leyden jar is so effective as a charge storage device, is the large surface area of the metal foils, and the small separation of the foils by the thickness of the glass.

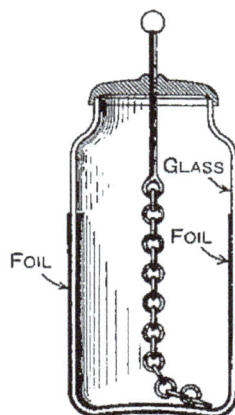

Figure 17.2: A Leyden jar [36].

A Leyden jar can be charged using an electrostatic generator, such as the Wimshurst machine. The Wimshurst machine itself has two Leyden jars, marked J_1 and J_2 in Figure 7.3(a), to store the charge generated by the machine. Connecting the terminals of the Wimshurst machine, marked S_1 and S_2 in Figure 17.3(a), to the outer and the inner metal foils, charge can be stored in the Leyden jar.

17.3 Proof that Electric Charges Flow in a Wire Connected to a Battery

A charged Leyden jar is discharged by a wire joining the outer and the inner metal foils, as shown in Figure 17.3(b). The wire joining the foils is routed over a compass, similar to the setup of Oersted's experiment. During the discharge, positive charges flow from the positively charged foil to the negatively charged foil. By convention, the positive charges are the ones that flow, as explained in Section 1.2. Once discharged, using an electroscope, it can be verified that the foils of the Leyden jar are no longer charged.

The discharge of the Leyden jar deflects the magnetic needle, and so does the battery with a wire connected across its terminals, as seen in Oersted's experiment. This is a conclusive evidence that charges flow in a wire connected to a battery. The discharge happens quickly, and the magnetic needle oscillates back and forth due to its rotational inertia, and eventually coming to rest.

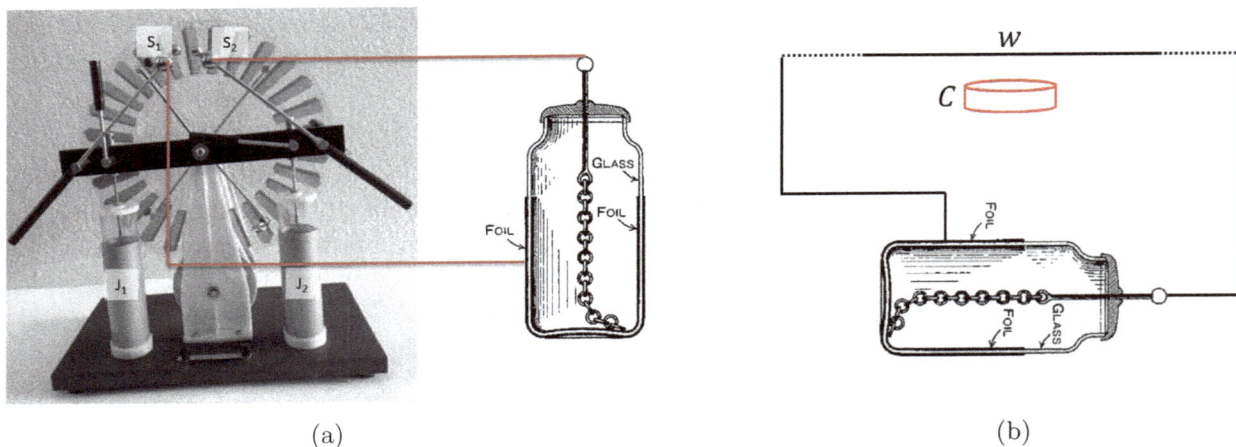

(a) (b)

Figure 17.3: (a) A Wimshurst machine used to charge a Leyden jar. (b) The Leyden jar is discharged by a wire routed over a magnetic compass, similar to Oersted's experiment setup.

17.4 Iron-Filings Pattern of a Long Current-Carrying Wire

Iron filings can be used to visualize the magnetic-field pattern around a magnet. This experiment will be described first in this section. This experiment will be repeated on a current-carrying wire do determine its magnetic-field pattern. If the current-carrying wire does not generate a magnetic field, then no pattern will be formed by the iron filings. This experiment, therefore, verifies the conclusion drawn from Oersted's experiment: a current-carrying wire generates a magnetic field.

Iron filings are sprinkled over a planar surface, beneath which a magnet is placed, shown in Figure 17.4(a). The iron filings, in the presence of the magnetic field of the magnet, get magnetized. The topic of certain objects becoming magnetized in the presence of a magnetic field will be looked at in detail in Chapter 41, during the discussion of magnetic materials. The magnetized iron filings attract their nearest neighbors, and rotate to align to the magnetic field of the magnet, similar to a magnetic compass aligning to Earth's magnetic field, shown in Figure 14.1(b). The end result is the formation of a pattern by the iron filings that will display the magnetic field of the magnet.

The direction of the magnetic field at any point can be found by placing a tiny magnetic compass at that point and the magnetic field is in the direction that the north pole of the compass points. By convention, the magnetic needle of compass is modeled with positive and negative magnetic charges $\pm m$, at its north and south pole end points, respectively, as shown in Figure 14.1(a). By definition of the magnetic field, Equation 14.1, the direction of the field is the same as the direction of the force exerted on a positive magnetic charge. The torque on the magnetic compass rotates the needle, aligning it to the magnetic field, and the north pole of the needle points in the direction of the field. It would be observed that the field of a magnet is in the direction from the north to the south poles of the magnet, as shown in Figure 17.4(b).

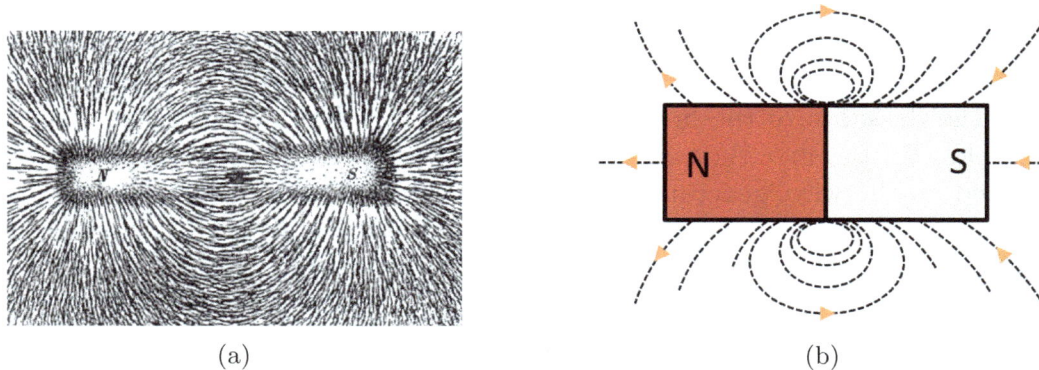

(a)   (b)

Figure 17.4: (a) The iron-filings pattern of a bar magnet [38]. (b) The direction of the magnetic-field pattern around a bar magnet.

A similar experiment can be done on a long and straight current-carrying wire to determine its magnetic-field pattern. In Chapter 21, Ampere's law will be used to derive the equation quantifying the magnetic field.

A long, straight wire exhibits circular symmetry. Any point on a circle centered on the wire will not be be any different from any other point on the circle. It is expected that the resulting strength of the magnetic field from the current carrying wire will also have circular symmetry. As shown in Figure 17.5, the magnetic filings, shown by the dashed lines, do form circular patterns

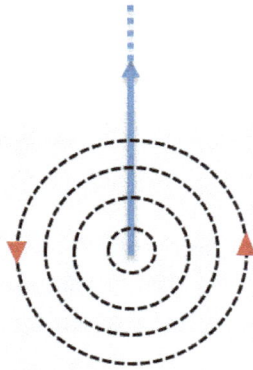

Figure 17.5: An illustration of the magnetic-filings pattern around a current-carrying wire.

around the wire [39].

The circular pattern also indicates that the direction of the magnetic field is either in the clockwise or in the counter-clockwise direction. This can be easily verified by placing a small magnetic compass at different points around the wire, and as explained earlier in the chapter, the direction of the magnetic field at the point is the same as the direction of the north pole of the magnetic compass.

How does the direction of the flow of positive charges in the wire, relate to the magnetic field generated? The convention followed is given by the first right-hand rule: if the thumb of the right hand points in the direction of the flow of positive charges, the fingers curl in the direction of the magnetic field. By the first right-hand rule, the positive charges flowing in the direction of the arrow marked on the wire, generate the magnetic field in the direction shown by the arrows on the field pattern. Several other right-hand rules will be presented in Chapter 21, during the discussion of Ampere's law.

There are many important consequences of the first right-hand rule. The first right-hand rule sets the positive and negative terminals of a battery. It also sets the polarity of a charged object to be positive or negative. These topics will be discussed in the following chapter.

17.5 Superposition of Magnetic Fields

In the experiment shown in Figure 17.5, the magnetic field at any point can be measured using Gauss's technique in Chapter 16. Instead of using one wire, two wires can be used, each with identical batteries, to double the current flow. It can be verified that the magnetic field is now $2\times$ greater.

This experiment shows that superposition applies to magnetic fields generated by current sources. Several other examples will be presented in Chapter 22 to show that superposition is applicable to magnetic fields.

17.6 Electrostatic Units (ESU) vs Electromagnetic Units (EMU)

Oersted's experiment clearly shows that electricity and magnetism are not independent of each other. This coupling of electricity and magnetism gives freedom to either

☐ define electrical quantities first, and define magnetic quantities from the electrical quantities. This system is called electrostatic units (ESU), or

☐ define magnetic quantities first, and define electrical quantities from the magnetic quantities. This system is called electromagnetic units (EMU).

The flowchart of definitions of electrical quantities in ESU and EMU will be discussed in detail in Chapter 25 and Chapter 26. It will become clear after this, why some of the equations are valid only in ESU, or EMU, or both. The flowchart of definitions will be revised as electromagnetic theory is developed, and the revised set of definitions, without the use of magnetic charges, will be presented in Chapter 58 and Chapter 59.

17.7 Detecting Magnetic Charges

If magnetic charges do exist, a similar experiment to Oersted's, can be used to detect magnetic charges. Assuming that magnetic charges exhibit a symmetric behavior to electric charges, similar to the electric battery, there would be a magnetic battery generating a flow of magnetic charges. As observed in Oersted's experiment, the flow of electric charges creates a magnetic field, and deflects a magnetic compass. Likewise, it can be hypothesized that the flow of magnetic charges creates an electric field, and deflects an "electric compass".

An "electric compass" may be created using two tiny metal spheres, joined together by a small dielectric rod, such that it is free to rotate on a plane, similar to the magnetic needle of a magnetic compass. The spheres are positively and negatively charged by electric charges. Using a setup similar to Oersted's experiment, a magnetic current-carrying wire will deflect the needle of the electric compass. Such a result has not been observed until now.

If magnetic charges do exist, perhaps, there may be the equivalent of a Wimshurst machine for charging objects with magnetic charges. The magnetic charge on the object would modify the direction of a magnetic compass, and can be used like an electroscope to detect magnetically charged objects. The discharging magnetic-charge flow between two oppositely charged objects can be detected using an electric compass, using a setup similar to Oersted's experiment. However, none of these effects have been observed, and no magnetic charges detected yet.

18

Definition of Current

Charges flow in a wire connected to a battery, as proven in Chapter 17. Current is the flow of charges. One can think of positive charges, or negative charges, as the ones that flow, as explained in Section 1.2. Current in electromagnetics, however, by convention, always means the flow of positive charges, because of the bias that was held towards the positive charge.

Although current is a scalar value, it has a direction associated with it. The direction of current flow in a wire is always one of the two possible directions.

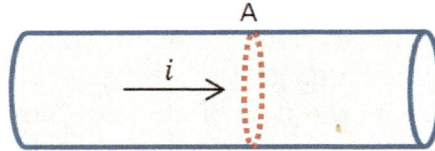

Figure 18.1: A cylindrical conductor carrying uniform current i.

A part of a current-carrying wire is shown in Figure 18.1. The cross-section of the conductor at some point on the wire is shown in the figure and marked A. Let q be the cumulative charge, beginning at some time, flowing across area A, in the direction marked by the arrow.

An example of such a waveform, as a function of time, is shown in Figure 18.2(a). As explained in Section 1.2, one could think of positive charges flowing in the direction of the arrow, or negative charges flowing in the opposite direction. The waveform in Figure 18.2(b) is equivalent to the waveform in Figure 18.2(a), but tracks the equivalent cumulative negative charge flowing in the opposite direction.

In Figure 18.2(a), the positive charges flow in the direction of the arrow, and the waveform increases. However, the waveform decreases after some time, which means that negative charges begin flowing in the direction of the arrow at this point, or equivalently, positive charges flowing in the opposite direction to the arrow (see Section 1.2).

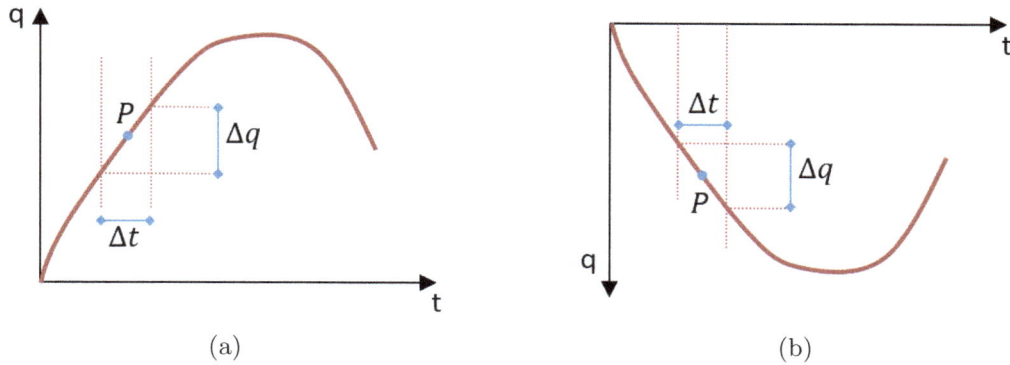

Figure 18.2: (a) A cumulative charge-time waveform, of charges flowing in the direction of the arrow in Figure 18.1, and (b) the equivalent waveform of cumulative charge flow in the opposite direction.

The slope of the cumulative charge-time waveform is current,

$$i = \frac{\Delta q}{\Delta t}, \tag{18.1}$$

and is the rate of flow of charges. In the limit as Δt becomes infinitesimal, current is the derivative of the cumulative charge-time waveform,

$$i = \frac{dq}{dt}. \tag{18.2}$$

The units of current in ESU and EMU will be discussed in Chapter 25-Chapter 26.

The derivative of the waveform in Figure 18.2(a) is negative of the derivative of the waveform in Figure 18.2(b). For example, the derivative at Point P is a positive number for the case shown in Figure 18.2(a), which means that positive current, or positive charges are flowing in the direction of the arrow in Figure 18.1, which is the direction chosen for tracking the cumulative charge flowing across A.

However, in Figure 18.2(b), the derivative at Point P is a negative number, which is a negative current, or negative charges flowing in the direction opposite to the arrow. Since the two waveforms are equivalent representations of the same current flow, as illustrated in Figure 18.3, i flowing from A to B is the same as $-i$ flowing from B to A.
Rearranging Equation 18.1,
$$\Delta q = i\,\Delta t, \tag{18.3}$$
where Δq is the charge flowing across A during time Δt, and i is approximated as a constant during a short time interval Δt.

Rearranging Equation 18.2,
$$dq = i\,dt, \tag{18.4}$$

121

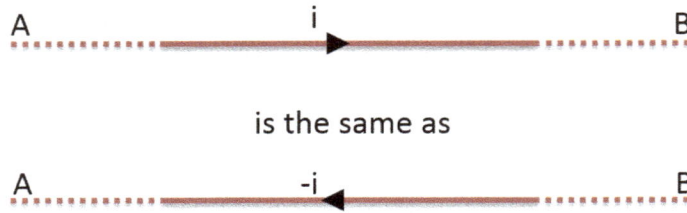

Figure 18.3: The current flow i in the top wire is equivalent to the current flow of $-i$ in the opposite direction, shown in the bottom wire.

and integrating both sides of the above equation over the time interval $[t_i, t_f]$,

$$\Delta q = \int_{t_i}^{t_f} i \, dt, \tag{18.5}$$

where Δq is the net charge flowing across the cross-section A during the time interval. Graphically, the above equation is the signed area under the curve of a current waveform as a function of time.

An example of a current-time waveform is shown in Figure 18.4, where the current flow is in the direction shown by the arrow in Figure 18.1. The shaded region represents the quantity of charge flowing across A during the time Δt.

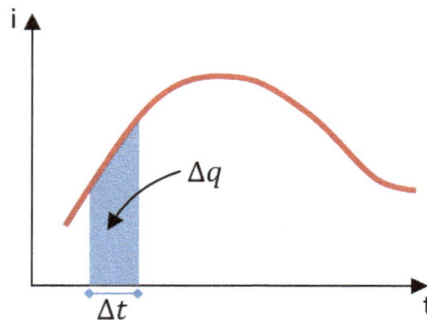

Figure 18.4: The charge contained in a current flowing for a time duration Δt, is the area under the curve of the current-time waveform, over the time interval Δt.

19

Volta's Invention of the Battery

The battery was invented by Alessandro Volta in the year 1800, also called as a voltaic pile in his honor. The reader is referred to other texts to understand the chemistry of the battery. A high-level overview of the structure of the battery will be presented in this chapter. Determining the polarity of the terminals of a battery, as well as a charged object, will be discussed.

The voltaic pile is a sandwich of two different metals, such as zinc and copper, separated by an acidic material. Two constructions of the voltaic pile are the penny battery and the lemon battery. These examples are very similar to the original construction of the battery, although the materials used in the construction are different.

A penny battery is shown in Figure 19.1(a) [40]. It consists of copper pennies and zinc washers sandwiched between paper soaked in vinegar. The stack is placed on a metal foil, which is one of the terminals of the battery. The other terminal is the first copper penny on top of the stack.

The torpedo fish was known at Volta's time to cause a sting, similar to the sting of an electric shock from the discharge of a Leyden jar. It is believed that Volta got the idea of the metal stack in a voltaic pile, by looking at the similar arrangement of the electric organs in a torpedo fish.

A lemon battery, shown in Figure 19.1(b), can be created by sticking two different metals in a lemon, and these two metals form the two terminals of the battery [41].

19.1   The Voltage of a Battery

Oersted's experiment shows that the behavior of a battery is similar to a charged Leyden jar. A Leyden jar can be discharged by connecting its two conductors with a wire. The current flow during the discharge of a charged Leyden jar generates a magnetic field, and modifies the direction of the magnetic needle in a compass. A similar behavior is observed in a battery. Unlike a Leyden jar, however, in the case of the battery, the deflection of the compass remains steady. This shows that a steady current flows in the wire connected to a battery.

(a)

(b)

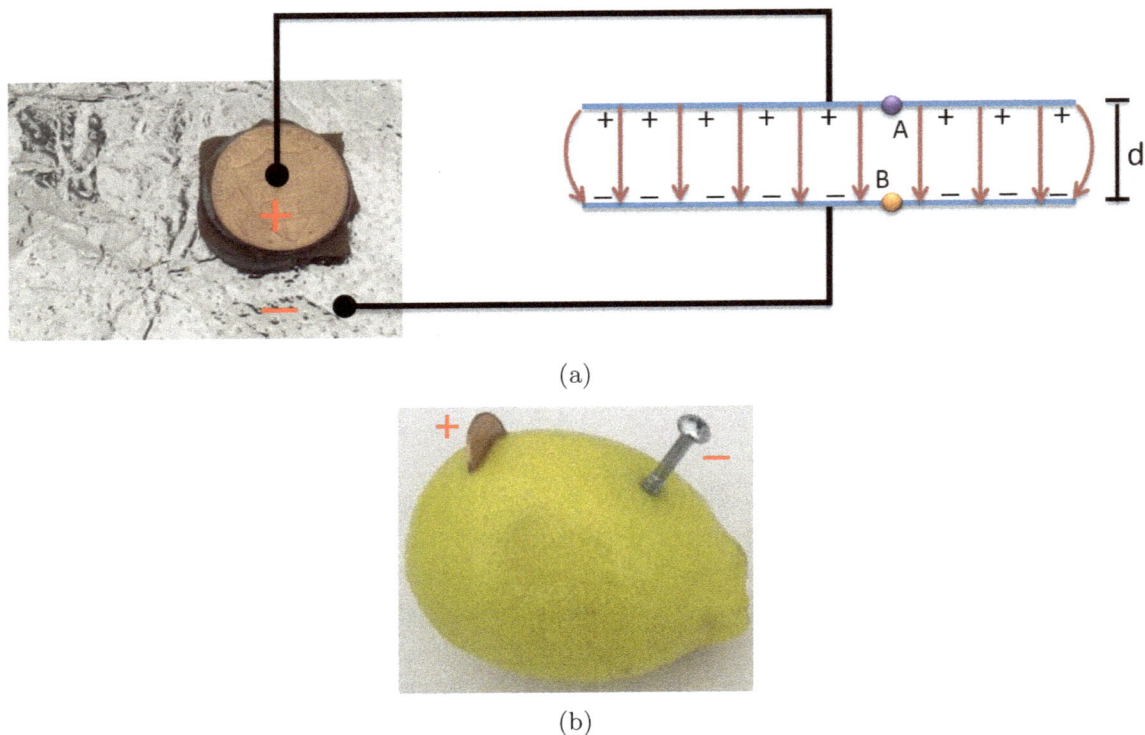

Figure 19.1: Batteries similar to Volta's invention: (a) a penny battery, (b) a lemon battery.

The two terminals of a battery are similar to the two terminals of a Leyden jar: one of the terminals is at a higher potential than the other terminal. This will be proven in Section 19.4. The terminal at the higher potential will be referred to as the positive terminal. The terminal at the lower potential is the negative terminal. A method to find the positive and negative terminals of a battery is presented in Section 19.3. The steady current in a wire connected to a battery, shows that the voltage difference between the terminals is a constant, unlike the voltage difference between the terminals of a Leyden jar that decays to 0 during the discharging process.

What does it mean when a battery is labeled as 1.5V or 9V? The voltage of a battery will be written as the path integral of the electric field. The terminals of a battery are connected to the two plates of a parallel-plate capacitor, which are separated by a distance d, as shown in Figure 19.1(a). Since a metal is an equipotential volume in electrostatics, the plate, the wire, and the terminal of the battery to which it is connected, are at the same potential in the steady state. Connecting the battery to the capacitor, charges the capacitor. The plate connected to the positive terminal of the battery charges the plate positive, and the plate connected to the negative terminal, negative. This will be proven by experiment in Section 19.4.

Voltage, from Equation 10.23, is the work done by the electric field in moving a unit positive charge from A to B, as marked in Figure 19.1(a). If V is the voltage of the battery, by definition of voltage,

$$V = \int_A^B \vec{E} \cdot \vec{dl},\tag{19.1}$$

where $\vec{E}$ is the electric field due to the charged plates. Since the electrostatic field is a conservative field, and a metal is an equipotential volume, A and B can be two points located anywhere on or within the plates.

19.2 Increasing the Battery Voltage With Series Connections

The voltage of a battery can be increased by connecting many cells in series, as shown in Figure 19.2(a). In a series connection, the negative terminal of a battery is connected to the positive terminal of a second battery. The negative terminal of the second battery is connected to the positive terminal of the third, and so on. If N cells of equal voltages are cascaded in series, the total voltage is $N\times$ the voltage of a single cell. This will be proved from the definition of voltage.

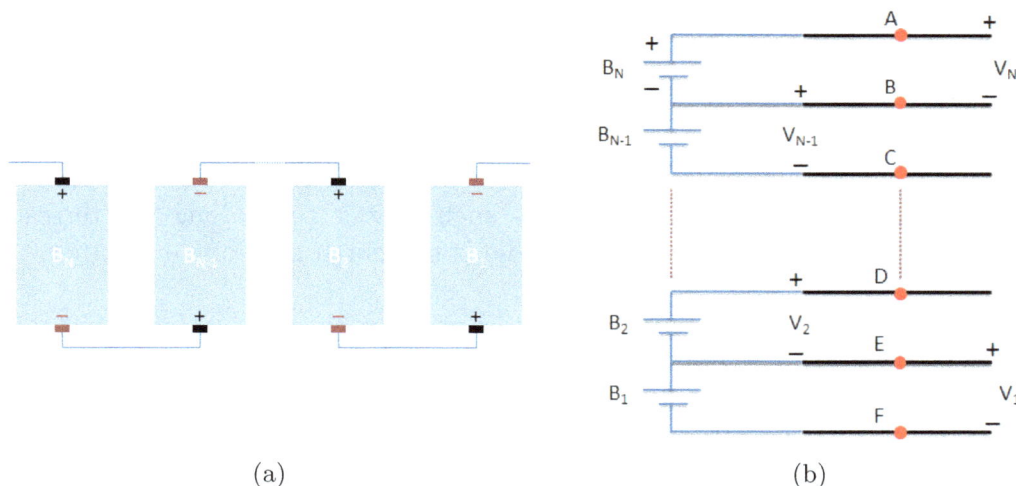

(a) (b)

Figure 19.2: (a) N batteries connected in series. (b) N batteries in series connected to parallel plates.

A series connection of N batteries $\{B_1, B_2, ..., B_N\}$, with voltages $\{V_1, V_2, ..., V_N\}$, connected to parallel plates marked A through F is shown in Figure 19.2(b). The symbol of a battery is shown by the parallel long and short dashes, with the long dash representing the positive terminal, and the short dash is the negative terminal. By the definition of voltage in Equation 10.23, the work done by the electric field in moving a charge q from A to B is qV_N, from B to C is qV_{N-1}, and so on. The work W_{AF} done by the electric field in moving a charge q from A to F is

$$W_{AF} = q\left(V_1 + V_2 + ... + V_N\right).\tag{19.2}$$

Since a metal is an equipotential volume, no work is done in moving the charge across the thickness of the metal.

The voltage difference V_{AF} between points A and F, by definition, is the work done by the electric field in moving a unit positive charge from A to F, and setting $q = 1$ in the above equation,

$$V_{AF} = V_1 + V_2 + ... + V_N. \tag{19.3}$$

This shows that the voltage difference between the positive terminal of B_N, and the negative terminal of B_1, which are the output terminals of the N batteries connected in series, is the sum of the voltage of each of the cells. This can be empirically verified using the measurement technique in Chapter 48 to characterize the voltage of a battery.

19.3   Determining the Positive and Negative Terminals of a Battery

By definition, positive charges flow out of the positive terminal of a battery. The terminal into which the positive charges flow is the negative terminal of the battery. It will be proven in Section 19.4, the positive terminal is at a higher potential than the negative terminal. An experiment technique to determine the positive and negative terminals of a battery will be presented in this section.

A wire running above a compass C is shown in Figure 19.3(a). The connection to the battery has been omitted for simplicity, indicated by the dotted lines. The top view is shown in Figure 19.3(b). The arrangement of the wire is such that it runs along the terrestrial $N - S$ direction. The darker shaded triangle of the needle is the north pole of the magnet.

Current in electromagnetics is strictly the flow of positive charges, a consequence of the bias that was held toward positive charges. By the first right-hand rule of Section 17.4, if the thumb of the right hand points in the direction of the flow of positive charges, the fingers curl in the direction of the magnetic field. In the experiment setup, the magnetic field generated by the current-carrying wire is in the plane of the magnetic needle, and perpendicular to it.

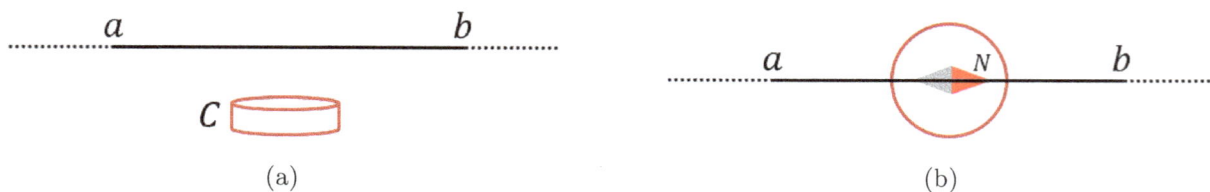

Figure 19.3: (a) The side view of a wire over a magnetic compass. (b) The top view of the wire routed over the compass.

If the needle deflects in the direction shown by the dotted arrow in Figure 19.4(a), using the model of the compass in Figure 14.1(a), the magnetic field generated by the wire is also in the direction of the dotted arrow. By the first right-hand rule, the direction of the current flow is from a to b. The positive terminal of the battery is the terminal connected to a, and the negative terminal is the terminal connected to b, as shown in the figure. If the needle deflects in the opposite direction, shown in Figure 19.4(b), the terminals are now reversed compared to the previous case.

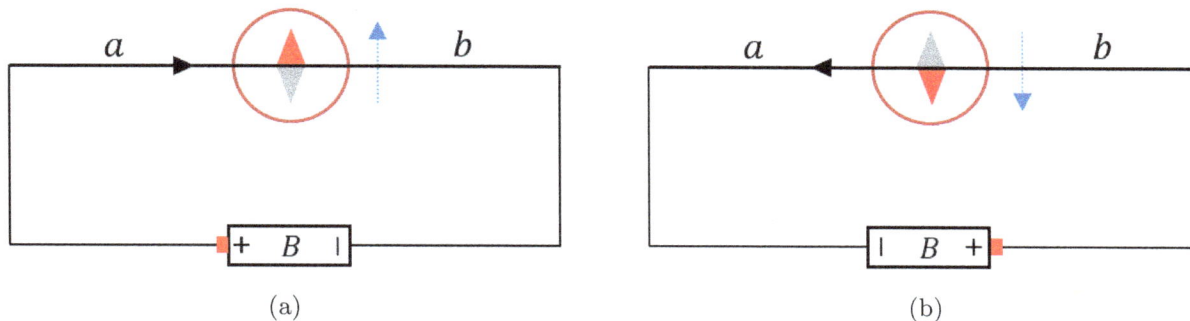

(a)

(b)

Figure 19.4: The deflection of a magnetic compass when current flows from (a) a to b, and (b) b to a.

19.4 Proof that the Positive Terminal of a Battery is at a Higher Potential than the Negative Terminal

In the discharge of a Leyden jar, positive charges flow from the positively charged conductor that is at a higher potential, to the negatively charged conductor at a lower potential. From Section 19.3, the positive terminal is the terminal from which the positive charges flow out of the battery, and the charges flow into the negative terminal. Since positive charges flow from the positive terminal to the negative terminal of a battery, in comparison to a Leyden jar, the positive terminal of the battery is at a higher potential than the negative terminal.

Charging a capacitor with a battery and discharging the capacitor, can also be used to prove that the positive terminal of a battery is at a higher potential than its negative terminal. The purpose of this experiment is to also show that a capacitor gets charged when connected to a battery, as illustrated in Figure 19.1(a).

A Leyden jar may be charged by connecting it to a battery. The charge stored in the metal foils, however, would be very small to be detected. More charge can be stored in the Leyden jar by using a battery of a higher voltage. Many batteries may be connected in series, as explained earlier, perhaps even several hundreds of them, to increase the voltage of the battery to a level, where the charge stored in the metal foils can be detected using an electroscope.

Rather than using a battery of a very high voltage, more charge may be stored by using a more effective capacitor. To create an effective capacitor, or to store more charge for a given voltage difference between the plates, it will be shown in Chapter 38, the spacing between the plates will have to be smaller, and the area of the plates larger. Using an electrolytic capacitor, which is formed using thin metal sheets, sandwiched between an extremely thin layer of insulator material, much thinner than that of a Leyden jar, sufficient charge can be stored, which can be detected

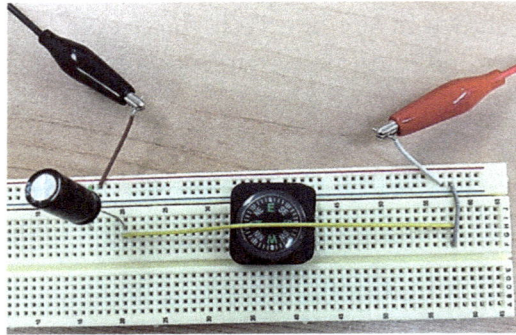

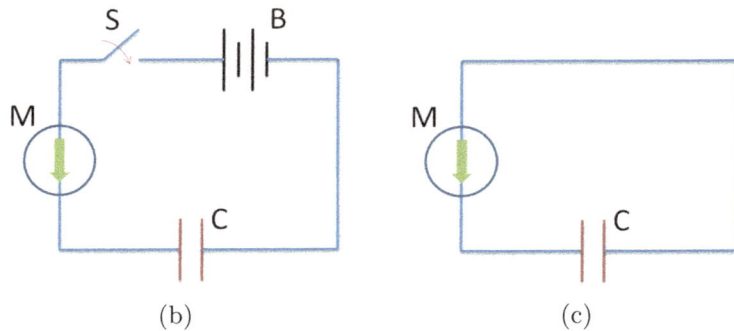

Figure 19.5: (a) An experiment setup to show that a capacitor gets charged when connected to a battery. (b) The circuit schematic of the charging capacitor. (c) The circuit schematic of the discharging capacitor.

using a setup similar to Oersted's experiment.

The experiment setup is shown in Figure 19.5(a). A magnetic compass M is placed over a wire connected to a capacitor C, such that the wire lies along the magnetic N-S direction. When the switch S is closed in Figure 19.5(b), a transient current flow charges the capacitor. The transient current generates a burst of magnetic field, which can be detected from the oscillation of the magnetic needle. Likewise, in Figure 19.5(c), when the capacitor is discharged by connecting its terminals together, an oscillation of the needle can be observed. The oscillations of the magnetic needle will be observed in opposite directions during the charging and the discharging phases, indicating the current flow in opposite directions.

From the experiment presented in Section 19.3, the positive and the negative terminals of the battery are known. The terminals of the capacitor connected to the positive and the negative terminals of the battery, when it is charged, are also known. When the capacitor is discharged, from the exercise in Section 19.3, using a capacitor instead of the battery, the positively and the negatively charged plates of the capacitor can be determined.

The direction of the electric field is from the positively charged plate of the capacitor to the negatively charged plate, as shown in Figure 19.1(a). The positively charged plate is at a higher potential than the negatively charged plate. From this exercise, it can be verified that the terminal of the capacitor connected to the positive terminal of the battery is positively charged, and the terminal of the capacitor connected to the negative terminal of the battery is negatively charged. Therefore, the positive terminal of the battery is at a higher potential than its negative terminal.

19.5   Finding the Absolute Polarity of a Charged Object

A Leyden jar has two metal foils separated by a thin dielectric material such as glass, shown in Figure 17.2. The metal foils can be charged using an electrostatic generator, such as a Wimshurst machine.

If the foils are discharged by a wire running over a compass, the discharge current in the wire generates a magnetic field, and deflects the magnetic compass, similar to Figure 19.4. Since the discharge current is not continuous, but a burst of charge, the discharge happens quick, before the needle deflects. As a result of the angular momentum imparted on the needle, it oscillates back and forth in a periodic motion.

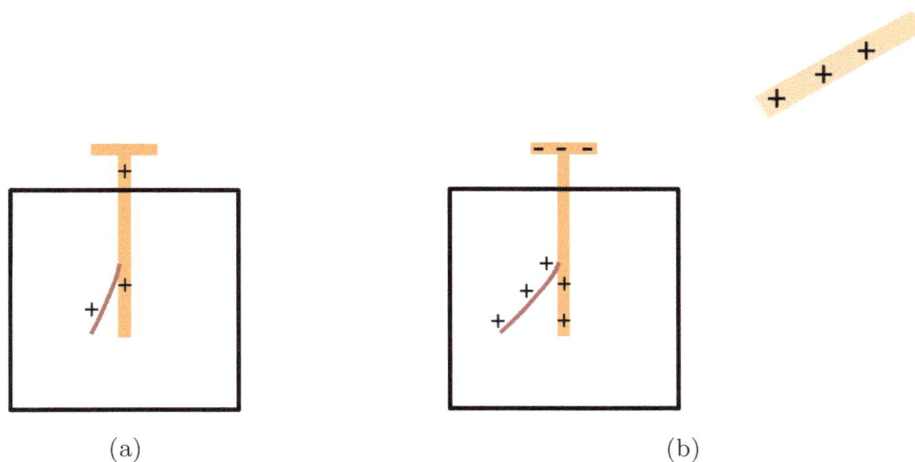

(a)   (b)

Figure 19.6: (a) A positively charged electroscope. (b) A positively charged object near the electroscope diverges the leaf more.

Depending on which direction the magnetic needle deflects first, the direction of flow of the positive charges can be determined, similar to how the polarity of the battery terminals was determined earlier. The foil of the Leyden jar that is charged positive is the foil from which the positive charges flow, and the other foil negative.

If one of the foils of the Leyden jar was put in contact with an electroscope, before discharging the Leyden jar to determine the polarity of the foils, the polarity of the charge on the electroscope is also known.

For example, lets assume that the charge on the electroscope has been determined to be positive, as shown in Figure 19.6(a). If a charged object, whose polarity is to be determined, is brought near the electroscope, if the leaf diverges further, then the charged object is also positively charged. The positively charged object would repel more positive charges to the leaf, thereby diverging the leaf further, as illustrated in Figure 19.6(b). However, if the leaf is less diverged, then the object must be negatively charged. The negatively charged object would attract the positive charges from the leaf of the electroscope, thereby making it less diverged [3].

There is flexibility in the formulation of the first right-hand rule of Chapter 18. It was chosen arbitrarily. A left-hand rule could have been used: if the thumb of the left hand points in the direction of the positive current flow, the fingers curl in the direction of the magnetic field. If the left-hand rule is used, the positive current flow direction will be the opposite of the right-hand rule. Charged objects that we call positive (negative) will be called negative (positive). The convention that is followed, however, is the right-hand rule.

20

Magnetic Potential

Coulomb's law of electric charges, Equation 4.1, and Coulomb's law of magnetic charges, Equation 13.1, have the same form. The definitions related to electric charges will also be applied to magnetic charges. It was proven earlier that the electrostatic field is conservative. The same proof can be applied to the magnetostatic field generated by stationary magnetic charges. It will be assumed that the magnetic field is also conservative. The results derived in this chapter will be used in the derivation of Ampere's law in Chapter 21.

A conservative field can be written as the gradient of a scalar field by the fundamental theorem of calculus for line integrals, as explained before. In the case of an electrostatic field, repeating Equation 10.1,

$$\vec{E} = -\nabla V, \tag{20.1}$$

where $\vec{E}$ is the electric field and V is the electric potential or voltage. Likewise, the magnetostatic field can be written as

$$\vec{H} = -\nabla V_m, \tag{20.2}$$

where $\vec{H}$ is the magnetic field and V_m is the magnetic potential. The negative sign makes the direction of the magnetic field point from higher magnetic potential to lower magnetic potential.

Rewriting Equation 10.6 for magnetic potential,

$$V_m^{AB} = V_m(A) - V_m(B) = \int_A^B \vec{H} \cdot \vec{dl}, \tag{20.3}$$

where V_m^{AB} is the magnetic potential difference points A and B, and written as the line integral of the magnetic field from Point A to Point B.

20.1 Magnetic Potential Near a Magnetic Point Charge

The force exerted by a magnetic charge of magnitude m on m_2, separated by a distance r, using Equation 13.1,

$$\vec{F} = \frac{m\, m_2}{r^2}\, \hat{r}. \tag{20.4, EMU}$$

The magnetic field generated by a positive point magnetic charge of strength m, from Equation 20.4 and Equation 14.1 is

$$\vec{H} = \frac{\vec{F}}{m_2} = \frac{m}{r^2}\hat{r}. \tag{20.5, EMU}$$

The magnetic potential V_m^{AB} between Point A at $r = r_0$ and Point B at $r = \infty$, applying

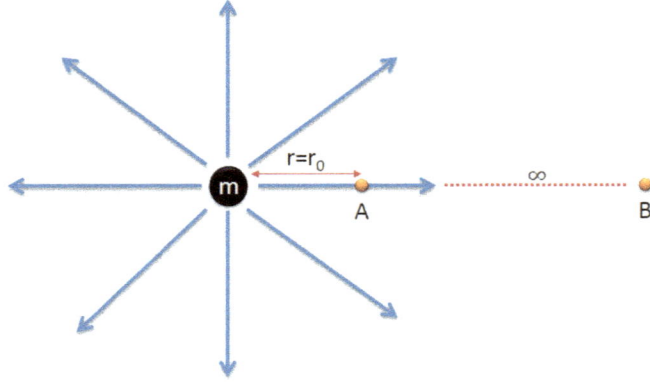

Figure 20.1: The magnetic field around a magnetic point charge m.

Equation 20.3,

$$V_m^{AB} = V_m(A) - V_m(B) = \int_A^B \vec{H} \cdot \vec{dl}. \tag{20.6}$$

Substituting Equation 20.5 in the above equation,

$$V_m^{AB} = V_m(A) - V_m(B) = \int_{r=r_0}^{r=\infty} \frac{m}{r^2} dr. \tag{20.7, EMU}$$

Evaluating the integral,

$$V_m(A) = V_m(B) + \frac{m}{r_0}. \tag{20.8, EMU}$$

Since only potential difference is defined, as explained in Chapter 10, a reference point can be assigned as 0 potential. The potential at all other points are calculated with respect to the reference point that is at 0 potential. Typically, the reference point at $r = \infty$ is assigned as 0 potential. Since the magnetic field decays to 0 at ∞, all points at ∞ are at the 0 potential. Therefore, $V_m(B)$ at $r = \infty$ is set to 0. The magnetic potential at a radial distance of $r = r_0$ from the point magnetic charge is

$$V_m(A) = \frac{m}{r_0}. \tag{20.9, EMU}$$

Similar to electric charges, it will be assumed that superposition is also applicable to magnetic charges. When N magnetic charges $m_0, m_1, m_2, \dots m_N$ are present at distances $r_0, r_1, \dots r_N$ from Point A,

$$V_m(A) = \frac{m_0}{r_0} + \frac{m_1}{r_1} + \frac{m_2}{r_2} + \dots + \frac{m_N}{r_N}. \tag{20.10, EMU}$$

20.2 Magnetic Potential Near a Magnetic Dipole

A magnetic dipole consists of a positive and a negative magnetic charge, $+m$ and $-m$, separated by a small distance ℓ, as shown in Figure 20.2(a). $+m$ is at Point Q and $-m$ at Point R. Let P be the point at which the magnetic potential is to be calculated, and O be the midpoint between the charges. Length OP is r, PQ is r_1, and PR is r_2. Let $\angle QOP$ be θ.

If

$$\ell \ll r, \tag{20.11}$$

segments QP, OP, and RP can be approximated as three parallel segments cut by a transversal QR. This is easier to see if the above equation is rewritten as

$$r \gg \ell, \tag{20.12}$$

and is captured in Figure 20.2(b). $\angle ORP$ and $\angle QOP$ are corresponding angles and are equal,

$$\angle ORP = \angle QOP = \theta. \tag{20.13}$$

The magnetic potential at Point P $V_m(P)$, generated by $+m$ and $-m$, using Equation 20.10, is

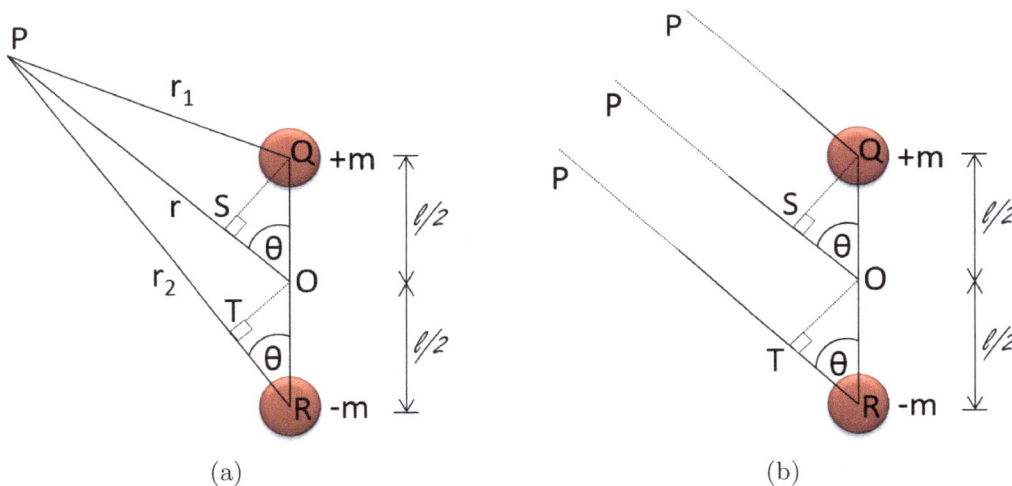

Figure 20.2: (a) A magnetic dipole. (b) The magnetic dipole with $r \gg \ell$.

$$V_m(P) = \frac{m}{r_1} - \frac{m}{r_2}. \tag{20.14, EMU}$$

r_1 and r_2 can be written in terms of r. Segments QS and OT are drawn perpendicular to segments OP and RP, respectively. Since $r \gg \ell$, by visual inspection of Figure 20.2(b), segments $PS \approx PQ$ and $OP \approx TP$. Therefore,

$$r_1 = r - \frac{\ell}{2} \cos \theta \tag{20.15}$$

$$r_2 = r + \frac{\ell}{2} \cos \theta. \tag{20.16}$$

Substituting the above expressions in Equation 20.14, and simplifying,

$$V_m(P) = \frac{m\ell \cos\theta}{r^2 - \frac{\ell^2}{4}\cos^2\theta}. \tag{20.17}$$

Assuming that $r \gg \ell$, the second term in the denominator of the above equation can be neglected,

$$V_m(P) \approx \frac{m\ell \cos\theta}{r^2}. \tag{20.18, EMU}$$

Eliminating variables r_1 and r_2, Figure 20.2(a) can be simplified as Figure 20.3, where $+m$ is located at the midpoint of the dipole charges, at Point O, and the magnetic potential at P is written as the above equation.

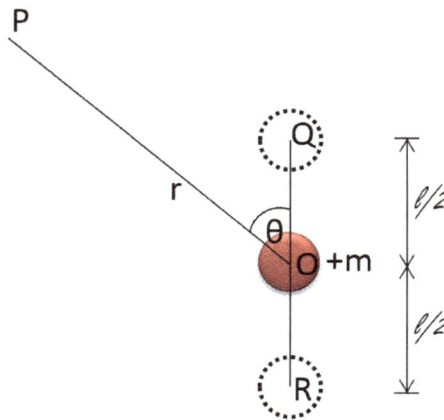

Figure 20.3: A different representation of the magnetic dipole in Figure 20.2(a).

The product of magnetic charge m and length ℓ in the above equation, appears often, such as Equation 15.4, and is called the magnetic moment ϕ_M,

$$\phi_M = m\ell. \tag{20.19}$$

Rewriting Equation 20.18 using the above expression,

$$V_m(P) = \frac{\phi_M \cos\theta}{r^2}. \tag{20.20, EMU}$$

20.3   Magnetic Potential Near a Magnetic Dipole Area

Equation 20.20 can be extended to calculate the potential resulting from a dipole area with magnetic charge density $\pm\sigma_M$. The unit of magnetic charge density is $[maxwell/cm^2]$. Two patches $P1$ and $P2$ of the same area, with magnetic charge density $\pm\sigma_M$, are shown in Figure 20.4(a).

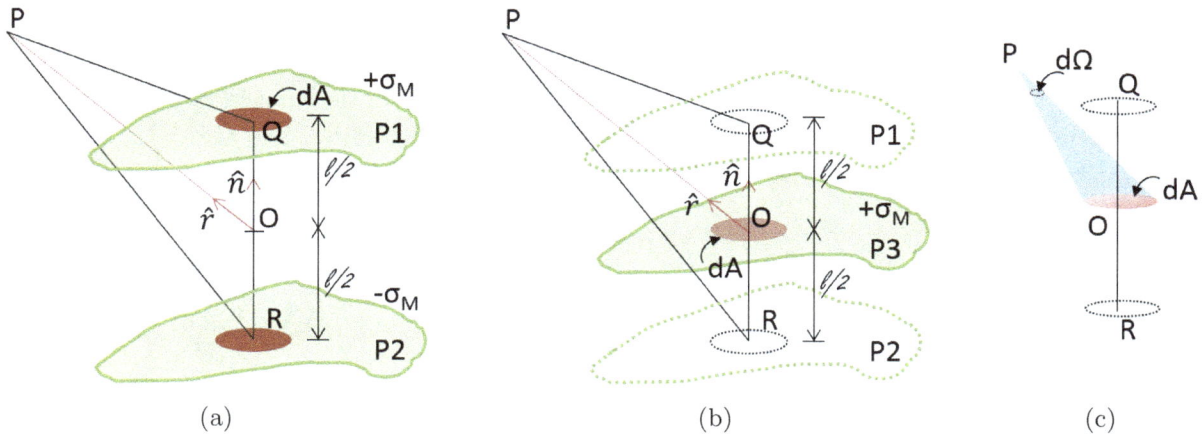

Figure 20.4: (a) A magnetic dipole area. (b) A different representation of the magnetic dipole area. (c) The solid angle $d\Omega$ subtended by the dipole area dA at Point P.

Element areas shaded within the patch are marked dA. The element areas form an element dipole with element magnetic charges,

$$dm = \pm \sigma_M dA. \tag{20.21}$$

The element magnetic moment $d\phi_M$, using Equation 20.19, is

$$\begin{aligned} d\phi_M &= dm\, \ell \\ &= \sigma_M\, \ell\, dA \\ &= \phi'_M\, dA, \end{aligned} \tag{20.22}$$

where $\phi'_M = \sigma_M\, \ell$ is the magnetic moment per unit area. It will be assumed that ϕ'_M is constant throughout the area of the patch. Substituting this equation into Equation 20.20 to calculate the element potential resulting from the element dipole,

$$dV_m(P) = \phi'_M \frac{\cos\theta}{r^2}\, dA. \tag{20.23, EMU}$$

Figure 20.4(a) can be simplified to Figure 20.4(b), similar to Figure 20.2(a) redrawn as Figure 20.3. Instead of two patches $P1$ and $P2$, a single patch $P3$, with magnetic charge density $+\sigma_M$, is located at the center of Patches $P1$ and $P2$. The potential at Point P can be calculated as

$$V_m(P) = \phi'_M \int_{P3} \frac{\cos\theta}{r^2}\, dA, \tag{20.24, EMU}$$

where ϕ'_M is assumed to be a constant, and moved out of the integral.

Note from Figure 20.4(b), if P lies above Point $P1$, the magnetic potential is positive at Point P, since $0° \le \theta < 90°$ result in positive values of $\cos\theta$ for all the element areas. If P lies below Point

135

$P2$, however, the potential is negative.

Equation 20.24 can be written in terms of the solid angle subtended at P by the area of the patch. A tutorial on solid angle is provided in Appendix $\mathcal{A}$. Solid angles are always non-negative. The sign can be included, depending on the location of P, once the magnitude of the potential has been calculated.

Let $\hat{r}$ be the unit vector along Segment OP, and $\hat{n}$ be the unit vector normal to the element area dA, shown in Figure 20.4(a). By definition of the dot product, $\angle QOP = \theta$, can be written as,

$$\cos\theta = \hat{r} \cdot \hat{n}. \tag{20.25}$$

Substituting this into Equation 20.24,

$$V_m(P) = \phi'_M \int_{P3} \frac{\hat{r} \cdot \hat{n}}{r^2} \, dA. \tag{20.26, EMU}$$

By definition of solid angle,

$$\frac{\hat{r} \cdot \hat{n}}{r^2} \, dA = d\Omega, \tag{20.27}$$

where $d\Omega$ is the element solid angle subtended by the element area dA at Point P.

The solid angle $d\Omega$ subtended by the element area dA at Point P, is illustrated by the cartoon in Figure 20.4(c). Accounting for all the area elements making up the patch, the potential at P is

$$V_m(P) = \phi'_M \int_{P3} d\Omega, \tag{20.28, EMU}$$

and

$$\Omega = \int_{P3} d\Omega \tag{20.29}$$

is the solid angle subtended by the entire patch $P3$ at Point P. This result will be used in the derivation of Ampere's law in the following chapter.

21

Ampere's Law

François Arago brought the news of Oersted's experiment to Paris in 1820. Within a short time after that, André-Marie Ampère devised new experiments, and theorized the force between currents. A distinguished scientist, Ampere earned the title "Newton of Electricity" from James Clerk Maxwell.

Ampere's law will be derived in this chapter, which will then be used to solve for the magnetic field around a long, straight, current-carrying wire. This result was verified by Jean-Baptiste Biot and Felix Savart, whose experiments will be presented in Chapter 24.

A current loop, drawn as the solid line, is shown in Figure 21.1, and arrows indicating the direction of the current flow. The battery is not shown in the loop for simplicity. Oersted's experiment showed that the current generates a magnetic field. Ampere's law relates a closed path integral

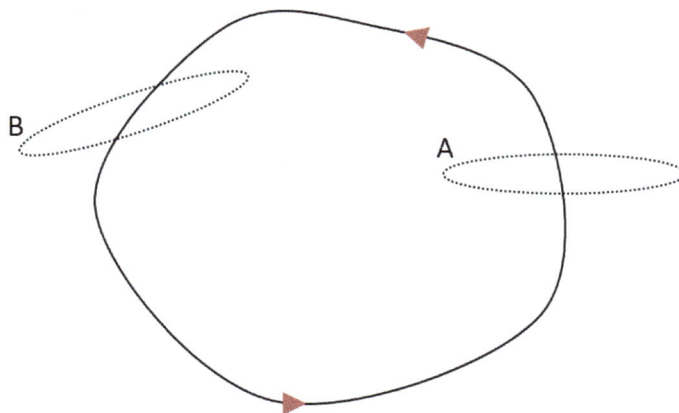

Figure 21.1: Path integrals around a current loop.

of the magnetic field, shown by the dotted line marked A or B in the figure, to the current enclosed by the closed integral path. Note that it is not necessary for the integral path to be in any particular angle with respect to the current it encloses, Ampere's law always holds true.

It will be shown that a current-carrying loop of wire behaves like a small bar magnet [42]. This model will be used to construct Ampere's law.

Two coils of wire, marked A and B in Figure 21.2, neither repel nor attract each other, when there is no current flowing in them. Depending on the direction of the current flow in the two coils, the coils can be made to attract or repel each other, similar to magnets. They attract each other when the currents flow in the coils in the same direction, shown in Figure 21.2(a). When the currents flow in the opposite directions, by reversing the connections to the battery terminals in one of the coils, the coils repel each other, as shown in Figure 21.2(b).

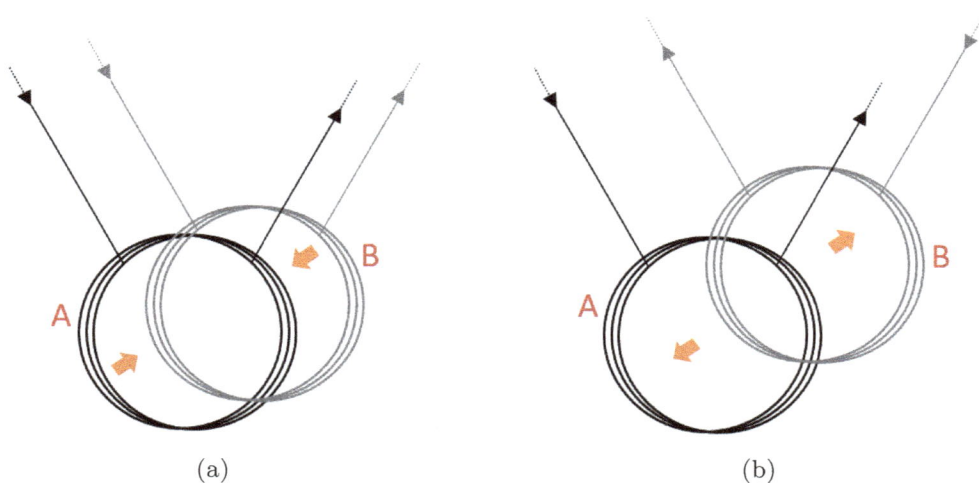

| (a) | (b) |

Figure 21.2: (a) The two coils attract each other when the currents flow in the same direction in both the coils, and (b) repel each other, when the currents flow in opposite directions.

A current-carrying coil is either attracted to, or repelled by a magnet. This too confirms the same behavior as a magnet. This is demonstrated in Figure 21.3 [43]. The coil is neither attracted to, nor repelled by a magnet, when there is no current flowing in the coil. It is attracted to the magnet, however, when current flows in the coil in one of the two directions, as shown in Figure 21.3(a). If the magnet is reversed with the current flowing in the same direction, or if the current flow is reversed with the magnet oriented the same way, then the coil would be repelled away from the magnet, as shown in Figure 21.3(b).

Furthermore, the magnetic-field pattern of a current-carrying coil is very similar to that of a magnet. The magnetic-field pattern can be observed using iron filings [44], similar to the experiment in Figure 17.4(a). A coil is wound around a planar surface, by drilling two holes on the surface to wrap the coil around, as shown in Figure 21.4(a). The two end points of the coil are marked a and b, which are then connected to a battery, not shown for simplicity. The top view of the coil is shown in Figure 21.4(b).

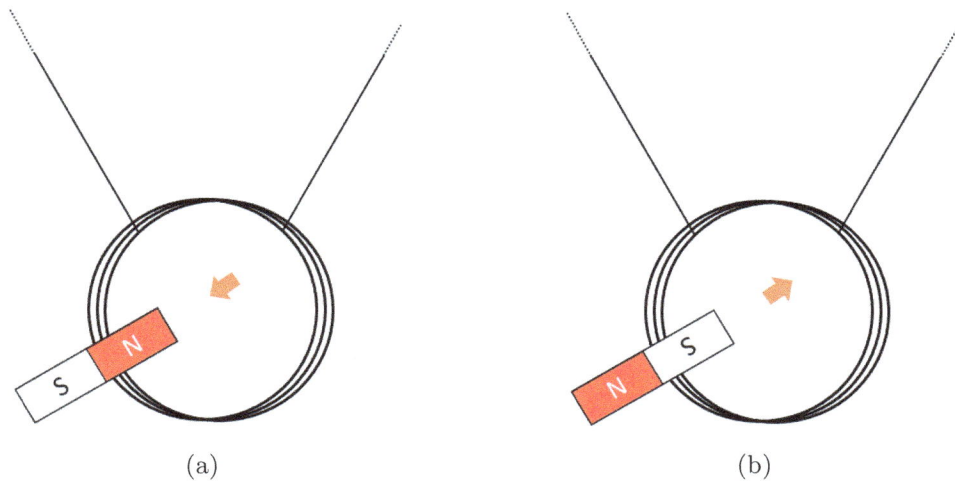

(a)　　　　　　　　　　　　　　(b)

Figure 21.3: (a) The current flowing in the coil in one of the directions is attracted to the magnet, and repelled in (b) when the magnet, or the current direction in the coil, is reversed [43].

By the same reasoning in Section 17.4, the iron filings sprinkled over the current-carrying coil in Figure 21.4(b), shown by the dashed lines, forms a pattern that captures the magnetic field. The field pattern of the coil is similar to a bar magnet in Figure 17.4. To show their similarity, the magnet, the coil, and their iron-filings patterns, are shown together in Figure 21.5.

An N-turn tightly wound current coil, with negligible thickness, can be viewed as made up of N current loops. If each of the current loops are identical, the magnetic-field pattern of each of the loops must also be identical. Assuming that superposition holds true, the magnetic field of each of the loops can be added together, resulting in the magnetic field that is $N\times$ greater in strength. A single current loop, having the same field pattern as a tightly wound coil, also behaves like a bar magnet.

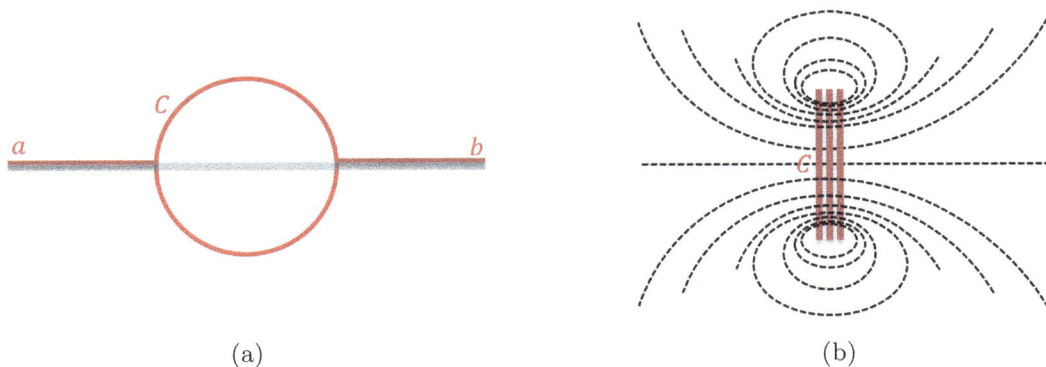

(a)　　　　　　　　　　　　　　(b)

Figure 21.4: (a) The side view of a coil wound around a planar surface. (b) The iron-filings pattern formed on the planar surface, caused by the magnetic field generated by the current-carrying coil.

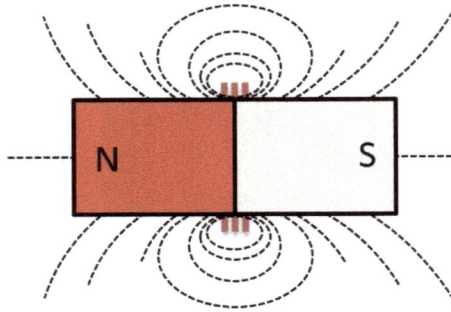

Figure 21.5: The iron-filings pattern of a bar magnet is similar to a current-carrying coil.

21.2 The Second Right-Hand Rule for Current-Carrying Loops and Coils

The conclusion from the experiments presented in the previous section is that a current loop behaves like a bar magnet, illustrated in Figure 21.6. But in which direction does the north pole of the magnet face? This question will be answered next.

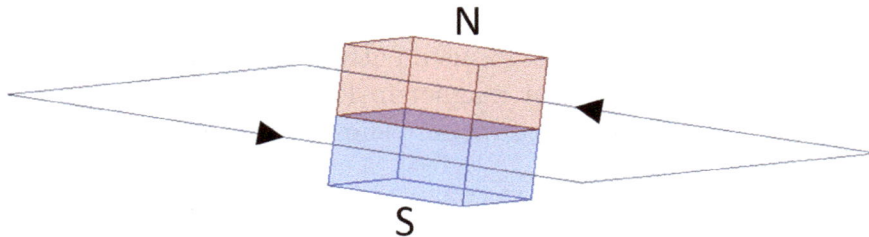

Figure 21.6: A current loop element viewed as a small bar magnet.

The side view of the current loop in Figure 21.6 is shown in Figure 21.7. The current loop is in-out of the page. The current directions are marked in the figure using the • and × convention: a • indicates current flowing out of the page, and × is current flowing into the page. The current directions are reversed between Figure 21.7(a) and Figure 21.7(b).

By the first right-hand rule of Chapter 18, if the thumb of the right hand points in the direction of the current, the fingers curl in the direction of the magnetic field. A magnet is shown in each of the figures, and oriented such that it is consistent with the magnetic-field lines traversing from the north to the south poles.

The first right-hand rule can be cast in a different form that is more suited for current loops and coils, which will be referred to as the second right-hand rule: if the fingers of the right hand curl in the direction of the current in a loop or coil, the thumb points in the direction of the north pole of the magnet, or in the direction of the magnetic field. Note that the orientation of the magnet in the figures is consistent with the second right-hand rule.

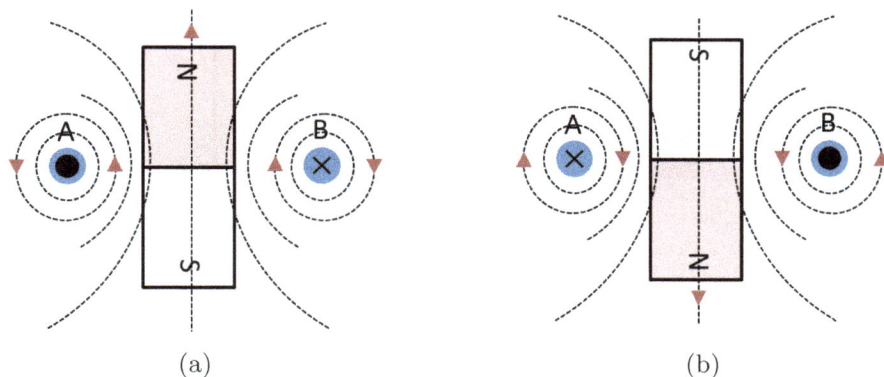

Figure 21.7: The orientation of a bar magnet that represents the field pattern of a current loop, when (a) the current flows in the counter-clockwise direction looking from above, and (b) the current flows in the clockwise direction looking from above.

The second right-hand rule can be verified from the experiment in Figure 21.3. The magnet that represents the coil, can be confirmed to be oriented such that like poles repel and unlike poles attract each other.

The second right-hand rule explains the experiment results observed in Figure 21.2. In the case where the currents flowing in the coils are in the same direction, by the second right-hand rule, the opposite poles of the magnets representing the coils are near each other. In this case, the coils attract each other, consistent with unlike poles attracting each other. The currents flowing in the opposite directions have like poles near each other, consistent with the coils repelling each other.

21.3   Ampere's Discretization of the Current Loop

The top view of a current loop is shown in Figure 21.8. The direction of the current flow, for example, is marked by the arrows in the figure. The area enclosing the current loop can be discretized into a fine mesh, as shown in the figure. Ampere observed that each of the mesh elements can be viewed as a tiny current loop, each flowing in the same direction as the outer current loop. When all the mesh elements are put together, the current on the edges common to adjacent mesh elements cancel each other, leaving the outer current loop.

To illustrate this, the zoom of the mesh elements, marked A, B, C, and D, are shown in the figure. The mesh current in each of the elements also flows in the counter-clockwise direction, which is the same direction as the outer current loop. The nodes of the elements are marked 1 through 9. Edge $2 - 5$ common to the elements B and C, has currents flowing in the opposite directions, and therefore cancel each other. This is also true for Edges $4 - 5$, $5 - 6$, and $5 - 8$, leaving the current path along the periphery $1 - 4 - 7 - 8 - 9 - 6 - 3 - 2 - 1$, in that direction. Considering the entire mesh, similarly, the currents on the inner edges cancel each other, resulting in the current flowing on the periphery.

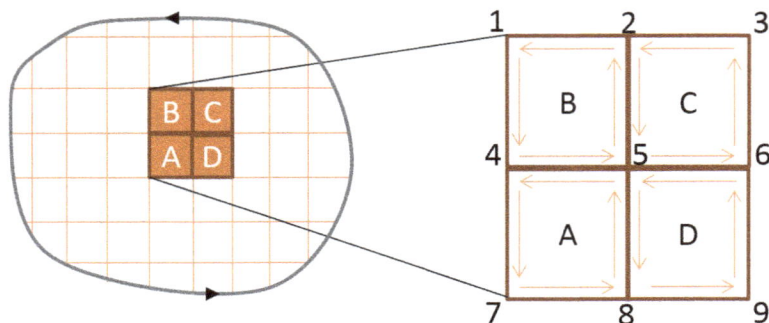

Figure 21.8: The top view of a current loop discretized into small rectangular current-loop elements.

Each of the mesh elements behaves like a tiny bar magnet, as seen earlier. With finer and finer mesh, in the limit, the current loop can be modeled as a "magnetic shell". The side view of the current loop, and its model as a magnetic shell, is shown in Figure 21.9. The current loop viewed from the side is in-out of the page, where the current flows out of the page at A and into the page at B. The current directions are marked in the figure using the $\bullet$ and $\times$ convention, as before. The magnetic shell is the parallel plate with a very small separation, having magnetic-charge density $\pm\sigma_M$ on the top plate and bottom plate, respectively.

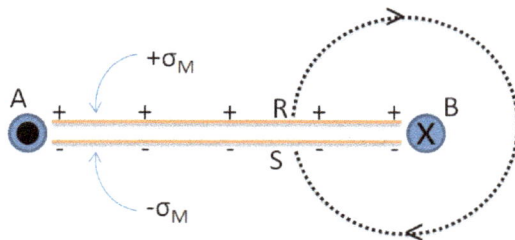

Figure 21.9: The side view of a current loop viewed as a magnetic shell.

As defined in Equation 20.22, let the magnetic moment per unit area of the shell be ϕ'_M, where the superscript $'$ is used to indicate that the variable represents a per unit area value. In Figure 21.9, let Point R and Point S be two points located somewhere within the current loop, however, very closely spaced above the positive plate and below the negative plate, respectively. If that is the case, Point R and Points S would appear as though they are located over an infinite plane.

As derived in Appendix $\mathcal{A}$, the solid angle subtended by an infinite plane at a point above is equal to 2π. Therefore, the potential at Point R in Figure 21.9, using Equation 20.28, is

$$V_m(R) = 2\pi\phi'_M. \qquad \text{(21.1, EMU)}$$

The potential at Point S will be negative, since it is located below the negatively charged plate, as explained in Section 20.3. Using Equation 20.28, the potential at Point S is

$$V_m(S) = -2\pi\phi'_M. \qquad \text{(21.2, EMU)}$$

From Equation 20.3,

$$V_m^{RS} = V_m(R) - V_m(S). \tag{21.3}$$

Substituting Equation 21.1 and Equation 21.2 in the above equation,

$$V_m^{RS} = 4\pi\phi'_M. \tag{21.4, EMU}$$

From Equation 20.3, V_m^{RS} can also be expressed as the path integral of the magnetic field,

$$V_m^{RS} = \int_R^S \vec{H} \cdot \vec{dl}, \tag{21.5}$$

along any path from Point R to Point S, one of which is shown in Figure 21.9 by the dotted lines. Figure 21.4(b) shows some of the magnetic-field patterns forming closed loops. The filings at the center form larger loops. Since the magnetic field becomes weaker farther away from the coil, this may not be visible from the pattern formed by the filings.

From the filings pattern, it can be noted that the magnetic field is continuous and well-defined within the area enclosing the current loop. The path integral in the above equation will be modified to a *closed* path integral,

$$\oint \vec{H} \cdot \vec{dl} = 4\pi\phi'_M. \tag{21.6, EMU}$$

Since the magnetic field is continuous within the area enclosing the current loop, and the magnetic shell has infinitesimal thickness, writing Equation 21.5 as a closed path integral, has a negligible change on the value of the integral. In Chapter 30, Ampere's law will be derived from Biot-Savart's law. This shows that the above equation is the correct formulation. In Section 21.6, the magnetic field around a long, straight current-carrying wire will be derived using Ampere's law. The result is in agreement with the magnetic-field pattern observed from iron filings. This is also a verification of the above formulation.

An N-turn tightly wound coil carrying current i, can be viewed as N single-turn loops, each carrying current i. Assuming superposition holds true, the magnetic field generated by each of the loops can be added together, resulting in a magnetic field that is $N\times$ greater, compared to one single-turn loop. This shows that a single-turn loop carrying $N\times$ more current generates $N\times$ greater magnetic field. The higher the strength of the current in a loop, the stronger is the magnetic field generated, and therefore, a stronger magnetic shell from which the magnetic field originates. It will be assumed that

$$\phi'_M \propto i$$
$$\phi'_M = Ci, \tag{21.7}$$

where i is the current in the loop, and C is a proportionality constant that is set to the dimensionless value of 1 in EMU, which is the simplest value. Substitution of the above equation in Equation 21.6, Ampere's law is

$$\oint \vec{H} \cdot \vec{dl} = 4\pi i. \tag{21.8}$$

This equation will be used to relate $\vec{H}$ and i in both ESU and EMU. The reason for this will become clear after the definitions flowchart in ESU and EMU are discussed in detail in Chapter 25 and Chapter 26.

Often times, i in the above equation is written as i enclosed i_{enc},

$$\oint \vec{H} \cdot \vec{dl} = 4\pi i_{enc},$$ (21.9)

to emphasize that i is the current enclosed by the closed integration path.

In the case where the integration path does not enclose a current, from the above equation,

$$\oint \vec{H} \cdot \vec{dl} = 0.$$ (21.10)

Although the derivation of Ampere's law in this chapter made many assumptions, using Biot-Savart's law, a rigorous mathematical derivation of Ampere's law will be presented in Chapter 30.

Points $\{R, S\}$ can be located anywhere within the current loop area. They will be viewed as being located over infinite planes. In addition, any closed path may be chosen for the integration, since Equation 20.2 implies path independence, similar to the definition of voltage in Equation 10.1.

From the derivation of Ampere's law, it can be noted that the angle a current-carrying wire makes with the integration path area, does not affect the result of the path integral. For example, two wires carrying the same current are at different angles to the identical enclosing integration paths ℓ, in Figure 21.10(a) and Figure 21.10(b). By Ampere's law, the path integral of the magnetic field is the same in each of the cases, since the current enclosed by the integration path is the same in each of the cases. The rigorous mathematical derivation of Ampere's law in Chapter 30, will confirm that this is indeed the case.

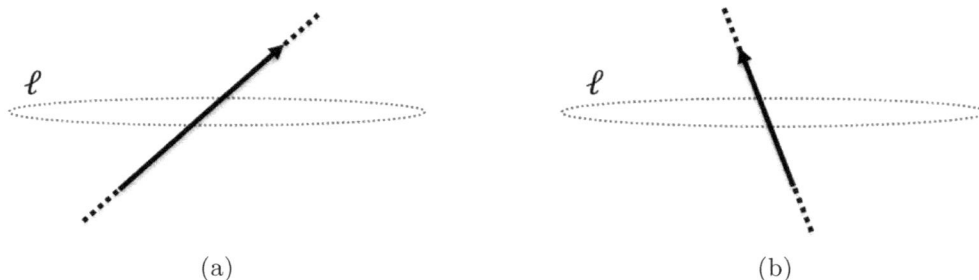

(a)   (b)

Figure 21.10: The result of the path integral of the magnetic field in Ampere's law is the same, although the wire is at an angle (a) towards the right, or (b) towards the left.

From Chapter 18, the positive direction of current flow in a wire may be chosen from one of the two possible directions along the wire. A positive value of the current means that positive charges are flowing in this direction. A negative value of current in the chosen positive direction is equivalent to the positive current flowing in the opposite direction. In Ampere's law, there are two possibilities for the direction of the integration path, either clockwise, or counter clockwise, and two possibilities for the positive current direction along the wire carrying current i_{enc}.

The direction of the path of integration cannot be chosen independent of the positive current direction, and vice versa. Once the direction of integration is chosen, the direction of the positive current gets fixed. Likewise, picking the direction of positive current, fixes the direction of the integration path.

In the case shown in Figure 21.11(a), the positive current direction is chosen as out of the page, marked by the • to indicate 'out of the page'. The current can be a positive or a negative value in the chosen current direction. If i_{enc} in Equation 21.9 is a positive number, then positive charges are flowing in the chosen direction of the positive current flow, which is out of the page. This results in the magnetic field in the counter-clockwise direction, using the first right-hand rule of Chapter 18.

If the path of integration is chosen in the clockwise direction c_1, the magnetic field is anti-parallel to the path of integration, resulting in a negative value for the path integral of the magnetic field. i_{enc} is a positive value, while the path integral of the magnetic field is negative. The left-hand side and the right-hand side of Equation 21.9 have inconsistent signs.

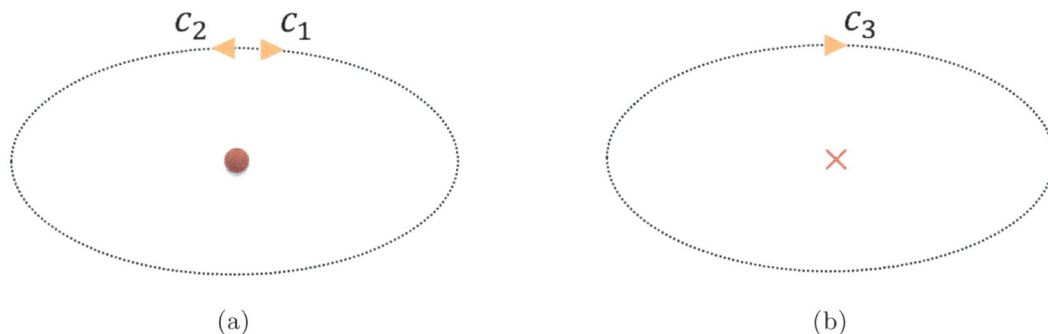

(a)   (b)

Figure 21.11: (a) The current in the wire is flowing out of the page, and the path of integration of the magnetic field is chosen in the clockwise direction c_1, or in the counter-clockwise direction c_2. (b) By the third right-hand rule, if the path of integration c_3 is chosen in the clockwise direction, the positive direction of current flow is into the page.

If the path of integration is chosen in the counter-clockwise direction c_2, the path integral of the magnetic field results in a positive value, since the magnetic field is then parallel to the path of integration. Both the sides of Equation 21.9 have the same signs, and are now consistent with each other. Picking the direction of the positive current flow, fixes the path of integration of the magnetic field.

Likewise, picking the direction of integration fixes the direction of positive current flow. Suppose c_3 in Figure 21.11(b) is chosen as the direction of integration, and the path integral of the magnetic field turns out to be a negative value. Therefore, i_{enc} is also a negative value in Equation 21.9. The direction chosen for the positive current flow must be such that i_{enc} is a negative value. The magnetic field generated by the current must be anti-parallel to the direction of c_3, in the counter clockwise direction, so that the path integral result is a negative value. By the first right-hand rule, the positive charges of the current must be flowing out of the page. Equivalently, the negative current must be flowing into the page. Therefore, the direction chosen for the positive current flow must be into the page, marked by the symbol $\times$.

A third right-hand rule is created to relate the direction of the positive current flow, and the direction of the path of integration in Ampere's law: if the thumb of the right hand points in the direction of positive current flow, the fingers curl in the direction of the path of integration. It is left as an exercise for the reader to verify that this rule does not contradict Ampere's law for the different possibilities of the paths of integration, and the directions of current flow, similar to how the cases in the figures were analyzed.

21.5   Superposition of Enclosed Currents

As explained in Section 17.5, it will be assumed that superposition of magnetic fields holds true. If N different current-carrying loops, $\{\ell_1, \ell_2, ..., \ell_N\}$, generate the magnetic fields $\{\vec{H}_1, \vec{H}_2, ..., \vec{H}_N\}$, it will be assumed that the net magnetic field $\vec{H}$ is

$$\vec{H} = \vec{H}_1 + \vec{H}_2 + ... + \vec{H}_N. \tag{21.11}$$

Making this assumption, Ampere's law can be extended to include multiple current loops in the formulation.

Two current loops carrying current i_1 and i_2 are shown in Figure 21.12, which generate the magnetic fields $\vec{H}_1$ and $\vec{H}_2$. The integration path in Ampere's law, shown by the dotted line, encloses both the currents i_1 and i_2. The direction of current flow is chosen to be in the $\hat{z}$ direction. By the third right-hand rule, the integration path is along the dotted line shown by the arrow. Applying Ampere's law on the loop carrying current i_1,

$$\oint \vec{H}_1 \cdot \vec{dl} = 4\pi i_1, \tag{21.12}$$

where current i_1 flows in the direction of $\hat{z}$. In the case of the loop carrying current i_2,

$$\oint \vec{H}_2 \cdot \vec{dl} = -4\pi i_2, \tag{21.13}$$

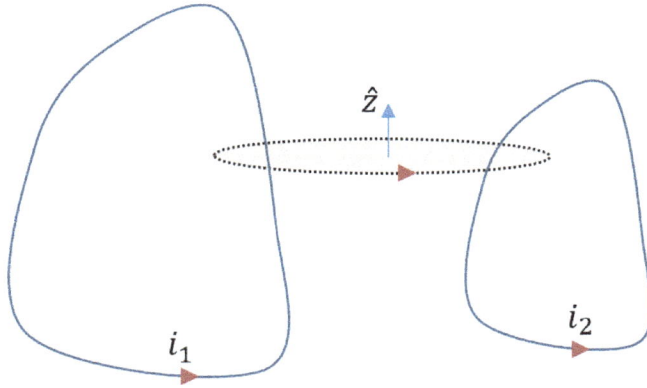

Figure 21.12: The path of integration enclosing two current loops.

since the current flowing in the $\hat{z}$ is $-i_2$ (see Chapter 18). The integration paths in the above two equations are in the same direction, along the arrow shown on the dotted line.

Adding the above equations, and using superposition of the magnetic fields,

$$\oint \vec{H} \cdot \vec{dl} = \oint \vec{H}_1 \cdot \vec{dl} + \vec{H}_2 \cdot \vec{dl}$$
$$= 4\pi i_{enc}, \tag{21.14}$$

where

$$\vec{H} = \vec{H}_1 + \vec{H}_2 \tag{21.15}$$

is the total magnetic field generated by the current loops, and

$$i_{enc} = i_1 - i_2 \tag{21.16}$$

is the net current flowing in the chosen direction $\hat{z}$, enclosed by the closed integration path. The above equation has the same form as Ampere's law for single current loops.

From the above result, Ampere's law can be generalized for any number of current loops,

$$\oint \vec{H} \cdot \vec{dl} = 4\pi i_{enc}, \tag{21.17}$$

where $\vec{H}$ is the total magnetic field generated by all the current loops, and i_{enc} is the net current flowing in the chosen direction of current flow, enclosed by the closed integration path. This result will also be derived from Biot-Savart's law in Chapter 30.

21.6   Magnetic Field of a Long, Straight Current-Carrying Wire

The magnetic field around a current-carrying wire can be visualized using iron filings, repeating Figure 17.5 in Figure 21.13(a). The magnetic field is circular and concentric around the wire. By

147

circular symmetry, no particular direction is favored over another, and such a pattern is expected. Ampere's law will be used to calculate the magnetic field around a long, straight current-carrying wire.

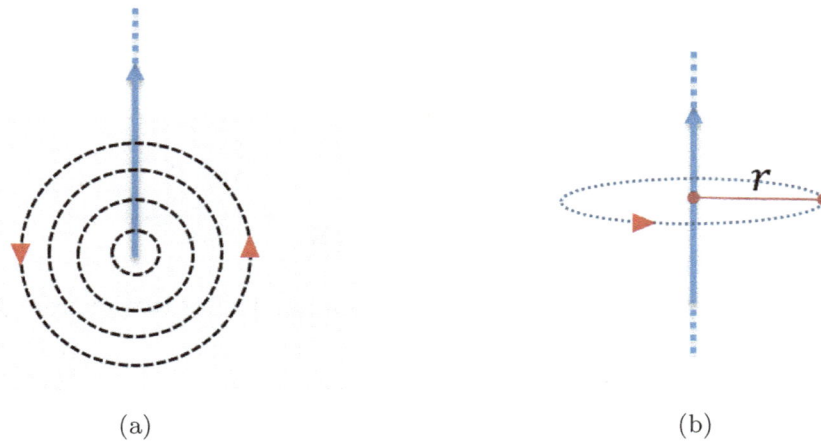

(a)   (b)

Figure 21.13: (a) An illustration of the magnetic-filings pattern around a current-carrying wire. (b) A circular path of integration of radius r around a current-carrying wire.

Assume that the direction of the current is shown by the arrow on the wire in Figure 21.13. The dotted line in Figure 21.13(b) indicates the circular path integral around the wire at radius r, and the direction of integration marked by the arrow, is consistent with the third right-hand rule. By circular symmetry, the magnetic field must have the same magnitude at radius r.

Using Ampere's law, the relation between the current i enclosed within the integration path, and the magnetic field generated by the current is

$$\oint \vec{H} \cdot \vec{dl} = 4\pi i. \tag{21.18}$$

By the first right-hand rule in Chapter 18, the magnetic-field direction is the same as the direction of the path integral. The magnetic field $\vec{H}$ is parallel to the differential segment $\vec{dl}$. Therefore, the dot product simplifies to unity, resulting in the equation

$$\oint H \, dl = 4\pi i. \tag{21.19}$$

By circular symmetry, the magnetic-field strength H is constant at a given radius r, and therefore, can be moved out of the integral, resulting in

$$H \oint dl = 4\pi i. \tag{21.20}$$

The closed path integral is simply the perimeter of the circle with radius r,

$$H (2\pi r) = 4\pi i. \tag{21.21}$$

148

Therefore,

$$H = \frac{2i}{r}. \tag{21.22}$$

The direction of the magnetic field is in the azimuth direction $\hat{\varphi}$,

$$\vec{H} = \frac{2i}{r}\hat{\varphi}, \tag{21.23}$$

where $\hat{\varphi}$ is the azimuth unit vector along the direction dictated by the first right-hand rule. The above equation of the magnetic field around a current-carrying wire is very important. In the next few chapters, this result will be used in the derivation of Biot-Savart's law.

22

Observations Leading to Biot-Savart's Law

The focus of this chapter is to present some of the experiments of Ampere, which show that a current-carrying wire can be divided into small vector elements. Each of the current elements behaves like a vector, whose direction is along the direction of the current. This view of a current-carrying wire as a set of vector elements will be used in the formulation of Biot-Savart's law.

Biot-Savart's law is named after Jean-Baptiste Biot and Felix Savart, who developed the relation to calculate the magnetic field of a current-carrying wire. Ampere's law was used to derive the magnetic field around a long and straight, current-carrying wire in Chapter 21. Biot-Savart's law is a more general relation, which can be used to calculate the magnetic field around a wire of any configuration, not limited to a long and straight wire. The derivation of Biot-Savart's law will be presented in the following chapter, and Biot-Savart's experiments to verify the formulation will also be presented later.

Several of Ampere's experiments are "null experiments". In this type of an experiment, the outcome of the experiment is no action. For example, the compass, or a wire, etc. in the experiment remains stationary. Two of Ampere's experiments will be presented in this chapter.

22.1 Force Between Current-Carrying Wires

Ampere showed experimentally that current-carrying wires exert a force upon each other. The force between current-carrying wires will be derived in a later chapter. Since a current-carrying wire creates a magnetic field, as explained earlier in Chapter 17, intuitively it can be seen that two current-carrying wires may either attract or repel each other, similar to magnets.

Ampere showed that currents in the same direction attract each other, while currents flowing in the opposite directions repel. This can be easily shown by observing the force between two parallel wires, illustrated in Figure 55.1(a) and Figure 55.1(b), respectively. The arrows on the wire show the current directions, and the block arrows are the forces of repulsion or attraction.

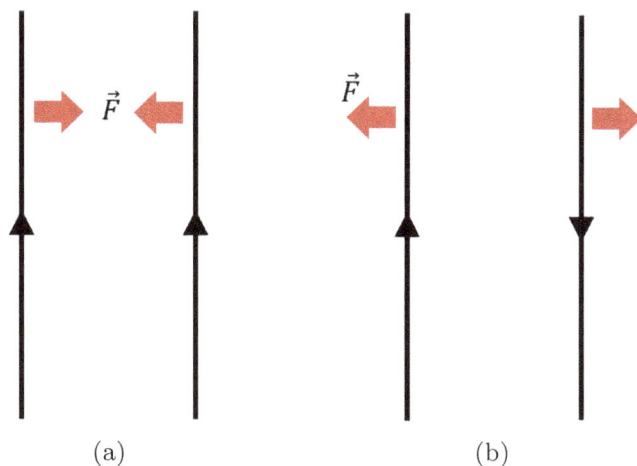

Figure 22.1: (a) The currents flowing in two parallel wires in the same direction attract each other, while (b) currents flowing in the opposite directions repel.

22.2   Astatic Balance

Ampere's astatic balance can be used to detect the force between current-carrying wires. Its special feature is that it is constructed such that it is immune to any force due to the terrestrial magnetic field. It will become clear shortly, factoring out Earth's magnetic field makes the balance more accurate to detect any force between wires.

A wire is twisted to form two loops, marked A and B in Figure 22.2(a). The two ends of the wires are each attached to a tiny thin metal, marked C, and are placed on metal bars attached to pillars E and F. The wire loop, A and B, is free to turn about C clockwise or counter clockwise, illustrated by the block arrows. Mercury, which is a metal in the liquid state at room temperature, is added to the tiny containers marked D, and C is immersed in it to increase the point of contact between C and D for better electrical conduction. Otherwise, since C is thin, there is little contact to D.

The reason for having two loops, A and B, instead of one, is to make the loop immune to terrestrial magnetism. As explained in Section 21.1, a loop of wire behaves like a small bar magnet. If there is only a single coil, it would turn and align towards magnetic north-south when free to rotate, similar to how a magnetic compass seeks the magnetic north-south direction. However, the astatic balance has two coils, A and B, with currents flowing in them in opposite loops, as marked by the arrows in Figure 22.2(a). The equivalent bar magnets of the coils are now in opposite directions, which would cancel any turning force on the coils due to terrestrial magnetism. Therefore, any turning action of the astatic balance is solely caused by other sources.

A current-carrying wire a near the astatic balance is shown in Figure 22.2(b). When placed in front of the current-carrying wire b of the astatic balance, a force is exerted on b causing the

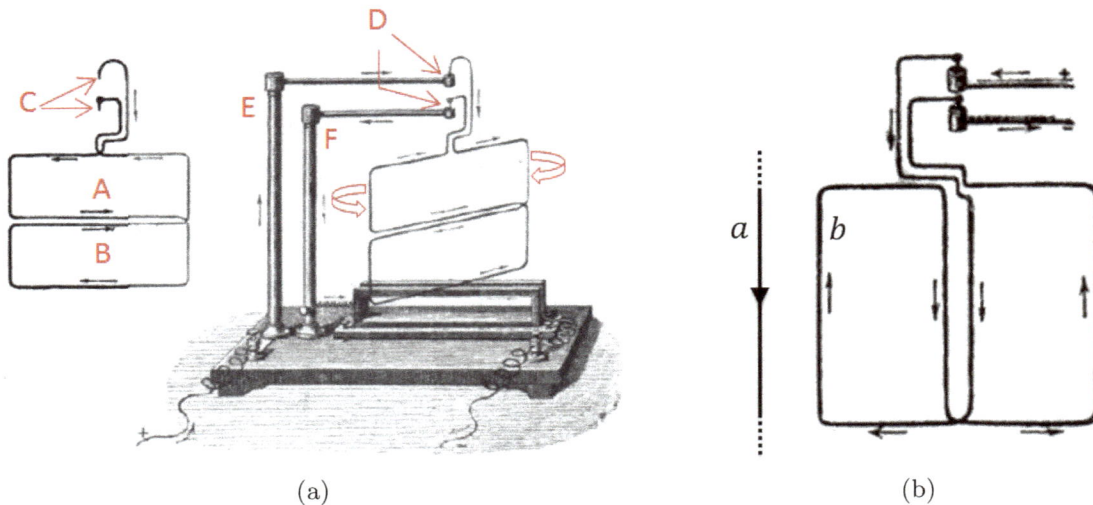

Figure 22.2: (a) An astatic current balance. (b) The magnetic field generated by the current-carrying wire a, exerting a force on an astatic balance causing it to rotate [45].

balance to rotate.

Ampere experimented with two types of wire: The first one, Figure 22.3(a), is a straight wire and looping back so that the currents are parallel and in opposite directions. The second one in Figure 22.3(b) is the sinuous wire type. The wire that loops back looks like a sinusoidal waveform. The current directions are shown by the arrows in the figure.

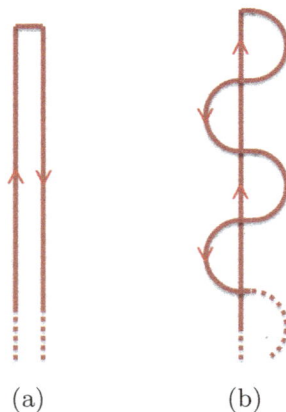

Figure 22.3: (a) A wire looping back. (b) A wire looping back in a sinusoid pattern.

Ampere noted that both these wire types behave in an identical way. Figure 22.4(a) shows the wire in the loop-back configuration near the astatic balance, and Figure 22.4(b) shows the sinuous wire near the astatic balance. Both the wire types have no effect on the astatic balance. Similar behavior is observed when the wire types are placed over a magnetic compass [46]. A wire looping

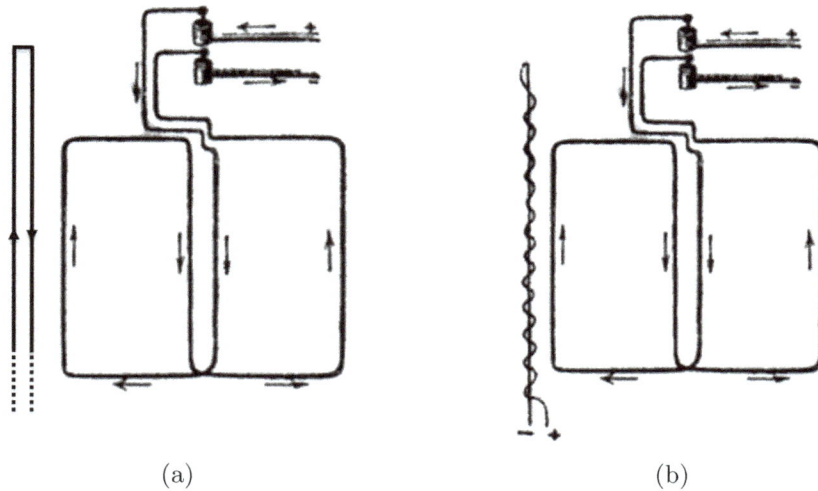

Figure 22.4: The wire in (a) Figure 22.3(a), and (b) Figure 22.3(b), placed near an astatic balance.

back over a magnetic compass is shown in Figure 22.5(a), and a sinuous wire in Figure 22.5(b). In both the cases, the needle does not deflect from the magnetic meridian. The experiments presented in this section are a few examples of Ampere's null experiments. The outcome of such experiments is no action is observed.

22.3 Meshing a Current-Carrying Wire into a Set of Vector Current Elements

The results of Ampere's experiments can be explained by discretizing the wire into small segments. Each of the segments can be treated as a vector, whose direction is the same as the current flowing in the segment. This is illustrated in Figure 22.6(a) and Figure 22.6(b), respectively, for the linear case and the sinuous case.

Using vector addition of the discretized elements, the resulting vector is the same for both the

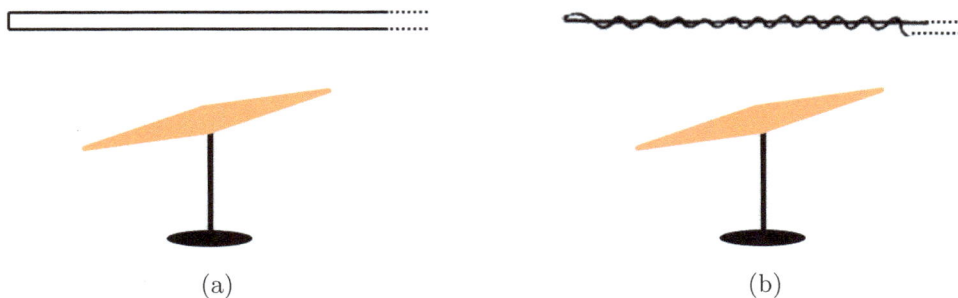

Figure 22.5: Ampere's experiments with different wire configurations. (a) A wire looping back over a magnetic compass. (b) A sinuous wire over a magnetic compass.

sinuous case and the typical case. The wire vector after the vector additions of the segments is shown by the dashed line, which are linear in both the wire types, hence exhibiting the same behavior on the compass and the astatic balance.

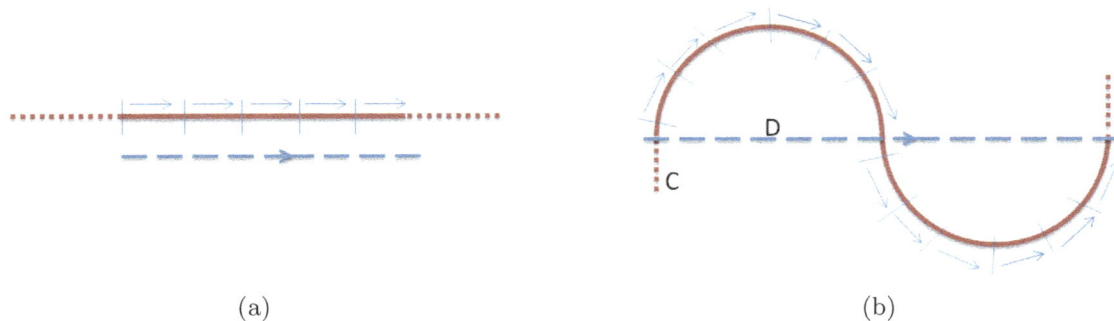

(a)   (b)

Figure 22.6: The vector addition of current elements in a (a) straight wire, and (b) sinuous wire.

In general, however, vector addition of wire segments would not be correct. For example, in the experiment shown in Figure 22.5(a), if the wire loop is a circle, although the vector addition of the elements is the zero vector in both cases, the circular loop will modify the direction of the magnetic compass.

Nevertheless, the idea of discretizing a wire into vector elements is very useful. The direction of a wire changes the magnetic-field pattern. In a long, straight wire, the magnetic field is concentric around the wire, as seen in Figure 21.13(a). If the wire is wound into loops, as shown in Figure 21.4(b), the magnetic-field pattern is completely different. A wire loop discretized into vector elements captures the direction of current flow in the wire. If the magnetic field of a wire element is quantified, using superposition, the contribution of all the elements can be added together to calculate the magnetic field of the wire loop. This will be explained in detail in the derivation of Biot-Savart's law, presented next.

23

Derivation of Biot-Savart's Law

Ampere studied the force between current-carrying wires, while Jean-Baptiste Biot, Felix Savart, and Laplace studied the magnetic field generated by a current-carrying wire. The magnetic field generated by a wire is named Biot-Savart's law in their honor, and is the focus of this chapter. The derivation of Biot-Savart's law is presented first, followed by Biot-Savart's two experiments to prove their formulation in the next chapter.

Biot-Savart's law is equivalent to Ampere's law, disguised in different forms. The derivation of Ampere's law from Biot-Savart's law will be presented in Chapter 30. Their equivalence is a verification of the formulations. Biot-Savart's law is valid only in magnetostatics, where there are no time-varying currents or fields. Modifying Ampere's law to include time-varying fields and currents will be presented later.

23.1   Biot-Savart's Law

There are sufficient clues available until now to theorize the magnetic field generated by a current-carrying wire segment. There are two parts to this problem: calculating (1) the direction of the magnetic field, and (2) the magnitude of the magnetic field, at any point.

23.1.1   Hypothesizing the Direction of the Magnetic Field

The magnetic-field patterns of a long wire in Figure 17.5, and a circular wire loop in Figure 21.4(b), will be used to surmise the direction of the magnetic field caused by an infinitesimal wire segment.

From Section 21.1, the iron-filings pattern provides a visual image of the magnetic field generated by the current-carrying wire, and its direction can be determined with a small magnetic compass. The direction of the magnetic field at a point is the same as the direction of the north pole of the compass.

In a linear wire, from Section 17.4, by circular symmetry, the magnetic field is concentric to the wire. From Figure 21.4(b), it can seen that the magnetic field at the center of a circular loop, should be perpendicular to the plane containing the loop. The direction of the magnetic field in the two cases can be determined from the first and second right-hand rules.

A long current-carrying wire, marked w, is shown in Figure 23.1(a), lying along the z-axis from $-\infty$ to $+\infty$. The direction of the current in the wire, of strength i, is marked by the arrow, as shown.

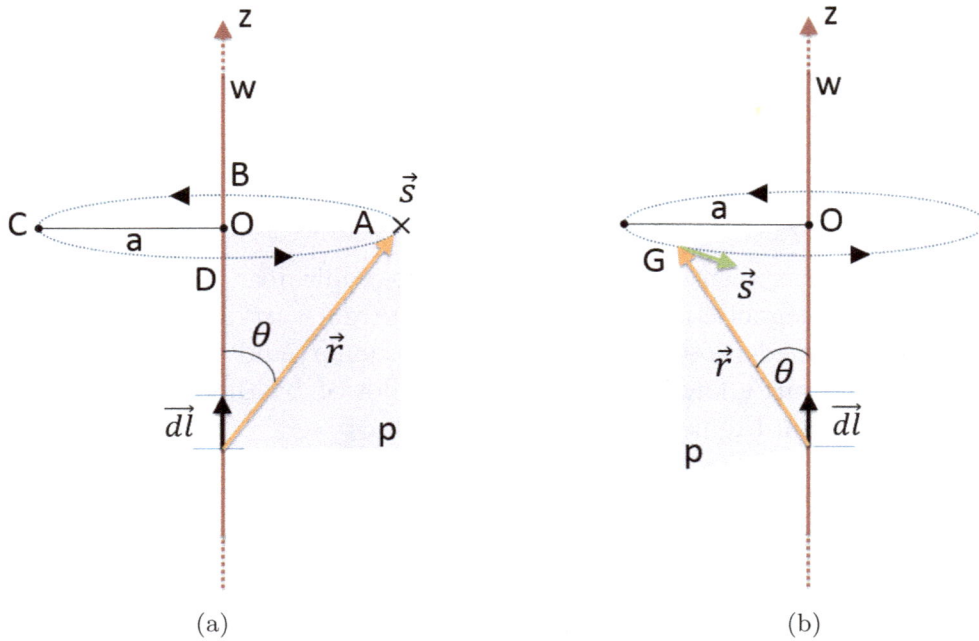

Figure 23.1: A long current-carrying wire on the z-axis. The direction of the magnetic field $\vec{s}$ at (a) Point A, and (b) Point G.

The wire is divided into tiny segments, one of which is shown, represented by the vector $\vec{dl}$. $\vec{dl}$ is a vector of magnitude dl, whose direction is the same as the direction of the current in the segment. Some of the experiments that led to the idea of modeling a wire by a set of vector elements was presented in Chapter 22.

Let $\vec{r}$ be the position vector from the current segment to Point A, where the magnetic field is to be calculated. Let Point A be located on the xy-plane with the wire extending to $\pm\infty$ above and below. A circle of radius a is drawn in the figure that shows the magnetic field around the wire in the direction marked by the arrows.

The two vectors $\vec{dl}$ and $\vec{r}$ must be combined in a way, such that the resulting vector, points in the correct direction of the magnetic field. Note that the magnetic field at any point on the circle is always in a direction perpendicular to the plane containing the vectors $\vec{dl}$ and $\vec{r}$. The plane containing $\vec{dl}$ and $\vec{r}$ is the shaded rectangle p. The direction of the magnetic field at Point A is into the page, marked by the symbol $\times$, and is perpendicular to the plane p.

156

Another example at a different point G is shown in Figure 23.1(b). Similar to the previous case, the magnetic field is perpendicular to the plane containing the vectors $\vec{dl}$ and $\vec{r}$. It would be easy to verify this for the other points B, C, or D, or any other point on the circle.

Looking at a menu of mathematical operations, the operator relating two vectors $\vec{dl}$ and $\vec{r}$, such that the resultant vector $\vec{s}$ is perpendicular to the plane containing the two vectors is the cross product,

$$\vec{s} = \vec{dl} \times \vec{r}. \tag{23.1}$$

Reversing the order of the cross-product operation would make $\vec{s}$ point in the opposite direction, which would be incorrect. Note that the cross-product operation would result in the correct direction of the magnetic field, for any of the wire segments making up the wire.

Making the assumption that superposition holds true, the direction of the magnetic field at a point is the sum of the direction vectors of the magnetic field generated by all the segments making up the wire. Considering all the segments of the wire in Figure 23.1, applying superposition, Equation 23.1 is consistent with the direction of the magnetic field observed in the iron-filings pattern.

The hypothesis can be tested on a circular current loop, whose magnetic-field pattern was shown earlier in Figure 21.4(b). The iron-filings pattern clearly shows that the field at the center of the loop is perpendicular to the loop area, and in the direction corresponding to the second right-hand rule of Section 21.2.

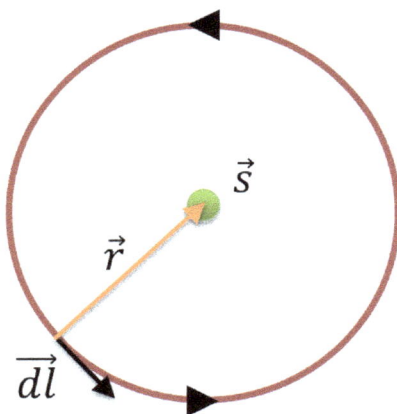

Figure 23.2: The magnetic field at the center of a circular current-carrying wire.

A circular current-carrying loop is shown in Figure 23.2, with current flowing in the direction marked by the arrows. The loop is assumed to be tightly wound with negligible thickness, and lying on the page. By the second right-hand rule, the magnetic-field direction at the center of the loop is out of the page, marked by the solid green dot. The direction of the cross product between $\vec{dl}$ at any point on the circular wire, and $\vec{r}$ to the center of the loop, does indeed point out of the page. Applying superposition on all the segments, the direction is consistent with the empirical observation.

23.1.2   Hypothesizing the Magnitude of the Magnetic Field

The method to calculate the magnetic-field direction caused by an infinitesimal current element was derived in the previous section. The magnetic field, however, is a vector quantity with both magnitude and direction. The magnitude can be included by scaling the direction vector by some scalar, say κ, to be determined in this section. Ampere's law was used to derive the magnetic field around a long current-carrying wire in Equation 21.23, and this result will be used to determine κ.

Let Point A in Figure 23.3, located at radius a cm from the wire, be the point at which the magnetic field is to be calculated. Assuming that superposition holds true, the magnetic field at Point A is the sum of the contributions of all the wire segments, written as the Riemann sum,

$$\vec{H} = \sum_{i=1}^{N} \kappa_i \vec{s}_i, \tag{23.2}$$

where N is the number of segments making up the wire, κ_i is a positive scalar multiplier of the direction vector $\vec{s}_i$. Substituting Equation 23.1 in the above equation,

$$\vec{H} = \sum_{i=1}^{N} \kappa_i \vec{\Delta z}_i \times \vec{r}_i, \tag{23.3}$$

where $\vec{\Delta z}_i$ is the ith wire element, and $\vec{r}_i$ is the position vector from the center of the ith element $(0, 0, z_i)$ to Point A. By definition of the cross product, the above equation can be written as

$$\vec{H} = \sum_{i=1}^{N} \kappa_i r_i \sin\theta_i \, \Delta z_i \, \hat{n}, \tag{23.4}$$

where θ_i is the angle between $\vec{r}_i$ and $\vec{\Delta z}_i$ as shown in the figure, r_i and Δz_i are the magnitudes of $\vec{r}_i$ and $\vec{\Delta z}_i$.

In this example, for any segment $\vec{\Delta z}_i$, the direction of the magnetic field,

$$\vec{s}_i = \vec{\Delta z}_i \times \vec{r}_i, \tag{23.5}$$

is always pointing into the page, as shown by the $\times$ symbol at Point A. Each of the direction vectors, corresponding to a differential element of the wire, is scaled by a positive multiplier. Therefore, the direction of the magnetic field, accounting for all the segments making up the wire, is also into the page, denoted as the unit vector $\hat{n}$ in Equation 23.4.

Equation 23.4 is written as

$$\vec{H} = \sum_{i=1}^{N} f(z_i) \Delta z_i \, \hat{n}, \tag{23.6}$$

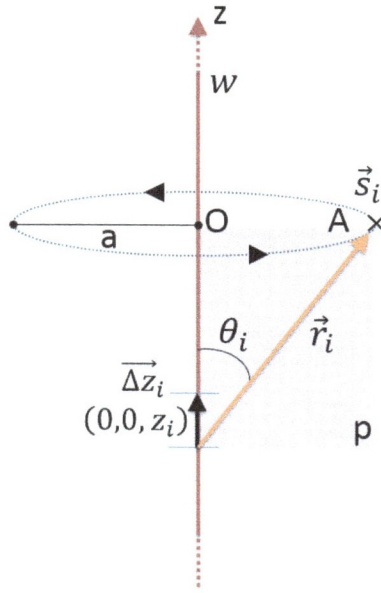

Figure 23.3: The magnetic field generated by a current element $\vec{\Delta z_i}$ at Point A.

to indicate that κ_i, r_i, $\sin\theta_i$ are functions of z_i. In the limit, as $N \to \infty$, the Riemann sum becomes the integral

$$\vec{H} = \int_\ell f(z)dz\,\hat{n}, \tag{23.7}$$

where the integral is calculated along the path of the wire ℓ. Substituting the expression for $f(z)$,

$$\vec{H}(A) = \int_{-\infty}^{+\infty} \kappa\,r\sin\theta\,dz\,\hat{n}, \tag{23.8}$$

and the magnitude of the field at Point A is

$$H(A) = \int_{-\infty}^{+\infty} \kappa\,r\sin\theta\,dz. \tag{23.9}$$

Using the technique of integration by substitution, the above integral will be evaluated. If the midpoint of a differential wire segment is $(0, 0, z)$, from Figure 23.3, θ satisfies the relation,

$$\tan\theta = \frac{-a}{z}$$
$$z = -a\cot\theta. \tag{23.10}$$

Differentiating the above equation with respect to θ,

$$\frac{dz}{d\theta} = a\csc^2\theta. \tag{23.11}$$

159

Making use of the relation

$$\sin \theta = \frac{a}{r}, \qquad (23.12)$$

Equation 23.11 can be written as

$$\sin \theta \, dz = r \, d\theta. \qquad (23.13)$$

Substituting the above expression in Equation 23.9, and rewriting the limits of the integral in terms of θ, by noting visually from the figure, as z varies from $-\infty$ to $+\infty$, θ varies from 0 to π rad,

$$H(A) = \int_0^\pi \kappa \, r^2 \, d\theta. \qquad (23.14)$$

The limits of the above integral can also be obtained from Equation 23.10. As z varies from $-\infty$ to $+\infty$, for any non-zero finite value of a, dividing both sides of the equation by a, $-\cot \theta$ also varies from $-\infty$ to $+\infty$. Looking at the plot of $-\cot \theta$, not presented here, θ varies from 0 to π, when $-\cot \theta$ varies from $-\infty$ to $+\infty$.

κ must be chosen such that the result of the above integral should equal $\frac{2i}{a}$, so that it is consistent with the result of Ampere's law in Equation 21.23. By trial and error, it can be guessed that κ should equal

$$\kappa = \frac{i}{r^3}. \qquad (23.15)$$

Substituting the above expression for κ in Equation 23.14,

$$H(A) = \int_0^\pi \frac{i}{r} \, d\theta. \qquad (23.16)$$

Rewriting Equation 23.12 as

$$r = a \csc \theta, \qquad (23.17)$$

and substituting the expression in Equation 23.16,

$$H(A) = \int_0^\pi \frac{i \sin \theta}{a} \, d\theta. \qquad (23.18)$$

i and a can be moved out of the integral, since they are constants independent of θ, and the result of the integral is

$$H(A) = \frac{2i}{a}, \qquad (23.19)$$

which is the same as the magnitude of the result in Equation 21.23, and therefore, consistent with Ampere's law.

The value of κ has been determined. The integration steps can be verified by substituting Equation 23.15 in Equation 23.9, and evaluating the integral without the use of integration by substitution.

Scaling the direction vector in Equation 23.1 by κ, Biot-Savart's law is formulated as

$$\vec{H} = \int_\ell \frac{i}{r^3} \left(\vec{dl} \times \vec{r} \right),$$
(23.20)

where the line integral is calculated along the path of the wire ℓ. Biot-Savart's law is usually written as an inverse square law,

$$\vec{H} = \int_\ell \frac{i}{r^2} \left(\vec{dl} \times \hat{r} \right),$$
(23.21)

where $\hat{r}$ is the unit vector of $\vec{r}$.

24

Experimental Verification of Biot-Savart's Law

Biot and Savart did two experiments to verify their formulation derived in Equation 23.21 [48][37]. In their first experiment, they showed that the magnetic field varies inversely as distance from a long current-carrying wire, as derived from Biot-Savart's law in Equation 23.19. In the second experiment, they verified the $\sin\theta$ relation, written as the cross-product operator.

The measurement of the magnetic-field strength in both of their experiments was done using Gauss's technique from Chapter 16. From Equation 16.10, the strength of a magnetic field H is inversely proportional to the square of the period of oscillation T,

$$H = \frac{K}{T^2}, \tag{24.1}$$

$$K = \frac{4\pi^2 I}{m_1 L_1}. \tag{24.2}$$

K is a constant that depends on the physical properties of the magnet, which are its moment of inertia I, length of the magnet L_1, and the strength of its poles $\pm m_1$.

They used relative measurements of the strength of the magnetic field in the experiments, and this does not require the value of K to be known. For example, if the magnetic field at Point A is

$$H_A = \frac{K}{T_A^2}, \tag{24.3}$$

and the magnetic field at Point B is

$$H_B = \frac{K}{T_B^2}, \tag{24.4}$$

the magnetic field at Point B is

$$H_B = \frac{T_A^2}{T_B^2} H_A, \tag{24.5}$$

relative to Point A.

24.1   Biot-Savart's First Experiment

A small magnet AB is suspended on a thin thread, and placed in front of a long current-carrying wire marked CMZ, shown in Figure 24.1. The distance between the magnet and the wire can be varied. AB is placed such that AB points in the terrestrial magnetic north-south direction, and the plane containing the wire and the midpoint of AB is perpendicular to the magnetic meridian. In this setup, by the first right-hand rule, the magnetic field generated by the current-carrying wire is either parallel or anti-parallel to the terrestrial magnetic field, depending on the current direction. Permanent magnet $A'B'$ is used to partially or completely cancel the effect of terrestrial magnetism on AB.

Biot-Savart did two variations of the first experiment: (a) in the first variation, they completely cancelled the effect of terrestrial magnetism, and therefore, the measurement made using AB is solely the effect of the current in the wire, and in the second variation (b) they only partially cancelled the effect of terrestrial magnetism. Both these variations led to the same confirmation: the magnetic-field strength of a current-carrying wire varies inversely as the distance from the wire.

The first variation of the experiment is presented next. The placement of $A'B'$ is such that the horizontal component of the terrestrial magnetic field at AB is cancelled. With no currrent flowing in the wire, $A'B'$ is brought little by little closer towards AB, until the oscillation of AB, when perturbed, becomes negligible. No oscillation of AB is equivalent to having an oscillation of period ∞. From Equation 24.1, the magnetic-field strength is 0.

Biot-Savart noted down the time required for ten oscillations of AB located $a =$15 mm, 20 mm, 30 mm, 40 mm, 50 mm, 60 mm, and 120 mm from the current-carrying wire CMZ, which was used to calculate the electrical magnetic-field strength using Equation 24.1. They used the measurement at 30 mm to determine the relative factor of the increase or decrease of the magnetic-field strength.

To cope with poor batteries that lose their energy quickly, they used an averaging technique: to measure the magnetic-field strength at 20 mm from the wire CMZ, for example, they collected 3 data points, each noting the duration of ten oscillations of AB. (1) first at 30 mm, which was used as the reference point for relative comparisons, (2) at 20 mm, and finally (3) again at 30 mm, and averaged the first and third measurements for comparison against the data at 20 mm. By doing this, they could reduce the error that could occur from the battery becoming weaker during the course of the experiment.

From Biot-Savart's law, using Equation 23.19, the ratio of the field strengths at distance 30 mm and distance a from the wire, is

$$\frac{H(a = 30\,mm)}{H(a)} = \frac{\frac{2i}{30\,mm}}{\frac{2i}{a}}$$

$$= \frac{a}{30\,mm}, \tag{24.6}$$

where $H(a = 30\,mm)$ and $H(a)$ are the magnetic-field strengths at $a = 30\,mm$ and at a.

The results are plotted in Figure 24.2. The square markers are the ratio of the measured magnetic-field strengths at distance 30 mm from CMZ, to the field at distance a. The diamond markers are the calculated results from the above equation. The x-axis is distance a from the wire. Both the plots show a good match, proving that the magnetic field from the wire varies inversely as distance.

In the second variation of the experiment, rather than completely cancel the effect of terrestrial magnetism, the magnet $A'B'$ was placed at a position that only decreased the effect of terrestrial magnetism at the location of the magnet AB. As before, the plane containing the wire and the mid point of AB is perpendicular to the magnetic meridian. The direction of the current in the wire was chosen such that the magnetic field generated is anti-parallel to the terrestrial magnetic field. This further decreased the terrestrial magnetic-field strength at AB. The magnetic-field strength of the current-carrying wire is not large enough to reverse the direction of AB, but only reduces the terrestrial magnetic field.

The data obtained from Biot-Savart's publication is summarized in Table 24.1 [48]. The measurement distance from the wire is shown in the column labeled *Distance a (mm)*. The duration of 40 oscillations with magnet AB of length *10 mm*, without and with current flowing in the wire, are noted in the columns T'_a and T'_b, respectively. The prime indicates that the value is not calculated or measured per oscillation, but 40 oscillations.

Since there is no current flow in the T'_a measurements, it captures the strength of the terrestrial magnetism alone. T'_b includes the effects of both terrestrial magnetism, as well as the magnetic field generated by the current-carrying wire, since the terrestrial magnetic field at AB has not been fully cancelled by $A'B'$.

From the data shown in the table, since $H \propto \frac{1}{T^2}$, $T'_b > T'_a$ indicates that the strength of the field that includes the effects of the current in the wire and terrestrial magnetism, is weaker than the strength due to terrestrial magnetism alone. Using Equation 24.1, the magnetic field H'_w due to the wire alone, can be calculated from

$$H'_w = \frac{K}{T'^2_a} - \frac{K}{T'^2_b}. \tag{24.7}$$

Since only relative field strengths are compared, the result of the ratio will be the same if T'_a, T'_b are used instead of the periods of a single oscillation. The measured ratio of the magnetic-field strengths due to the wire at distances 32.9 mm and 62.9 mm is shown in the last column. Similar to the calculation in Equation 24.6, the theoretical value using Biot-Savart's law is

$$\frac{1}{32.9mm} \div \frac{1}{62.9mm} \approx 1.91. \tag{24.8}$$

The measured value of 1.84 is approximately the same as the calculated value of 1.91, which

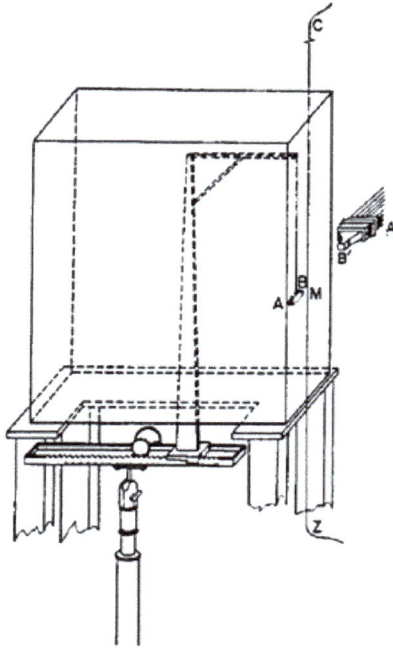

Figure 24.1: Biot-Savart's experiment setup to demonstrate the inverse distance variation of magnetic-field strength, caused by a current-carrying wire [49].

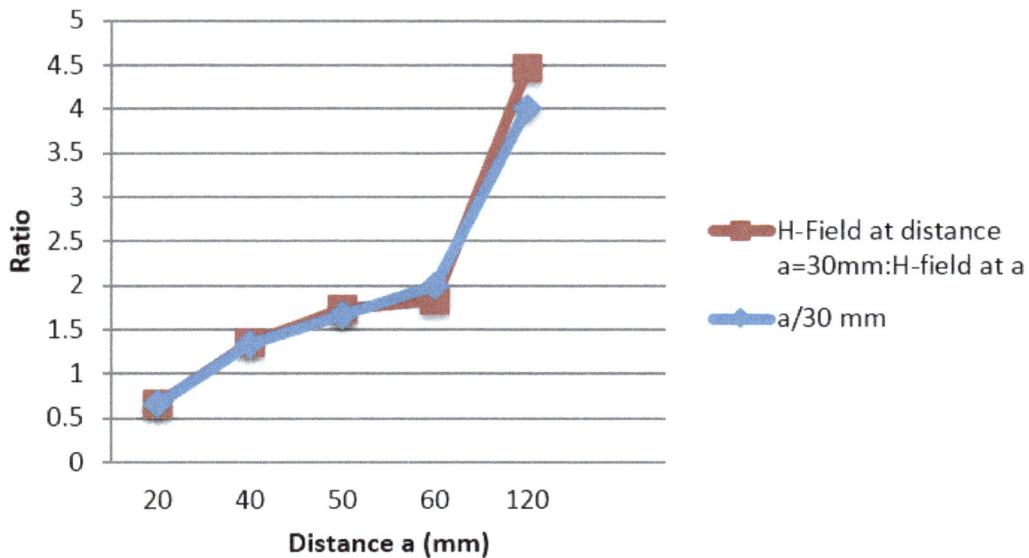

Figure 24.2: The experiment results of Figure 24.1, showing that the magnetic-field strength varies inversely as the distance from the wire.

Second Variation of Experiment 1 Data from Biot-Savart's Publication [48]					
Trial	Distance a (mm)	T'_a (sec)	T'_b (sec)	H'_w due to only the wire $(K \times 10^{-4})$	Ratio of $\left|\vec{H}_w\right|$ at 32.9 to 62.9 mm
1	32.9	67.5	101	1.21449	
2	62.9	66	78.5	0.67290	1.84239
3	32.9	66	98.5	1.26500	

Table 24.1: Experiment 1 data of Figure 24.1, using a magnetic needle of length 10 mm to record its time duration of 40 oscillations.

confirms that the magnetic-field strength varies inversely as the distance from the wire.

Instruments to measure current in electromagnetic units will be discussed in Chapter 27. Using a similar experiment setup, it is possible to show that the magnetic field H generated by a current-carrying wire, is proportional to the current i. This can be accomplished by changing the current in the wire, while keeping the location of the magnet AB from the wire fixed. The measured values of the magnetic field at the location of AB, using similar calculations to the ones shown above, can be used to show that $H \propto i$.

24.2 Biot-Savart's Second Experiment

Biot-Savart's first experiment shows that the magnetic-field strength varies inversely as distance. They constructed their second experiment to prove the $\sin\theta$ term, written as the cross product in Biot-Savart's law.

The experiment setup is shown in Figure 24.3. They compared the magnetic-field strength ratios between two wires, oblique and vertical, marked w_O and w_V, as shown. By doing this, as shown in the derivation presented next, they were able to validate the $\sin\theta$ term in their formulation. Biot-Savart doubled wire w_O, marked $\times 2$ in the figure, to double its field effect, which would make measurement easier, and more conclusive. The analytical derivation of the ratio of the magnetic-field strengths of the two wire types will be presented first, followed by their experiment results.

The current direction in the two wire types are marked by the arrows. The direction of the magnetic field generated by any differential wire element $\vec{dl}$, using Equation 23.1, is out of the page at Point P in either of the wire types. The total magnetic field accounting for all the segments, therefore, is also out of the page at Point P, in both the cases.

The magnetic field of the vertical wire has already been analyzed previously, written in Equation 23.19. The magnetic-field strength at Point P is

$$H_{w_V}(P) = \frac{2i}{a}, \tag{24.9}$$

166

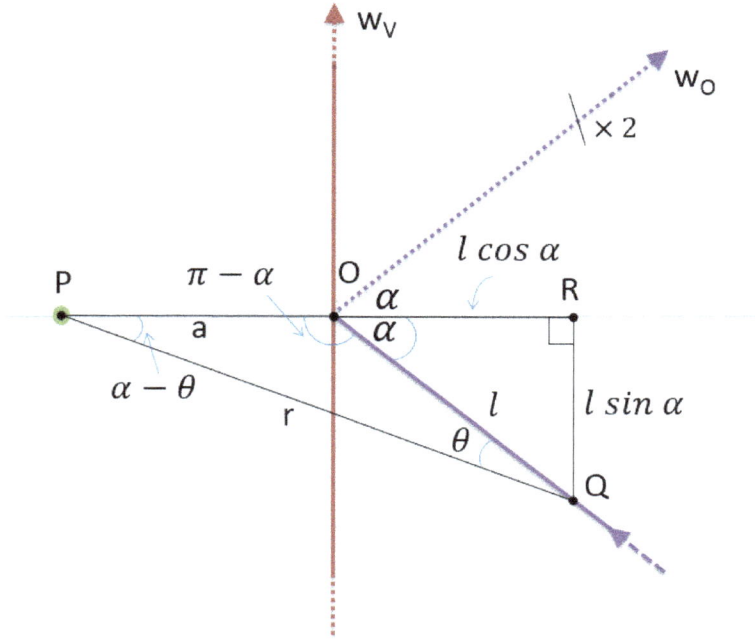

Figure 24.3: The second experiment setup of Biot-Savart to empirically prove the $\sin \theta$ relationship of their hypothesis.

where i is the current in the wire, and a is the distance between the wire and Point P. Only the ratios of the fields between the oblique and the vertical wires are compared, and therefore, it is not necessary to know the value of i, since it gets cancelled.

The magnetic-field strength at Point P, generated by the oblique wire w_O, will be derived next. Half of w_O is shown dotted, and the other half is a solid line. The wire marked by the solid line will be analyzed. By symmetry, this result is multiplied by 2 to model the complete wire. Wire w_O is doubled to increase the field strength. To account for this effect, the magnetic field of w_O should be multiplied by 2 again, to calculate the final value.

Lengths OP, OQ, and QP are a, l, and r, respectively, as marked in the figure. The oblique wire is inclined at $\angle ROQ = \alpha$. In Biot-Savart's experiment, they used $\alpha = \frac{\pi}{4} \, rad$. It follows that $\angle QOP = \pi - \alpha$, and lengths

$$OR = l \cos \alpha, \tag{24.10}$$
$$RQ = l \sin \alpha. \tag{24.11}$$

Let $\angle OQP = \theta$ and therefore, $\angle OPQ = \alpha - \theta$. By the law of sines applied to $\triangle POQ$,

$$\frac{\sin (\alpha - \theta)}{l} = \frac{\sin \theta}{a}. \tag{24.12}$$

167

Expanding $\sin(\alpha - \theta)$, and simplifying the relation,

$$l = a \sin \alpha \cot \theta - a \cos \alpha. \tag{24.13}$$

Differentiating the above expression with respect to θ

$$\frac{dl}{d\theta} = \frac{-a \sin \alpha}{\sin^2 \theta} \tag{24.14}$$

Applying the law of sines to $\triangle POQ$,

$$\frac{\sin(\pi - \alpha)}{r} = \frac{\sin \theta}{a}. \tag{24.15}$$

Using the identity

$$\sin(\pi - \alpha) = \sin \alpha, \tag{24.16}$$

Equation 24.15 can be simplified as

$$\frac{\sin \alpha}{r} = \frac{\sin \theta}{a}. \tag{24.17}$$

Simplifying Equation 24.14 using the above expression,

$$dl \sin \theta = -r \, d\theta. \tag{24.18}$$

By symmetry, the strength of the magnetic field at Point P due to wire w_O, can be calculated by analyzing only the portion of the semi-infinite solid wire using Equation 23.21,

$$H_{w_O}(P) = 2 \int_0^\infty \frac{i \sin \theta}{r^2} \, dl, \tag{24.19}$$

and doubling the result to account for the semi-infinite dotted portion. As $l \to 0$, $\angle OPQ = \alpha - \theta \to 0$. Therefore, $\theta \to \alpha$, as $l \to 0$. As $l \to \infty$, it can be visually seen from the figure, OQ and PQ can be viewed as parallel lines, cut by the transversal RP. In this case, $\angle OPQ = \alpha - \theta$ and $\angle ROQ = \alpha$ are corresponding angles, and are equal to each other. Therefore, as $l \to \infty$, $\alpha - \theta \to \alpha$, or $\theta \to 0$. To summarize, as l changes from 0 to ∞, θ changes from α to 0 radians.

Using the technique of integration by substitution, substituting Equation 24.18 in the above integral, and changing the limits of the integral,

$$H_{w_O}(P) = 2 \int_\alpha^0 -\frac{i}{r} \, d\theta. \tag{24.20}$$

Solving for r in Equation 24.17, and substituting this expression in the above equation,

$$H_{w_O}(P) = \frac{2i}{a \sin \alpha} \int_0^\alpha \sin \theta \, d\theta. \tag{24.21}$$

Evaluating the above integral, the strength of the magnetic field generated by w_O at P is

$$H_{w_O}(P) = \frac{2i}{a} \tan\left(\frac{\alpha}{2}\right). \tag{24.22}$$

Biot-Savart doubled the oblique wire to enhance its effect. Taking this into account, the ratio of the field strengths at P of the oblique wire to the vertical wire, Equation 24.22 to Equation 24.9, is

$$2 \times \frac{H_{w_O}(P)}{H_{w_V}(P)} = 2 \times \tan\left(\frac{\alpha}{2}\right). \tag{24.23}$$

For $\alpha = 45°$ that was used in Biot-Savart's experiment,

$$2 \times \frac{H_{w_O}(P)}{H_{w_V}(P)} = 2 \times \tan\left(\frac{45°}{2}\right)$$
$$= 0.828427 \tag{24.24}$$

is the expected result from their experiment. The above ratio is independent of a, which is the distance from the wire where the measurement is made. Biot-Savart's experiment result is indeed very close to this value, as discussed next.

In Biot-Savart's experiment setup in Figure 24.3, the plane containing the oblique and the vertical wires is perpendicular to the magnetic meridian. In such a setup, the direction of the current in the wires can be chosen such that the magnetic field generated by the wires is parallel or anti-parallel to the terrestrial magnetic field at Point P, in-out of the page. The terrestrial magnetism can be subtracted from the measurement.

Magnet AB, similar to Experiment 1, is placed at Point P, where the magnetic-field strength is to be measured. Similar to the second variation of Experiment 1, the terrestrial magnetism is partially cancelled by the magnet $A'B'$.

Biot-Savart's results are summarized in Table 24.2 and Table 24.3, and measurements of the magnetic-field strengths are made at Point P, located 28.5 mm and 33 mm from the wire, respectively. The length of Magnet AB in Table 24.2 is 20 mm, and 10 mm for the results in Table 24.3.

The columns marked T'_a and T'_b are the time taken for the number of oscillations noted in the tables. T'_a is used in the calculation of terrestrial magnetism, where no current flows in either of the wires. T'_b includes the effects of terrestrial magnetism, and the current in either the oblique wire w_O, or the vertical wire w_V.

When current flows in the wire, it either adds to terrestrial magnetism, the additive case, or subtracts from it, the subtractive case, depending on the direction of the current flowing in the wire. The direction of the field generated by the current can be determined from Biot-Savart's law in Equation 23.21. Trials 1 and 2 in both the tables are the subtractive cases, while Trials 3 and 4 are additive.

From Equation 24.1, since $H \propto \frac{1}{T^2}$, $T_b' > T_a'$ is the subtractive case, which indicates that the strength of the field that includes the effects of the current in the wire and terrestrial magnetism, is weaker than the strength caused only by terrestrial magnetism. In this case, the magnetic field of the current-carrying wire partially cancels the terrestrial magnetic field. Similarly, $T_b' < T_a'$ is the additive case, which indicates that the field strength caused by the current-carrying wire and Earth's magnetism, is greater than the strength only due to Earth's magnetism.

The terrestrial magnetic-field strength can be subtracted from the measurement in the subtractive case using Equation 24.7. In the additive case, the terrestrial magnetic field can be subtracted from the measurement, using the equation

$$H_w' = \frac{K}{T_b'^2} - \frac{K}{T_a'^2}. \tag{24.25}$$

To reduce the error caused by poor batteries at the time, which ran out of energy relatively quick, as before, for each trial they collected 3 data points. They averaged the first and the third data points for comparison with the second. The magnetic-field strength due to the current-carrying wire alone H_w', can be obtained from Equation 24.25 and Equation 24.7 for the additive and the subtractive cases, respectively.

The prime indicates that the value is not calculated or measured per oscillation, rather, many oscillations, as noted in the tables. Since only the ratios of the field strengths are compared with the theoretical value, the results of the ratio will be the same regardless of whether per oscillation data, or many oscillations data are used in the calculation. Comparison of the ratios also has the added advantage that the physical properties of the magnet, K in Equation 24.2, is not required, since it gets cancelled.

The magnetic-field strength due to the wire is normalized with respect to $K \times 10^{-4}$, as shown in the column label. The last column is the ratio of the magnetic-field strengths due to the current-carrying oblique to vertical wires, without the effect of terrestrial magnetism. Biot-Savart averaged these values for the different trials, obtaining the empirical value of 0.821, which is almost the same as the theoretical ratio of 0.828 in Equation 24.24, thereby confirming their hypothesis, now known as Biot-Savart's law in their honor.

T_a' (sec)	Trial	T_b' (sec)	Average of Trials a & c	H_w' due to only the wire $(K \times 10^{-4})$	Ratio of H_{w_O} to H_{w_V}
			Experiment 2 Data from Biot-Savart's Publication [48]		
58.375	1a 1b 1c	w_O: 80.25 w_V: 87.75 w_O: 80.00	w_O: 80.125 w_V: 87.75	w_O: 1.37695 w_V: 1.63589	0.841714
	2a 2b 2c	w_V: 87.75 w_O: 80 w_V: 87	w_O: 80 w_V: 87.375	w_O: 1.37208 w_V: 1.62472	0.844504
58.875	3a 3b 3c	w_O: 48.19 w_V: 47.08 w_O: 48.68	w_O: 48.435 w_V: 47.08	w_O:1.37772 w_V:1.62661	0.846985
	4a 4b 4c	w_V: 47.08 w_O: 48.68 w_V: 47.27	w_O: 48.68 w_V: 47.175	w_O:1.33492 w_V:1.60846	0.829935
Average					0.840785

Table 24.2: Experiment 2 data of Figure 24.3 using a magnetic needle length of 20 mm to record its time duration of 20 oscillations. The measurement location, Point P in Figure 24.3, is 28.5 mm from the wire.

T_a' (sec)	Trial	T_b' (sec)	Average of Trials a & c	H_w' due to only the wire $(K \times 10^{-4})$	Ratio of H_{w_O} to H_{w_V}
			Experiment 2 Data from Biot-Savart's Publication [48]		
67	1a 1b 1c	w_V: 102 w_O: 90.5 w_V: 100	w_O: 90.5 w_V: 101	w_O: 1.00670 w_V: 1.24737	0.807060
	2a 2b 2c	w_O: 90.5 w_V: 100 w_O: 89.5	w_O: 90 w_V: 100	w_O: 0.99310 w_V: 1.22767	0.808932
	3a 3b 3c	w_V: 54.0 w_O: 56 w_V: 53.75	w_O: 56 w_V: 53.875	w_O:0.96111 w_V:1.21762	0.789334
	4a 4b 4c	w_O: 56 w_V: 53.75 w_O: 55.5	w_O: 55.75 w_V: 53.75	w_O:0.98977 w_V:1.23366	0.802303
Average					0.801907
Average from Table 24.2					0.840785
Final Average					0.821346

Table 24.3: Experiment 2 data of Figure 24.3 using a magnetic needle length of 10 mm to record its time duration of 40 oscillations. The measurement location, Point P in Figure 24.3, is 33 mm from the wire.

Definitions in Electrostatic Units

As explained in Section 17.6, the observations made in Oersted's experiment gives rise to either define electrical quantities first, or magnetic quantities first, which are named ESU and EMU. The flowchart of definitions in ESU will be presented in this chapter, and the definitions in EMU will be presented in the following chapter. The definitions will be revised in the future chapters, without the use of magnetic charges.

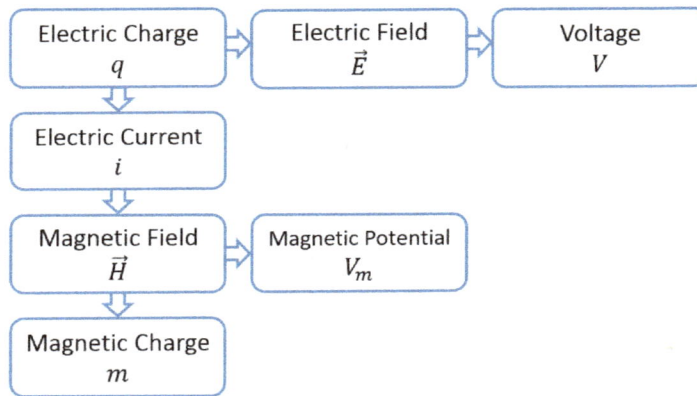

Figure 25.1: The definitions flowchart in electrostatic units.

Each of the definitions will be summarized in this chapter. Note that there are no circular definitions of the same quantity, and therefore, the electrical quantities are well defined. It will become clear, why some of the equations presented in the previous chapters are valid only in ESU, or EMU, or common to both ESU and EMU. It will become clear that some of the equations, although the units of the variables are different between ESU and EMU, are common to both the systems of units.

25.1 Electric Charge q

The electric charge is defined first in ESU using Coulomb's law,

$$\vec{F} = \frac{q_1\, q_2}{r^2}\, \hat{r},$$

$$(25.1, \text{ESU})$$

where q_1 and q_2 are two points charges, r is the distance between them, and $\hat{r}$ is the unit vector indicating the direction of the force. The unit electric charge is defined as two equal unit point charges, separated by 1 cm, exerting a force of 1 dyne on each other.

25.2 Electric Field $\vec{E}$

Once the electric charge has been defined, it follows that the electric field $\vec{E}$ can be defined from the force $\vec{F}$ exerted on charge q,

$$\vec{E} = \frac{\vec{F}}{q}. \tag{25.2}$$

25.3 Voltage V

The voltage V between two points A and B is the path integral of the electric field $\vec{E}$,

$$V = \int_A^B \vec{E} \cdot \vec{dl}. \tag{25.3}$$

It was proven that the electric field is a conservative field, and the above value is independent of the path from A to B.

25.4 Electric Current i

Since electric current is the flow of charges, using Equation 18.1,

$$i = \frac{\Delta q}{\Delta t}, \tag{25.4}$$

current i is defined as the flow of charge Δq in time Δt.

25.5 Magnetic Field $\vec{H}$

Biot-Savart's law relates magnetic field $\vec{H}$ and electric current i, by the path integral along a current-carrying wire ℓ,

$$\vec{H} = \int_\ell \frac{i}{r^2} \left(\vec{dl} \times \hat{r} \right), \tag{25.5}$$

where r is the distance from $\vec{dl}$ to the point where the magnetic field is calculated, and $\hat{r}$ is the unit vector in that direction.

The relation between magnetic field $\vec{H}$ and i is the same in ESU and EMU. In ESU, Biot-Savart's law is used to define magnetic field from the definition of current. In EMU, however, it will be shown in the following chapter, current is defined from the definition of magnetic field.

The magnetic field around a long straight wire, carrying current i, was derived in Equation 23.19,

$$\vec{H} = \frac{2i}{a}\,\hat{\varphi},$$

(25.6)

where $\hat{\varphi}$ is in the azimuth direction, and set by the first right-hand rule. Using the above equation, the magnetic field can be defined as the field of 2 units, generated at a unit distance from a long straight wire, which carries a unit current. The units of the electrical quantities will be summarized towards the end of the chapter.

25.6 Magnetic Potential V_m

The magnetic potential V_m, between two points A and B, is the path integral of the magnetic field $\vec{H}$,

$$V_m = \int_A^B \vec{H} \cdot \vec{dl}.$$

(25.7)

Assuming symmetry to electric charges, magnetic field due to magnetic charges are also conservative, and any path may be chosen from A to B, to calculate the magnetic potential.

25.7 Magnetic Charge m

The magnetic charge m, can be defined from the force $\vec{F}$ exerted on the charge in a magnetic field $\vec{H}$,

$$\vec{F} = m\vec{H}.$$

(25.8)

Note that the above definition of m has no relation to the Coulomb's law of magnetic charges in Equation 13.1. The force between two magnetic charges m_1 and m_2 separated by distance r, can only be written as

$$\vec{F} \propto \frac{m_1 m_2}{r^2}\,\hat{r},$$

(25.9)

where $\hat{r}$ is the unit vector in the direction of the force. The proportionality constant in ESU is to be determined. However, since the definitions will be revised in the future chapters without the need of magnetic charges, determining the proportionality constant in the above equation will not be pursued.

25.8 Units in ESU

Table 25.1 summarizes the units in ESU. It is left as an exercise for the reader to derive them from dimensional analysis, similar to the examples in the earlier chapters.

Electrical Quantity (Symbol)	Unit (ESU)	Unit (ESU)
Charge (q)	statcoulomb	$gm^{1/2}cm^{3/2}s^{-1}$
Electric field $(\vec{E})$	–	$gm^{1/2}cm^{-1/2}s^{-1}$
Voltage (V)	statvolt	$gm^{1/2}cm^{1/2}s^{-1}$
Electric current (i)	statampere	$gm^{1/2}cm^{3/2}s^{-2}$
Magnetic field $(\vec{H})$	–	$gm^{1/2}cm^{1/2}s^{-2}$
Magnetic potential (V_m)	–	$gm^{1/2}cm^{3/2}s^{-2}$
Magnetic charge (m)	–	$gm^{1/2}cm^{1/2}$

Table 25.1: A summary of the units in ESU.

26

Definitions in Electromagnetic Units

The flowchart of definitions in EMU is summarized in Figure 26.1. The definitions will be revised in the future chapters. Each of these definitions will be summarized next.

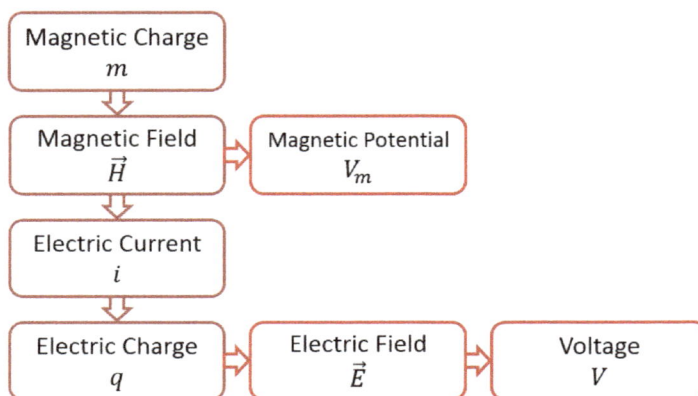

Figure 26.1: The definitions flowchart in electromagnetic units.

26.1 Magnetic Charge m

Unlike the definitions in ESU, where electric charge is defined first in ESU, the magnetic charge is defined first in EMU. Making the assumption that magnetic charges follow a similar behavior as Coulomb's law of electric charges, the magnetic charge is defined from the force $\vec{F}$ governing their behavior,

$$\vec{F} = \frac{m_1 \, m_2}{r^2} \, \hat{r},$$ (26.1, EMU)

where m_1 and m_2 are two point magnetic charges separated by distance r, $\hat{r}$ is the direction of the force between them. Unit magnetic charges repel each other with a unit force, when separated by a unit distance.

26.2 Magnetic Field $\vec{H}$

Similar to the definition of the electric field, the magnetic field can be defined from the force $\vec{F}$ exerted on a magnetic charge m,

$$\vec{H} = \frac{\vec{F}}{m}. \tag{26.2}$$

26.3 Magnetic Potential V_m

The magnetic potential V_m, between two points A and B, is the path integral of the magnetic field $\vec{H}$,

$$V_m = \int_A^B \vec{H} \cdot \vec{dl}. \tag{26.3}$$

Assuming symmetry to electric charges, magnetic field due to magnetic charges are also conservative, and any path may be chosen from A to B, to calculate the magnetic potential.

26.4 Electric Current i

Biot-Savart's law relates the current in a wire to the magnetic field generated,

$$\vec{H} = \int_\ell \frac{i}{r^2} \left(\vec{dl} \times \hat{r} \right), \tag{26.4}$$

where r is the distance from $\vec{dl}$ to the point where the magnetic field is calculated, and $\hat{r}$ is the unit vector in that direction. Since the magnetic field has been defined in Equation 26.2, it follows from Biot-Savart's law, the electric current is also defined.

For example, one way to define current is from the calculation of the magnetic field $\vec{H}$ around a long straight wire, at radius a, derived in Equation 23.19,

$$\vec{H} = \frac{2i}{a} \, \hat{\varphi}, \tag{26.5}$$

where $\hat{\varphi}$ is the azimuth direction that is in agreement with the first right-hand rule. Using the above equation, current i is of the value that generates the above magnetic field $\vec{H}$ around the wire.

26.5 Electric Charge q

Current is the flow of charges. It follows that charge is defined from Equation 18.1,

$$\Delta q = i \, \Delta t, \tag{26.6}$$

where Δq is the charge contained in current i flowing for time Δt. The unit charge is the charge contained in a unit current flowing for a unit time duration.

The above definition of the unit charge in EMU, has no relation to the definition of the unit electric charge using Coulomb's law in ESU. The force between two charges q_1 and q_2, separated by a distance r, can be written in EMU as

$$\vec{F} \propto \frac{q_1 q_2}{r^2} \hat{r}, \tag{26.7}$$

where $\hat{r}$ is the unit vector in the direction of the force, and the proportionality constant in EMU is yet to be determined.

ESU and EMU are not consistent with each other. The unit electric charge, unit electric current, etc. in ESU, are not the same as the unit electric charge, unit electric current, etc. in EMU. This will become clear in Chapter 57. Determining the ratio of the unit electrostatic charge to the unit electromagnetic charge was one of the greatest problems that was confronted in the 19th century. Weber and Kohlrausch first determined this ratio in 1856, and their experiment will be discussed in considerable detail later.

26.6   Electric Field $\vec{E}$

Electric field $\vec{E}$ is defined from the force $\vec{F}$ exerted on a charge q,

$$\vec{E} = \frac{\vec{F}}{q}. \tag{26.8}$$

26.7   Voltage V

Voltage V between two points A and B is the path integral of the electric field $\vec{E}$,

$$V = \int_A^B \vec{E} \cdot \vec{dl}. \tag{26.9}$$

Since the electrostatic field is a conservative field, the path integral is independent of the path from A to B.

26.8   Units in EMU

Table 26.1 summarizes the units in EMU. It is left as an exercise for the reader to derive them from dimensional analysis, similar to the examples in the earlier chapters.

Electrical Quantity (Symbol)	Unit (EMU)	Unit (EMU)
Charge (q)	abcoulomb	$gm^{1/2}cm^{1/2}$
Electric field ($\vec{E}$)	–	$gm^{1/2}cm^{1/2}s^{-2}$
Voltage (V)	abvolt	$gm^{1/2}cm^{3/2}s^{-2}$
Electric current (i)	abampere	$gm^{1/2}cm^{1/2}s^{-1}$
Magnetic field ($\vec{H}$)	oersted	$gm^{1/2}cm^{-1/2}s^{-1}$
Magnetic potential (V_m)	–	$gm^{1/2}cm^{1/2}s^{-1}$
Magnetic charge (m)	maxwell	$gm^{1/2}cm^{3/2}s^{-1}$

Table 26.1: A summary of the units in EMU.

27

Measurement of Current in Electromagnetic Units

A galvanometer is an instrument to measure current. Different types of galvanometers exist. The simplest is the tangent galvanometer, and will be described in detail in this chapter.

The magnetic field generated at the center of a current-carrying circular loop, and the terrestrial magnetic field, are required for the absolute measurement of current using a tangent galvanometer. The terrestrial magnetic field can be measured using Gauss's technique presented earlier. Since this technique is applicable only in EMU, for now, a tangent galvanometer will be used to measure current in absolute measure in EMU. The electrical magnetic field at the center of current-carrying loop can be calculated using Biot-Savart's law, presented next.

27.1 Magnetic Field at the Center of a Circular Loop

A circular current-carrying wire is shown in Figure 27.1, and without loss of generality, the current flows in the direction as marked. From Biot-Savart's law, the magnetic field at the center of the

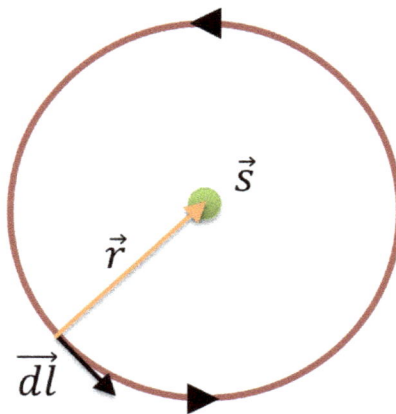

Figure 27.1: A circular current-carrying wire.

loop points out of the page. The direction vector of the magnetic field is $\vec{dl} \times \vec{r}$, and points out of the page for any value of $\vec{dl}$. $\vec{dl}$ is always perpendicular to $\hat{r}$, since the radius is perpendicular to the tangent,

$$\vec{dl} \times \hat{r} = dl\ (\sin 90°)\ \hat{n}$$
$$= dl\ \hat{n}, \tag{27.1}$$

where $\hat{n}$ is the unit vector pointing out of the page.

Applying Biot-Savart's law on the circular loop ℓ of radius r, the magnetic field at the center of the circle marked A is

$$\vec{H}(A) = \int_\ell \frac{i}{r^2}\ dl\ \hat{n}, \tag{27.2}$$

where i is the current flowing in the wire. From geometry, an arc length l is related to the radius r, and the angle θ in radians subtended by the arc, by the relation

$$l = r\theta. \tag{27.3}$$

Differentiating the above expression with respect to θ, and since r is a constant that does not depend on θ,

$$dl = r\ d\theta. \tag{27.4}$$

Using the technique of integration by substitution, substituting the above expression in Equation 27.2, and integrating over the complete circle $[0, 2\pi]$,

$$\vec{H}(A) = \frac{i}{r} \int_0^{2\pi} d\theta\ \hat{n}. \tag{27.5}$$

Since i and r are constants in the circular loop, these variables can be moved out of the integral. Solving the integral, the magnetic field at the center of the loop is

$$\vec{H}(A) = \frac{2\pi\ i}{r}\ \hat{n}. \tag{27.6}$$

The unit of magnetic field in EMU is the oersted (Oe), defined in Equation 14.2, and r is in cm. Using dimensional analysis, the unit of current in EMU is

$$[i] = Oe \cdot cm$$
$$= \frac{gm^{\frac{1}{2}}\ cm^{\frac{1}{2}}}{s}$$
$$= abampere. \tag{27.7, EMU}$$

The unit of current in EMU is the abampere, as noted in Table 26.1.

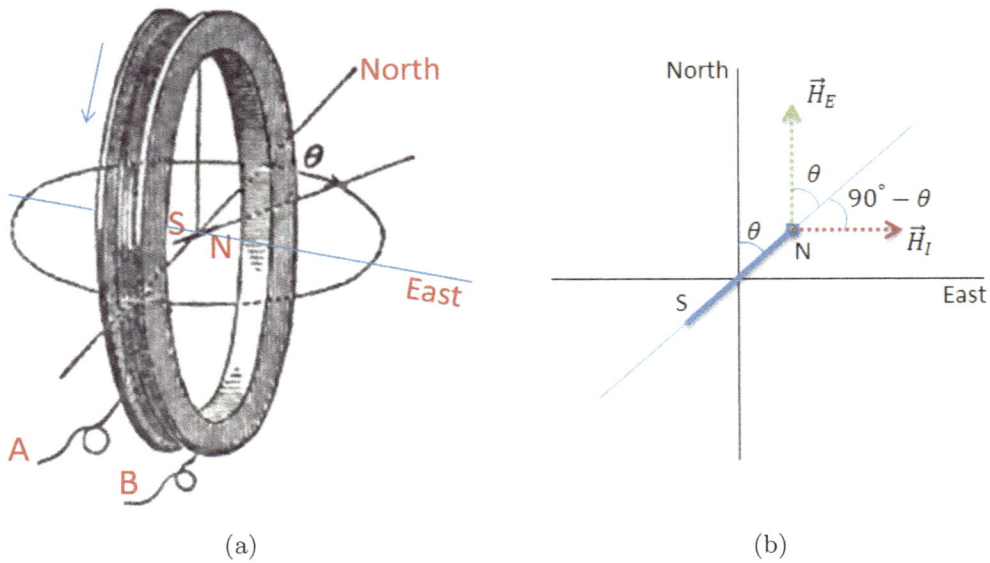

Figure 27.2: (a) A tangent galvanometer [50]. (b) The top view of the magnetic needle in the tangent galvanometer.

27.2   Tangent Galvanometer

The simplest galvanometer is the tangent galvanometer. The reason for calling it a *tangent* galvanometer will become clear in a moment. It consists of a small magnetic needle, suspended at the center of a tightly wound coil with many turns, as shown in Figure 27.2(a). The north and south poles of the magnetic needle are marked N and S.

With no current flowing in the wire, the north pole of the magnet points towards the north, as marked in the figure. Assume that the current flows in the direction of the arrow marked near the coil. From Section 21.2, using the second right-hand rule, the magnetic field generated by the circular coil at its center, points perpendicular to the loop area, and towards east for the current direction assumed.

The top view of the magnetic needle is shown in Figure 27.2(b). The torque on the magnetic needle caused by the terrestrial magnetic field $\vec{H}_E$, is balanced by the torque due to the magnetic field $\vec{H}_I$ generated by the current. The equilibrium angle of the magnetic needle is θ.

From Equation 27.6, the magnetic field generated by a circular coil of radius a with N tightly wound turns is

$$H_I = N\frac{2\pi i}{a}. \tag{27.8}$$

Equation 27.6 has been scaled by N to account for the N turns of the coil. The above equation is usually written as

$$H_I = G\,i$$
$$G = N\frac{2\pi}{a}, \tag{27.9}$$

where G is called the galvanometer constant. From Equation 14.1, the force exerted on a magnetic needle at its north and south poles by a magnetic field $\vec{H}$ is

$$\vec{F} = \pm m\,\vec{H}, \tag{27.10}$$

where $\pm m$ are the magnetic charges at the north and the south poles of the magnetic needle (see Chapter 14), respectively.

If ℓ is the length of the magnetic needle, similar to Equation 15.5, equating the torque on the needle due to $\vec{H}_E$ and $\vec{H}_I$, the rotational equilibrium condition of the magnetic needle is

$$\ell m H_E \sin\theta = \ell m H_I \sin\left(90° - \theta\right)$$
$$= \ell m H_I \cos\theta, \tag{27.11}$$

where the trigonometric identity $\sin\left(90° - \theta\right) = \cos\theta$ has been used to simplify the equation. Substituting Equation 27.8 in the above equation and simplifying,

$$i = \frac{H_E\,a}{N\,2\pi}\tan\theta. \tag{27.12}$$

Since the current i is proportional to $\tan\theta$, this type of a galvanometer is called the tangent galvanometer. The terrestrial magnetic-field strength H_E, can be determined using Gauss's method presented earlier.

Rewriting the above equation in terms of the galvanometer constant in Equation 27.9,

$$i = \frac{H_E}{G}\tan\theta. \tag{27.13}$$

The relative comparison of current measurements, does not require H_E and G to be known, since the constants cancel when calculating the ratio of the currents.

27.3   The Unit Charge in Electromagnetic Units: abcoulomb

From the definition of an abampere of current, it follows that the unit charge in EMU can be defined as the quantity of charge contained in a unit current flowing for a unit time. The unit of charge in EMU is the abcoulomb. Applying dimensional analysis on Equation 18.3,

$$abcoulomb = abampere \cdot s$$
$$= gm^{\frac{1}{2}}\,cm^{\frac{1}{2}}, \tag{27.14, EMU}$$

as noted in Table 26.1.

28

Volume Charge Density ρ_v and Volume Current Density $\vec{J}$

The focus of this chapter is to introduce a new scalar and a vector quantity, the volume charge density ρ_v, and the volume current density $\vec{J}$, which will be used to derive the current-continuity equation in the following chapter. These quantities are useful in writing electromagnetic equations in the differential form, such as the differential form of Ampere's law, which will be presented in this chapter.

Numerical techniques, such as the finite-element method or the finite-difference time-domain method, can be used to solve for electric and magnetic fields from electromagnetic equations in the differential form. Numerical methods, however, are not covered in this book. Electromagnetic equations in the differential form are easier to modify than the integral form, and derive or prove new conclusions. An example of this will be presented in Chapter 51, where Ampere's law will be modified to include the effects of time-varying fields and sources, and proven.

Numerical techniques using the integral form of electromagnetic equations are also used. It is useful to write the equations in both the integral and the differential forms.

28.1   Volume Charge Density

A volume V is discretized into tiny cells, one of which marked Cell i, is shown in Figure 28.1. The volume charge density ρ_v is defined at the center of each of the cells, such that the charge Q_i contained within Cell i is given by

$$
\begin{aligned}
Q_i &= \rho_{v,i}\,\Delta V_i \\
&= \rho_{v,i}\,\Delta x_i\,\Delta y_i\,\Delta z_i,
\end{aligned}
\tag{28.1}
$$

where ΔV_i is the volume of Cell i, $\rho_{v,i}$ is the volume charge density of Cell i, and Δx_i, Δy_i, Δz_i are the dimensions of the cell in the x, y, z directions.

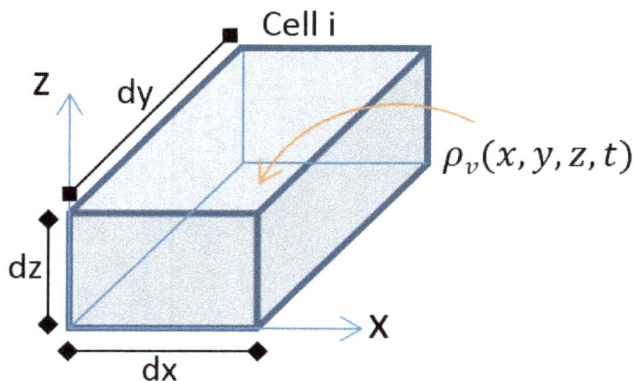

Figure 28.1: The volume charge density at the center of an infinitesimal cell.

From the above equation, the unit of volume charge density $[\rho_v]$ in EMU is

$$[\rho_v] = \frac{abcoulomb}{cm^3}. \qquad \text{(28.2, EMU)}$$

The charge Q contained in the entire volume V can be written as the Riemann sum

$$Q = \sum_{i=1}^{N} \rho_{v,i} \, \Delta V_i, \qquad (28.3)$$

where the summation is done on all the N cells contained in the volume V. From calculus, in the limit, as each cell becomes infinitesimal, and as $N \to \infty$, the charge Q contained within the volume V becomes the integral

$$Q = \int_V \rho_v \, dV. \qquad (28.4)$$

28.2   Volumetric Flow Rate in Fluid Mechanics

Volume current density $\vec{J}$ can be better understood from the volumetric flow rate in fluid mechanics [51]. Let the vector field $\vec{v}(x, y, z, t)$ be the velocity field of a fluid flowing across the surface S, shown in Figure 28.2(a). The surface is meshed into tiny planar elemental areas as shown.

Suppose the cumulative volume of the fluid flowing across the surface is kept track of. The arrows in the figure are examples of unit normal vectors to the surface on some of the mesh elements. The unit normal vectors of the mesh elements are all either outward normal or inward normal to the surface. The two regions above and below the surface have been marked R_2 and R_1. In this example, all the vectors are pointing towards R_2.

The convention that will be followed for the sign of the volume flow is as follows: if the fluid is flowing across the surface into the region that the normal vectors point towards, from R_1 to R_2, the cumulative volume increases, and decreases otherwise.

185

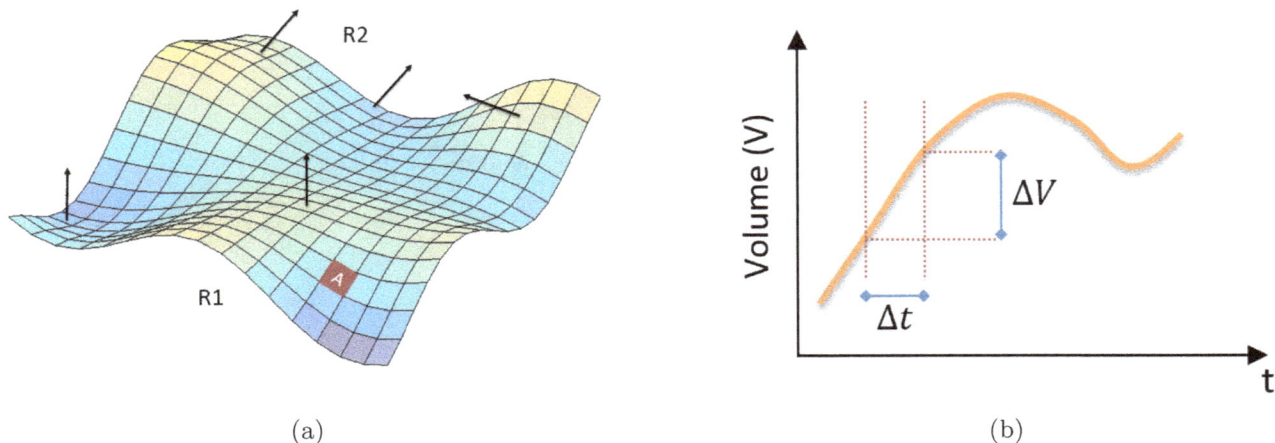

(a)   (b)

Figure 28.2: (a) A fluid flowing across a surface. (b) The cumulative volume flow across the surface as a function of time.

An example of a cumulative volume flow across the surface as a function of time, is shown in Figure 28.2(b). The slope of the cumulative volume-time waveform $\frac{\Delta V}{\Delta t}$, similar to Figure 18.2(a), is the average rate of flow of the fluid across the surface. As $\Delta t \to 0$, the slope of the tangent is the instantaneous rate of flow of the fluid across the surface.

One of the mesh elements, marked A in the figure, is shown in Figure 28.3. It is convenient to write the normal vector to the surface element as $\vec{\Delta A_i}$, where the area of the surface element ΔA_i, is the magnitude of the normal vector.

The mesh elements are sufficiently small that the velocity of the fluid within each of the mesh elements is approximated as being uniform, but may be different between the elements. In the temporal domain, Δt is sufficiently small, so that the velocity field $\vec{v}$ can be considered to be a constant during the time interval.

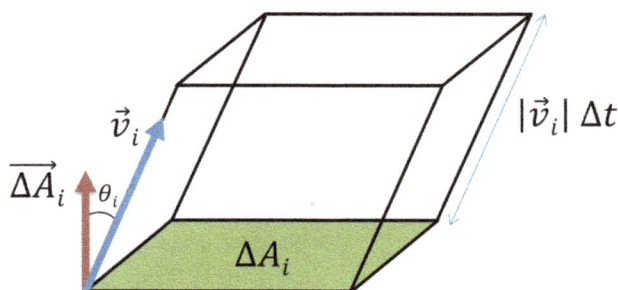

Figure 28.3: A blob of fluid with velocity $\vec{v}_i$ flowing across a small surface area element $\vec{\Delta A_i}$.

After time Δt, the blob of fluid would be of length $|\vec{v}_i|\Delta t$, as shown in the figure, where $\vec{v}_i$ is the velocity of the fluid in the ith mesh element. Since the velocity is uniform within the surface element, during time Δt, the blob forms a parallelepiped shape as shown. The volume of the parallelepiped is

$$\Delta V_i = |\vec{v}_i| \Delta t \cos\theta_i \, \Delta A_i, \tag{28.5}$$

where θ_i is the angle between $\vec{v}_i$ and $\Delta \vec{A}_i$. ΔA_i can be of any planar shape, and the above equation to calculate the volume would still be applicable.

By definition of the dot product, the above equation is written as

$$\Delta V_i = \vec{v}_i \cdot \Delta \vec{A}_i \, \Delta t. \tag{28.6}$$

If θ_i lies in the range $[0, 90°]$, ΔV_i is positive, and the fluid is flowing into the region that $\Delta \vec{A}_i$ is pointing towards. If, however, θ_i is over $90°$, ΔV_i is negative, and the fluid is flowing into the region that is opposite to where $\Delta \vec{A}_i$ is pointing towards. This is consistent with the convention used for the sign of the volume flowing across the surface, as stated at the beginning.

The volume ΔV flowing across the surface during time Δt, is obtained by adding the volume of the fluid ΔV_i, flowing across of each of the mesh elements making up the surface. If the surface S is meshed into N planar elements,

$$\Delta V = \Delta t \sum_{i=1}^{N} \vec{v}_i \cdot \Delta \vec{A}_i. \tag{28.7}$$

Δt has been moved outside of the summation, since it is independent of the index of summation i. As $N \to \infty$, the above Riemann sum becomes the surface integral,

$$\frac{\Delta V}{\Delta t} = \int_S \vec{v} \cdot d\vec{A}, \tag{28.8}$$

where the surface integral is the rate of flow across the surface, also called the volumetric flow rate.

28.3   Volume Current Density

Multiplying both sides of Equation 28.6 by $\rho_{v,i}$, the volume charge density in ΔV_i, the equation can be transformed from volume flow to charge flow,

$$\begin{aligned} \Delta Q_i &= \rho_{v,i} \, \Delta V_i \\ &= \rho_{v,i} \left(\vec{v}_i \cdot \Delta \vec{A}_i \, \Delta t \right), \end{aligned} \tag{28.9}$$

where ΔQ_i is the charge contained in the parallelepiped volume ΔV_i.

Although electricity can be viewed as the flow of positive or negative charges, the convention followed is that positive charges are the ones that flow. $\rho_{v,i}$ is restricted to being non-negative.

If θ_i lies in the range $[0, 90°]$, ΔQ_i is positive, and the positive charges are flowing into the region that $\vec{\Delta A_i}$ points towards. If, however, θ_i is over $90°$, ΔQ_i is negative, and the positive charges are flowing opposite to $\vec{\Delta A_i}$. The plot in Figure 28.2(b) tracks the cumulative charge flow instead of volume flow, and according to this convention.

A new vector quantity is defined, the volume current density at the ith surface element $\vec{J_i}$, replacing $\rho_{v,i}\,\vec{v_i}$. By definition, positive charges flow in the direction of $\vec{J_i}$. Its magnitude is defined such that it satisfies the equation,

$$\Delta Q_i = \vec{J_i} \cdot \vec{\Delta A_i}\,\Delta t, \tag{28.10}$$

ΔQ_i is the charge flowing across ΔA_i, and is equal to the volume of the parallelepiped formed from length $|\vec{J_i}|\Delta t$ and area ΔA_i, as shown in Figure 28.3.

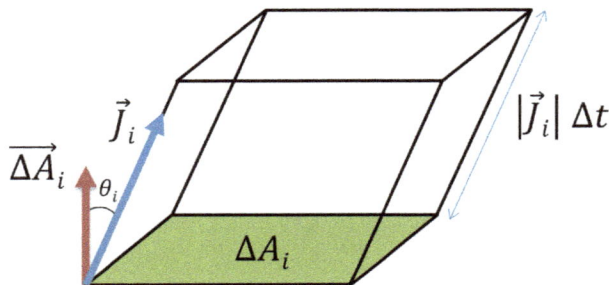

Figure 28.4: A blob of charge flowing across a small surface area element $\vec{\Delta A_i}$.

As explained in Chapter 18, one could think of positive charges flowing in the direction of $\vec{J_i}$, or as negative charges flowing in the opposite direction $-\vec{J_i}$.

The total charge ΔQ flowing across all the area elements can be calculated by adding the charge ΔQ_i flowing across each of the area elements. In the limit, as the area elements are infinitesimal, the flow of charges across all the area elements is calculated by the surface integral,

$$\Delta Q = \Delta t \int_S \vec{J} \cdot d\vec{A}. \tag{28.11}$$

Rearranging the above equation,

$$\frac{\Delta Q}{\Delta t} = \int_S \vec{J} \cdot d\vec{A}. \tag{28.12}$$

In the limit, as $\Delta t \to 0$, the above equation is the slope of the tangent of the cumulative charge vs time waveform at time t. The slope of the cumulative charge vs time waveform is the rate of flow of the charges. From Chapter 18, this is the current i flowing across the surface. Although i is a scalar value, its direction is along the direction in which the positive charges flow $\vec{J_i}$.

Rewriting the above equation using current i,

$$i = \int_S \vec{J} \cdot d\vec{A}. \tag{28.13}$$

In each of the mesh elements, the current flowing across the surface $\vec{J_i} \cdot \Delta\vec{A_i}$, can either be a positive or a negative value. If $\vec{J_i} \cdot \Delta\vec{A_i}$ is positive, such as the case shown in Figure 28.5(a), where $\vec{J_i}$ and $\Delta\vec{A_i}$ form an acute angle, the current flows into the region that $\Delta\vec{A_i}$ is pointing towards. If $\vec{J_i} \cdot \Delta\vec{A_i}$ is a negative value, where the angle between the vectors is an obtuse angle, as shown in Figure 28.5(b), current flows into the region opposite to the direction of $\Delta\vec{A_i}$.

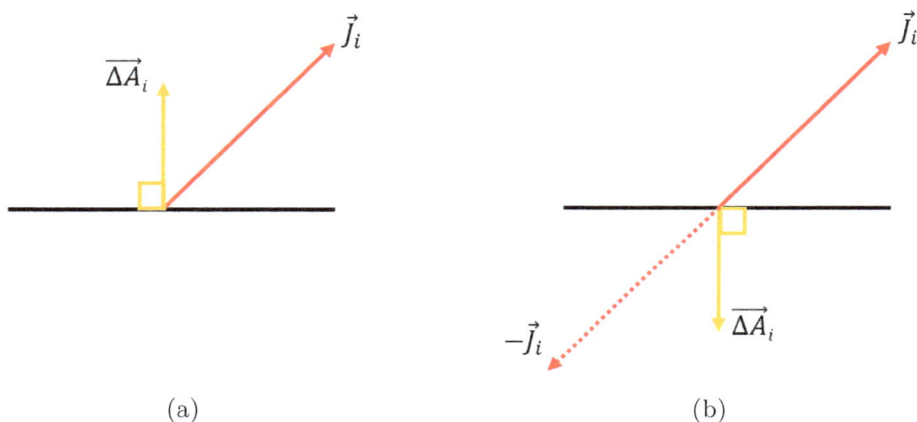

(a)   (b)

Figure 28.5: An example of a (a) positive current flow, and (b) negative current flow, into the region that $\Delta\vec{A_i}$ is pointing towards.

$\vec{J_i} \cdot \Delta\vec{A_i}$ can be viewed as the current flowing into the region that $\Delta\vec{A_i}$ is pointing towards. A negative value of $\vec{J_i} \cdot \Delta\vec{A_i}$ is negative current, or negative charges, flowing into the region that $\Delta\vec{A_i}$ is pointing towards, anti-parallel to $\vec{J_i}$, shown by the dotted line in Figure 28.5(b). A positive value of $\vec{J_i} \cdot \Delta\vec{A_i}$ is positive current, or positive charges, flowing into the region that $\Delta\vec{A_i}$ is pointing towards. In both cases, positive or negative current, is the current flowing into the region that $\Delta\vec{A_i}$ is pointing towards.

A special case where $\vec{J_i}$ is parallel or anti-parallel to $\Delta\vec{A_i}$, is shown in Figure 28.6. In such cases, the resulting current of the dot product is flowing in the same direction as $\Delta\vec{A_i}$. This observation will be used in Chapter 51, during the discussion of displacement current in a capacitor.

In Figure 28.6(a), $\vec{J} \cdot \Delta\vec{A_i}$ is positive, and the current flows in the direction of $\Delta\vec{A_i}$, since it is in the same direction as $\vec{J_i}$. Likewise, in Figure 28.6(b), $\vec{J_i} \cdot \Delta\vec{A_i}$ is negative, since the two vectors are at an angle of $180°$. The negative current flows in the direction of $\Delta\vec{A_i}$.

Applying dimensional analysis on Equation 28.13, the unit of current density $[\vec{J}]$, in electro-

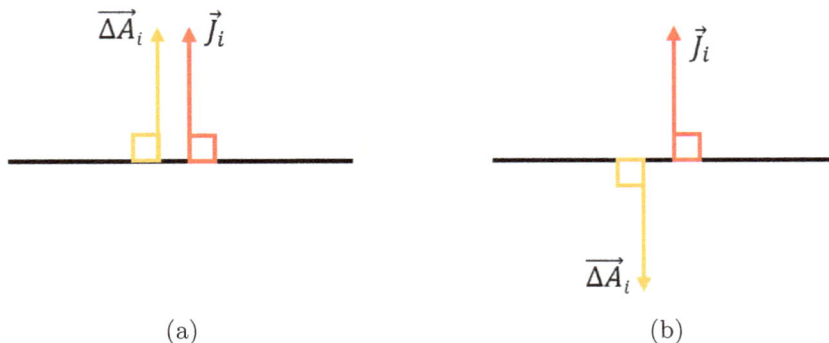

Figure 28.6: (a) $\vec{J_i}$ is parallel to $\vec{\Delta A_i}$. (b) $\vec{J_i}$ is anti-parallel to $\vec{\Delta A_i}$.

magnetic units is

$$[\vec{J}] = \frac{abampere}{cm^2}. \tag{28.14, EMU}$$

28.4   A Volume Current Density Example

Uniform current i is flowing across the cylindrical conductor in Figure 28.7, in the direction of the arrow. The direction of $\vec{J}$ is the same as the direction of current flow, perpendicular to the cross-section area A. The direction of the normal area vector for calculations using $\vec{J}$, will also be chosen in the direction of the arrow.

If A is discretized into small elements of equal areas $d\vec{A}$, the rate of flow of charges, or current, across each of the mesh elements $\vec{J} \cdot d\vec{A}$, must be the same in each of the elements, since the current is uniform across A. This shows that $\vec{J}$ must also be uniform across A, otherwise resulting in non-uniform currents flowing across the mesh elements.

Using Equation 28.13, the volume current density $\vec{J}$ is

$$\vec{J} = \frac{i}{A}\hat{x}, \tag{28.15}$$

where $\hat{x}$ is the unit vector along the direction of current flow. The same exercise can be repeated

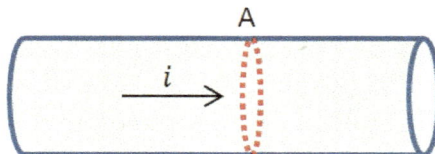

Figure 28.7: A cylindrical conductor carrying uniform current i.

on a surface A tilted at an angle θ, resulting in the same value of $\vec{J}$ in Equation 28.15. A similar example will be presented in Section 29.4.

28.5 Differential Form of Ampere's Law

Repeating Ampere's law in integral form,

$$\oint_\ell \vec{H} \cdot \vec{dl} = i_{enc}, \tag{28.16}$$

where ℓ is the path of integration, and i_{enc} is the enclosed current. By the third right-hand rule for Ampere's law in Section 21.4, the direction of the path of integation and the direction of positive current flow are not independent of each other. If the thumb of the right hand points in the direction of positive current flow, the fingers curl in the direction of the path of integration. Ampere's law in the integral form will be converted into the differential form.

The path of integration ℓ is shown by the dotted line in Figure 28.8. The direction of the path of integration, for example, is shown by the arrows on the path. The block arrow in the $\hat{z}$ direction at the center of the loop is the direction of positive current flow, by the third right-hand rule.

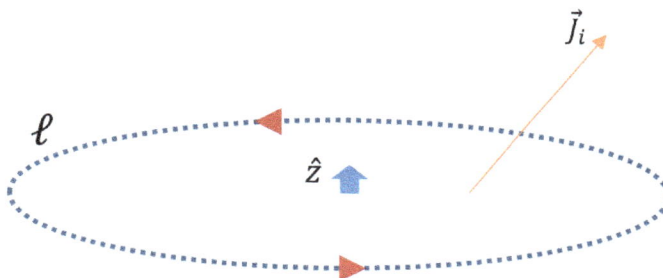

Figure 28.8: An example of the path of integration ℓ in the integral form of Ampere's law.

Using Equation 28.13, the current flowing across the loop area can be written using volume current density $\vec{J}$ as

$$i_{enc} = \int_A \vec{J} \cdot \vec{dA}, \tag{28.17}$$

where A is the area enclosing the loop, and i_{enc} is the current flowing into the region that $\vec{dA}$ is pointing towards. The normal vector to the loop area $\vec{dA}$, can be either in the $+\hat{z}$, or in the $-\hat{z}$ direction, but only one of them is correct. An example of a positive value of i_{enc}, where the current flow marked as $\vec{J}_i$ is flowing into the region that $\hat{z}$ is pointing towards, is shown in the figure. If the direction of the normal area vector is chosen in the $-\hat{z}$ direction, i_{enc} using the above equation is a negative value. This is an incorrect sign of i_{enc}, since it must be a positive value for the direction of $\vec{J}_i$ shown. Therefore, $\vec{dA}$ must be chosen in the $+\hat{z}$ direction.

Using Kelvin-Stokes theorem, the path integral of the magnetic field can be written as a surface integral over the loop area A,

$$\oint_\ell \vec{H} \cdot \vec{dl} = \int_A \left(\nabla \times \vec{H} \right) \cdot \vec{dA}. \tag{28.18}$$

In Kelvin-Stokes theorem, the direction of the normal area vector $d\vec{A}$ is determined by the right-hand rule: if the fingers of the right hand curl in the direction of the path of integration, the thumb points in the direction of the normal area vector. In this example, the direction of $d\vec{A}$ in Kelvin-Stokes theorem is also in the $+\hat{z}$ direction.

From the above equations, rewriting Ampere's law,

$$\int_A \left(\nabla \times \vec{H} \right) \cdot d\vec{A} = \int_A \vec{J} \cdot d\vec{A}. \tag{28.19}$$

Rearranging the above equation, and using the distributive property of the dot product,

$$\int_A \left(\nabla \times \vec{H} - \vec{J} \right) \cdot d\vec{A} = 0. \tag{28.20}$$

The above equation must be satisfied for any loop area A, and $\vec{H}$ generated by any value and distribution of $\vec{J}$. The solution that is guaranteed to satisfy the above equation is

$$\nabla \times \vec{H} = \vec{J}. \tag{28.21}$$

This equation is the differential form of Ampere's law. The difference between the integral and the differential forms is that the differential equation operates at a point, while the integral equation describes the behavior of fields over a region or volume [52]. However, both the forms are equivalent.

The above equation is valid at any point. Any operation may be applied on both sides of the above equation, and the equality still holds true. If both sides of the above equation are integrated over a loop area, and applying the above steps in reverse, the integral form of Ampere's law can be obtained from the differential form.

29

Current Continuity Equation

By the law of conservation of charge, as explained before, charge can neither by created nor destroyed, but only transferred from one object to another. The current continuity equation is a mathematical description of this law, and will be derived in this chapter.

It will be proven in Chapter 51 that the electromagnetic equations are consistent with the current continuity equation. When applied to current flow in circuits, this equation is called Kirchoff's current law, which will be discussed in a later chapter.

29.1 Current Continuity Equation

A closed surface S is shown in Figure 29.1. Let $\vec{J}$ be the volume current density that is a function of position and time. $d\vec{A}$ is the elemental surface area, as illustrated by the tiny shaded patch in the figure, whose magnitude is the elemental surface area. Its direction is *outward* normal to the surface, as shown in the figure.

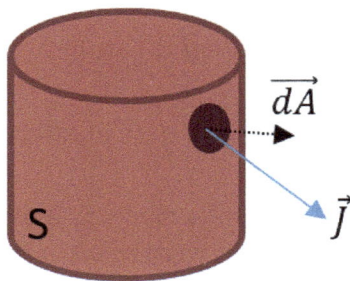

Figure 29.1: A current flow across a closed surface.

Repeating Equation 28.13, the current flowing across the surface S, in the direction of $d\vec{A}$, is given by the surface integral

$$i = \oint_S \vec{J} \cdot d\vec{A}. \tag{29.1}$$

The reader is referred to a calculus textbook for more details on surface integrals and flux.

By definition of $\vec{J}$ and the direction chosen for $d\vec{A}$ as the outward normal to the surface, the above equation is the net current, considering the entire surface, flowing *out* of the closed surface. A positive value of $\vec{J} \cdot d\vec{A}$, indicates that a net current is flowing *out* of the surface, and a negative value means that a net current is flowing *into* the surface. If the net current i flowing outside the surface is positive, then current is flowing out of the surface, and the volume enclosed by the surface is losing positive charge. If the net current is negative, then positive charges are flowing into the surface, and the enclosed surface is gaining positive charge.

Using Equation 18.1, the gain or loss of positive charge by the enclosed surface is

$$\Delta t \oint_S \vec{J} \cdot d\vec{A} = -\Delta Q, \tag{29.2}$$

where ΔQ is the change in the quantity of charge within the enclosed volume during time Δt. The negative sign arises from a positive current (negative current), resulting in a decrease (increase) in the charge contained within the volume.

Rearranging the terms, and in the limit, as Δt becomes infinitesimal,

$$\oint_S \vec{J} \cdot d\vec{A} = -\frac{\partial Q}{\partial t}. \tag{29.3}$$

This equation is the integral form of the current continuity equation.

29.2   Differential Form of the Current Continuity Equation

The differential form of the current continuity equation can be derived using Gauss's divergence theorem. The reader is referred to a calculus textbook for more details on the divergence theorem. Using the divergence theorem, the surface integral can be written as a volume integral,

$$\oint_S \vec{J} \cdot d\vec{A} = \int_V \nabla \cdot \vec{J} \, dV, \tag{29.4}$$

where V is the volume enclosed by the closed surface S, and $\nabla \cdot$ is the divergence operator. The right-hand side of Equation 29.3 can also be written as a volume integral using volume charge density ρ_v. Substituting Equation 29.4 and Equation 28.4 in Equation 29.3,

$$\int_V \nabla \cdot \vec{J} \, dV = -\int_V \frac{\partial \rho_v}{\partial t} \, dV. \tag{29.5}$$

Note that the partial derivative with respect to time can be moved inside the integral, since it is independent of the variable of integration. If the volume integral is viewed as a Riemann sum of tiny volume elements, the time derivative of the sums, if the partial time derivative is outside the integral, is equal to the sum of the time derivatives, if the partial time derivative is inside the integral, since the differentiation operation is distributive.

Rearranging the above equation,

$$\int_V \left(\nabla \cdot \vec{J} + \frac{\partial \rho_v}{\partial t} \right) dV = 0. \tag{29.6}$$

The above equation must be satisfied at any time t, for any volume integral, as well as any $\vec{J}$ and the resulting ρ_v. The solution that is guaranteed to satisfy the above equation is

$$\nabla \cdot \vec{J} = -\frac{\partial \rho_v}{\partial t}, \tag{29.7}$$

which is the differential form of the current continuity equation.

The divergence operation can also be written as

$$\nabla \cdot \vec{J} = \lim_{\Delta V \to 0} \frac{\oint_S \vec{J} \cdot d\vec{A}}{\Delta V}, \tag{29.8}$$

and is useful to understand the definition. In the above equation, $\nabla \cdot \vec{J}$ at any point is the ratio of the flux across the surface S to the enclosing volume ΔV, as ΔV enclosing the point at which the divergence is calculated approaches 0.

29.3 Current Continuity Equation in Magnetostatics

In magnetostatics, where there are no time-varying currents, the volume current density vector may have a spatial variation, but does not vary with time. Analogous to water flow in pipes, it will be assumed that charges in current flow do not accumulate within a closed surface. For example, a closed surface S is shown in Figure 29.2. Assume that the current flows uniformly across the faces marked A and B, whose current-density values are $\vec{J}_A$ and $\vec{J}_B$, and 0 everywhere else.

If more charges flow into the surface than flowing out, or vice versa, the total charge within the closed surface will be time varying. It is not possible for the current flow to not vary in time, but the charge within the closed surface to be time varying. As stated in the law of conservation of charge, it is non-physical for charges to get destroyed or created within the surface. In magnetostatics, therefore, the charges flowing into a closed surface must balance the charges flowing out of it.

If ΔS is the area of the faces A and B, using Equation 28.10, the charge flowing into face A ΔQ_A, during time interval Δt, is

$$\Delta Q_A = J_A \Delta S \Delta t. \tag{29.9}$$

Similarly, the charge flowing out of face B ΔQ_B, is

$$\Delta Q_B = J_B \Delta S \Delta t. \tag{29.10}$$

Since the above two values must be equal in magnetostatics,

$$\vec{J}_A = \vec{J}_B. \tag{29.11}$$

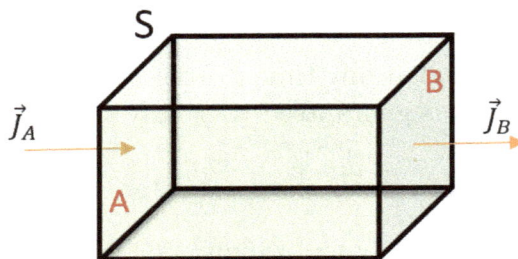

Figure 29.2: A current flow across a closed surface.

The time derivative is 0 in Equation 29.3, which then reduces to

$$\oint_S \vec{J} \cdot d\vec{A} = 0, \tag{29.12}$$

or equivalently in the differential form,

$$\nabla \cdot \vec{J} = 0. \tag{29.13}$$

This result will be revisited again during the discussion of Kirchoff's current law, and will also be used in the following chapter to derive Ampere's law from Biot-Savart's law.

29.4   Verifying the Current Continuity Equation in Magnetostatics

Here's an example that shows that charges do not accumulate within a closed surface in magnetostatics. A conductor with a rectangular cross-section is shown in Figure 29.3. Two planar surfaces S_1 and S_2, together with the top and bottom faces of the conductor form a closed surface. No current flows across the top and bottom faces of the conductor. It will be shown that the charge that flows across S_1 is equal to the charge that flows across S_2.

The current flow i_{S1} is uniform across the cross-section of the conductor S_1. $\vec{J}$ is the same as the direction of the current flow, and is perpendicular to S_1, as marked in the figure. $d\vec{A}_1$ and $d\vec{A}_2$ are the area normal vectors to S_1 and S_2, whose directions are marked in the figure. Similar to the example presented in Section 28.4, if w and l are the width and length of S_1, the value of $\vec{J}$ is

$$\vec{J} = \frac{i_{S1}}{wl}\,\hat{x}, \tag{29.14}$$

where $\hat{x}$ is the unit vector along the direction of current flow. Since S_1 may be located anywhere along the wire, the above value of $\vec{J}$ is the volume current density at any point in the wire. Rewriting the above equation,

$$i_{S1} = Jwl, \tag{29.15}$$

196

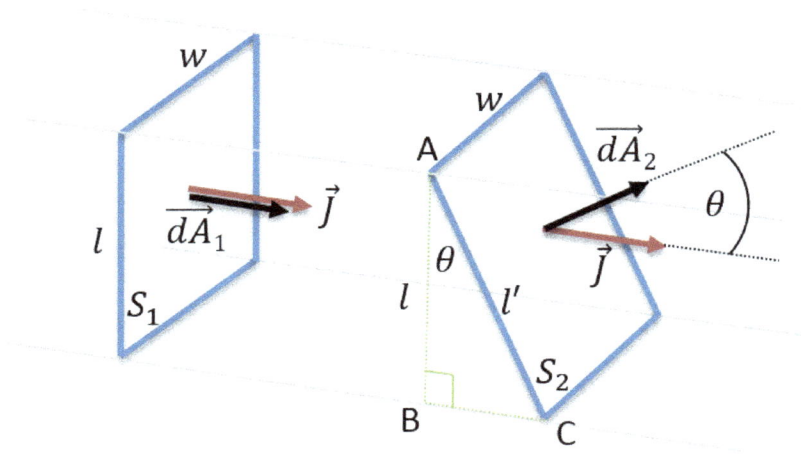

Figure 29.3: An example of current flow calculation across S_1 and S_2 using volume current density.

where J is the magnitude of the current-density vector. During time Δt, the charge ΔQ_{S1} that flows across S_1 is

$$\begin{aligned} \Delta Q_{S1} &= i_{S1}\Delta t \\ &= Jwl\Delta t. \end{aligned} \tag{29.16}$$

In S_1, vectors $\vec{J}$ and $d\vec{A}_1$ are in the same direction. S_2, however, is tilted with vectors $\vec{J}$ and $d\vec{A}_2$ at an angle θ. In the case of S_2, if the width and length of S_2 are w and l', the total current i_{S2} flowing across S_2, using Equation 28.13, is

$$i_{S2} = Jwl'\cos\theta. \tag{29.17}$$

In $\triangle ABC$, from trigonometry, $\angle BAC$ is also equal to θ. In $\triangle ABC$,

$$l' = \frac{l}{\cos\theta}. \tag{29.18}$$

Substituting the above value of l' in the previous equation and simplifying,

$$i_{S2} = Jwl, \tag{29.19}$$

which is the same result as Equation 29.15. The charge ΔQ_{S2} flowing across S_2 in time Δt is

$$\begin{aligned} \Delta Q_{S2} &= i_{S2}\Delta t \\ &= Jwl\Delta t, \end{aligned} \tag{29.20}$$

which is the same as ΔQ_{S1}. The shows that the charge flowing across S_1 is the same as the charge flowing across S_2 in time Δt. There is no charge accumulation within the enclosed surface, and therefore, there is no time variation of the total charge enclosed within the closed surface in magnetostatics.

30

Derivation of Ampere's Law from Biot-Savart's Law [Optional]

The focus of this chapter is to derive Ampere's law from Biot-Savart's law. The laws being consistent with each other, gives assurance on their validity and their formulations to be on the correct track. This chapter is optional and can be ignored without any loss in continuity. However, the reader is encouraged to read and appreciate the exquisite intricacies involved in the mathematical transformation.

30.1 Rewriting Biot-Savart's Law With Volume Current Density $\vec{J}$

A current loop is divided into tiny segments, one of which shown in Figure 30.1. In Biot-Savart's law, the current element is written as $i\vec{\Delta l}$, and the wire is assumed to be of negligible cross section. Δl is the length of the wire segment, and $\vec{\Delta l}$ points in the same direction as the current flow in the segment.

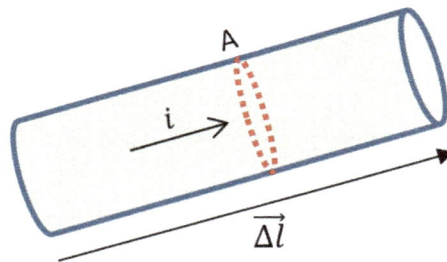

Figure 30.1: A uniform current flow across a cylindrical conductor.

The volume current density vector is in the same direction as the current flow, whose magnitude is

$$J = \frac{i}{A},\tag{30.1}$$

where i is the current flow in the segment, and A is the cross-section area. Using this equation

$i\Delta l$ can be written as

$$i\Delta l = JA\Delta l. \tag{30.2}$$

Since $A\Delta l$ is the volume of the current element, the above equation can be written as

$$i\Delta l = J\Delta V, \tag{30.3}$$

where ΔV is the volume of the current element. Since the direction of $\vec{J}$ is the same as the direction of $\vec{\Delta l}$, the above equation in vector form is

$$i\vec{\Delta l} = \vec{J}\Delta V. \tag{30.4}$$

The region containing the current element can be meshed into tiny cells. One such cell is shown in Figure 30.2. The current-density vector of the cell is defined at its center, and the volume of the cell is ΔV_i.

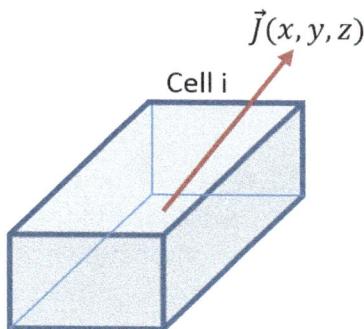

Figure 30.2: The volume current density in Cell i.

Biot-Savart's law was presented for a line current in Equation 23.21, rewritten using s instead of r to avoid confusion with the variables in the next section,

$$\vec{H} = \int_L \frac{i}{s^2}\left(\vec{dl} \times \hat{s}\right), \tag{30.5}$$

where $\vec{s}$ is a vector of magnitude s, between the infinitesimal current element to the point at which the field is calculated, and $\hat{s}$ is its corresponding unit vector.

Using Equation 30.4, Biot-Savart's law can be written using volume current density as

$$\vec{H} = \int_V \frac{\vec{J}dV \times \hat{s}}{s^2}, \tag{30.6}$$

where V is the volume of the source region. The variables in the equation have been marked in Figure 30.3. As before, $\hat{s}$ is the unit vector from an infinitesimal cell marked A, to the point B at which the field is calculated, s is the distance between A and B.

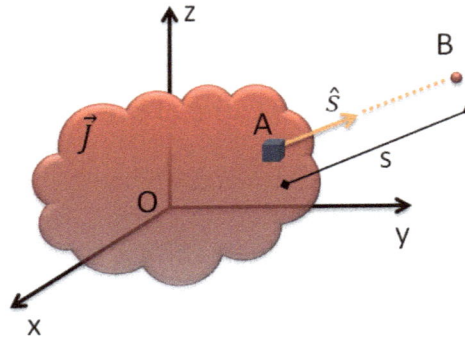

Figure 30.3: A source region with volume current density $\vec{J}$.

30.2 Rewriting Biot-Savart's Law in the Primed and Unprimed Coordinates

Biot-Savart's law written using volume current density in Equation 30.6, can be viewed as an operation in two coordinate systems: the source region, where the volume current density is present, is in the primed coordinate system, and the point at which the field is calculated is in the unprimed coordinate system, captured in Figure 30.4.

O and O' overlap, but shown separated. A vector in the unprimed coordinate space, say $\vec{r} = (1, 2, 3)$, is equal to $\vec{r}' = (1, 2, 3)$ in the primed coordinate space, since they have the same magnitude and direction, although the interpretation of the vectors are very different. $\vec{r} = (1, 2, 3)$ in the unprimed space is the point at which the magnetic field is calculated, but $\vec{r}' = (1, 2, 3)$ is the location of the volume current density in the primed coordinate space. The unit vectors of the coordinate systems are equal, and are not distinguished.

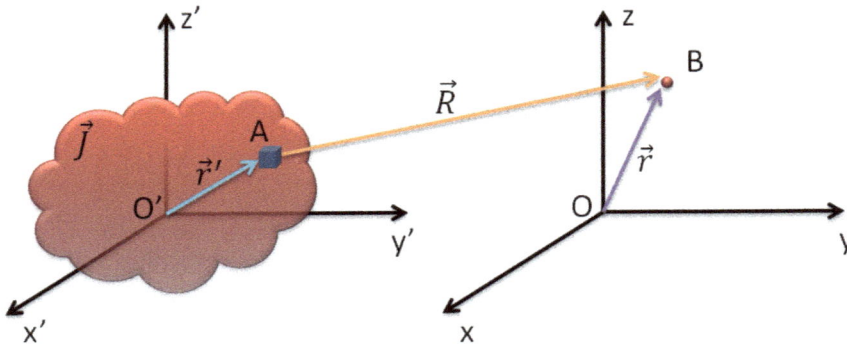

Figure 30.4: The source region, where the volume current density is present, is in the primed coordinate system, and the point at which the magnetic field is calculated is in the unprimed coordinate system.

The region of integration where the volume current density is present, is a function of (x', y', z'), and the magnetic field is calculated at some point (x, y, z). The two coordinate systems distinguish between the source region and the point at which the magnetic field is calculated.

Rewriting Biot-Savart's law in Equation 30.6,

$$\vec{H}\left(\vec{r}\right) = \int_{V'} \frac{\vec{J}\left(x', y', z'\right) dV' \times \hat{R}}{R^2}, \tag{30.7}$$

where $\vec{R} = \vec{r} - \vec{r}'$ is the vector between the volume current density volume element at $\vec{r}'$ and the point $\vec{r}$ at which the magnetic field is calculated, as shown in the figure. $\hat{R}$ and R are the unit vector and the magnitude of $\vec{R}$, respectively. The integration is carried out over the volume V', which is the region of the volume current density. Note that the integrand is a function of both (x, y, z) and (x', y', z'), and the integration is with respect to the primed variables x', y', and z'. The result of the integral is a function of the unprimed variables, and is the magnetic field at (x, y, z).

Variables and operators in the two coordinate systems are differentiated with a prime superscript. For example, the ∇ operator is defined as

$$\nabla = \frac{\partial}{\partial x} + \frac{\partial}{\partial y} + \frac{\partial}{\partial z}, \tag{30.8}$$

while ∇' operates on the primed variables,

$$\nabla' = \frac{\partial}{\partial x'} + \frac{\partial}{\partial y'} + \frac{\partial}{\partial z'}. \tag{30.9}$$

By these definitions, the unprimed variables are treated as constants, or as being fixed, in the primed coordinate system. Likewise, the primed variables are treated as constants in the unprimed coordinate system.

To get an understanding of how these operators will be used in the proof, an example would be

$$\nabla \left(\frac{1}{|\vec{r} - \vec{r}'|} \right), \tag{30.10}$$

compared to

$$\nabla' \left(\frac{1}{|\vec{r} - \vec{r}'|} \right). \tag{30.11}$$

In the first operation using ∇, if $\vec{r}'$ is fixed at a certain point in the source coordinate system, how would the inverse distance $\frac{1}{|\vec{r} - \vec{r}'|}$ vary if the observation point $\vec{r}$ changes. The second operation with ∇' is the reverse, if the observation point $\vec{r}$ is fixed, and if $\vec{r}'$, the location of the source changes, how would the inverse distance vary.

Alternate to the viewpoint of Figure 30.4, one can think of a single coordinate system containing primed and unprimed variables representing the source, and the point at which the field is calculated, respectively. When applying an unprimed operator, such as ∇ described above, the primed variables are treated as constants or fixed. Similarly, in a primed operation, the unprimed variables are treated as constants or fixed.

30.3 Laplacian of Inverse Distance

Identities to transform Biot-Savart's law into Ampere's law will be presented in this section, and the following. In this section, it will be shown that

$$\nabla^2 \left(\frac{1}{|\vec{r} - \vec{r}'|} \right) = -4\pi \delta^3(\vec{r} - \vec{r}'), \tag{30.12}$$

where $\delta^3(\vec{r} - \vec{r}')$ is the 3D Dirac-delta function that satisfies

$$\int_{V'} \delta^3(\vec{r} - \vec{r}') \, dV' = 1, \quad (\vec{r} \in V'), \tag{30.13}$$

if $\vec{r}$ lies within the volume of integration V', and the result of the integration is 0, otherwise.

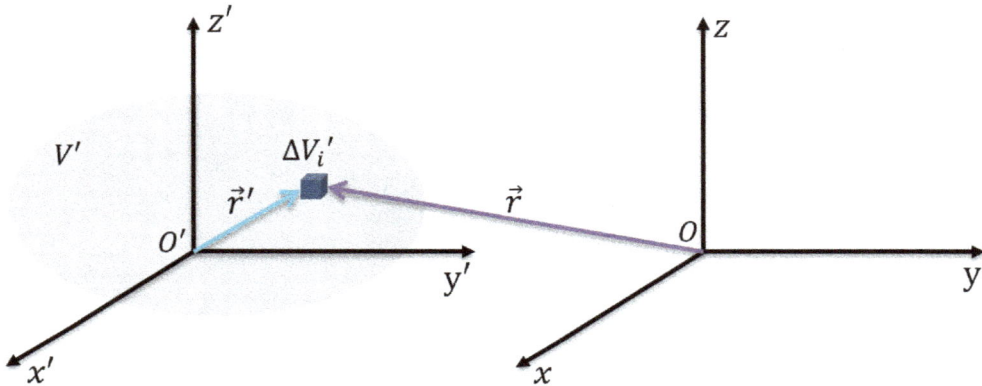

Figure 30.5: A figure to illustrate $\vec{r}$ lying within V'.

The integral in Equation 30.13 can be viewed as a Riemann sum,

$$\int_{V'} \delta^3(\vec{r} - \vec{r}') \, dV' = \lim_{N \to \infty} \sum_{i=1}^{N} \delta^3(\vec{r} - \vec{r_i}') \, \Delta V_i', \tag{30.14}$$

where the volume V' is divided into tiny volume elements $\Delta V_i'$, $\vec{r_i}'$ is located at the center of the ith volume element. Given a vector $\vec{r}$, if this vector lies within the volume of integration V', as shown in Figure 30.5, $\vec{r}$ must equal $\vec{r_i}'$ for one of the values of i in the Riemann sum. By definition of the Dirac-delta function, the result of $\delta^3(\vec{r} - \vec{r_i}' = \vec{0}) \, \Delta V_i'$ is 1, and 0 otherwise. If, however, $\vec{r}$ is located outside of V', $\vec{r}'$ will never equal $\vec{r}$, and the result of the integral is 0.

Integrating both sides of Equation 30.12 over a volume V', Equation 30.12 will be proven by showing that

$$\int_{V'} \nabla^2 \left(\frac{1}{|\vec{r} - \vec{r}'|} \right) dV' = 0, \quad (\vec{r} \notin V') \tag{30.15}$$

$$= -4\pi, \quad (\vec{r} \in V'). \tag{30.16}$$

In the first part of the proof, Equation 30.15 will be shown to be true, and in the second part, Equation 30.16.

30.3.1 Part 1

If $\vec{r} = (x, y, z)$ and $\vec{r}' = (x', y', z')$, then

$$|\vec{r} - \vec{r}'| = \sqrt{(x - x')^2 + (y - y')^2 + (z - z')^2}. \tag{30.17}$$

If $\vec{r} \neq \vec{r}'$,

$$\nabla^2 \left(\frac{1}{|\vec{r} - \vec{r}'|} \right) \tag{30.18}$$

is well defined. Applying the Laplacian operator ∇^2 in Cartesian coordinates, which operates on the unprimed variables x, y, z, it is left as an exercise for the reader to verify that

$$\nabla^2 \left(\frac{1}{|\vec{r} - \vec{r}'|} \right) = 0, \quad (\vec{r} \neq \vec{r}'). \tag{30.19}$$

If $\vec{r} \notin V'$, then $\vec{r} \neq \vec{r}'$ is satisfied, since $\vec{r}'$ extends only over the region V', where $\vec{r}$ is not present. Integrating the above equation over a non-zero volume V',

$$\int_{V'} \nabla^2 \left(\frac{1}{|\vec{r} - \vec{r}'|} \right) dV' = 0, \quad (\vec{r} \notin V'), \tag{30.20}$$

since the integrand is 0 at all points in V'. This completes the first part of the proof.

30.3.2 Part 2

In the second part of the proof, it will be shown that

$$\int_{V'} \nabla^2 \left(\frac{1}{|\vec{r} - \vec{r}'|} \right) dV' = -4\pi, \quad (\vec{r} \in V'). \tag{30.21}$$

V' can be any volume shape that includes the point $\vec{r}$. V' can be divided into two regions: a spherical volume centered at $\vec{r}$, marked B' in Figure 30.6, and the region outside the sphere that does not include $\vec{r}$, marked A'. The volume integral in the region outside the sphere A' is 0, as written in Equation 30.20, leaving only the integration over the spherical region B', redrawn in Figure 30.7. It will be shown next that the result of integrating over the volume B' is -4π.

For a fixed $\vec{r}$, and $\vec{R} = \vec{r}' - \vec{r}$, a value of constant magnitude of $\vec{R}$,

$$\begin{aligned} R &= |\vec{r}' - \vec{r}| \\ &= |\vec{r} - \vec{r}'| \\ &= \sqrt{(x - x')^2 + (y - y')^2 + (z - z')^2}, \end{aligned} \tag{30.22}$$

is a sphere of radius R centered at $\vec{r}$, marked Point P in Figure 30.7. If $R = 0$, Point P coincides with Point Q and $\vec{r}' = \vec{r}$. Few methods can be used to derive Equation 30.16, and the easiest one uses the divergence theorem. The divergence theorem is not valid, however, if a singularity

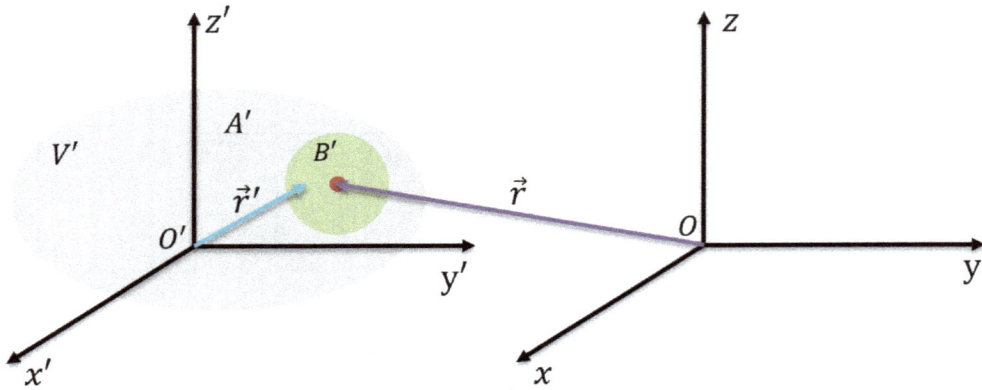

Figure 30.6: The volume V' is divided into A', and a spherical region B', with $\vec{r}$ located at the center of B'.

is present within the volume of integration [53]. The left-hand side of Equation 30.16 is singular when $\vec{r} = \vec{r}'$. For mathematical rigor, the method outlined in [54] will be used instead.

The workaround for the singularity in the term

$$\nabla^2 \left(\frac{1}{R} \right) = \nabla^2 \left(\frac{1}{|\vec{r}' - \vec{r}|} \right), \tag{30.23}$$

which is not defined for $\vec{r} = \vec{r}'$, is to replace it with the equivalent expression

$$\nabla^2 \left(\frac{1}{R} \right) = \lim_{a \to 0} \nabla^2 \left(\frac{1}{\sqrt{R^2 + a^2}} \right). \tag{30.24}$$

The argument of the Laplacian operator is now non-singular for $\vec{r} = \vec{r}'$, since the denominator is a non-zero number when $R = 0$. Let the spherical volume B' be a sphere of radius $R_\circ$.

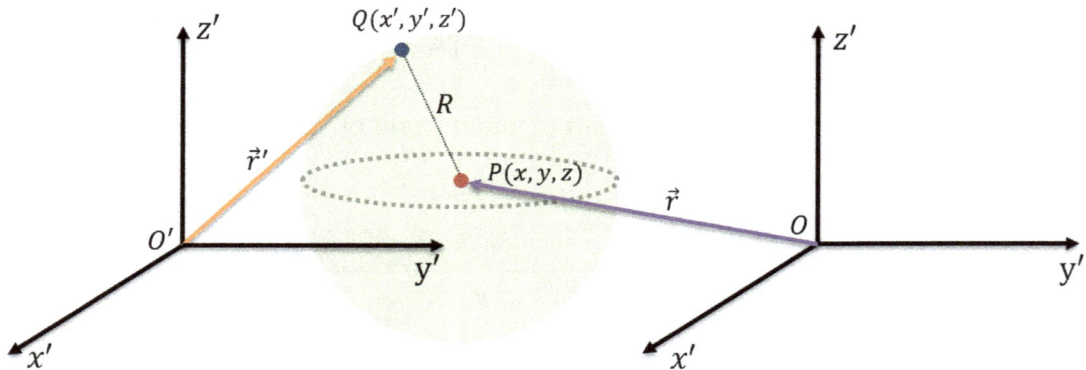

Figure 30.7: Constant values of R tracing out a sphere.

Using Equation 30.22, and applying ∇^2 in Cartesian coordinates on the unprimed variables, it is left as an exercise for the reader to verify that

$$\lim_{a \to 0} \int_{B'} \nabla^2 \left(\frac{1}{\sqrt{R^2 + a^2}} \right) dV' = \lim_{a \to 0} \int_{B'} \frac{-3a^2}{(R^2 + a^2)^{5/2}} \, dV'. \tag{30.25}$$

The elemental volume dV' of a spherical shell of radius R and thickness dR is $dV' = 4\pi R^2 \, dR$. The above integral evaluates to

$$\lim_{a \to 0} \int_{B'} \frac{-3a^2}{(R^2 + a^2)^{5/2}} \, dV' = \lim_{a \to 0} \int_0^{R_\circ} \frac{-3a^2}{(R^2 + a^2)^{5/2}} \, (4\pi R^2) \, dR$$

$$= -4\pi. \tag{30.26}$$

The above result is valid for any non-zero value of $R_\circ$. This completes the second part of the proof,

$$\int_{V'} \nabla^2 \left(\frac{1}{|\vec{r} - \vec{r}'|} \right) dV' = -4\pi, \quad (\vec{r} \in V'), \tag{30.27}$$

completing the derivation of Equation 30.12.

30.4 Other Identities

Several identities needed for the derivation are presented in this section. The reader is referred to a calculus textbook for more details.

30.4.1 Gradient of Inverse Distance

Noting from Figure 30.4,

$$\vec{R} = \vec{r} - \vec{r}', \tag{30.28}$$

its corresponding unit vector is

$$\hat{R} = \frac{\vec{R}}{R}, \tag{30.29}$$

where the magnitude of $\vec{R}$ is given by Equation 30.17,

$$R = |\vec{r} - \vec{r}'|. \tag{30.30}$$

Substituting $\vec{r} = (x, y, z)$ and $\vec{r}' = (x', y', z')$ into the above set of equations in Cartesian coordinates, it is left as an exercise for the reader to show that

$$\frac{\hat{R}}{R^2} = -\nabla \left(\frac{1}{R} \right), \tag{30.31}$$

where ∇ is the gradient operator that operates on the unprimed variables, x, y, z. Similarly, it can also be shown that

$$\nabla \left(\frac{1}{R} \right) = -\nabla' \left(\frac{1}{R} \right), \tag{30.32}$$

205

where ∇' operates on the primed variables x', y', z', and treating the unprimed variables as fixed or constants.

30.4.2   Product Rules

Given a scalar field f and a vector field $\vec{g}$,

$$\nabla \times (f\vec{g}) = f\nabla \times \vec{g} - \vec{g} \times \nabla f, \tag{30.33}$$

where $\nabla\times$ is the curl operator, ∇ is the gradient operator, $\vec{g} \times \nabla f$ is the cross product of $\vec{g}$ and ∇f. Similar identity is

$$\nabla \cdot (f\vec{g}) = f\nabla \cdot \vec{g} + \vec{g} \cdot \nabla f, \tag{30.34}$$

and $\nabla\cdot$ is the divergence operator, ∇ is the gradient operator, and $\vec{g} \cdot \nabla f$ is the dot product of $\vec{g}$ and ∇f.

30.4.3   Curl of a Vector Field

Given a vector field $\vec{A}$,

$$\nabla \times \left(\nabla \times \vec{A}\right) = \nabla \left(\nabla \cdot \vec{A}\right) - \nabla^2 \vec{A}, \tag{30.35}$$

where $\nabla\times$ is the curl, $\nabla\cdot$ is the divergence, ∇ is the gradient, and ∇^2 is the Laplacian operation on the unprimed variables x, y, z. The primed variables x', y', z' are treated as constants in the unprimed coordinate system, and vice versa.

30.4.4   One Last Identity

It will be proven in this subsection,

$$\int_{V'} \nabla \cdot \left(\frac{\vec{J}}{R}\right) dV' = 0, \tag{30.36}$$

where volume V' encloses the volume current density $\vec{J}(x', y', z')$, shown in Figure 30.8(a), R is the distance between an infinitesimal volume element in V' and a fixed point $\vec{r} = (x, y, z)$, as shown in Figure 30.4. Although a localized current-density region will be assumed to begin with, where the current density exists over a finite region, it will be shown later in this subsection that this identity also holds true for non-localized current, in which case V' extends over the entire primed-coordinate space.

Applying the product rule in Equation 30.34 on the integrand of the above equation,

$$\nabla \cdot \left(\frac{\vec{J}}{R}\right) = \frac{1}{R}\nabla \cdot \vec{J} + \vec{J} \cdot \nabla \left(\frac{1}{R}\right). \tag{30.37}$$

(a)   (b)

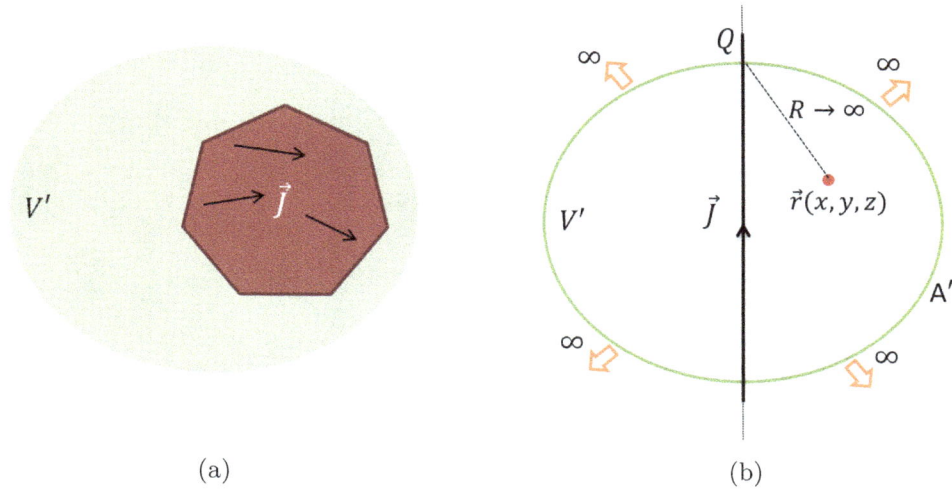

Figure 30.8: (a) A localized volume current density $\vec{J}$ lying within V'. (b) An example of a non-localized current density, which is a line current between $[-\infty, +\infty]$, and V' including all of the primed-coordinate space.

$\vec{J}$ is a function of only the primed variables x', y', z', and since ∇ treats primed variables as constants,

$$\nabla \cdot \vec{J} = 0, \tag{30.38}$$

which then reduces the equation to

$$\nabla \cdot \left(\frac{\vec{J}}{R} \right) = \vec{J} \cdot \nabla \left(\frac{1}{R} \right). \tag{30.39}$$

The above equation can also be written as, using the identity in Equation 30.32,

$$\nabla \cdot \left(\frac{\vec{J}}{R} \right) = -\vec{J} \cdot \nabla' \left(\frac{1}{R} \right). \tag{30.40}$$

Invoking the product rule in Equation 30.34 again on the term on the right-hand side of the above equation,

$$\nabla \cdot \left(\frac{\vec{J}}{R} \right) = -\nabla' \cdot \left(\frac{\vec{J}}{R} \right) + \frac{1}{R} \left(\nabla' \cdot \vec{J} \right). \tag{30.41}$$

In magnetostatics, since the currents are in their steady state, using the current-continuity equation in Equation 29.13,

$$\nabla' \cdot \vec{J} = 0. \tag{30.42}$$

In the current-continuity equation, the primed divergence operator $\nabla' \cdot$ must be used, since current is a function of the primed variables, and is present in the primed coordinate system. Equation 30.41, therefore reduces to

$$\nabla \cdot \left(\frac{\vec{J}}{R}\right) = -\nabla' \cdot \left(\frac{\vec{J}}{R}\right). \tag{30.43}$$

Integrating over the region of the volume current density V',

$$\int_{V'} \nabla \cdot \left(\frac{\vec{J}}{R}\right) dV' = -\int_{V'} \nabla' \cdot \left(\frac{\vec{J}}{R}\right) dV'. \tag{30.44}$$

To apply Gauss's divergence theorem in the primed coordinate space, the primed divergence operator $\nabla' \cdot$ must be used in the volume integral, as written on the right-hand side of the above equation. The proof is that the unprimed variables are treated as constants in the primed coordinate space, and the proof of the divergence theorem in a typical coordinate system is also applicable here. Applying Gauss's divergence theorem, the volume integral can be written as a surface integral,

$$\int_{V'} \nabla \cdot \left(\frac{\vec{J}}{R}\right) dV' = -\oint_{A'} \left(\frac{1}{R}\right) \vec{J} \cdot d\vec{A}', \tag{30.45}$$

where A' is the surface area enclosing the volume V'. As mentioned in Section 30.3.2, the divergence theorem can be applied only if there is no singularity in the volume V'. If $\vec{r} \in V'$, $R = 0$ when $\vec{r}' = \vec{r}$, and $\frac{\vec{J}}{R}$ is singular. Similar to the exercise in Section 30.3.2, the right-hand side of Equation 30.44 can be written as

$$-\int_{V'} \nabla' \cdot \left(\frac{\vec{J}}{R}\right) dV' = -\lim_{a \to 0} \int_{V'} \nabla' \cdot \left(\frac{\vec{J}}{\sqrt{R^2 + a^2}}\right) dV'. \tag{30.46}$$

The integrand in the right-hand side of the above equation is no longer singular when $R = 0$, and applying the divergence theorem,

$$-\lim_{a \to 0} \int_{V'} \nabla' \cdot \left(\frac{\vec{J}}{\sqrt{R^2 + a^2}}\right) dV' = -\lim_{a \to 0} \oint_{A'} \left(\frac{1}{\sqrt{R^2 + a^2}}\right) \vec{J} \cdot d\vec{A}'$$

$$= -\oint_{A'} \left(\frac{1}{R}\right) \vec{J} \cdot d\vec{A}'. \tag{30.47}$$

Using this trick to overcome the singularity, Equation 30.45 is obtained.

The case where the volume current density is localized, or located in a finite space, is shown in Figure 30.8(a). V' can always be extended beyond the region where the non-zero volume current density is present, by assuming that current is 0 in the extended region. The current lies within V', and there is no current flowing across the surface A'. Therefore, the surface integral evaluates to 0, resulting in

$$\int_{V'} \nabla \cdot \left(\frac{\vec{J}}{R}\right) dV' = 0. \tag{30.48}$$

208

Even if non-localized current is dealt with, where the current extends over an infinite region in space, the integral will still evaluate to 0. An example of such a case is shown in Figure 30.8(b), where a line current on the z-axis extends over $[-\infty, +\infty]$. For a given $\vec{r} = (x, y, z)$, as V' balloons to ∞, as shown in the figure, the value of $R \to \infty$, and $\frac{1}{R} \to 0$. Therefore,

$$-\oint_{A'} \left(\frac{1}{R}\right) \vec{J} \cdot d\vec{A}' \to 0, \tag{30.49}$$

and Equation 30.48 still holds true for a non-localized current.

30.5 Derivation of Ampere's Law from Biot-Savart's Law

The methodology in [55] will be used for the derivation. Biot-Savart's law written using the volume current density, repeating Equation 30.7,

$$\vec{H}(\vec{r}) = \int_{V'} \frac{\vec{J}(x', y', z') \, dV' \times \hat{R}}{R^2}. \tag{30.50}$$

Using the relation in Equation 30.31,

$$\frac{\hat{R}}{R^2} = -\nabla\left(\frac{1}{R}\right), \tag{30.51}$$

Biot-Savart's law is rewritten as

$$\vec{H}(\vec{r}) = -\int_{V'} \vec{J} \times \nabla\left(\frac{1}{R}\right) dV'. \tag{30.52}$$

Using the product rule in Equation 30.33,

$$\nabla \times (f\vec{g}) = f\nabla \times \vec{g} - \vec{g} \times \nabla f, \tag{30.53}$$

and setting $\vec{g} = \vec{J}$ and $f = \frac{1}{R}$,

$$\vec{H}(\vec{r}) = \int_{V'} \left[\nabla \times \left(\frac{\vec{J}}{R}\right) - \frac{1}{R}\nabla \times \vec{J}\right] dV'. \tag{30.54}$$

Since $\vec{J}$ is a function of only the primed variables, and depicted in Figure 30.4 as residing only in the primed coordinate system, the result of the unprimed curl $\nabla\times$ operator on $\vec{J}$ will be 0, and this term can be ignored. The resulting expression is

$$\vec{H}(\vec{r}) = \nabla \times \int_{V'} \left(\frac{\vec{J}}{R}\right) dV', \tag{30.55}$$

and the unprimed curl operator has been moved out of the integral, since the integration dV' is with respect to the primed variables. A mild verification of this is to try a few simple integral examples, and simple operators, such as the $\frac{d}{dx}$ derivative operator, and show that indeed, an unprimed operator can be moved out of an integral with respect to a primed variable. As a checkpoint to make sure that the derivation until now is correct, the equivalence of Equation 30.52 and Equation 30.55 is verified in Appendix $\mathcal{B}$.

Applying the curl operation on both sides of the equation,

$$\nabla \times \vec{H}(\vec{r}) = \nabla \times \nabla \times \int_{V'} \left(\frac{\vec{J}}{R}\right) dV'. \tag{30.56}$$

Applying Equation 30.35 on the right-hand side of the above expression,

$$\nabla \times \vec{H}(\vec{r}) = \nabla \int_{V'} \nabla \cdot \left(\frac{\vec{J}}{R}\right) dV' - \int_{V'} \nabla^2 \left(\frac{\vec{J}}{R}\right) dV', \tag{30.57}$$

where $\nabla\cdot$, and ∇^2 are brought inside the integral on the right-hand side, by the same argument the $\nabla\times$ operator was moved out of the integral in Equation 30.55. It was shown in Equation 30.48 that the first integral on the right-hand side is 0. The above expression reduces to

$$\nabla \times \vec{H}(\vec{r}) = -\int_{V'} \vec{J} \nabla^2 \left(\frac{1}{R}\right) dV'. \tag{30.58}$$

Substituting Equation 30.12 into the above equation,

$$\nabla \times \vec{H}(\vec{r}) = 4\pi \int_{V'} \vec{J} \delta^3(\vec{r} - \vec{r}') dV'. \tag{30.59}$$

Note that V' can extend to all of the primed space, each of x', y', z' varying between $[-\infty, +\infty]$, and $\vec{J}$ would be just be equal to 0 other than the location where the current exists. By the sifting property of the Dirac-delta function, the integral is zero everywhere, except when $\vec{r}' = \vec{r}$. The resulting equation is

$$\nabla \times \vec{H}(\vec{r} = \vec{r}') = 4\pi \vec{J}(\vec{r}' = \vec{r}). \tag{30.60}$$

In words, this means that $\nabla \times \vec{H}$ at a point is equal to the volume current source at that point, scaled by 4π.

Reverting back to the single coordinate system, where both the source and the point at which the field is calculated are present, as shown in Figure 30.3, the above equation can be written as

$$\nabla \times \vec{H}(\vec{r}) = 4\pi \vec{J}(\vec{r}), \tag{30.61}$$

which is the differential form of Ampere's law that was derived earlier in Equation 28.21.

31

Measurement of the Charge Stored in a Capacitor Using a Ballistic Galvanometer

The capacitor was introduced in Section 17.2, and was known as the Leyden jar in the past. It is made up of two closely spaced metal foils separated by a thin layer of non-conductive dielectric material, or an insulator, as shown in Figure 31.1. Michael Faraday experimented with different dielectric materials, and studied how the quantity of charge stored in a Leyden jar varied between them. This will be explained in detail in Chapter 35.

The focus of this chapter is to present a measurement technique to quantify the charge stored in a capacitor in electromagnetic units. The instrument used in the measurement is known as the ballistic galvanometer, described next.

31.1   Ballistic Gavanometer

A ballistic galvanometer is very similar to the working of a tangent galvanometer from Section 27.2. Unlike the tangent galvanometer to measure current, a ballistic galvanometer is used to measure the quantity of charge in a current impulse that exists for a short time duration. The definition of a *short* time duration will be explained in a moment.

Similar to the tangent galvanometer, the magnetic needle of the ballistic galvanometer, while at rest, lies on the plane of the coil. The north pole of the needle points North, as marked in Figure 31.1. In this arrangement, by the second right-hand rule, the magnetic field generated by the current flow is perpendicular to the magnetic needle.

A Leyden jar charged using a Wimshurst machine, for example, is discharged by connecting the inner and outer metal foils to the two terminals of the ballistic galvanometer, as shown in the figure. This will cause the jar to discharge almost instantaneously, and a current impulse flows in the wire for a short time duration. This creates a sudden burst of magnetic field that exerts a torque on the magnetic needle, causing it to oscillate back and forth, and finally coming to rest.

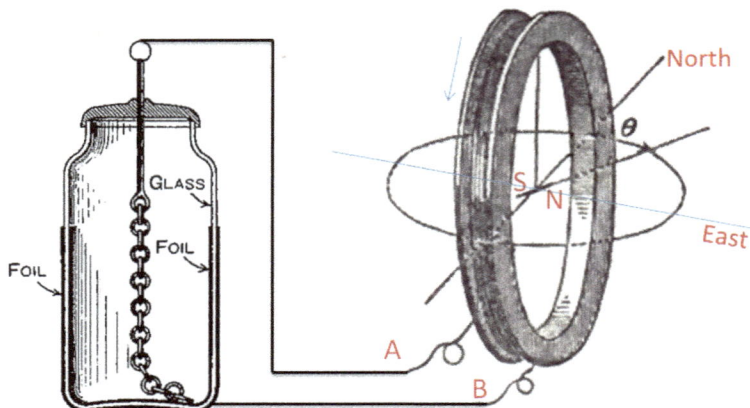

Figure 31.1: Discharging a Leyden jar through a ballistic galvanometer (figure not to scale).

It will be assumed that the discharge of the current occurs in a time duration, short enough to assume that the magnetic needle of the galvanometer has not moved appreciably, and can be assumed to be stationary.

The construction of the ballistic galvanometer is almost identical to the tangent galvanometer, except for the magnetic needle of a ballistic galvanometer, which has a higher moment of inertia. This slows down the oscillations and angular acceleration, for easier reading of the angle of oscillation, and its period, from which the charge in the current burst can be calculated, as explained later in the chapter. From the rotational equivalent of Newton's 2nd law,

$$\tau = I\alpha, \tag{31.1}$$

where I is the moment of inertia of the magnetic needle, α is the angular acceleration, and τ is the applied torque. From the above equation, since

$$\alpha \propto \frac{1}{I}, \tag{31.2}$$

the magnetic needle is made heavier to increase its moment of inertia, reducing its angular acceleration, and making it easier to read the "throw of the needle". The throw of the needle is a phrase to refer to the maximum angle of oscillation of the magnetic needle. The period of oscillation will be derived in Equation 31.29, which will also confirm that a higher moment of inertia results in a higher period of oscillation.

31.2 Angular Velocity of the Magnetic Needle at the End of the Current Burst

The angular velocity of the magnetic needle at the end of the current burst ω_f, will be derived in this section. This result will be used in the derivation of the charge in the current burst. ω is commonly used to represent both angular frequency, as in Equation 31.37, and angular velocity, $\omega = \dot{\theta}$, both of which are used in this chapter. To avoid confusion with the variables, subscripts/superscripts will be used to differentiate between them in this chapter, and from the context, which of the two is used will be clear. Table 31.1 summarizes the physical quantities represented by ω.

Variable	Usage
ω_o	undamped angular frequency
ω'	damped angular frequency
ω_i	angular velocity of the needle before the current burst
ω_f	angular velocity of the needle at the end of the current burst
ω	angular velocity

Table 31.1: The variables used to represent angular frequency and angular velocity in this chapter.

The burst of magnetic field generated by the discharging current impulse of the Leyden jar, exerts a torque on the magnetic needle to begin the oscillations. In Figure 31.2, $\vec{H}$ denotes the terrestrial magnetic field that is tangential to the Earth's surface. The setup of the ballistic galvanometer, similar to the tangent galvanometer, is such that the magnetic field generated by the current burst $\vec{h}$ is perpendicular to the needle. When no current flows in the galvanometer, the needle lies along $\vec{H}$. The direction of $\vec{h}$ can be determined by the second right-hand rule. The directions of the magnetic fields are indicated by the arrows.

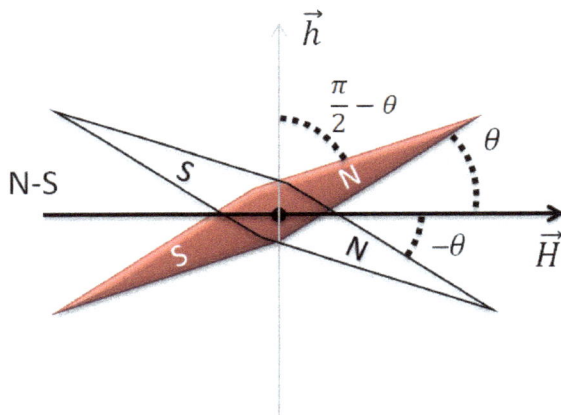

Figure 31.2: The top view of the magnetic needle in the ballistic galvanometer.

In the beginning, $\theta = 0\,rad$, and the magnetic needle points in the magnetic $N-S$ direction. It is assumed that the magnetic needle stays at $\theta = 0\,rad$ during the short time duration the burst of current occurs. Since the magnetic needle is approximately lying in the direction of the magnetic $N-S$ direction during the duration of the current burst, the restoring torque exerted by $\vec{H}$ (see Equation 16.4) can be neglected, and the only torque impulse on the needle is that caused by the magnetic field generated by the current burst.

The magnetic needle is modeled with magnetic charges $\pm m$ at its end points, as shown in Figure 14.1(a). The force on the magnetic charges caused by $\vec{h}$ is

$$\vec{F} = \pm m\vec{h}. \tag{31.3}$$

The magnitude of the magnetic field h generated by current i is

$$h = Gi, \tag{31.4}$$

where G is the galvanometer constant.

It will be assumed that counter-clockwise angles of rotation θ are positive, and clockwise angles are negative. Therefore, as explained in Section 3.2, $\dot{\theta}$, $\ddot{\theta}$, torque τ, are also positive in the counter-clockwise direction, and negative in the clockwise direction. The torque exerted on the needle by $\vec{h}$, using Equation 31.1, is

$$I\ddot{\theta} = 2\left[mGi\left(\frac{\ell}{2}\right)\sin\left(\frac{\pi}{2}-\theta\right)\right]$$
$$= MGi\cos\theta, \tag{31.5}$$

where I is the moment of inertia of the magnetic needle, $\ddot{\theta}(t)$ is the angular acceleration, G is the galvanometer constant from Equation 27.9, $i(t)$ is the burst of current in the coil, the magnetic moment $M = m\ell$ is the product of m and the length of the needle ℓ. The multiplying factor 2 in the above equation arises from the force acting on both the magnetic charges $\pm m$.

If $[0, T]$ is the time interval during which the current in the coil exists, integrating the above equation over this time interval,

$$\int_0^T I\ddot{\theta}\,dt = \int_0^T MGi\cos\theta\,dt. \tag{31.6}$$

It is assumed that $\theta \approx 0\,rad$, or $\cos\theta \approx 1$, during the time interval $[0, T]$, which is a mathematical way of stating that the current discharge happens quickly, before the needle moves appreciably. Angular acceleration is the time-derivative of angular velocity,

$$\ddot{\theta} = \frac{d\omega}{dt}. \tag{31.7}$$

Substituting the above equation, and integrating over this interval,

$$\int_0^T I\frac{d\omega}{dt}\,dt = MG\int_0^T i\,dt. \tag{31.8}$$

The left-hand side can be simplified to

$$\int_0^T I\frac{d\omega}{dt}\,dt = I\int_{\omega_i}^{\omega_f} d\omega$$
$$= I\left(\omega_f - \omega_i\right), \tag{31.9}$$

and I has been moved out of the integral, since the moment of inertia is a constant for all time, ω_i and ω_f are the angular velocities at time $t = 0$ and $t = T$, respectively. The above equation can be better understood by writing the integral as a Riemann sum,

$$I \int_0^T \frac{d\omega}{dt}\, dt = I \lim_{N \to \infty} \sum_{k=1}^N \frac{\Delta\omega_k}{\Delta t_k} \Delta t_k, \tag{31.10}$$

and cancelling Δt_k, the result of the summation

$$I \lim_{N \to \infty} \sum_{k=1}^N \Delta\omega_k = I\left[(\omega_2 - \omega_i) + (\omega_3 - \omega_2) + ... + (\omega_f - \omega_{N-1})\right], \tag{31.11}$$

is the same as Equation 31.9.

From Equation 18.5, the integral on the right-hand side of Equation 31.8 is the charge Q contained in the current burst $i(t)$ that exists during the interval $[0, T]$,

$$Q = \int_0^T i\, dt. \tag{31.12}$$

From the above equations, assuming that the needle is at rest at $t = 0$, and therefore $\omega_i = 0$,

$$I\omega_f = MGQ. \tag{31.13}$$

The above equation can be interpreted as the conservation of angular momentum. Angular momentum is $I\omega$, and the above equation states that the current burst results in the angular momentum of the needle. In the above equation, ω_f will be replaced by quantities which can be measured, such as the period of oscillation, or the maximum amplitude of oscillation, etc.

31.3   Undamped Oscillations of the Magnetic Needle

The calculation of charge in the current impulse can be simplified by neglecting damping in the oscillations of the magnetic needle, or assuming that the oscillations continue indefinitely. The calculation of the charge with damped oscillations is derived in the following section.

At the end of the current burst, the needle is assumed to be still at $\theta = 0\,rad$. The angular velocity imparted to the magnetic needle by the burst of magnetic field, causes the needle to swing to $\theta = \theta_1\,rad$, at which angle, the angular velocity ω becomes 0. The needle then swings to $-\theta_1\,rad$, and back and forth. The resulting waveform of $\theta(t)$ is shown in Figure 31.3. Since undamped oscillations are assumed, the peaks and valleys of the waveform are $\pm\theta_1\,rad$, and do not decrease in amplitude over time.

Before the discharge of the Leyden jar, the needle is at rest. The burst of magnetic field imparts kinetic energy to the magnetic needle. When the needle reaches its two extreme amplitudes

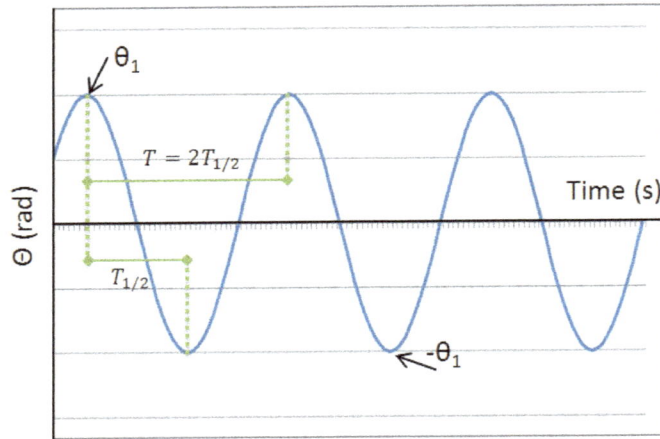

Figure 31.3: The undamped oscillations $\theta(t)$ of the magnetic needle shown in Figure 31.2.

at $\theta = \pm\theta_1 \, rad$, and comes to rest momentarily, the kinetic energy of the needle is transformed to potential energy. The law of conservation of energy will be used to derive an expression for ω_f that can be measured, which can be used to calculate the charge in the current burst Q in Equation 31.13.

The kinetic energy KE of the needle at the end of the current burst is

$$KE = \frac{1}{2}I\,\omega_f^2,\tag{31.14}$$

where I is the moment of inertia of the needle. The work W done on a rigid body is given by

$$W = \int_{\ell_1} \vec{F}_1 \cdot d\vec{s} + \int_{\ell_2} \vec{F}_2 \cdot d\vec{s} + ... + \int_{\ell_i} \vec{F}_i \cdot d\vec{s},\tag{31.15}$$

where $\{\vec{F}_1, \vec{F}_2, ..., \vec{F}_i\}$ are all the forces acting on the rigid body, $\{\ell_1, \ell_2, ..., \ell_i\}$ are the paths of the points on the rigid body where the force acts.

The only force acting on the needle after the current impulse is the restoring force of the terrestrial magnetic field $\vec{H}$ on the magnetic charges $\pm m$ at its end points,

$$\vec{F} = \pm m\vec{H}.\tag{31.16}$$

As shown in Figure 31.4, the differential distance ds the needle moves is related to θ by

$$ds = \left(\frac{\ell}{2}\right) d\theta,\tag{31.17}$$

where ℓ is the length of the needle.

216

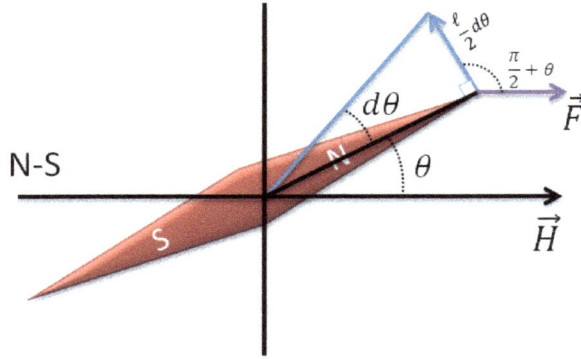

Figure 31.4: The top view of the magnetic needle to explain the work-energy calculations.

As θ changes from $0\,rad$ to $\theta_1\,rad$, the work done by the terrestrial magnetic field $\vec{H}$ is

$$W = 2 \times \int_0^{\theta_1} mH\frac{\ell}{2} \cos\left(\frac{\pi}{2} + \theta\right) d\theta, \tag{31.18}$$

where the source of the multiplier 2 is the terrestrial magnetism acting on the two poles of the magnet. Using the trigonometric identity

$$\cos\left(\frac{\pi}{2} + \theta\right) = -\sin\theta \tag{31.19}$$

to simplify and evaluate the integral,

$$W = -MH\left(1 - \cos\theta_1\right), \tag{31.20}$$

where $M = m\ell$. Note that W is negative work done by the restoring force, since its perpendicular component to the needle that contributes to work, acts opposite to the direction of motion of the needle.

As the needle swings from $\theta = 0$ to $\theta = \theta_1$, the kinetic energy of the needle is transformed to potential energy ΔPE, which is a positive number, or $-W$,

$$\Delta PE = MH\left(1 - \cos\theta_1\right). \tag{31.21}$$

The change in KE as the needle moves from $\theta = 0\,rad$ to its maximum amplitude $\theta = \theta_1\,rad$, when its KE is 0, is

$$\Delta KE = -\frac{1}{2}I\,\omega_f^2. \tag{31.22}$$

By the law of conservation of energy, assuming undamped oscillations, and therefore no loss in energy,

$$\Delta KE + \Delta PE = 0. \tag{31.23}$$

From the above equations,

$$-\frac{1}{2}I\omega_f^2 + MH(1 - \cos\theta_1) = 0. \tag{31.24}$$

Solving for ω_f,

$$\omega_f = 2\sqrt{\frac{MH}{I}}\sin\left(\frac{\theta_1}{2}\right), \tag{31.25}$$

simplified by making use of the trigonometric identity,

$$\sin\left(\frac{\theta_1}{2}\right) = \frac{1}{\sqrt{2}}\sqrt{1 - \cos\theta_1}, \ \theta_1 \geq 0. \tag{31.26}$$

Equating the expressions for ω_f in Equation 31.25 and Equation 31.13,

$$2\sqrt{\frac{MH}{I}}\sin\left(\frac{\theta_1}{2}\right) = \frac{MGQ}{I}. \tag{31.27}$$

Solving for Q,

$$Q = 2\sin\left(\frac{\theta_1}{2}\right)\frac{H}{\pi G}\left[\pi\sqrt{\frac{I}{MH}}\right]. \tag{31.28}$$

Substituting Equation 16.10 for the term within the square brackets,

$$T_{1/2} = \frac{T}{2}$$
$$= \pi\sqrt{\frac{I}{MH}}, \tag{31.29}$$

where $T_{1/2}$ is half the period of oscillation of the magnetic needle, as shown in Figure 31.3. From the above equation, it can be noted that the period of oscillation can be increased by using a magnetic needle with a higher moment of inertia. Q can also be written as

$$Q = 2\sin\left(\frac{\theta_1}{2}\right)\frac{T_{1/2}}{\pi}\frac{H}{G}. \tag{31.30}$$

Maxwell uses half the period $T_{1/2}$ in his treatise [56], instead of the period T. Half period is used in the above equation to be consistent with Maxwell's treatise.

The right-hand side of the above equation are measurable values from experiments. θ_1 can be observed from the galvanometer, $T_{1/2}$ can be averaged over many oscillations of the magnetic needle, measurement of H was explained in detail in Chapter 16, and G can be calculated from Equation 27.9. In this manner, the charge Q contained in the current burst can be measured.

Gauss's technique to measure the magnetic-field strength in Chapter 16, is valid only in electromagnetic units. If the measurement result of Q is obtained using Equation 16.10, then the result would be valid only in EMU.

31.4 Spring-Mass System With Damping [Optional]

The spring-mass system, shown in Figure 31.5, provides a good introduction to oscillations with damping. A brief overview of the spring-mass system will be followed by its rotational counterpart, the magnetic needle. When the mass in the figure is stretched or compressed, and released, it oscillates back and forth about $x = 0$, marked by the dotted vertical line. The amplitude of the

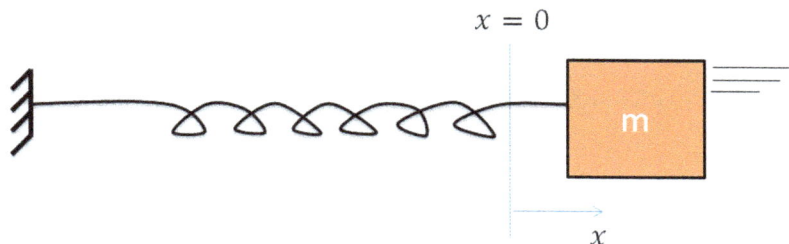

Figure 31.5: A spring-mass system.

oscillations decay as energy is lost as heat due to friction, referred to as damping, and eventually the mass comes to rest at $x = 0$.

The forces acting on the mass are the restoring $\vec{F}_r$ and the damping forces $\vec{F}_d$. The restoring force of the spring is always acting opposite to the direction of the displacement of the mass, "restoring" the mass to $x = 0$. The damping force is like friction, reducing the amplitude of the oscillations.

As explained in detail in Section 3.1, vectors can be avoided in 1D problems, and the vector information of the position of the mass x, velocity $\dot{x}$, acceleration $\ddot{x}$, and the forces acting on it, are captured in the sign of the values. Positive values of these quantities are vectors pointing to the right, and negative values are to the left.

Applying Newton's 2nd law,

$$\sum F = ma, \tag{31.31}$$

to the spring-mass system,

$$F_r + F_d = ma. \tag{31.32}$$

Expressions for the restoring force F_r and the damping force F_d can be guessed, and verified to be correct if the solution of the differential equation is meaningful, and models the physical behavior.

From Hooke's law,

$$F_r = -kx, \tag{31.33}$$

where k is called the spring constant, x is the position of the mass, and the positive direction of x has been marked by the horizontal arrow in Figure 31.5. Hooke's law states that the restoring force is proportional to the displacement of the mass, and the negative sign makes the direction of the restoring force always opposite to the direction of the displacement. This is consistent with the restoring force "pulling" the mass back to $x = 0$.

F_d is the damping force due to friction, given by

$$F_d = -b\dot{x}, \tag{31.34}$$

and b is called the damping coefficient.

From the above equations, and since the acceleration of the mass m is $a = \ddot{x}$, Equation 31.32 is written as

$$-kx - b\dot{x} = m\ddot{x}. \tag{31.35}$$

If $b = 0$, the lossless case without damping, the solution to the above equation is

$$x(t) = a \cos(\omega_o t + \phi) \tag{31.36}$$

$$\omega_o = \sqrt{\frac{k}{m}}, \tag{31.37}$$

where a is the amplitude of the oscillation or the maximum displacement of the mass from $x = 0$, ϕ is the phase of the waveform that can be used to set the position of the mass at time $t = 0$, and ω_o is the angular frequency of the oscillation, which is related to the period of the oscillation as written in Equation 3.31.

In the solution to the differential equation shown above, the argument of the cosine function is in radians, and not degrees, as explained in Section 3.2. The equation of the restoring force in Equation 31.33, gives rise to a sinusoidal function as the solution to the partial differential equation, which makes physical sense, as it represents the oscillation of the mass back and forth. This verifies the restoring force relation in Equation 31.33.

With damping, the solution to the partial differential equation is

$$x(t) = a \exp\left(-\frac{b}{2m}t\right) \cos\left(\omega' t + \phi\right)$$

$$\omega' = \sqrt{\omega_o^2 - \left(\frac{b}{2m}\right)^2}, \tag{31.38}$$

where ω' is the angular frequency with the effect of damping, and ω_o is the angular frequency without damping in Equation 31.37. The damping force in Equation 31.34 gives rise to the decaying exponential term that models the reduction in amplitude over time. b can be set to best fit the decaying amplitude of the oscillating mass, thereby accurately representing the physical behavior. This verifies the assumption of the damping force made in Equation 31.34.

The unit of $[\omega_o]$, as mentioned in Chapter 3, is rad/s. Applying dimensional analysis in Equation 31.38, the unit of the argument of the decaying expontential $\left[\frac{b}{2m}\right]$, must have the same unit as rad/s. As expected, the above expression reduces to Equation 31.37 for $b = 0$, the lossless case.

The rotational equivalent of the spring-mass system, the oscillating magnetic needle with damping, exhibits similar behavior, and therefore, a similar set of equations will be derived next.

31.5 Oscillations of the Magnetic Needle With Damping [Optional]

A magnetic needle, shown in Figure 31.6, if rotated by some angle θ and released, oscillates back and forth, eventually coming to rest in the magnetic $N - S$ direction. The north pole of the needle

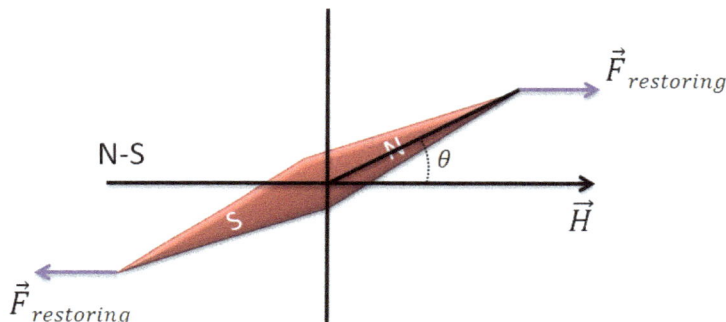

Figure 31.6: The restoring torque on a magnetic needle.

finally comes to rest pointing in the direction of $\vec{H}$, as marked by the arrow in the figure, which is the component of Earth's magnetic field that is tangential to the surface of Earth. The solution to θ, including the effect of damping to model the decreasing amplitude of the oscillations over time, will be derived.

As before, counter-clockwise angles of rotation θ are set as positive, and clockwise negative. As a result, as explained in Section 3.2, angular velocity $\dot{\theta}$, angular acceleration $\ddot{\theta}$, and torques τ, are all positive in the counter-clockwise direction, and negative clockwise.

The rotational equivalent of Newton's 2nd law is

$$\sum \tau = I\ddot{\theta}, \tag{31.39}$$

where τ are the torques acting on the rotating body, I is the moment of inertia of the rotating object, and using the dot convention as before, $\ddot{\theta}$ is the angular acceleration.

The restoring force $\vec{F}_r$ is caused by the terrestrial magnetic field acting on the magnetic needle. Using Equation 14.1,

$$\vec{F}_r = \pm m\vec{H} \tag{31.40}$$

are the forces acting at the two endpoints of the needle, modeled with magnetic charges $\pm m$. The restoring torque on the rigid body is therefore,

$$\tau_r = -2 \times mH \left(\frac{\ell}{2}\right) \sin\theta, \tag{31.41}$$

where θ is the angle of twist of the magnetic needle from its rest position, ℓ is the length of the magnetic needle, the multiplying factor $2\times$ arises from the restoring force acting on the two poles $\pm m$. For small angles of θ, $\sin\theta \approx \theta$, and the above equation can be approximated to be

$$\tau_r \approx -2 \times mH \left(\frac{\ell}{2}\right) \theta$$
$$= -\kappa\,\theta, \tag{31.42}$$

similar in form to Equation 31.33. Since the direction of the restoring torque τ_r is always opposite to the angle of rotation θ, the negative sign is included in the above equation.

A similar relation to Equation 31.34 is used to include the effect of damping,

$$\tau_d = -b\,\dot{\theta}, \tag{31.43}$$

where b is the damping coefficient. As before, the above relation can be verified, if the solution models the decaying amplitude of the oscillations, and represents the physical behavior.

Substituting the above equations in Equation 31.39,

$$-m\ell\,H\,\theta - b\,\dot{\theta} = I\,\ddot{\theta}. \tag{31.44}$$

The solution to the partial differential equation is

$$\theta(t) = a\exp\left(-\frac{b}{2I}t\right)\cos\left(\omega't + \phi\right), \tag{31.45}$$

$$\omega' = \sqrt{\omega_o^2 - \left(\frac{b}{2I}\right)^2}, \quad \omega_o = \sqrt{\frac{MH}{I}}, \tag{31.46}$$

where $M = m\ell$ is the magnetic moment. The period of the oscillation, similar to the derivation of Equation 3.31, is

$$T = \frac{2\pi}{\omega'}. \tag{31.47}$$

This result will be used to derive the charge stored in a capacitor in the following section, taking the effect of damping of the magnetic needle into account.

31.6 Calculation of the Charge Stored, Including the Effect of Damping of the Magnetic Needle [Optional]

The equation for the charge stored in a capacitor will be derived, taking the damping of oscillations of the ballistic galvanometer into account. Equation 31.45 is of the form

$$\theta(t) = a \exp\left(-\alpha t\right) \cos\left(\omega' t + \phi\right),$$
(31.48)

$$\omega' = \sqrt{\omega_o^2 - \alpha^2}, \ \omega_o = \sqrt{\frac{MH}{I}}.$$
(31.49)

Making use of trigonometric identities, the expression for the charge stored in a capacitor can be considerably simplified if α is written as

$$\alpha = \omega' \tan \beta,$$
(31.50)

for $\beta \in [0, \frac{\pi}{2})$, and ω' is the angular frequency of oscillation of the magnetic needle with damping, as illustrated in Figure 31.7. For the assumed interval of β, $0 \leq \tan \beta < \infty$, and therefore, any

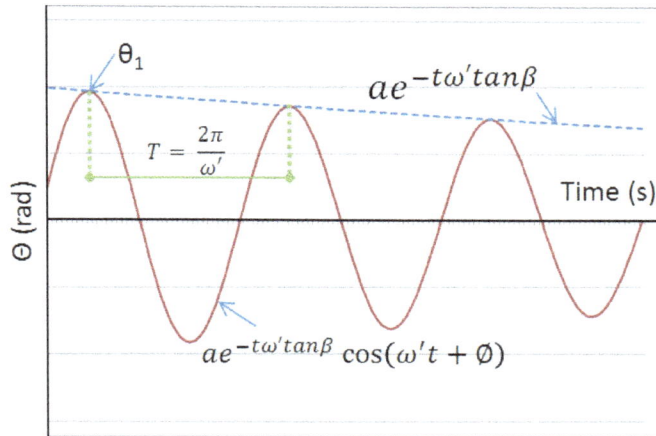

Figure 31.7: $\theta(t)$ of the magnetic needle in Figure 31.2 with damped oscillations.

non-negative value of α can be represented by the product $\omega' \tan \beta$.

ϕ in Equation 31.48 is set to $-\frac{\pi}{2}$, so that $\theta(t) = 0$ at time $t = 0$. Making use of the above equation of α,

$$\theta(t) = a \exp\left(-\omega' t \tan \beta\right) \sin\left(\omega' t\right).$$
(31.51)

Using the identity

$$\cos\left(\alpha + \beta\right) = \cos(\alpha)\cos(\beta) - \sin(\alpha)\sin(\beta),$$
(31.52)

and differentiating Equation 31.51 with respect to time, the angular velocity is

$$\frac{d\theta}{dt} = a\omega' \sec \beta \exp\left(-\omega' t \tan \beta\right) \cos\left(\omega' t + \beta\right).$$
(31.53)

223

The angular velocity is 0 when $\cos(\omega't + \beta) = 0$, which occurs when $\omega't + \beta$ is an integer multiple of $\frac{\pi}{2}$. The first occurrence when the angular velocity is 0 is when

$$\omega't + \beta = \frac{\pi}{2}. \tag{31.54}$$

Let θ_1 be the *first* occurrence of the throw of the needle, or the first maximum angle of swing. The subsequent angles where the angular velocity is 0 will be smaller than θ_1 because of the damping of the oscillations. Solving the above equation for $\omega't$,

$$\omega't = \frac{\pi}{2} - \beta. \tag{31.55}$$

Substituting this equation in Equation 31.51, and using the identity

$$\sin\left(\frac{\pi}{2} - \beta\right) = \cos\beta \tag{31.56}$$

to simplify the resulting expression,

$$\theta_1 = a \exp\left(-\left(\frac{\pi}{2} - \beta\right)\tan\beta\right)\cos\beta. \tag{31.57}$$

At time $t = 0$, the angular velocity from Equation 31.53 is

$$\frac{d\theta}{dt}\bigg|_{(t=0)} = a\omega'. \tag{31.58}$$

It will be assumed that the current burst occurs instantaneously at time $t = 0$, and as assumed earlier $\theta = 0$ at time $t = 0$. From Equation 31.13,

$$\begin{aligned}\frac{d\theta}{dt}\bigg|_{(t=0)} &= \omega_f \\ &= \frac{MGQ}{I}.\end{aligned} \tag{31.59}$$

Equating the above two equations, and solving for a,

$$a = \frac{MGQ}{I\omega'}. \tag{31.60}$$

Substituting this in Equation 31.57,

$$\theta_1 = \frac{MGQ}{I\omega'}\exp\left(-\left(\frac{\pi}{2} - \beta\right)\tan\beta\right)\cos\beta. \tag{31.61}$$

Using the trigonometric identity

$$1 + \tan^2\beta = \sec^2\beta, \tag{31.62}$$

and combining Equation 31.49 and Equation 31.50, and simplifying,

$$\omega_o^2 = \omega'^2 \sec^2\beta. \tag{31.63}$$

224

Relating the undamped angular frequency ω_o of the oscillating needle, to the magnetic field of Earth in Equation 16.9,

$$\omega_o^2 = \frac{MH}{I}.$$

(31.64)

Equating the above two equations, solving for ω', and using Equation 3.31,

$$\omega' = \sqrt{\frac{MH}{I}} \cos \beta$$
$$= \frac{\pi}{T_{1/2}},$$

(31.65)

where $T_{1/2} = \frac{T}{2}$ is the half period of a full oscillation T. $T_{1/2}$ is used rather than the full period T to be consistent with Maxwell's treatise. Substituting the above equation in Equation 31.61,

$$\theta_1 = \frac{MGQT_{1/2}}{I\pi} \exp\left(-\left(\frac{\pi}{2} - \beta\right)\tan\beta\right)\cos\beta.$$

(31.66)

The above expression can be further simplified, and will be expressed as the same equation in Maxwell's treatise. Solving for M using Equation 31.65,

$$M = \frac{I\pi^2}{HT_{1/2}^2 \cos^2\beta},$$

(31.67)

and eliminating M in the expression for θ_1,

$$\theta_1 = \frac{\pi GQ \sec\beta}{HT_{1/2}} \exp\left(-\left(\frac{\pi}{2} - \beta\right)\tan\beta\right).$$

(31.68)

Further simplifying by writing

$$\tan\beta = \frac{\lambda}{\pi},$$

(31.69)

where λ is a real non-negative number that satisfies the above relation. Since $\beta \in [0, \frac{\pi}{2})$ as mentioned earlier, $\tan\beta$ lies in the interval $[0, \infty)$, and therefore, λ is a real non-negative number that can represent the range of $\tan\beta$. Using the above two equations, and the identities,

$$\sec\beta = \sqrt{1 + \tan^2\beta}, \quad \beta \in \left[0, \frac{\pi}{2}\right)$$

(31.70)

$$\arctan\left(\frac{1}{x}\right) = \frac{\pi}{2} - \arctan x, \quad x > 0,$$

(31.71)

θ_1 is rewritten as

$$\theta_1 = \frac{QG}{HT_{1/2}}\sqrt{\pi^2 + \lambda^2} \exp\left(\frac{-\lambda}{\pi}\arctan\left(\frac{\pi}{\lambda}\right)\right).$$

(31.72)

Rearranging the above equation,

$$Q = \frac{HT_{1/2}\theta_1}{G}\frac{1}{\sqrt{\pi^2 + \lambda^2}} \exp\left(\frac{\lambda}{\pi}\arctan\left(\frac{\pi}{\lambda}\right)\right).$$

(31.73)

For small values of λ, $\lambda^2 \approx 0$, and

$$\frac{1}{\sqrt{\pi^2 + \lambda^2}} \approx \frac{1}{\pi}, \tag{31.74}$$

and

$$\arctan\left(\frac{\pi}{\lambda}\right) \approx \frac{\pi}{2}. \tag{31.75}$$

For $x \ll 1$, e^x can be approximated as

$$\exp x \approx 1 + x. \tag{31.76}$$

Using the above approximations, the charge stored in the capacitor is

$$Q = \frac{H\,T_{1/2}\,\theta_1}{\pi\,G}\left(1 + \frac{\lambda}{2}\right), \tag{31.77}$$

which is the same equation in Maxwell's treatise [56]. The right-hand side are measurables values: θ_1 and $T_{1/2}$ are data collected from the experiment, measurement of H in EMU was explained in detail in Chapter 16, and the galvanometer constant can be calculated from Equation 27.9.

32

Gauss's Law – Part I

Gauss's law will be discussed in Chapter 32-Chapter 38. Two proofs of Gauss's law will be presented, one that uses solid angle in this chapter [57], and one that does not in the following chapter. Both the results are consistent with each other. This also verifies the solid-angle definition in Equation $\mathcal{A}$.19, which was informally derived.

Gauss's law presented in this chapter is not the most general form of the law. It is valid only in the special case, where air and metal are present. The results of Faraday's experiments will be used to modify Gauss's law in the upcoming chapters, which will then be valid in any material medium.

32.1   Calculation of Electric Field Using Coulomb's Law

The assumption made in this section will be used in the derivation of Gauss's law. Coulomb's law to calculate the force between two point charges was presented in Equation 4.1. A charged metal object is shown in Figure 32.1. As concluded from the experiments in Chapter 8, the charges lie on the outer surface of the metal object. In the steady state, they arrange themselves in a manner that the electric field is 0 within the metal.

The electric field at Point Q, by definition in Equation 5.2, is the force exerted by the charges on the metal surface, on a unit positive point charge at Q. Superposition of forces, see Section 4.3, can be used to calculate the electric field at Point Q. The surface is discretized into tiny area elements. The field at Point Q can be calculated by applying Coulomb's law on each of the area elements, and assuming that the charge distributed in an area element, can be lumped into a point charge located at the center of the area element. Two such area elements have been marked dA_1 and dA_2 in the figure.

The dotted lines are the lines joining the center of the area elements and Point Q. As shown in the figure, the space between dA_1 and Point Q is air, while the space between dA_2 and Point Q is a combination of metal and air. Coulomb's law, however, in Equation 4.1 is defined between two point charges in air. Is Coulomb's law still applicable in the case of dA_2 to calculate the

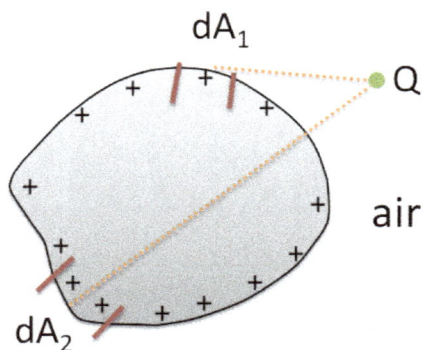

Figure 32.1: A charged conductor.

electric field at Point Q, where metal and air exist between dA_2 and Point Q? It will be assumed that Coulomb's law is still applicable, whether the medium between the charges is metal, or air, or a combination of both.

It was shown earlier in Section 8.2 that a Coulomb force exists between charges in a metal medium. The charges lying on the outer surface of a metal object, and none within the metal volume is a consequence of the Coulomb force. The like charges repel each other, trying to move away from each other. The maximum that they can repel each other is until they get to the outer surface of the metal. This experiment result shows that a Coulomb force does exist between the charges in dA_2 and Point Q.

There are many checkpoints that will be presented along the way during the course of development of electromagnetic theory. These checkpoints give assurance that such assumptions do not lead the theory on a wrong track.

32.2   Gauss's Law in Electrostatic Units

The cross section of an arbitrary closed 3D surface S is shown in Figure 32.2, with charges $\{q_1, ..., q_N\}$ somewhere within the surface, and q_0 outside. M is a charged metal object, and air present everywhere else. For simplicity, the cross section of the 3D surface is drawn as 2D in the figure. If $\vec{E}$ is the electric field generated by all of the charges, including the ones outside of S, Gauss's law relates the electric-field flux across the surface to the net charge contained within the surface. This is written as the surface integral,

$$\oint_S \vec{E} \cdot d\vec{A} = 4\pi \, q_{enc},$$
(32.1, ESU)

where q_{enc} is the charge enclosed within the surface, $d\vec{A}$ is the outward normal area vector of a surface element. In this example,

$$q_{enc} = q_1 + q_2 + ... + q_N.$$
(32.2)

228

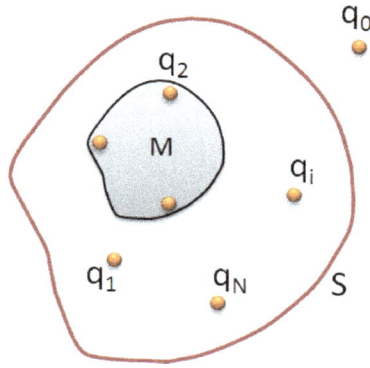

Figure 32.2: The cross section of a closed 3D surface S, enclosing charges $\{q_1, ..., q_N\}$.

q_0 is outside of the closed surface, as shown in the figure, and is not part of q_{enc}.

The electric-field flux across S has a very special property, and Equation 4.1 will be used to calculate the flux. As stated in Section 32.1, Coulomb's law will be assumed to be applicable everywhere, although a metal may be present. Gauss's law in the form presented in this chapter, is valid only if the Gaussian surface S is present in a medium with air and metal. Gauss's law will be augmented, and this restriction will be removed in the next chapter.

32.3   Proof 1 of Gauss's Law Using Solid Angle

Two proofs of Gauss's law will be presented. The proof in this section uses solid angle, and the one in the following chapter does not. Since the two proofs are consistent with each other, this indirectly verifies the general definition of solid angle in Equation $\mathcal{A}.19$, which has been informally derived. A tutorial on solid angle is presented in Appendix $\mathcal{A}$.

32.3.1 q_i Lies Within S

Point charge q_i lies somewhere within the surface S, as shown in Figure 32.3(a). From Coulomb's law in electrostatic units, and by definition of the electric field in Equation 5.2, the electric field caused by q_i is

$$\vec{E} = \frac{q_i}{r^2}\,\hat{r},$$

(32.3, ESU)

where r is the distance from the point at which q_i is located, $\hat{r}$ is the unit vector from q_i to the point at which the electric field is calculated.

An infinitesimal area marked dA_1 on S is shown in Figure 32.3(a), with the unit normal vector $\hat{n}$. From calculus, the electric-field flux through dA_1 is

$$\vec{E} \cdot \hat{n}\, dA_1 = q_i\, \frac{dA_1\,\hat{r} \cdot \hat{n}}{r^2}.$$

(32.4, ESU)

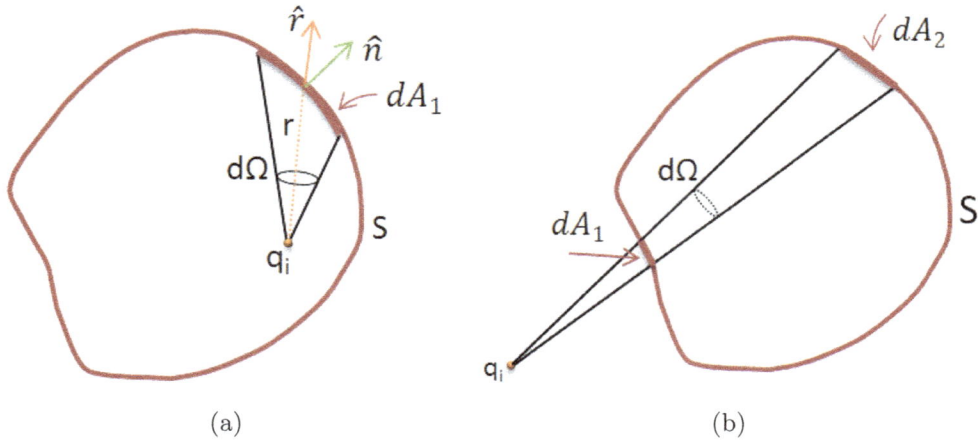

Figure 32.3: A point charge q_i (a) lying within S, and (b) outside of S.

From Equation $\mathcal{A}$.19, recognizing that the term

$$\frac{dA_1 \, \hat{r} \cdot \hat{n}}{r^2} = d\Omega, \tag{32.5}$$

is the solid angle subtended by dA_1 at the point where the point charge q_i is located. The net flux through S is

$$\oint_S \vec{E} \cdot d\vec{A} = q_i \oint_S d\Omega. \tag{32.6, ESU}$$

From Section $\mathcal{A}$.4, the solid angle subtended by any closed surface at any point within the surface is $4\pi \ sr$, and therefore,

$$\oint_S \vec{E} \cdot d\vec{A} = 4\pi \, q_i. \tag{32.7, ESU}$$

32.3.2 q_i Lies Outside of S

The second part of the proof is to show that if q_i lies outside the surface S,

$$\oint_S \vec{E} \cdot d\vec{A} = 0, \tag{32.8}$$

since q_i is not part of the enclosed charge, and the right-hand side of Equation 32.1 is 0.

A closed 3D surface S is shown in Figure 32.3(b), drawn as 2D for simplicity, with q_i lying outside the surface. Solid angle $d\Omega$ is drawn in the figure so that dA_1 and dA_2 subtend the same solid angle. Surface dA_1 is redrawn in Figure 32.4(a). The electric-field flux through dA_1 is

$$\vec{E} \cdot d\vec{A}_1 = q_i \frac{dA_1 \, \hat{r}_1 \cdot \hat{n}_1}{r_1^2}$$

$$= -q_i \frac{dA_1 \, \hat{r}_1 \cdot \hat{n}_1'}{r_1^2}, \tag{32.9, ESU}$$

230

where $\hat{n}_1$ is the outward unit normal to the surface S, $\hat{n}'_1 = -\hat{n}_1$ is the unit vector opposite to $\hat{n}_1$, or the inward unit normal to the surface. Noting that

$$d\Omega = \frac{dA_1\,\hat{r}_1 \cdot \hat{n}'_1}{r_1^2},\qquad(32.10)$$

the flux through dA_1 is

$$\vec{E} \cdot d\vec{A}_1 = -q_i\,d\Omega.\qquad(32.11,\text{ESU})$$

Repeating the same exercise for dA_2 in Figure 32.4(b),

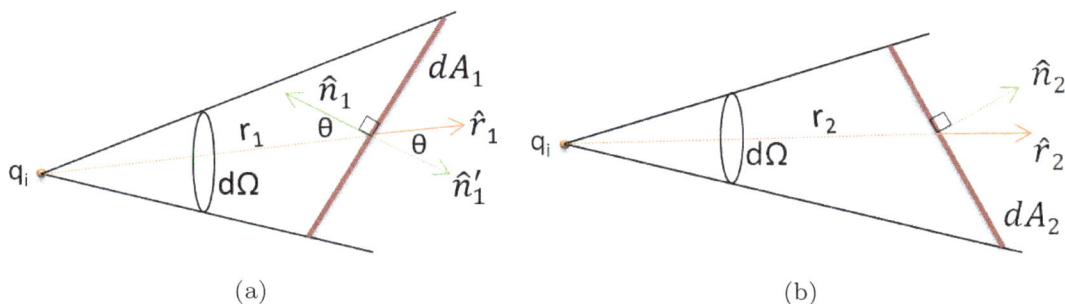

Figure 32.4: (a) dA_1 and (b) dA_2 in Figure 32.3(b).

$$\vec{E} \cdot d\vec{A}_2 = q_i\,\frac{dA_2\,\hat{r}_2 \cdot \hat{n}_2}{r_2^2},\qquad(32.12,\text{ESU})$$

where $\hat{n}_2$ is the outward unit normal to the surface S. From Equation $\mathcal{A}.19$, the term

$$\frac{dA_2\,\hat{r}_2 \cdot \hat{n}_2}{r_2^2} = d\Omega,\qquad(32.13)$$

is the solid angle subtended by dA_2 at the point where the charge q_i is located. The flux through dA_2 is therefore,

$$\vec{E} \cdot d\vec{A}_2 = q_i\,d\Omega.\qquad(32.14,\text{ESU})$$

The flux through dA_2 is equal and opposite to dA_1, and cancel each other. Considering the entire surface S, the flux through infinitesimal pairs of areas, such as dA_1 and dA_2, are equal and opposite, and the net flux is 0. Therefore, if q_i is outside the surface S,

$$\oint_S \vec{E} \cdot d\vec{A} = 0.\qquad(32.15)$$

Although the above proofs assumed a single point charge, using this result, and by superposition, Gauss's law can be shown to be true for any arbitrary charge distribution, as explained next.

231

32.3.3   Gauss's Law for a Charge Distribution

Applying the principle of superposition in Section 4.3 to the calculation of electric field, if $\{q_0, q_1, \ldots, q_N\}$ generate individually the electric fields $\{\vec{E}_0, \vec{E}_1, \ldots, \vec{E}_N\}$, then the charges generate the field

$$\vec{E} = \vec{E}_0 + \vec{E}_1 + \ldots + \vec{E}_N. \tag{32.16}$$

Calculating the flux through a closed surface S on both sides of the above equation,

$$\oint_S \vec{E} \cdot d\vec{A} = \oint_S (\vec{E}_0 + \vec{E}_1 + \ldots + \vec{E}_N) \cdot d\vec{A}$$

$$= \oint_S \vec{E}_0 \cdot d\vec{A} + \oint_S \vec{E}_1 \cdot d\vec{A} + \ldots + \oint_S \vec{E}_N \cdot d\vec{A}, \tag{32.17}$$

where the distributive property of the dot product, and the equivalence of integral of sums and the sums of integrals, have been used to derive the above equation. The above equation can be concisely written as

$$\oint_S \vec{E} \cdot d\vec{A} = \sum_{i=0}^{N} \Phi_i, \tag{32.18}$$

where Φ_i is the flux caused by the charge q_i.

Let charges $\{q_0, q_1, \ldots, q_N\}$, some lying within a surface S and some outside of it, generate the flux through the surface $\{\Phi_0, \Phi_1, \ldots, \Phi_N\}$. By Gauss's law, as proved earlier,

$$\Phi_i = 4\pi\, q_i, \text{ if } q_i \text{ lies within S}$$

$$= 0, \text{ if } q_i \text{ is outside of S.} \tag{32.19, ESU}$$

Applying the above result to Equation 32.18,

$$\oint_S \vec{E} \cdot d\vec{A} = 4\pi\, q_{enc}, \tag{32.20, ESU}$$

where q_{enc} is the total enclosed charge within the surface.

Instead of discrete point charges, for example, a region with a volume charge density exists. The region containing the charge can be discretized into small volume elements, small enough that the volume charge density can be assumed to be uniform within each of the elements. The charge within a volume element can be approximated as a point charge at the center of the element. In this way, volume charge density can be treated as a set of discrete point charges. Gauss's law, as proved above, is therefore valid for any charge distribution.

33

Alternate Proof of Gauss's Law [Optional]

In this chapter, unlike the proof of Gauss's law in Chapter 32, Gauss's law will be proved without the use of solid angles. The two proofs are consistent with each other. This also implicitly verifies the solid-angle definition in Equation $\mathcal{A}.19$, which was informally stated.

Using a similar setup of the coordinate system as Chapter 30, the primed coordinate system contains the source, which are the charges generating the electric field, and the unprimed coordinate system contains the point at which the electric field is calculated. As shown in Figure 33.1, the source are the charges $\{q_0, q_1, ..., q_N\}$ located at $\{\vec{r_0}', \vec{r_1}', \vec{r_2}', ..., \vec{r_N}'\}$ in the primed coordinate space. As assumed in Chapter 32, although metal may be present in addition to point charges, it will be assumed that Coulomb's law to calculate the electric field is still applicable.

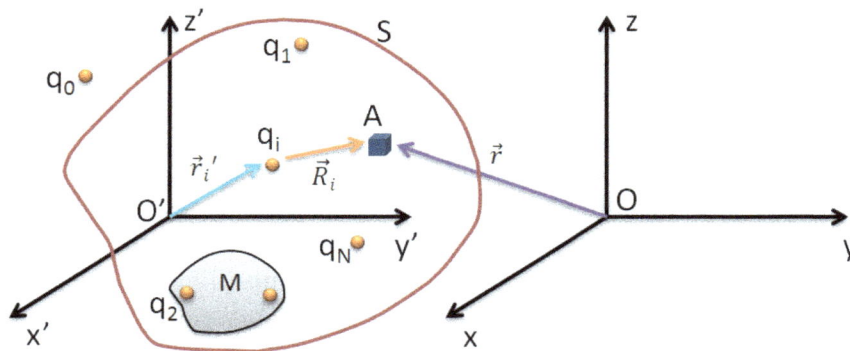

Figure 33.1: The charges $\{q_0, q_1, q_2, ..., q_N\}$ lie in the primed coordinate system, and $\vec{r}$ in the unprimed coordinate system is the point at which the electric field is calculated.

$\{\vec{r_0}', \vec{r_1}', \vec{r_2}', ..., \vec{r_N}'\}$ are fixed at the position of the point charges, and $\vec{r}$, the point at which the electric field is calculated, is a variable. Origin O coincides with Origin O', but drawn separately in the figure to illustrate the two coordinate systems. S is a closed 3D surface in the unprimed coordinate system, and enclosing the charges $\{q_1, q_2, ..., q_N\}$, but not q_0, as an example of some charges lying within S, and others outside of it.

Given a charge distribution, Equation 4.1 and Equation 5.2 can be used to calculate the electric field $\vec{E}$ at any point. The electric field at $\vec{r}$ generated by the charges, using superposition, is

$$\vec{E}(\vec{r}) = \sum_{i=0}^{N} q_i \frac{\vec{r} - \vec{r}_i'}{|\vec{r} - \vec{r}_i'|^3}$$

$$= \sum_{i=0}^{N} q_i \frac{\hat{R}_i}{|\vec{R}_i|^2}, \qquad\qquad (33.1, \text{ESU})$$

where the summation is over all the charges present, both within and outside of S, $\vec{R}_i = \vec{r} - \vec{r}_i'$ is shown in the figure, and $\hat{R}_i$ is the corresponding unit vector.

Applying Gauss's divergence theorem to calculate the flux across S,

$$\oint_S \vec{E} \cdot d\vec{A} = \int_V \nabla \cdot \vec{E} \, dV, \qquad\qquad (33.2)$$

where V is the volume enclosing S, and ∇ is the del operator defined in Equation 30.8, operating in the unprimed coordinate system. From the above equations,

$$\int_V \nabla \cdot \vec{E} \, dV = \sum_{i=0}^{N} q_i \int_V \nabla \cdot \left(\frac{\hat{R}_i}{|\vec{R}_i|^2} \right) dV, \qquad\qquad (33.3, \text{ESU})$$

where the summation and q_i have been moved out of the integral, since integral of sums is the same as sum of integrals, and q_i is a constant.

The above integral can be viewed as a Riemann sum to get a better understanding. The volume V is discretized into tiny volume elements, one of which centered at Point A is shown in the figure. With $\vec{r}_i'$ fixed at q_i, the integrand is evaluated and the Riemann sum calculated for $\vec{r}$ at the center each of the volume elements making up the volume. This operation is repeated for all the charges in the primed coordinate system, including the ones outside of S.

It is left as an exercise for the reader to show, similar to Equation 30.31,

$$\frac{\hat{R}_i}{|\vec{R}_i|^2} = -\nabla \left(\frac{1}{|\vec{R}_i|} \right). \qquad\qquad (33.4)$$

This can be easily shown in the Cartesian coordinates by setting $\vec{r} = (x, y, z)$ and $\vec{r}_i' = (x_i', y_i', z_i')$. Therefore,

$$\frac{1}{|\vec{R}_i|} = \frac{1}{\sqrt{(x - x_i')^2 + (y - y_i')^2 + (z - z_i')^2}}, \qquad\qquad (33.5)$$

234

and applying the gradient operation, Equation 33.4 can be shown to be true. Substituting Equation 33.4 in Equation 33.3,

$$\int_V \nabla \cdot \vec{E} \, dV = \sum_{i=0}^{N} q_i \int_V -\nabla^2 \left(\frac{1}{|\vec{R_i}|} \right) dV, \qquad \text{(33.6, ESU)}$$

where ∇^2 is the Laplacian operation on the unprimed coordinates. Using the identity in Equation 30.12 on the Laplacian of inverse distance,

$$\int_V \nabla \cdot \vec{E} \, dV = \sum_{i=0}^{N} q_i \int_V 4\pi \delta^3(\vec{r} - \vec{r_i}') \, dV, \qquad \text{(33.7, ESU)}$$

where $\delta^3(\vec{r} - \vec{r_i}')$ is the 3D Dirac delta function. The integral on the right-hand side of the above equation is 0 if $\vec{r} \neq \vec{r_i}'$, and 4π when $\vec{r} = \vec{r_i}'$.

V is discretized into infinitesimal volume elements, and the right-hand side of the above integral is viewed as a Riemann sum, with $\vec{r}$ pointing at the center of each of the infinitesimal elements during the sum operation, for each $\vec{r_i}'$ staying fixed at charge q_i. $\vec{r}$ must equal $\vec{r_i}'$ during the Riemann summation, as $\vec{r}$ makes an excursion to the center of each of the volume elements, if the charge q_i lies within V. By definition of the Dirac-Delta function, when $\vec{r} = \vec{r_i}'$,

$$4\pi \delta^3(\vec{r} - \vec{r_i}') \, dV = 4\pi, \qquad \text{(33.8)}$$

and 0, otherwise.

However, if q_i does not lie within the surface S, $\vec{r_i}'$ is outside the volume V. In this case, as $\vec{r}$ traverses each of the infinitesimal volume elements within V during the Riemann sum, $\vec{r}$ would never be equal to $\vec{r_i}'$, and the result of the integration is 0. As illustrated in the figure, q_0 located at $\vec{r_0}'$ is outside of S. $\vec{r}$ always lies within S during the integration, and will never be equal to $\vec{r_0}'$.

Equation 33.7, therefore, reduces to

$$\oint_S \vec{E} \cdot d\vec{A} = \int_V \nabla \cdot \vec{E} \, dV$$
$$= 4\pi \, q_{enc}, \qquad \text{(33.9, ESU)}$$

where q_{enc} is the sum of the charge enclosed within the surface S, completing the proof of Gauss's law. This result is the same as Equation 32.1. The two proofs of Gauss's law, with and without the use of solid angle, result in the same conclusion. This proof also verifies the general definition of the solid angle in Equation $\mathcal{A}$.19.

34

Verifying the Formulation of Gauss's Law

To verify the formulation of Gauss's law, the electric field of an infinite sheet of charge, solved using Gauss's law, will be compared with the result calculated using Coulomb's law.

34.1   Electric Field Generated by an Infinite Sheet of Charge

A positive uniformly charged infinite sheet of charge lying on the xy-plane is shown in Figure 34.1(a). The electric field at $Q(0, 0, z)$ will be solved using Coulomb's law.

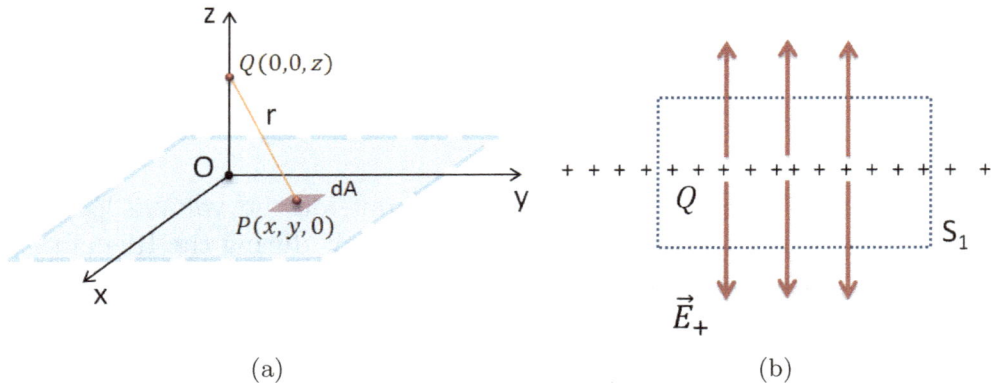

(a)   (b)

Figure 34.1: (a) An infinite sheet of positive charges with uniform charge density. (b) The calculation of electric field using Gauss's law.

An infinitesimal sheet area $dA = dx\, dy$, whose center is at $P(x, y, 0)$ is shown in the figure. The electric field $d\vec{E}_+$ generated by the infinitesimal area is

$$d\vec{E}_+ = \frac{\sigma\, dA}{r^2}\, \hat{r}, \tag{34.1, ESU}$$

where σ is the uniform charge density of the sheet, r is the distance between Point P and Point Q, and $\hat{r}$ is the unit vector in the direction from Point P to Point Q,

$$\hat{r} = \frac{\vec{PQ}}{|\vec{PQ}|}. \tag{34.2}$$

The field generated by the entire sheet is

$$\vec{E}_+ = \int_{-\infty}^{+\infty} \int_{-\infty}^{+\infty} \frac{\sigma}{(x^2 + y^2 + z^2)} \frac{(-x, -y, z)}{\sqrt{x^2 + y^2 + z^2}} \, dx \, dy \qquad \text{(34.3, ESU)}$$

It is left as an exercise for the reader to show that the result of the above integral is

$$\vec{E}_+ = 2\pi\sigma \, \hat{z}. \qquad \text{(34.4, ESU)}$$

The same result is obtained at any point over the sheet of positive charge. The negatively charged sheet will be the same magnitude as above, except for the sign.

By planar symmetry, the electric field is in the vertical direction at any point over the infinite sheet of charge. At any point over the sheet, the horizontal components cancel, leaving only the vertical component, as explained earlier in Figure 5.3(b).

The same result can be derived using Gauss's law. The use of Gauss's law to calculate the electric field requires the electric-field pattern to be known. This can be known from the symmetry of the problem, such as planar or spherical symmetry. Several examples will be presented in the future sections.

A Gaussian surface in the shape of a rectangular box is shown in Figure 34.1(b). By planar symmetry, as explained earlier, the electric field generated by the positive sheet of charge, denoted as $\vec{E}_+$, is vertical as shown. Let Q be the charge enclosed by the surface S. The non-zero electric-field flux is only across the top and the bottom faces of S, and the flux across the side faces is 0, since $\vec{E}_+$ is parallel to those faces.

Applying Gauss's law,

$$2\,E_+A = 4\pi\,Q, \qquad \text{(34.5, ESU)}$$

where A is the area of the top and bottom rectangular faces, and E_+ is the magnitude of the electric field. Recognizing that

$$\sigma = \frac{Q}{A} \qquad \text{(34.6)}$$

is the uniform surface charge density,

$$E_+ = 2\pi\sigma. \qquad \text{(34.7, ESU)}$$

In vector form, the electric field above the sheet is

$$\vec{E}_+ = 2\pi\sigma \, \hat{z}, \qquad \text{(34.8)}$$

which is the same result obtained earlier in Equation 34.4.

34.1.1 Electric Field Between Positive and Negative Sheets of Charges

The side view of an infinite sheet of positive charge and negative charge is shown in Figure 34.2(a), and of uniform surface charge density $\pm\sigma$. The region above the positively charged sheet, between the charge sheets, and below the negatively charged sheet, have been marked A, B, and C.

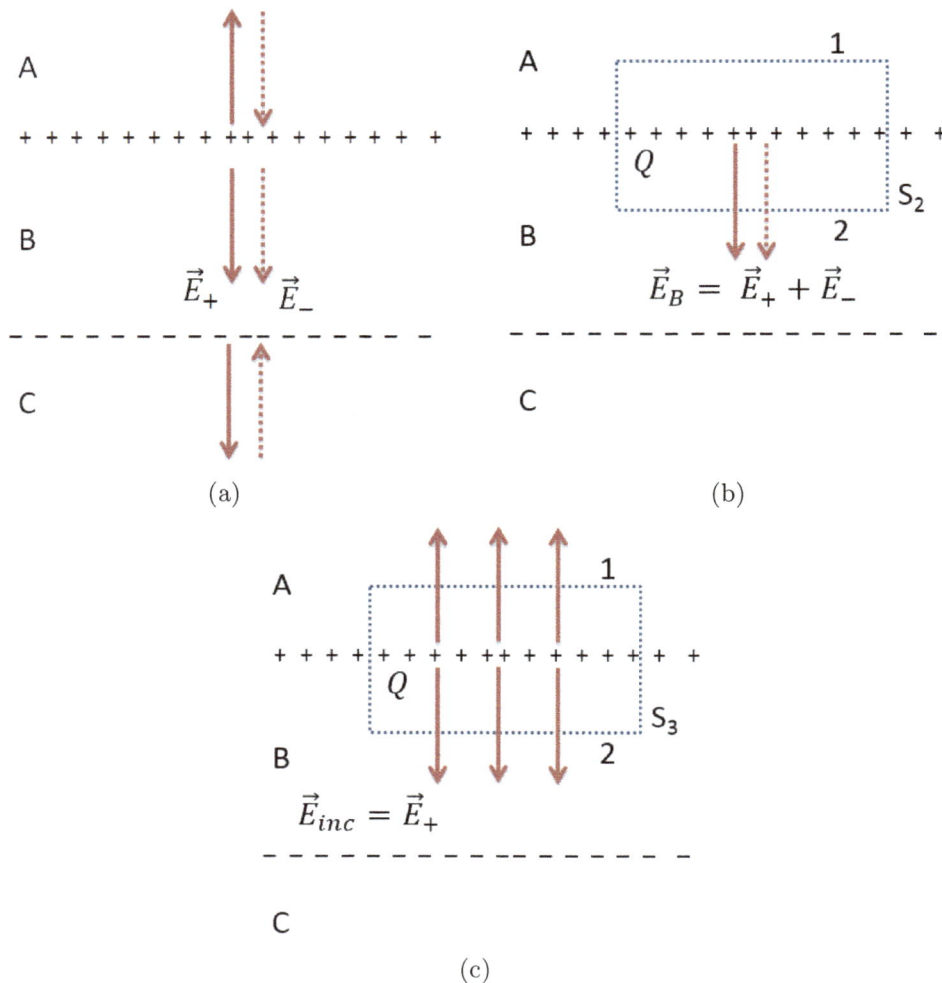

Figure 34.2: (a) The electric field generated by two infinite sheets of positive and negative charges. (b) The net electric field in (a). (c) The electric-field pattern considering only the positively charged sheet.

The dotted arrows represent the electric field generated by the negatively charged sheet $\vec{E}_-$, and the solid arrows by the positively charged sheet $\vec{E}_+$. The net electric field in the regions A, B, and C, are shown in Figure 34.2(b). Using superposition, the fields cancel in Regions A and C, and the net field in Region B is the sum of the fields generated by the positively and the negatively charged sheets, marked $\vec{E}_B$.

Using Equation 34.4, the field between the charged sheets is

$$\vec{E}_B = -4\pi\sigma\,\hat{z}. \tag{34.9, ESU}$$

The same result for the field between the plates can be obtained by applying Gauss's law to the surface S_2 shown in Figure 34.2(b). For the field pattern shown in the figure, only the flux through the face marked 2 is non-zero, and zero for all the other faces of S. Therefore,

$$E_B = 4\pi\frac{Q}{A}$$
$$= 4\pi\sigma. \tag{34.10, ESU}$$

In vector form, the above result is

$$\vec{E}_B = -4\pi\sigma\,\hat{z}, \tag{34.11}$$

and the results are consistent.

Suppose one does not include the contribution of the field generated by the negative charges, with the reasoning that the negative charges lie outside of S_2, and therefore, does not contribute to the net flux, the result would be incorrect. If the field contribution of the negatively charged sheet is ignored, the resulting field pattern is shown in Figure 34.2(c).

Applying Gauss's law to the surface S_3 in Figure 34.2(c),

$$2E_{inc}A = 4\pi Q, \tag{34.12, ESU}$$

and as before, Q is the enclosed charge within S_3, A is the surface area of the faces marked 1 and 2, and E_{inc}, where the subscript stands for *incorrect*, is the magnitude of the electric field due to the positive charges only. From the above equation,

$$E_{inc} = 2\pi\sigma, \tag{34.13, ESU}$$

which is an incorrect value for the field between the charged sheets, and is inconsistent with the results obtained without the use of Gauss's law.

The correct result for the field between the plates is Equation 34.9. This example shows that in applying Gauss's law, the net field pattern due to *all* the charges present, both within and outside the Gaussian surface must be used. If only a part of the charges is accounted for in the electric-field calculation using Gauss's law, superposition will have to be used to account for the electric field due to the remaining charges.

Faraday's Experiments With Spherical Capacitors

Michael Faraday's experiments with spherical capacitors, to study the effect of a dielectric material on the charge stored in a capacitor, will be demonstrated in this chapter. A spherical capacitor will be described first, followed by Faraday experiments, and the conclusions drawn from his experiments.

35.1   Spherical Capacitor

Gauss's law will be applied on a spherical capacitor. A spherical capacitor consists of two closely spaced concentric metal spheres, separated by a cavity between them. The 3D view is shown in Figure 35.1(a), and the 2D view of the cross section in Figure 35.1(b). The radius of the inner sphere is a, and the radius of the inner surface of the outer sphere is b, as marked in the figure. The cross section of two spherical Gaussian surfaces S_1 and S_2 are shown by the dotted lines.

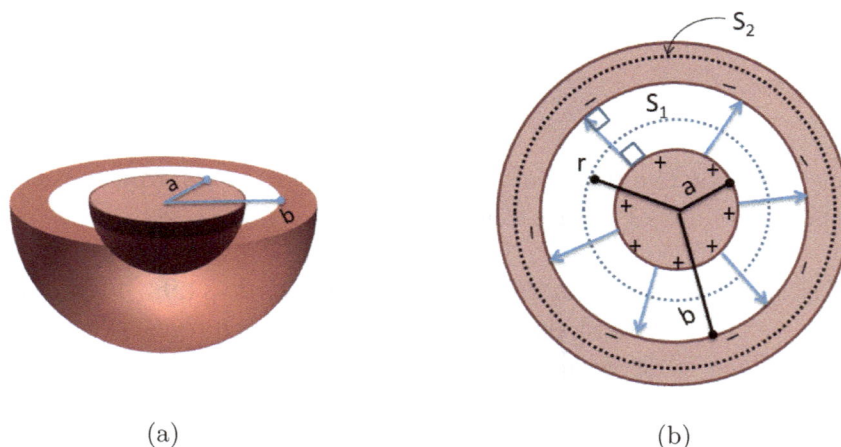

(a)   (b)

Figure 35.1: (a) A 3D view, and (b) a 2D view of the cross section of a spherical capacitor.

Without loss of generality, let the inner sphere be charged positive and the outer, negative. The charges on the inner sphere would entirely lie on its outer surface, as shown by the experiments in Chapter 8. The positive charges on the inner sphere would attract the negative charges to the inner surface of the outer sphere, as marked in the figure.

It was demonstrated by experiments in Chapter 8, the charges distribute themselves in a way that the electric field is 0 within a metal. Therefore, the electric-field flux across S_2 is 0, since S_2 lies within the metal. From Gauss's law,

$$\oint_{S_2} \vec{E} \cdot d\vec{A} = 4\pi \, q_{enc}$$

$$= 0. \tag{35.1}$$

The charges enclosed by S_2 must be equal and opposite, so that the net charge is 0, as derived from Gauss's law in the above equation. The charge stored in the concentric metal spheres will be related to the electric field in the cavity between the spheres.

The cross section of a spherical Gaussian surface S_1 at radius r, is shown by the dotted line in Figure 35.1(b). Assuming perfectly concentric spheres, by spherical symmetry, the magnitude of the electric field would be constant at any point on the Gaussian surface S_1 at radius r in the cavity. The electric field is perpendicular to the surface S_1, since the field is normal to the surfaces of the metal spheres, as proved in Section 8.5.

The flux across S_1 is

$$\oint_{S_1} \vec{E} \cdot d\vec{A} = 4\pi \, r^2 \, E, \tag{35.2}$$

where $4\pi \, r^2$ is the surface area of S_1. By Gauss's law,

$$4\pi \, r^2 \, E = 4\pi \, Q, \tag{35.3, ESU}$$

where Q is the positive charge in the inner sphere that is enclosed by S_1. Rearranging the above equation,

$$\vec{E} = \frac{Q}{r^2} \, \hat{r}, \tag{35.4, ESU}$$

where $\hat{r}$ is the unit vector in the radial direction, and the origin is at the center of the concentric spheres. Using Equation 10.6, the voltage difference V between the inner and the outer spheres is

$$V = \int_{r=a}^{r=b} \vec{E} \cdot d\vec{r}$$

$$= \int_{r=a}^{r=b} \frac{Q}{r^2} \, dr. \tag{35.5, ESU}$$

Evaluating the above equation, rearranging the result, and simplifying,

$$Q = \frac{ab}{b-a} \, V. \tag{35.6, ESU}$$

The above equation is of the form

$$Q = CV, \tag{35.7}$$

where C is called the capacitance. Rearranging,

$$C = \frac{Q}{V}, \tag{35.8}$$

and from this equation, capacitance is defined as the amount of charge stored per unit potential difference between the two conductors making up the capacitor. The capacitance of a spherical capacitor is therefore,

$$C = \frac{ab}{b-a}, \tag{35.9, ESU}$$

and note that the value of the capacitance is purely a function of the geometry of the capacitor. $b - a$ is the length of the cavity, and the smaller the length of the cavity, the greater is the capacitance, and more charge is stored per unit potential difference between the spheres. Greater values of a and b also result in a higher capacitance. To create an effective capacitor, the spacing between the conductors making up the capacitor must be smaller, and the surface areas of the conductors must be greater. This conclusion will be verified when the parallel-plate capacitor is analyzed in Section 38.2.

Michael Faraday studied the relation between the charge stored in a spherical capacitor, and the dielectric material filling the cavity of the capacitor. The results of the experiments show that Gauss's law, and the capacitance value in the above equation need to be modified, if a material other than air fills the cavity. Gauss's law presented in the previous chapter will be modified in the following chapter, to account for the effect of a dielectric material on the electric field.

35.2 Faraday's Experiment Setup

The spherical capacitor used by Faraday is shown in Figure 35.2. It has an opening at the top by which dielectric materials, such as distilled water or melted wax, can be filled in the cavity between the spheres.

The experiment setup is shown in Figure 35.3. Faraday used two identical capacitors, A and B, with the two inner metal spheres, p and p', connected together by a wire, as well as the two outer metal spheres q and q'. It was shown in Chapter 10, a metal is an equipotential volume, and so are metal objects connected together. Therefore, p and p' are at the same potential, and so are q and q'. Since the inner spheres are connected, as well as the outer spheres, the potential difference between the inner sphere and the outer sphere in A, V_{pq}, is equal to that of B, $V_{p'q'}$,

$$V_{pq} = V_{p'q'}. \tag{35.10}$$

This is true, independent of the dielectric material that fills the cavities of A and B.

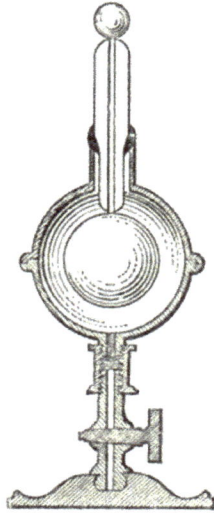

Figure 35.2: The spherical capacitor used by Faraday [58].

35.3 Experiment Results

In the cavity of A, the material is always kept as air, but the material in B is varied. Capacitor A with the unfilled cavity acts as a reference for the experiment. The capacitors are charged simultaneously by connecting the inner and the outer spheres to the terminals of a Wimshurst machine.

Faraday studied the ratio of the charge stored in A, Q_A, and B, Q_B. The quantity of charge stored in a capacitor can be measured using a ballistic galvanometer, as explained in detail in Chapter 31. The results of the charge stored will be in electromagnetic units. The ratio of the charges, however, would be independent of the units, since the units cancel in the ratio, resulting in a dimensionless number.

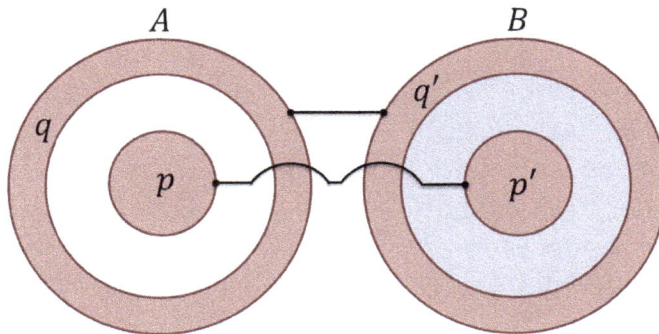

Figure 35.3: The experiment setup to study the variation of the charge stored in a capacitor with different dielectric materials.

If the dielectric material in both A and B is air, by symmetry, the charge stored in the two identical capacitors are equal. The ratio of the charge stored $Q_B : Q_A$ is 1. However, in the case when B is filled with a dielectric material other than air, Faraday observed that the ratio is greater than 1. This ratio has a special name and is called the relative permittivity of the dielectric material, denoted by the symbol ϵ_r, where r stands for *relative*, and it means the permittivity of a material relative to air.

To be precise, the cavity in Capacitor A must be vacuum, which is the absence of any material, including air. The reference dielectric material of air will be assumed for simplicity. Table 35.1 summarizes the results of ϵ_r observed for different materials.

Permittivity of Dielectric Materials	
Dielectric Material in Capacitor B	Relative Permittivity ϵ_r
Paraffin Oil	1.92
Distilled Water	76
Alcohol	26

Table 35.1: The permittivity of a few dielectric materials, as documented in Reference [58] in the year 1909.

If the material in the cavity of Capacitor B is a conductive material, charges would flow between the outer and the inner conductors, discharging them, and being unable to characterize the permittivity of the material. To prevent this, the walls of the cavity in Capacitor B can be coated with a thin layer of a non-conductive material, and then filling the cavity with the conductive material. In this way, the permittivity of a conductive material can be measured. If the conductive material is a metal, for example, intuitively, one could expect the permittivity to be 1.0.

35.4 Explanation of Faraday's Experiment Results

Faraday observed that Capacitor B, whose cavity is filled with a dielectric material, stores $\epsilon_r \times$ more charge than Capacitor A, whose cavity is unfilled, or contains air.

Gauss's law was used to derive the electric field in the cavity in Equation 35.4. The electric-field strength is $\propto Q$. If this same equation is applied to Capacitor B, the field strength is $\epsilon_r \times$ greater. By definition, voltage is the path integral of the electric field, and therefore, $V_{p'q'}$ is $\epsilon_r \times$ greater than V_{pq}. Since the inner spheres of Capacitors A and B are connected by a wire, as well as the outer spheres, it was shown in Section 35.2 that $V_{pq} = V_{p'q'}$. This contradicts the experiment results. This can be resolved with the explanation presented next.

If the dielectric material in the cavity of Capacitor B reduces the electric-field strength by ϵ_r, then the electric fields in the cavities of A and B are equal. If the electric fields are equal, this means that the voltage $V_{pq} = V_{p'q'}$. This explains the reason that more charge is present in Ca-

244

pacitor B: the additional charge is present to overcome the reduction in the electric field caused by the dielectric material, so that the electric fields in the cavities of A and B are equal. This explanation can be captured as

$$\vec{E}_{material} = \frac{\vec{E}_{air}}{\epsilon_r},$$ (35.11)

where $\vec{E}_{material}$ is the electric field in any material, and $\vec{E}_{air}$ is the electric field in air. This observation will be used to reformulate Gauss's law in the next chapter, taking into account the reduction of electric-field strength in a dielectric material.

36

Gauss's Law – Part II

To summarize the conclusions drawn in Chapter 35, Faraday discovered that for the same potential difference between the conductors of a spherical capacitor, the charge stored depends on the dielectric material filling the cavity. The dielectric material reduces the strength of the electric field, and more charge is stored in the capacitor to compensate for the reduction in the field strength.

In the derivation of Gauss's law in Equation 32.1, the electric-field flux through the surface S was calculated assuming that S is in air, so that Coulomb's law in Equation 4.1 can be used to calculate the flux. Faraday's experiments with spherical capacitors show how Gauss's law needs to be modified, if the Gaussian surface S lies in a dielectric material other than air, which is the focus of this chapter. It will also be shown that Coulomb's law, in general, cannot be applied to calculate the electric field in a non-uniform dielectric medium.

36.1   Electric-Flux Density $\vec{D}$

Faraday's experiment setup is repeated in Figure 36.1. If $\vec{E}_A$ is the electric field in capacitor A, applying Gauss's law in Equation 32.1,

$$\oint_{S_A} \vec{E}_A \cdot d\vec{A} = 4\pi \, Q_A, \qquad \text{(36.1, ESU)}$$

where $\pm Q_A$ is the charge stored in the spheres of A, and S_A is the spherical Gaussian surface in the cavity, as shown by the dotted line.

The ratio of the charges stored in A and B, empirically measured using the technique in Chapter 31, is the relative permittivity ϵ_r,

$$Q_A = \frac{Q_B}{\epsilon_r}. \qquad \text{(36.2)}$$

The electric fields in the cavities of A and B are equal,

$$\vec{E}_A = \vec{E}_B, \qquad \text{(36.3)}$$

246

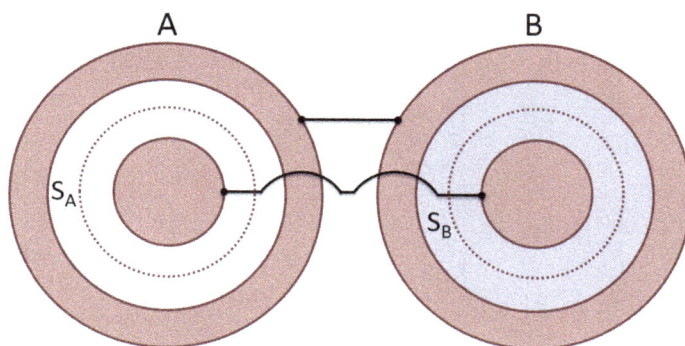

Figure 36.1: The setup of Faraday's experiment with spherical capacitors.

since the potential difference between the outer and the inner conductors are the same in both the identical capacitors, as noted in Section 35.2–Section 35.3. Therefore,

$$\oint_{S_B} \vec{E}_B \cdot d\vec{A} = \oint_{S_A} \vec{E}_A \cdot d\vec{A}, \tag{36.4}$$

where S_B is a spherical Gaussian surface lying in the cavity of B, and of the same radius as S_A. From the above equations,

$$\oint_{S_B} \vec{E}_B \cdot d\vec{A} = 4\pi \frac{Q_B}{\epsilon_r}. \tag{36.5, ESU}$$

Rearranging the above equation,

$$\oint_{S_B} \epsilon_r \vec{E}_B \cdot d\vec{A} = 4\pi Q_B, \tag{36.6, ESU}$$

is the general form of Gauss's law.

In this example, the Gaussian surface is present in an uniform dielectric material. ϵ_r is moved inside the integral, which will account for the variation in the dielectric material over S. Gauss's law is also valid in this case. However, the proof of Gauss's law in a non-uniform dielectric material medium requires special attention, and will be discussed in more detail in the later sections of this chapter.

The integrand in the above equation is assigned a new vector-field quantity $\vec{D}$, and is called electric-flux density,

$$\vec{D} = \epsilon_r \vec{E}. \tag{36.7, ESU}$$

$\vec{D}$ and $\vec{E}$ are related at any point by the above equation. In words, $\vec{D}$ is the electric field that would be present at a point, if the dielectric material at that point does not modify the electric field.

ϵ_r has no units as explained before, and therefore, by dimensional analysis $\vec{D}$ and $\vec{E}$ have the same units in ESU. However, $\vec{D}$ and $\vec{E}$ have different units in EMU and SI units, as explained in later chapters. From Equation 5.4, and applying dimensional analysis on Equation 36.7,

$$[\vec{D}] = \frac{statcoulomb}{cm^2} = \frac{gm^{1/2}}{cm^{1/2}\,s}. \tag{36.8, ESU}$$

From the above equations, the general form of Gauss's law is written as

$$\oint_S \vec{D} \cdot d\vec{A} = 4\pi\,q_{enc}, \tag{36.9}$$

where S is the Gaussian surface enclosing charge q_{enc}.

36.2 Capacitance of a Spherical Capacitor With a Dielectric Material

In Section 35.1, the capacitance of a spherical capacitor for the special case of air as the dielectric material filling the cavity was derived. In this section, the capacitance value is derived using the general form of Gauss's law, for any dielectric material filling the cavity of the spherical capacitor. The 2D cross section of the spherical capacitor is shown in Figure 36.2. As before, a and b are the radii of the inner sphere, and the inner surface of the outer sphere, respectively. The cross section of the Gaussian surfaces S_1 and S_2 are shown by the dotted lines.

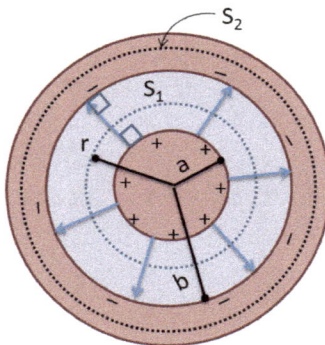

Figure 36.2: A 2D view of the cross section of a spherical capacitor filled with a dielectric material.

Without loss of generality, lets assume the inner sphere is charged positive and the outer, negative. The electric-field lines are shown by the solid arrows, and the field lines are perpendicular to the surfaces of the spheres, as explained in Section 8.5.

Applying the general form of Gauss's law to the spherical surface S_1 in Figure 36.2,

$$\oint_{S_1} \vec{D} \cdot d\vec{A} = 4\pi\,q_{enc}. \tag{36.10}$$

By spherical symmetry, the magnitude of $\vec{D}$ is the same anywhere on S_1, and the direction of the vector is radially outward. Using this result, the above equation can be simplified as

$$4\pi r^2 D = 4\pi\,Q, \tag{36.11}$$

where Q is the charge stored in the inner sphere of the capacitor, and r is the radius of S_1. Assuming the dielectric material has permittivity ϵ_r, substituting Equation 36.7 in the above expression, and simplifying,

$$\vec{E} = \frac{1}{\epsilon_r} \frac{Q}{r^2} \hat{r}, \qquad \text{(36.12, ESU)}$$

where $\hat{r}$ is the unit vector in the direction of the electric field in the cavity. Following the same steps outlined in Equation 35.5–Equation 35.9, the capacitance is

$$C = \epsilon_r \frac{ab}{b-a}. \qquad \text{(36.13, ESU)}$$

With the dielectric material present, the capacitance has increased by the factor ϵ_r, compared to the air-filled cavity in Equation 35.9. By the definition of capacitance in Equation 35.7, for the same potential difference between the conductors with air and dielectric material in the cavity, the capacitor with the dielectric material holds ϵ_r times more charge. As expected, this is consistent with the results shown in Table 35.1.

36.3 Differential Form of Gauss's Law

The differential form of Gauss's law can be derived from the integral form using Gauss's divergence theorem. Applying the divergence theorem on Equation 36.9,

$$\int_V \nabla \cdot \vec{D}\, dV = \int_V \rho_v\, dV, \qquad \text{(36.14)}$$

where V is the volume enclosing a closed Gaussian surface, with volume charge density ρ_v. Rearranging the above equation,

$$\int_V \left(\nabla \cdot \vec{D} - \rho_v \right) dV = 0. \qquad \text{(36.15)}$$

The above equation is satisfied for any volume V, any charge distribution ρ_v in a material medium that generates $\vec{D}$, if the integrand is 0,

$$\nabla \cdot \vec{D} = \rho_v. \qquad \text{(36.16)}$$

This is the differential form of Gauss's law.

Coulomb's law is singular at the exact location of the point charge. Therefore, the electric field calculated using Coulomb's law is also singular at the location of the point charge. However, by using Gauss's law and volume charge density ρ_v, instead of Coulomb's law and point charges, the singularity can be avoided. As a result, Gauss's law is more convenient to solve for electromagnetic fields in numerical and analytical methods. The reader is referred to other books for more details on numerical techniques in electromagnetics.

36.4 Non-Uniform Dielectric Materials and Coulomb's Law

The application of Coulomb's law in the case of a non-uniform dielectric material to calculate $\vec{D}$ and $\vec{E}$ will be analyzed in this section. The point of the exercise presented in this section is to show that in general, Coulomb's law cannot be applied to calculate the electric field when non-uniform dielectric regions are present.

A non-uniform dielectric region is shown in Figure 36.3. Coulomb's law will be used to calculate the electric field in the non-uniform dielectric region. It will be shown that this leads to a non-physical result.

A horizontal boundary separates the dielectric regions with relative permittivity ϵ_{r1} and ϵ_{r2}. A rectangle P across the boundary is shown in the figure, over which the path integral of the electric field will be calculated. The sides of the rectangle have been marked ℓ_1 and ℓ_2, and the direction in which the closed path integral is calculated is shown by the arrows. It will be assumed that $\Delta\ell$ is a very small length, and its contribution to the path integral is negligible.

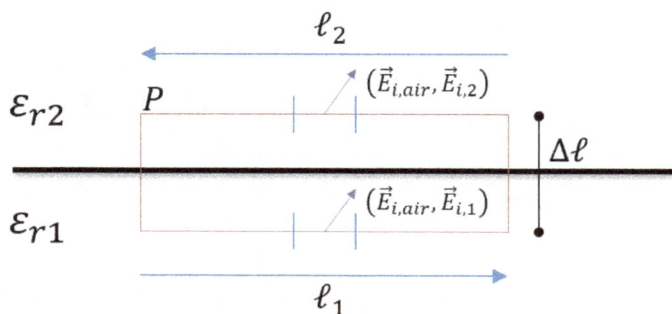

Figure 36.3: The electric field at the boundary between two different dielectric materials.

Given some charge distribution (not shown), Coulomb's law is used to calculate the electric field on P. In the application of Coulomb's law, it will be assumed that air is present everywhere. The effect of the dielectric material is then taken into account using Equation 35.11.

ℓ_1 and ℓ_2 are each divided into N segments, small enough that the electric field can be assumed to be uniform in the segment. The ith segment on ℓ_1 and ℓ_2 is shown in the figure. The electric field at the ith segment, assuming that the dielectric material is air everywhere, is marked as $\vec{E}_{i,air}$. Since $\Delta\ell$ is very small and negligible length, $\vec{E}_{i,air}$ at the ith segment on ℓ_1, is the same as $\vec{E}_{i,air}$ at the opposite i^{th} segment on ℓ_2, as shown in the figure.

The effect of a dielectric material, using Equation 35.11, reduces the strength of the electric field by ϵ_r. The electric field at the ith segment in the region with permittivity ϵ_{r1} is

$$\vec{E}_{i,1} = \frac{\vec{E}_{i,air}}{\epsilon_{r1}}, \tag{36.17}$$

and

$$\vec{E}_{i,2} = \frac{\vec{E}_{i,air}}{\epsilon_{r2}},$$ (36.18)

in the region with permittivity ϵ_{r2}.

The path integral along the path P, written as a Riemann sum is

$$\sum_P \vec{E} \cdot \vec{dl} = \sum_{i=1}^N \vec{E}_{i,1} \cdot \vec{\Delta\ell}_{i1} + \sum_{i=1}^N \vec{E}_{i,2} \cdot \vec{\Delta\ell}_{i2},$$ (36.19)

where N is the number of segments making up ℓ_1 and ℓ_2, $\vec{\Delta\ell}_{i1}$ is the ith segment on ℓ_1, and $\vec{\Delta\ell}_{i2}$ is the ith segment on ℓ_2. From the above equations,

$$\sum_P \vec{E} \cdot \vec{dl} = \frac{1}{\epsilon_{r1}} \sum_{i=1}^N \vec{E}_{i,air} \cdot \vec{\Delta\ell}_{i1} + \frac{1}{\epsilon_{r2}} \sum_{i=1}^N \vec{E}_{i,air} \cdot \vec{\Delta\ell}_{i2}.$$ (36.20)

The direction of integration along ℓ_1 is opposite to that of ℓ_2, and therefore,

$$\vec{\Delta\ell}_{i2} = -\vec{\Delta\ell}_{i1}.$$ (36.21)

Using this equation, the path integral is

$$\sum_P \vec{E} \cdot \vec{dl} = \frac{1}{\epsilon_{r1}} \sum_{i=1}^N \vec{E}_{i,air} \cdot \vec{\Delta\ell}_{i1} - \frac{1}{\epsilon_{r2}} \sum_{i=1}^N \vec{E}_{i,air} \cdot \vec{\Delta\ell}_{i1}.$$ (36.22)

In general, if $\epsilon_{r1} \neq \epsilon_{r2}$, the above result is not equal to 0.

However, in electrostatics, it was shown in Chapter 5 that the closed path integral of the electric field must always be 0,

$$\oint \vec{E} \cdot \vec{dl} = 0,$$ (36.23)

otherwise violating the law of conservation of energy. Since solving for the electric field in a non-uniform dielectric region using Coulomb's law, does not guarantee meeting the conservative property of the electric field, this exercise shows that in general, Coulomb's law cannot be used to solve for the electric field. Coulomb's law can only be applied in cases where the solution of the electric field also satisfies the above equation. Otherwise, the result is an incorrect solution.

In Equation 36.22, if a uniform dielectric material is present, $\epsilon_{r1} = \epsilon_{r2}$. The terms cancel, reducing the expression to 0, thereby satisfying the conservative property of the electric field. Coulomb's law can therefore be used to calculate the electric field in the case where a uniform dielectric material of permittivity ϵ_r is present. Coulomb's law will be strictly defined only for air as the dielectric medium. The electric field in air can then be scaled by $\frac{1}{\epsilon_r}$ to account for the reduced field strength in the material. In the case of a non-uniform dielectric medium, $\vec{D}$ and $\vec{E}$ can be solved using Poisson's equation, as explained in Section 36.6.

This example shows that the electric field component tangential to a material boundary must be continuous across the boundary. Otherwise, Equation 36.23 is not satisfied, violating the law of conservation of energy.

36.5   Non-Uniform Dielectric Materials and Gauss's Law

Gauss's law was proved using Coulomb's law in Chapter 32, in the case where the dielectric medium is air, and assuming that Coulomb's law is also valid when conductors are present. The general form of Gauss's law was derived in the case of a uniform dielectric material of permttivity ϵ_r. Is Gauss's law also valid in a non-uniform dielectric medium? The answer is yes. It is always valid!

If Coulomb's law can always be used to solve for the electric field, including the non-uniform dielectric medium case, following the same steps in Chapter 32, it would be easy to show that Gauss's law is always valid. As seen in Section 36.4, however, Coulomb's law in general, cannot be used to solve for the electric field in a non-uniform dielectric region.

For now, it will be assumed that Gauss's law is always valid, including the case of a non-uniform dielectric material. The proof, however, will be deferred until Chapter 51. It will be shown using the modified Ampere's law for time-varying fields, Gauss's law must always be valid. Gauss's law is a much more powerful equation: it will also be shown that Gauss's law must also be satisfied in the case of time-varying fields,

$$\oint_S \vec{D}(t) \cdot d\vec{A} = 4\pi \, q_{enc}(t), \tag{36.24}$$

or in the differential form,

$$\nabla \cdot \vec{D}(t) = 4\pi \, \rho_v(t), \tag{36.25}$$

where $\vec{D}(t)$, $q_{enc}(t)$, and $\rho_v(t)$ may be time-varying values. In other words, Gauss's law is always valid, regardless of whether the fields and charges are time-varying or non time-varying, whether the charges are moving or not, or the dielectric medium is uniform or non-uniform!

36.6   Numerical Solution of $\vec{D}$ and $\vec{E}$ in Non-Uniform Dielectric Materials

If Coulomb's law is not always applicable to solve for the electric fields, and since Gauss's law is always valid, given a charge distribution, Gauss's law can be used to solve for the electric fields. Substituting the definition of $\vec{D}$ in Equation 36.7 in Gauss's law,

$$\nabla \cdot \left(\epsilon_r \vec{E} \right) = 4\pi \, \rho_v. \tag{36.26, ESU}$$

If ϵ_r is not uniform, it cannot be moved out of the $\nabla\cdot$ operator. This is only half the story to solve for the electric field $\vec{E}$ [59]. As proved in Chapter 5, electric fields must also satisfy

$$\oint \vec{E} \cdot \vec{dl} = 0. \tag{36.27}$$

From calculus, the above equation is equivalent to saying that the electric field can be written as the gradient of a scalar potential V,

$$\vec{E} = -\nabla V. \tag{36.28}$$

From the above equations,

$$\nabla \cdot (\epsilon_r \nabla V) = -4\pi \, \rho_v. \tag{36.29, ESU}$$

This equation can be numerically solved for V, and the field values can be obtained from the solution of V using Equation 36.28 [60].

36.7   Avoiding Discontinuity of Fields at a Material Boundary

If a material boundary is an abrupt transition, fields may be discontinuous at the boundary. One such example was seen in Section 8.5. The electric field just outside a charged conductor surface is perpendicular to the surface, and its tangential component is 0. Just within the conductor surface, however, the electric field is 0. The normal component of the electric field is discontinuous *at* the boundary of the metal surface.

A different example is presented in Figure 36.4. A material boundary between two different interfaces is shown in Figure 36.4(a). $\vec{E}_{1t}$ and $\vec{E}_{2t}$ are electric-field components close to the two sides of a boundary, and tangential to the boundary. $\vec{D}_{1t}$ and $\vec{D}_{2t}$ are the tangential components of electric-flux density, and $\vec{D}_{1n}$ and $\vec{D}_{2n}$ are the normal components to the boundary.

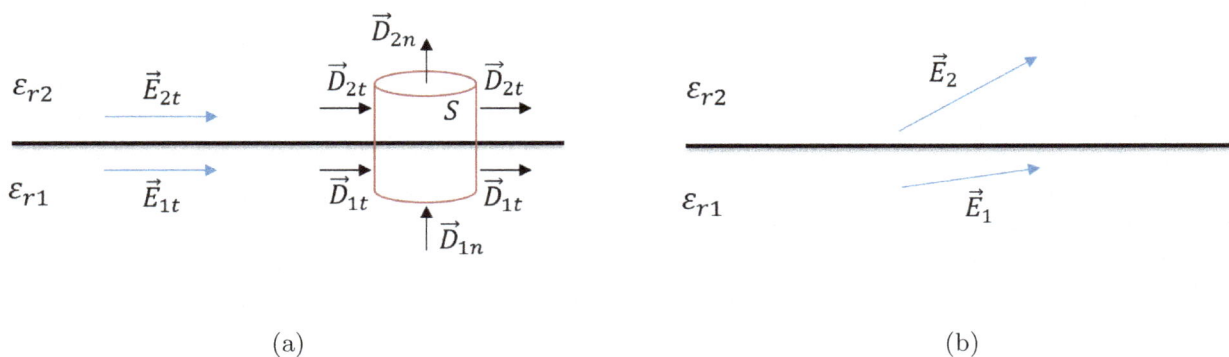

(a)   (b)

Figure 36.4: (a) The tangential component of $\vec{E}$ must be continuous across a material boundary. The normal component of $\vec{D}$ must also be continuous, assuming that there are no charges at the material boundary. (b) Net electric field $\vec{E}$ is discontinuous at the material boundary.

It was proven in Section 36.4, the tangential component of an electric field across a material boundary must be continuous. A similar boundary condition exists for electric-flux density $\vec{D}$. It will be proven next, the normal component of $\vec{D}$ must be continuous across a boundary, if there are no charges present at the boundary.

A Gaussian cylindrical surface S straddling across a material boundary is shown in the figure. S is assumed to be small enough that the electric flux due to $\vec{D}_{1t}$ and $\vec{D}_{2t}$ is negligible. If S does not enclose any charge, by Gauss's law,

$$\vec{D}_{1n} = \vec{D}_{2n}. \tag{36.30}$$

Therefore, the tangential component of $\vec{E}$, and the normal component of $\vec{D}$ must be continuous across a material boundary. This implies that the net electric field or electric-flux density may be discontinuous across, and therefore, discontinuous *at* the material boundary, as shown in Figure 36.4(b). The discontinuity results in undefined values, either of a field, or any mathematical operation, such as Equation 36.16, on a field at the boundary.

To avoid this type of discontinuity at material interfaces, instead of a material boundary making an abrupt transition, one could imagine the boundary as having a very tiny negligible thickness, across which the material property continually varies, such that the fields are continuous. In the first example presented at the beginning of this section, one could think of the charge as being distributed over a tiny negligible thickness on the surface of the conductor. The charge density continually decreases over the thickness, and the electric field continually diminishing to 0 over the tiny thickness on the surface. This results in a continuous field at all points, and any mathematical operation on the field at any point would be well defined.

37

Dielectric Materials

A simplistic view of atoms in a dielectric material provides an understanding of how a dielectric material reduces the strength of an electric field. In the presence of an electric field, the positive and the negative charges in the atoms become distorted, as illustrated in Figure 37.1.

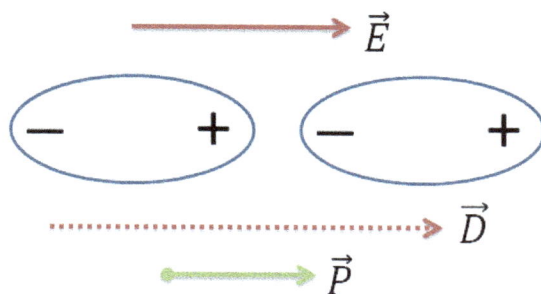

Figure 37.1: The reduction in electric-field strength due to polarization.

$\vec{D}$ was defined in Chapter 36 as the electric field that exists at a point, if the material at that point does not modify the electric field. From Faraday's experiment results, it can be reasoned that the distortion of atoms creates an electric field in the dielectric material, in a direction that opposes $\vec{D}$ and reducing its field strength. The net electric field is marked by the solid arrow $\vec{E}$.

$\vec{D}$ is called the electric-flux density, since it is used to calculate the electric flux across a closed surface in Gauss's law. Its early name, however, is the electric-displacement vector, so named to represent the displacement of the charges in the atoms of a dielectric material in the presence of an electric field. Although $\vec{D}$ may be called different names, it is still an *electric field*. $\vec{D}$ and $\vec{E}$ have the same units in the electrostatic system of units. The polarization vector, the different types of dielectric materials, and bound charges will be discussed next.

37.1   Polarization Vector

As shown in Figure 37.1, the dielectric material reduces the strength of the electric field from $\vec{D}$ to $\vec{E}$. Equation 36.7 relates $\vec{D}$ and $\vec{E}$ with a scaling factor ϵ_r. An alternate way is to relate $\vec{D}$ and $\vec{E}$ as a difference,

$$\vec{D} = \vec{E} + 4\pi\vec{P}, \qquad \text{(37.1, ESU)}$$

where $\vec{P}$ is called the polarization vector. It will become clear in a moment the rationale for including the factor 4π.

By the rules of dimensional analysis, 4π has no units, and it only scales the value of $\vec{P}$. The units of the left-hand side and the right-hand side must match, and therefore, the units of $\vec{P}$ must be the same as those of $\vec{D}$ and $\vec{E}$ in electrostatic units. The above equation is written in a different form in the electromagnetic system of units. This will be discussed in a later chapter. Nature sometimes gives flexibility in choosing scaling factors, such as the factor 4π in the above equation, as long as it is consistent with every other equation. Other examples of such scenarios will be presented in later chapters.

Calculating the flux across a closed 3D surface S on both sides of the above equation,

$$\oint_S \vec{D} \cdot d\vec{A} = \oint_S \vec{E} \cdot d\vec{A} + 4\pi \oint_S \vec{P} \cdot d\vec{A}. \qquad \text{(37.2, ESU)}$$

Substituting Gauss's law in the left-hand side of the above equation,

$$4\pi\, q_{enc} = \oint_S \vec{E} \cdot d\vec{A} + 4\pi \oint_S \vec{P} \cdot d\vec{A}, \qquad \text{(37.3, ESU)}$$

where q_{enc} is the charge enclosed by S. Rearranging the above equation,

$$\oint_S \vec{E} \cdot d\vec{A} = 4\pi \left(q_{enc} - \oint_S \vec{P} \cdot d\vec{A} \right) \qquad \text{(37.4, ESU)}$$

By dimensional analysis, the flux across the surface S due to the polarization vector $\vec{P}$ must have units of charge, since it is subtracted from q_{enc}. The above equation is similar to Gauss's law, except that the left-hand side is written as the flux due to $\vec{E}$, rather than $\vec{D}$. The flux due to the polarization vector is the quantity of charge to be subtracted from q_{enc}, to write Gauss's law in terms of $\vec{E}$, instead of $\vec{D}$. The reason for including the factor 4π in Equation 37.1 is to be able to cast it in a form similar to Gauss's law, as written in the above equation.

To be able to write Equation 37.1 in the form shown in Equation 36.7, the polarization vector is written as

$$\vec{P} = \chi_e \vec{E}, \qquad \text{(37.5, ESU)}$$

where χ_e is called the electric susceptibility. Since the units of $\vec{P}$ and $\vec{E}$ are the same in electrostatic units, χ_e is a dimensionless quantity. Substituting the above equation in Equation 37.1 and factoring $\vec{E}$,

$$\vec{D} = (1 + 4\pi\chi_e)\, \vec{E}, \qquad \text{(37.6, ESU)}$$

where

$$\epsilon_r = 1 + 4\pi\chi_e,\tag{37.7}$$

and is now of the same form shown in Equation 36.7. For permittivity values nearly equal to 1.0, such as 1.0004, it is convenient to specify the value of χ_e, rather than ϵ_r.

37.2 Dielectric Material Types

Using the theory developed until now, dielectric materials can be classified into different types. The electric-flux density and the electric field are related by permittivity ϵ_r, as written in Equation 36.7. By analyzing the permittivity, which is an electrical property of the dielectric material, different material types can be theorized.

37.2.1 Linear and Non-Linear Materials

In a linear material, the value of the permittivity does not depend on the strength of the electric field. Likewise, the permittivity is a function of the electric-field strength in a non-linear material.

37.2.2 Isotropic and Anisotropic Materials

Equation 36.7 can be rewritten as

$$\begin{pmatrix} D_x \\ D_y \\ D_z \end{pmatrix} = \begin{bmatrix} \epsilon_r & 0 & 0 \\ 0 & \epsilon_r & 0 \\ 0 & 0 & \epsilon_r \end{bmatrix} \begin{pmatrix} E_x \\ E_y \\ E_z \end{pmatrix},\tag{37.8, ESU}$$

where $\vec{D}$ and $\vec{E}$ are resolved into its components in the x, y, and z directions. This type of a material with the relation, as shown in the above equation, is known as an isotropic material, where the permittivity does not change with the x, y, or z components of the field. Permittivity expressed in the form above is called a permittivity tensor [61].

On the other hand, a material may exhibit the relation,

$$\begin{pmatrix} D_x \\ D_y \\ D_z \end{pmatrix} = \begin{bmatrix} \epsilon_r^x & 0 & 0 \\ 0 & \epsilon_r^y & 0 \\ 0 & 0 & \epsilon_r^z \end{bmatrix} \begin{pmatrix} E_x \\ E_y \\ E_z \end{pmatrix},\tag{37.9, ESU}$$

where the permittivity changes depending on the x, y, or z component of the field. This type of a material is called anisotropic. In the most general case, the permittivity tensor may also have non-zero off-diagonal elements.

The above equations may be written in short form as

$$\vec{D} = \bar{\bar{\epsilon}}_r \vec{E},\tag{37.10, ESU}$$

where $\bar{\bar{\epsilon}}_r$ is the permittivity tensor.

An ideal material is linear and isotropic. It is linear because the permittivity does not depend on the electric-field strength. It is isotropic because the permittivity is independent of the direction of the electric field in the material.

The susceptibility may also be expressed in the form of a tensor. In Equation 37.1, if $\vec{P}$ is written as

$$\vec{P} = \bar{\bar{\chi}}_e \vec{E}, \tag{37.11, ESU}$$

where $\bar{\bar{\chi}}_e$ denotes the susceptibility tensor, then Equation 37.1 becomes

$$\vec{D} = \vec{E} + 4\pi \bar{\bar{\chi}}_e \vec{E}. \tag{37.12, ESU}$$

Factoring $\vec{E}$,

$$\vec{D} = \left(I + 4\pi \bar{\bar{\chi}}_e \right) \vec{E}, \tag{37.13, ESU}$$

where I is the 3×3 identity matrix. Equating the above equation with Equation 37.10, the relation between the permittivity tensor and the susceptibility tensor is

$$\bar{\bar{\epsilon}}_r = I + 4\pi \bar{\bar{\chi}}_e. \tag{37.14}$$

37.3   Bound Charge [Optional]

Analyzing Equation 37.4 results in a different interpretation of the polarization vector. By dimensional analysis, the surface integral of the polarization vector must have the unit of charge, denoted as

$$\left[\oint_S \vec{P} \cdot d\vec{A} \right] = [q], \tag{37.15}$$

since it is subtracted from q_{enc}.

Since the electric field $\vec{E}$ is smaller than the electric-flux density $\vec{D}$ in a dielectric material, the surface integral of $\vec{E}$ is smaller than the surface integral of $\vec{D}$. Equation 37.4 states that the reduced electric-field flux across a surface S, translates to a reduction in the enclosed charge q_{enc} by the amount

$$\oint_S \vec{P} \cdot d\vec{A}. \tag{37.16}$$

An infinitesimal surface-area element dA_i of a surface S is shown in Figure 37.2(a). The polarization vector $\vec{P}_i$ is uniform in the area dA_i, and its direction is at an angle θ_i to the outward unit-normal vector of the surface element $\hat{n}_i$. $\vec{P}_i$ and the surface dA_i form a parallelepiped, as shown by the box with the solid outline in the figure. The flux across dA_i due to $\vec{P}_i$,

$$\vec{P}_i \cdot d\vec{A}_i = P_i \, dA_i \, \cos\theta_i, \tag{37.17}$$

is also the volume of the parallelepiped. The volume must have the unit of charge, as noted in Equation 37.15. The parallelepiped can be viewed as a blob of charge marked $+dq_i$,

$$dq_i = P_i \, dA_i \, \cos\theta_i. \tag{37.18}$$

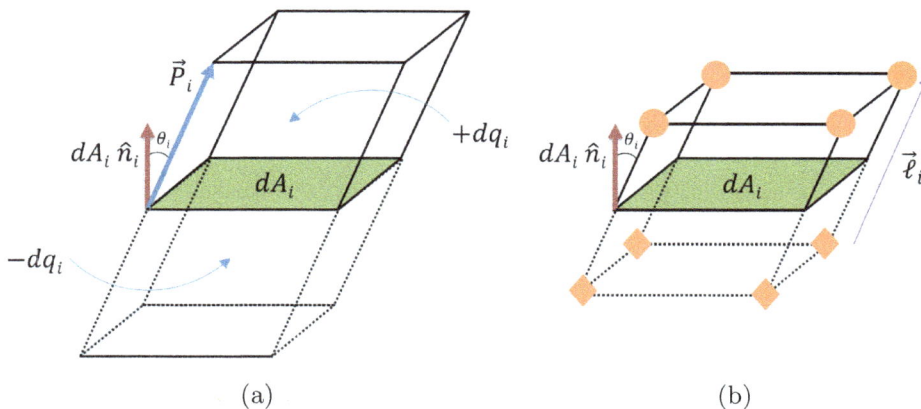

Figure 37.2: (a) The polarization vector at a surface element dA_i. (b) The flux due to the polarization vector across dA_i, viewed as an electric dipole moment.

Since this blob of charge is subtracted from q_{enc} in Equation 37.4, a mirror image of the blob of charge, shown by the dotted parallelepiped lying *within* the surface, which represents $-dq_i$ is drawn. This blob of charge $-dq_i$, reduces the enclosed charge within the surface S, q_{enc}, to satisfy the flux across the surface caused by $\vec{E}$ in Equation 37.4.

The charge dq_i associated with the flux caused by the polarization vector is known as bound charge, since it is viewed as the dipole charges of the dielectric material that are bound, and not free to move. q_{enc} is known as free charge, which is free to move, such as the ones in a charged metal object. Note that q_{enc} in Gauss's law in Equation 36.9 is the free charge, since the electric-flux density $\vec{D}$, by definition, that is used to calculate the flux across a Gaussian surface, is the electric field unaltered by a material. Hence, the bound charges of a material are not part of Gauss's law.

As shown in Figure 37.1, $\vec{P}_i$ will be related to the dipoles formed in a dielectric material by the presence of an electric field. Perhaps, this may only be a mathematical representation, rather than a physical one, but it provides a different perspective of the polarization vector.

The bound charge $\pm dq_i$ viewed as an electric dipole is shown in Figure 37.2(b). The positive and negative charges of an electric dipole are distinguished by circles and diamonds. $\vec{\ell}_i$ is a small length vector straddling across the boundary of the surface, and separating the dipole charges. $\vec{\ell}_i$ is in the direction as $\vec{P}_i$, whose magnitude is small enough to assume that $\vec{P}_i$ is a constant within the volume formed by ℓ_i and dA_i. Multiplying both sides of Equation 37.18 by length ℓ_i, the magnitude of $\vec{\ell}_i$, the magnitude of the electric dipole moment is

$$dq_i \, \ell_i = \vec{P}_i \cdot d\vec{A}_i \, \ell_i. \tag{37.19}$$

But

$$\ell_i \, dA_i \cos \theta_i = dV_i, \tag{37.20}$$

the volume of the parallelepiped formed by ℓ_i and dA_i. Rewriting the above equation,

$$P_i = \frac{dq_i\, \ell_i}{dV_i}. \tag{37.21}$$

Since the direction of $\vec{P_i}$ is the same as $\vec{\ell_i}$, the above equation can be written as a vector,

$$\vec{P_i} = \frac{dq_i\, \vec{\ell_i}}{dV_i}, \tag{37.22}$$

where $dq_i\, \vec{\ell_i}$ is the electric dipole moment, and the right-hand side is the electric dipole moment per unit volume. Note that the value of the right-hand side does not vary with ℓ_i. From the above equation, the polarization vector $\vec{P_i}$ can be viewed as an electric dipole moment per unit volume.

A volume is discretized into volume elements as shown in Figure 37.3. Each of the volume

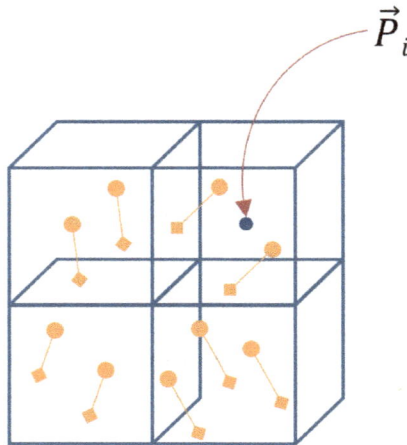

Figure 37.3: The polarization vector viewed as the electric dipole moment per unit volume.

elements contains many many atoms, and each of the atoms forming a dipole with an applied electric field. Lets say that there are N of them, each with a dipole moment $\vec{p_i}$. $\vec{P_i}$ is defined using the average of all the dipole moments in a volume element with volume ΔV as

$$\vec{P_i} = \frac{\frac{1}{N}\sum_{i=1}^{N} \vec{p_i}}{\Delta V}. \tag{37.23}$$

Examples of Gauss's Law

Several examples of Gauss's law will be presented in this chapter. The electric field in a parallel-plate capacitor will be calculated by several methods using Gauss's law, and the results are all consistent with each other. These examples provide more insight into how Gauss's law can be used to solve for the electric field.

38.1 Applying Gauss's Law on a Conductor

A charged metal object is shown in Figure 38.1, and without loss of generality, assume that it is positively charged. The charges lie on the surface of the conductor. As explained in Section 8.5, the electric field generated by the charges is perpendicular to the conductor surface, marked by the solid arrows. Assume that a linear, isotropic dielectric material with permittivity ϵ_r surrounds the metal object.

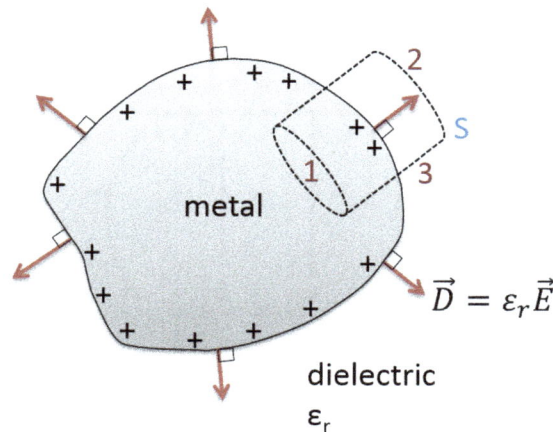

Figure 38.1: A charged metal object.

A cylindrical Gaussian surface S, with its areas marked 1 through 3, and enclosing the charge dq is shown in the figure. S is chosen small enough, and oriented in a direction that the electric field is perpendicular to its surface area 2, and parallel to the surface 3. The charges are distributed on the surface such that the electric field is 0 within the metal surface. From the relation in Equation 36.7, $\vec{D} = \vec{E} = 0$ within the metal surface, and therefore, the flux through the area of S marked 1 is 0. Also, since the electric field is parallel to the area 3, the flux is 0.

Applying Gauss's law,

$$
\begin{aligned}
D &= 4\pi \frac{dq}{dA} \\
&= 4\pi \sigma
\end{aligned}
\tag{38.1}
$$

where dA is the end cap area of the cylinder marked 2, and σ is the surface charge density, whose unit is charge per unit area. This result will be used to solve for $\vec{D}$ in a parallel-plate capacitor, in the following section.

The electric field just outside the metal surface, can be calculated using Equation 36.7,

$$
\vec{E} = \frac{4\pi \sigma}{\epsilon_r} \hat{n},
\tag{38.2, ESU}
$$

where $\hat{n}$ is the unit normal vector to the metal surface.

38.2   Parallel-Plate Capacitor

The cross section of a parallel-plate capacitor is shown in Figure 38.2, where the metal conductors are planar, and separated by distance d. Without loss of generality, let the top plate be charged positive, and the bottom negative.

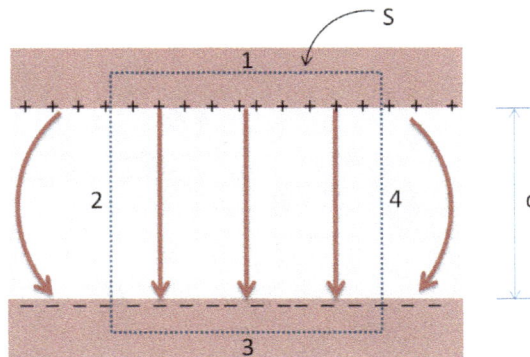

Figure 38.2: A charged parallel-plate capacitor.

The charges on the plates distribute themselves on the outer surface of the metal plates, such that the electric field within the metal is 0. Assuming the metal plates are at certain voltages, the charge distribution on the metal plates can be solved numerically using the methodology outlined in Reference [62], but this is not covered in this book. If the spacing between the plates is small, it can be expected that most of the charges would be on the *inner* surfaces of the metal plates, as shown by the '+' and '-' signs, since the charges would attract each other and will be closest to each other. It will be assumed that the charges on the plates are distributed as shown in the figure.

It will be shown using Gauss's law that the surface charge density on the plates must be equal and opposite. Following this, the electric field between the plates will be solved using several methods.

38.2.1 Properties of the Charge Density on the Metal Plates

The cross section of a Gaussian surface S, in the shape of a rectangular box, is shown by the dotted lines in Figure 38.2, with its faces marked 1 to 4. As explained in Section 5.1, in the center region of the plates, away from the edges, the electric field is vertical, but close to the edges, however, the fringing fields are as shown. The electric-field flux across the faces 1 and 3 is 0, since the field is 0 within the metal surface. The flux across the faces 2 and 4 is also 0, as the field is parallel to these faces. The flux resulting from $\vec{E}$ and $\vec{D}$ across S is therefore 0,

$$\oint_S \vec{D} \cdot d\vec{A} = 0. \tag{38.3}$$

By Gauss's law, the charge enclosed within S must be equal and opposite, so that the charge enclosed is 0. If charge Q lies on the inner surface of the positive plate, then $-Q$ would exist on the inner surface of the negative plate. Since the charges are equal and opposite, the surface charge density $\pm\sigma$, which is the charge per unit area, are also equal and opposite. In the remainder of this section, the electric field between the plates will be calculated in several ways.

38.2.2 Method 1: With Superposition of Fields

The parallel plates are redrawn in Figure 38.3, with the plate thicknesses exaggerated. Three regions A, B, and C have been marked that are the regions above the positive sheet of charge, between the plates, and below the negative sheet of charge, respectively.

By definition, the electric-flux density $\vec{D}$ is electric field $\vec{E}$, without the influence of a dielectric material. Using the results from Section 34.1, the electric-flux density $\vec{D}_+$ and $\vec{D}_-$ due to the positive and negative sheet of charges, are drawn as solid and dotted lines in Figure 38.3.

Considering the electric field generated by only the positive charges or only the negative charges, the electric field can no longer be assumed to be 0 within the metal plates. The field will be assumed to be uniform, and from the figure, $\vec{D}_+$ and $\vec{D}_-$ cancel in Regions A and C, but add up in Region B. This is consistent with $\vec{D}$ being 0 within the metal regions.

A rectangular Gaussian surface S_+ will be used to calculate $\vec{D}_+$, which is the electric-flux density generated by the positive charges. The flux across the faces marked 1 and 2 are nonzero, and $\vec{D}_+$ is perpendicular to those faces. The electric-flux density is parallel to the other two faces, and

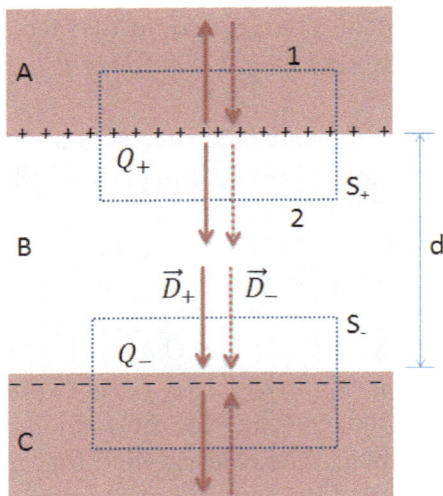

Figure 38.3: The use of superposition to calculate the electric field between the plates.

therefore, does not contribute to the flux. Applying Gauss's law,

$$\oint_{S_+} \vec{D}_+ \cdot d\vec{A} = 2 D_+ A$$
$$= 4\pi \, Q_+, \tag{38.4}$$

where A is the area of the faces 1 or 2, and Q_+ is the positive charge enclosed by S_+. The factor 2 in the above equation is to account for the flux across the two faces 1 and 2 of S_+. Rearranging the above equation,

$$D_+ = 2\pi \, \frac{Q_+}{A}$$
$$= 2\pi \, \sigma, \tag{38.5}$$

where σ is the surface charge density of the positive charges. By symmetry, the electric-flux density generated by the negative charges is

$$D_- = 2\pi \, \sigma, \tag{38.6}$$

where $-\sigma$ is the charge density in the negatively charged plate. It was shown in Section 38.2.1, the positive and the negative charge densities must be equal $\pm\sigma$. Using superposition, the fields $\vec{D}_+$ and $\vec{D}_-$ add between the plates, as they are vectors that are in the same direction,

$$D = D_+ + D_-$$
$$= 4\pi \, \sigma. \tag{38.7}$$

Using the relation in Equation 36.7, the electric-field strength E between the plates is

$$E = \frac{4\pi\sigma}{\epsilon_r}, \tag{38.8, ESU}$$

264

assuming a linear, and an isotropic dielectric material between the plates of permittivity ϵ_r. It is left as an exercise for the reader to calculate the capacitance of the parallel-plate capacitor from the electric field, using the steps outlined in Section 36.2,

$$C = \epsilon_r \frac{A}{4\pi d}. \tag{38.9, ESU}$$

38.2.3 Method 2: Without Superposition of Fields

In the second method, Gauss's law will be applied to the surface S in Figure 38.4, and without the use of superposition. Since the field is 0 within the metal surface, the flux through the face marked 1 is 0. $\vec{D}$ is the total field between the plates contributed by all the charges present.

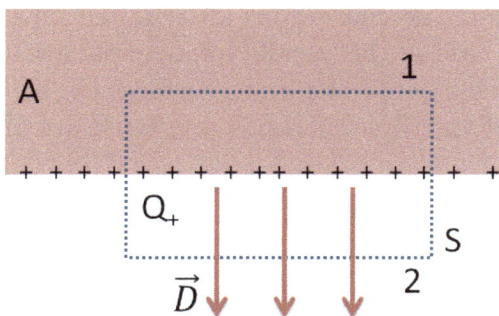

Figure 38.4: The calculation of electric field between the plates without the use of superposition.

Applying Gauss's law on the dotted Gaussian surface S,

$$\oint_S \vec{D} \cdot d\vec{A} = 4\pi\, Q_+. \tag{38.10}$$

As before, the resulting electric-flux density is

$$D = 4\pi \frac{Q_+}{A}$$
$$= 4\pi\, \sigma, \tag{38.11}$$

where A is the area of the face 2 of the rectangular box S, and σ is the surface charge density. The value of $\vec{D}$ is the same as the result in Equation 38.7, obtained using superposition.

38.2.4 Method 3

The third method is the quickest way to calculate the electric field between the plates. Applying Equation 38.1 on the positive plate, or on the negative plate, is the electric-flux density between the plates, and we are done! Without loss of generality, lets pick the positive plate with the surface charge density σ,

$$D = 4\pi\, \sigma, \tag{38.12}$$

and the direction of the field is normal to the conductor, from the positively charged plate to the negatively charged plate.

How is the above result in a parallel-plate capacitor, where there is a positively-charged conductor and a negatively-charged conductor, different from the conductor in Figure 38.1, where there is only a single conductor, if the expressions of $\vec{D}$ are the same in both cases?

The difference between the two cases is captured in the value of the surface charge density σ, by which the presence of the second plate is implicitly accounted for. Although the expressions of $\vec{D}$ are the same in both the cases, the values of σ are different.

The charge distribution is different between a charged conductor and a parallel-plate capacitor. In the case of a single conductor, the charges are distributed on the *entire* surface, as shown in Figure 38.1, but this is not the case in the parallel-plate capacitor. The oppositely charged plates in close proximity, attract the charges to the inner surfaces of the plates, thereby influencing the charge density σ. The results of the three methods presented are all consistent with each other.

39

Ohm's Law

In the year 1826, George Simon Ohm formulated the relation between the current flowing in a wire to the voltage of a battery connected to it, which would become one of the most famous equations, known as Ohm's law in his honor. Although he performed many experiments to arrive at the equation, only his most important work will be presented in this chapter, from which Ohm's law will be derived [63][64].

A difficulty that Ohm initially faced in his experiments was that the battery source didn't last long enough as desired. The current and therefore, the voltage of the battery, continually decreased during the course of the experiment, making it difficult to draw any conclusions. He overcame this problem, at the suggestion of Poggendorf, by replacing his battery with a thermoelectric source. This works on the principle of Seebeck effect, as explained next, followed by the derivation of Ohm's law.

39.1 Seebeck Effect

In 1821, Thomas Seebeck discovered that when two different metals, marked $M1$ and $M2$ in Figure 39.1, joined together and when subject to a temperature gradient, acts as a voltage source, similar to a battery [65]. This type of a source is known as a thermoelectric source, and is called the Seebeck effect in his honor. When it is connected by a wire, and if the temperature gradient is held constant, a steady current i flows in the wire. A temperature gradient is created using a heat source, such as a candle or immersing the junction in boiling water, and the other junction cooled using ice, for example.

The current flow in the wire can be detected using a magnetic needle, similar to Oersted's experiment, and the current flow is steady if the needle remains stationary. Using a thermoelectric source, Ohm was able to generate a steady voltage and current, with which accurate results could be obtained.

39.2 Ohm's Apparatus

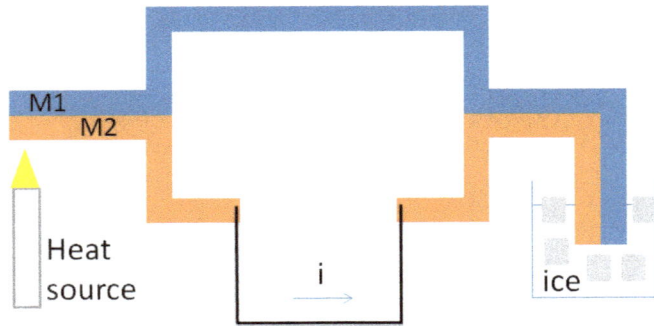

Figure 39.1: A thermoelectric source.

Ohm's experiment setup is shown in Figure 39.2. The schematic of the setup is the same as Figure 39.1. To measure the *relative* strength of the current, Ohm used a simplified galvanometer, marked G in Figure 39.2. A magnetic needle is attached to a torsion head using a wire or a string, similar to Coulomb's torsion balance. The magnetic needle lies above the current-carrying wire, and parallel to it, when no current flows in the wire, similar to the setup in Oersted's experiment. This state will be referred to as the 0 position of the magnetic needle.

The 0 position of the needle is along the magnetic N-S direction, and the terrestrial magnetic field is marked $\vec{H}$, as shown in Figure 39.3. The current-carrying wire is routed parallel to the needle, so that the magnetic field generated by the current flow, marked $\vec{h}$, is perpendicular to the magnetic needle. $\vec{h}$ causes the needle to deflect.

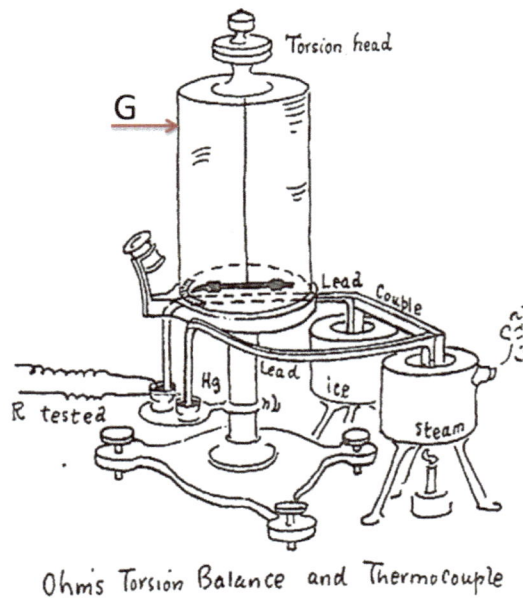

Ohm's Torsion Balance and Thermocouple

Figure 39.2: Ohm's experiment setup [64].

268

The torsion head is turned, which may be many turns or angles over 360°, until the needle returns back to its 0 position. The angle that the torsion head needs to be turned θ, or the twist in the wire, is used as the relative measure of the strength of the current flow. This will be proven next.

Since the needle is at the 0 position, the force caused by the terrestrial magnetic field on the magnetic needle is along the direction of the magnetic needle, and therefore, it exerts no torque on the needle. The restoring torque τ_r, which is the torque exerted on the needle by the string untwisting itself, is balanced by the torque on the needle caused by $\vec{h}$, denoted as τ_h. Repeating Equation 3.21,

$$\tau_r = -\kappa\theta, \tag{39.1}$$

where κ is a constant that depends on the string material and physical properties. The torque caused by the magnetic field of the current flow $\vec{h}$, and magnitude h, is

$$\tau_h = 2 \times mh \left(\frac{\ell}{2}\right) \sin 90°, \tag{39.2}$$

where $\pm m$ is the magnetic charge of the needle at its endpoints, as explained in Chapter 14, ℓ is the length of the magnetic needle, and the factor $2\times$ arises from the force acting on both the magnetic charges of the needle.

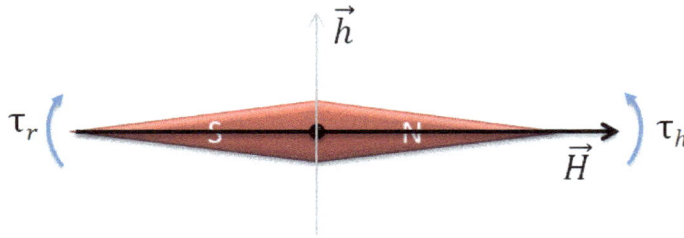

Figure 39.3: The top view of a magnetic needle.

By the rotational equivalent of Newton's 2nd law of motion, since the needle is in equilibrium, the sum of the torques acting on the needle must be 0,

$$\tau_r + \tau_h = 0. \tag{39.3}$$

Since h is proportional to the current i flowing in the wire, as seen in Biot-Savart's law, from the above equations,

$$\theta \propto i. \tag{39.4}$$

From the above equation, the angle the torsion head is turned, can be used as a relative measure of the strength of the current flow.

Before presenting the experiment results, Ohm's law will be intuitively derived. The expected results can be verified using experiments. For a given voltage source, it can be expected that the current flow depends on the wire dimensions and the material of the wire.

As the length of the wire increases, the wire would resist the flow of current more, as the charges would have to traverse a longer path, colliding with the atoms of the wire along the way, and generating heat in the process caused by friction. The experiment in Reference [28] proves that a current-carrying wire generates heat. If the cross section of the wire is greater, this offers more space in the wire for more current to flow. The material property of the wire would also determine how much current flows. If the material is a perfect insulator then no current flows, and high current, in the case of a conductor. If the voltage source has a higher strength, achieved by connecting the batteries in series, and using a tangent galvanometer, it can be shown that more current flows in the wire. By the above reasoning, it can be expected that

$$i \propto \frac{VA}{L},\tag{39.5}$$

where i is the current flowing in a wire of cross-section area A, length L, and V is the voltage of the battery. Ohm empirically verified the above relation.

Ohm used 8 copper strips, and cascaded them one by one. He noted how the strength of the current varied as a function of the length of the wire, using the thermoelectric voltage source, and the current measurement device explained earlier. He observed that

$$\theta = \frac{a}{b + L},\tag{39.6}$$

where L is the length of the wire, a and b are constants independent of the wire dimensions, to be explained shortly, and $\theta \propto i$ is the angle of the torsion head, which measures the relative current strength.

Ohm repeated the above experiment with a reduced temperature difference of the voltage source, thereby reducing its strength. He observed that b remained the same, but a reduced by 10×. This shows that a depends only on the voltage source, and b is the term accounting for the reduction in the current caused by the rest of the circuit other than L.

Using wires of different diameter, but keeping the length to its cross-section area ratio $L : A$ the same, the above experiment can be repeated to show that the same current flows, thereby confirming the relation

$$\theta = \frac{a}{b + \frac{L}{A}}.\tag{39.7}$$

The above equation may also be verified by keeping the length of the wire the same, but varying its cross-section area. The above equation will be rewritten in the present day notation later in

Equation 39.20.

b is the term that accounts for the reduction in the current in the rest of the circuit other than L, since its value stays constant when L is varied. Assuming that the rest of the circuit other than L is ideal, or nearly ideal, keeping the terms related only to L,

$$\theta = \frac{a}{\left(\frac{L}{A}\right)}. \tag{39.8}$$

a depends only on the voltage source, as noted earlier. The voltage source, such as a battery, is well defined using the definition of voltage, as explained in Section 19.1. a can be written as some scaling factor of the battery voltage V, or

$$a \propto V. \tag{39.9}$$

The voltage of a battery has a very specific definition. As shown earlier in Figure 19.1(a), by definition of voltage, a battery with voltage V satisfies the electric-field path integral relation in Equation 19.1. The definitions of the different electrical quantities may seem confusing, which quantity defines which? However, a flowchart summarizing the definitions of all the different electrical quantities will be presented in Chapter 58 and Chapter 59.

Since $\theta \propto i$, and from the above equations,

$$i \propto \frac{V}{\left(\frac{L}{A}\right)}. \tag{39.10}$$

In Equation 39.9, note that a is a linear function of V, and not any other relation such as V^2. This can be easily shown to be true by a thought experiment. In Figure 39.4(a), a battery B of voltage V, is connected to a capacitor C using a switch S. When S is closed, the battery charges the capacitor. The charge Q stored in the capacitor of capacitance C, is given by the relation,

$$Q = CV. \tag{39.11}$$

When the capacitor is discharged, as shown in Figure 39.4(b), current i flows in the wire, where

$$Q = \int_0^\infty i \, dt. \tag{39.12}$$

If the capacitor was to be charged using a battery of twice the voltage $2V$, the charge stored in the capacitor, using Equation 39.11, would be $2\times$ greater. When discharged, $2\times$ the current i has to flow in the wire, so that

$$2Q = \int_0^\infty 2i \, dt. \tag{39.13}$$

When the capacitor is charged to voltage V, the discharge current is i. When the capacitor is charged to $2V$, the discharge current is $2i$. A charged capacitor is similar to a battery, but unlike

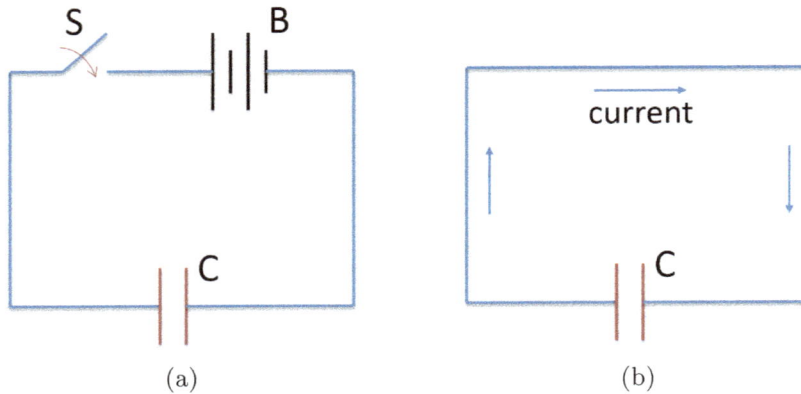

Figure 39.4: The (a) charging, and the (b) discharging, of a capacitor.

the battery, the voltage across the capacitor is not constant. This experiment shows that $i \propto V$, as captured in Equation 39.10.

The constant of proportionality in Equation 39.10 is σ,

$$i = \sigma \frac{V}{\left(\frac{L}{A}\right)}, \tag{39.14}$$

and is called the conductivity. Since L and A account for the dimensions of the wire, the only parameter unaccounted for is the material that the wire is made up of. σ is therefore, only a property of the material of the wire. Alternately, the above equation is written as

$$i = \frac{1}{\rho} \frac{V}{\left(\frac{L}{A}\right)}, \tag{39.15}$$

where

$$\rho = \frac{1}{\sigma} \tag{39.16}$$

is defined as the resistivity of the material. From the above equations, note that the conductivity of a metal must be high, but low for a non-metal, since current flows readily in a metal wire connected to a battery.

In the final form, Ohm's law is written as

$$i = \frac{V}{R} \tag{39.17}$$

$$R = \rho \frac{L}{A}. \tag{39.18}$$

In the above equation, measurement of the voltage of a battery V, resistivity of the wire ρ, and the resistance of a wire R, will be presented in later chapters.

In the case of time-varying voltages $V(t)$ and currents $i(t)$, Ohm's law will be assumed to be valid at any time t,

$$i(t) = \frac{V(t)}{R}.$$

(39.19)

39.4 Internal Resistance of a Voltage Source

The empirical result in Equation 39.7 can be explained by noting that the total resistance in a circuit, is the sum of the resistances cascaded in series. This can be proved after the discussion of Kirchoff's circuit laws in Chapter 47. In Equation 39.7, since b is a term that depends on the circuit other than L, in comparison to Equation 39.15, the equation is of the form,

$$i = \frac{V}{r + R},$$

(39.20)

where r is the internal resistance of the voltage source, and R is the resistance of the wire, as illustrated in Figure 39.5. The dotted line represents the current path in the thermoelectric voltage source of resistance r, and the solid line is the resistance R of the wire connecting the terminals of the voltage source. The total resistance of the series connection of the two is $r + R$.

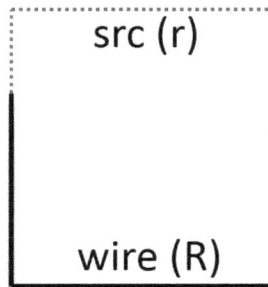

src (r)

wire (R)

Figure 39.5: The internal resistance of the voltage source, and the resistance of the wire in Ohm's experiment.

In the special case of two series wires of lengths L_1 and L_2 in Figure 39.6, but equal cross-section area and material, it will be shown that the total resistance of the wire is the sum of the individual resistances of the wire. Intuitively, it can be seen that this should also be valid for unlike wires cascaded together.

Using Equation 39.18, the resistance of a wire of length $L_1 + L_2$ is

$$\begin{aligned}
R &= \rho\frac{L_1 + L_2}{A} \\
&= \rho\frac{L_1}{A} + \rho\frac{L_2}{A} \\
&= R_1 + R_2,
\end{aligned}$$

(39.21)

where R_1 and R_2 are the resistances of the individual wires L_1 and L_2.

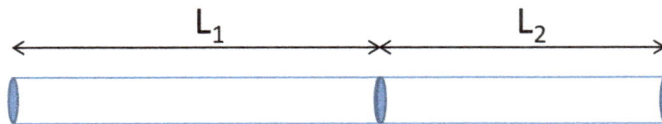

Figure 39.6: A wire of length L_1+L_2.

39.5 Alternate Form of Ohm's Law

It will become clear after the discussion of Faraday's law, that voltage is not well defined in time-varying fields. Fortunately, Ohm's law can be cast in a different form, without the use of voltage, that is also valid for time-varying fields.

A wire segment of length L, with uniform current i flowing in the direction shown by the arrow, is shown in Figure 39.7. The electric field $\vec{E}$ in the wire exerts a force on the charges, propelling the charges, creating a current. The direction of $\vec{E}$ is the same as the direction of the current flow. If $\vec{E}$ is uniform along the wire, the voltage between the end points of the segment

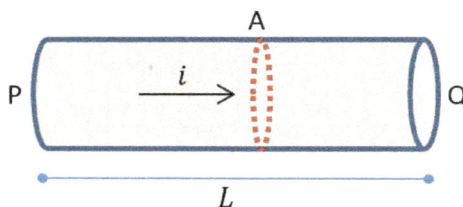

Figure 39.7: A conductor carrying uniform current i.

V, between P and Q, is

$$V = EL, \tag{39.22}$$

where E is the electric-field strength in the wire. If A is the cross-section area of the wire, i can be written in terms of volume current density J, repeating Equation 28.15,

$$J = \frac{i}{A}. \tag{39.23}$$

From Ohms' law,

$$i = \sigma \frac{VA}{L}. \tag{39.24}$$

From the above equations,

$$J = \sigma E. \tag{39.25}$$

The direction of the electric field $\vec{E}$, and the direction of current flow, and therefore, the direction of $\vec{J}$, are all the same. Therefore, writing the above equation in vector form,

$$\vec{J} = \sigma \vec{E}. \tag{39.26}$$

The above equation is Ohm's law cast in a different form, without the use of voltage, and written as vector fields that is valid at any point. For time-varying fields, it will be assumed that

$$\vec{J}(t) = \sigma \vec{E}(t), \tag{39.27}$$

where $\vec{J}(t)$ and $\vec{E}(t)$ are time-varying fields.

40

Time-Varying Magnetic Fields

Electrostatics, the study of stationary charges, and magnetostatics, the study of constant current flow, have been studied until now. The electric and magnetic fields caused by stationary charges and steady currents do not change over time. Time-varying magnetic fields will be studied in this chapter.

The experiment observations of time-varying magnetic fields were independently discovered by Michael Faraday and Joseph Henry during the years 1831-1832. Similar experiments will be presented in this chapter [66].

The mathematical construct formulated by Maxwell that describe the experiment results is the famous Faraday's law, and will be gradually developed in the upcoming chapters. A way by which the formulation of Faraday's law can be verified using the Kelvin-Ampere current balance, will be explained in Chapter 56.

40.1   Experiments on Time-Varying Magnetic Fields

Two circuit loops, shown in Figure 40.1(a), are arranged such that the loop areas are parallel to each other, and in close proximity. Each of the coils has hundreds of tightly wound turns. The coil connected to the battery B, marked P, is called the primary coil. The other coil marked S is the secondary coil. A galvanometer G is connected to the secondary coil to indicate if there exists an induced current flow.

When the switch W is open, no current flows in the primary and the secondary. However, when the switch W is closed, a transient current momentarily exists in the secondary coil. When the switch W is opened from its closed position, a transient current would be detected momentarily again in the secondary coil, but now flowing in the opposite direction. The induced current is stronger when there are more number of turns in the primary and/or the secondary coils.

Current flowing in a wire generates a magnetic field. A time-varying current therefore, creates a time-varying magnetic field. This experiment shows that a time-varying magnetic field generated by the time-varying current flow in the primary, induces a current in the secondary. Since no current flow is observed in the secondary when a steady current flows in the primary, it can be concluded that a time-varying magnetic field in the primary is a necessary condition to induce a current flow in the secondary.

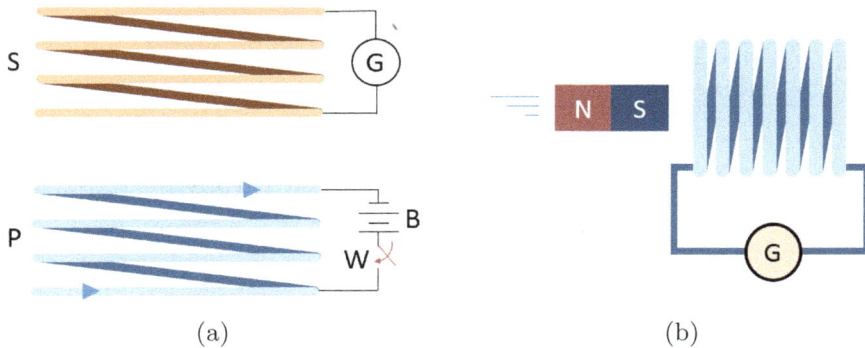

Figure 40.1: Experiments to demonstrate an induced current flow caused by a time-varying magnetic flux: (a) two parallel current coils, P and S, where each of the coils has hundreds of turns, and (b) a magnet moved in and out of a tightly-wound coil.

A classic experiment to demonstrate similar results is by using a magnet to induce a current flow in a coil of wire. A coil of wire with N-turns, where N is tens or hundreds of turns, connected to a galvanometer G is shown in Figure 40.1(b). If the south pole of the magnet, marked S, is moved in and out of the loop, a current flow is observed in the galvanometer. The direction of the current flow when the magnet is moved into the coil is opposite to that of the case, when the magnet is pulled out of the coil. If, however, the magnet is kept stationary within the coil, no current flow is observed. More current flows in the coil if the magnet is moved faster. This experiment confirms that a time-varying magnetic field is required to induce a current flow.

It would be observed that more number of turns in the coil amplifies the induced current. If the magnet is reversed so that the north pole, marked N, is moved in-out of the coil, similar result is observed, but the direction of the current flow is reversed.

40.2 Magnetic Flux

The observations made from the experiments is summarized in Table 40.1. Observation I in Table 40.1 states that a time-varying magnetic field within a loop of coil is a requirement to generate an induced current flow. The magnetic-field vector present at the loop area can be decomposed into a component parallel to the loop area, and the other perpendicular to the loop area. The experiment in Figure 40.2 can be used to show that the component of $\vec{H}$ perpendicular to the loop area is what that matters to induce a current in the secondary loop.

The magnetic-field pattern around a bar magnet was discussed in Figure 17.4(a). In Figure 40.2(a), the bar magnet is oriented so that its magnetic field is mostly perpendicular to the loop area.

	Observation
I	A changing magnetic field within a loop or coil is required to generate an induced current.
II	The more the number of turns in the primary coil, the higher the induced current.
III	The more the number of turns in the secondary coil, the higher the induced current.
IV	The faster the change in the magnetic field, the greater is the induced current.

Table 40.1: A summary of the conclusions that can be drawn from the experiments described in this section.

On the other hand, the magnetic field is mostly parallel to the loop area in Figure 40.2(b). The magnet is moved vertically in and out of the loop area.

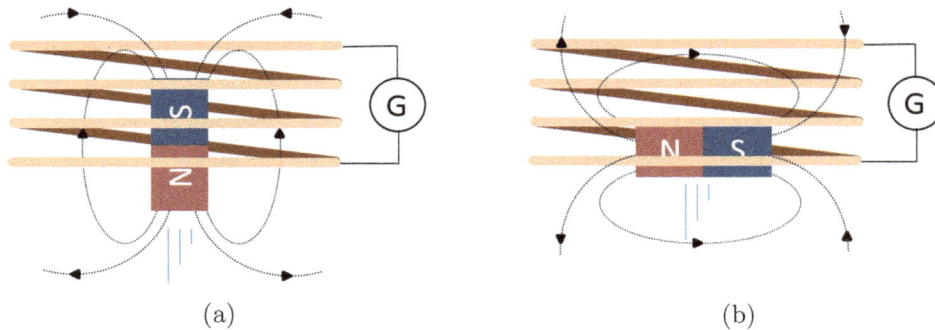

(a)   (b)

Figure 40.2: A magnet oriented so that its magnetic field component is mostly (a) perpendicular, and (b) parallel, to the loop area.

This experiment, performed using the list of items in Figure 40.3, shows that the current in Figure 40.2(a) is greater than the case in Figure 40.2(b) by over 50%. This result shows that the magnetic-field component perpendicular to the loop area is the contributing element to the induced current flow, and not the total magnitude of the magnetic field in the loop area, or the field component parallel to the loop area. The induced current flow in the case shown in Figure 40.2(b) is not 0, because magnetic field may not be perfectly horizontal to the loop area.

Another example to show that the magnetic-field component perpendicular to the loop area is what that matters to generate an induced current flow, can be demonstrated using the primary and secondary coils, as shown in Figure 40.4. In the case shown in Figure 40.4(a), the coils are parallel to each other, and arranged perpendicular in Figure 40.4(b).

In the first case, the magnetic field generated by the current in the primary coil (see Section 21.1) is mostly perpendicular to the loop area of the secondary, as shown. As a consequence, the induced current in the secondary coil is greater than the case in Figure 40.4(b), where the magnetic field

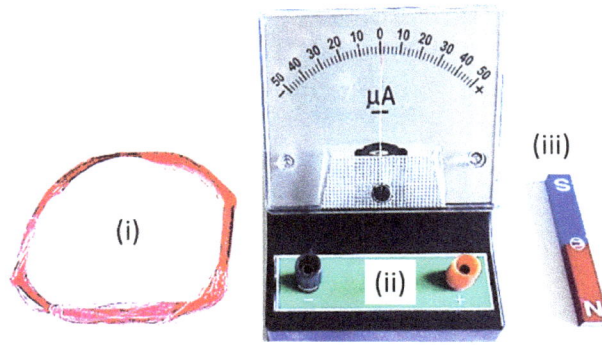

Figure 40.3: The articles (i) a 100-turn thin 36 AWG Copper wire, (ii) a galvanometer, and (iii) a bar magnet, to verify the experiment in Figure 40.2.

is mostly parallel to the secondary coil. This experiment can also be used to show that a larger coil area of the primary and secondary, results in a larger coupling of the field, and therefore, a higher induced current.

Biot-Savart's law was used to derive the magnetic field at the center of a 1-turn loop. If M turns are present, for the same current flowing in the coil, the magnetic field generated would be $M\times$ greater, neglecting the thickness of the coil. From Observation II in Table 40.1, increasing the number of turns in the primary coil strengthens the magnetic field generated by the primary coil, and coupling to the secondary coil. This shows that the stronger the magnetic field coupling, the higher the induced current flow.

A mathematical representation of the magnetic field within the loop area can be formulated. This can then be equated to the induced current, which is a mathematical description of the physical behavior.

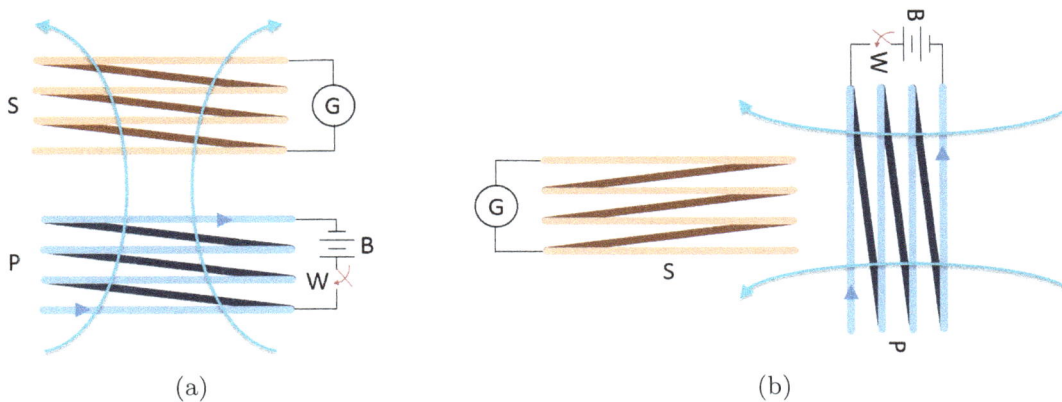

Figure 40.4: The arrangement of two coils P and S, such that the magnetic field generated by P is mostly (a) perpendicular, and (b) parallel, to S.

Looking at a menu of mathematical definitions, the definition of flux is ideally suited to capture the strength of the perpendicular component of the magnetic-field component through a loop area. By definition of flux, both the magnitude of the field strength, and the contribution of the perpendicular component of the magnetic field are taken into account, as explained next. This formulation will be verified using the Kelvin-Ampere current balance in Chapter 56.

The secondary loop in Figure 40.4(a) is drawn in Figure 40.5, with its loop area discretized into tiny area elements. The magnetic field at the center of the shaded element of area dA is marked $\vec{H}$, and $\hat{n}$ is the unit-normal vector to dA. The area elements are small enough to assume that $\vec{H}$ is uniform within dA. The component of $\vec{H}$ perpendicular to dA is $\vec{H} \cdot \hat{n}$.

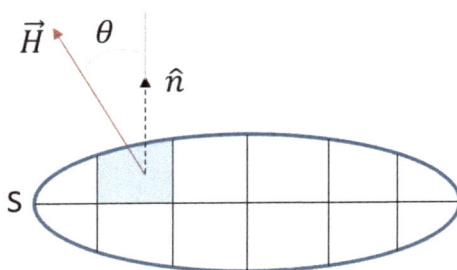

Figure 40.5: The secondary loop is discretized into tiny area elements to calculate the magnetic flux across the loop area.

The magnetic flux Φ caused by $\vec{H}$, over the entire loop area A is the surface integral,

$$\Phi = \int_A \vec{H} \cdot \hat{n} \, dA. \tag{40.1}$$

Although flux is not a vector, it is defined with respect to the direction of the normal vector to the loop area $\hat{n}$. In mathematical terms, using the above equation and from Observation I, it can be hypothesized that

$$i_{ind} \propto \frac{\partial \Phi}{\partial t}. \tag{40.2}$$

From Observation II, if M turns are present in the primary coil, the strength of the magnetic field that gets coupled to the secondary coil gets amplified $M\times$. This $M\times$ increase in the strength is captured in the calculation of Φ across the secondary coil.

If there are N turns in the secondary coil, the flux across the secondary coil Φ gets amplified $N\times$,

$$\Phi = N\Phi_s, \tag{40.3}$$

compared to the flux across a single turn Φ_s, since there are N turns that the magnetic field now couples to. Using the above relation, a higher flux Φ would result in a higher induced current, which is in agreement with Observation III.

The above equation also states that the induced current is higher with a faster rate of change of flux, which is the same as Observation IV. The definition of flux is ideally suited to capture all of the observations made in the experiments. This equation will be used as a starting point to formulate Faraday's law in the next few chapters.

40.3 Lenz's Law

A closer look at the experiments presented, shows that the direction of the induced current flow is such that it opposes the changing magnetic flux in the loop area. In other words, Lenz's law states that the induced current flows in a direction to "fight" against the time-varying magnetic field altering it. Lets revisit the example in Figure 40.4(a) to illustrate this point.

When the switch is closed, the direction of the current in the primary coil is marked as shown in Figure 40.6(a). The current increases from 0 to its steady-state value in a short-time duration. The magnetic field generated by the primary coil increases in the $+\hat{z}$ direction, as shown in the figure. The induced current flows in the secondary coil in the direction shown in Figure 40.6(b) to "fight" back this change in the flux. Once the steady-state value of the current has been reached in the primary coil, the magnetic field and the magnetic flux become constant, and the induced current in the secondary loop ceases to exist.

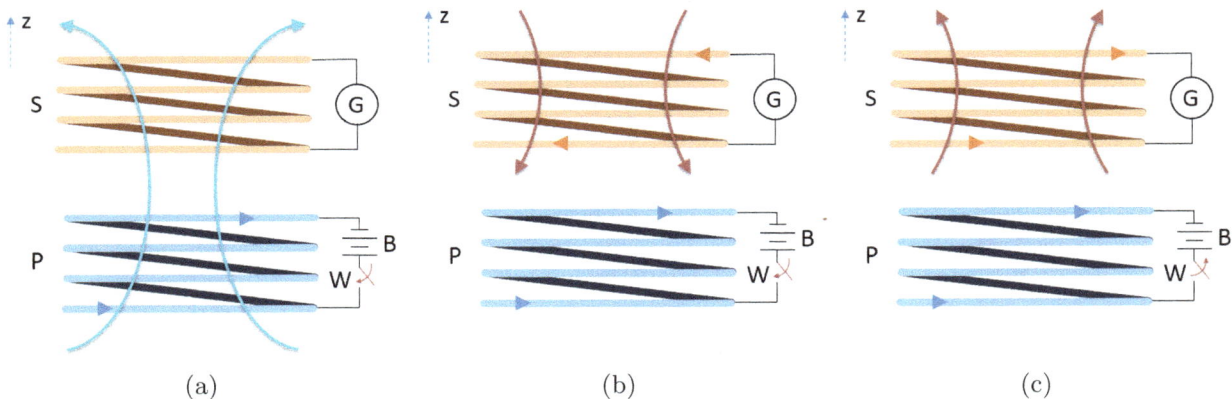

(a) (b) (c)

Figure 40.6: (a) The coupling of magnetic field from the primary to the secondary coil. (b) The induced current in the secondary coil when the switch W is closed. (c) The induced current flow direction is reversed when the switch is opened.

When the switch in the primary coil is opened, the current diminishes rapidly from its steady-state value to zero. The magnetic field and flux decrease in the $\hat{z}$ direction. The current in the secondary coil flows in the direction in Figure 40.6(c), thereby counteracting the decrease in the magnetic flux. This explains the reason for the induced current flow in the opposite direction compared to the case in Figure 40.6(b), when the switch in the primary is closed from open.

Similarly, Lenz's law can be used to explain the results of the experiment in Figure 40.1(b). When the magnet is moved into the coil, the magnetic field, illustrated by the dotted lines in Figure 40.7(a), is increasing within the coil in the direction marked a. The field directions may be opposite, as shown in Figure 40.7(b). However, the magnetic-field strength is strongest at the poles. Iron filings attracted to the magnet are mostly concentrated at the poles, showing that the magnetic-field strength is strongest at the poles. This can also be verified using Gauss's technique to measure the strength of a magnetic field. Overall, the flux increases within the coil in the direction marked a.

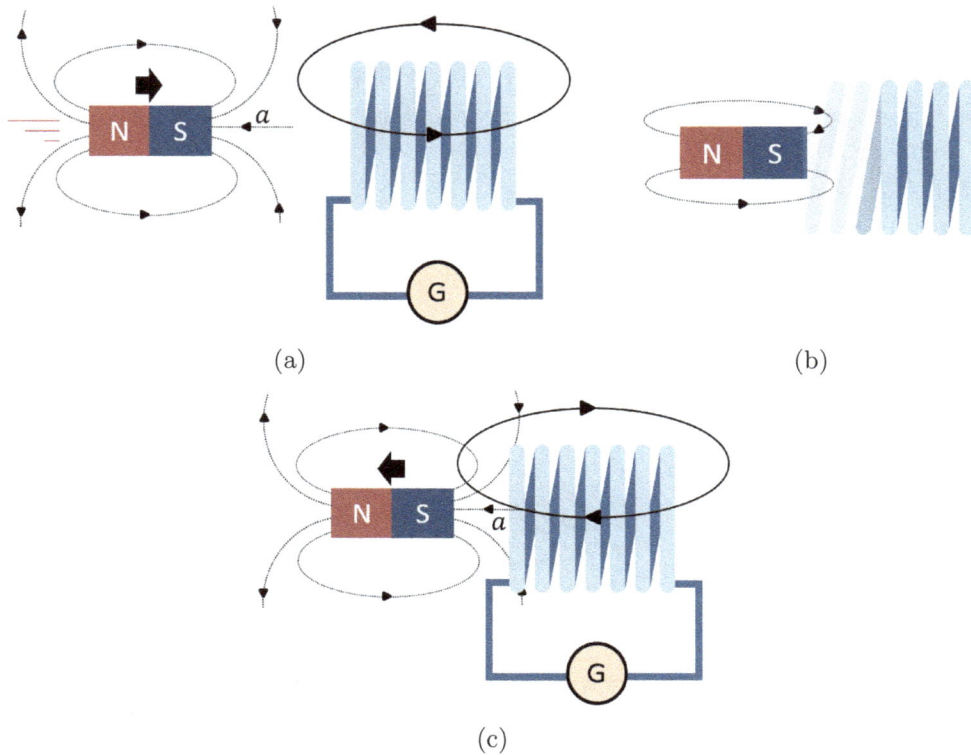

(a)   (b)

(c)

Figure 40.7: Lenz's law to determine the direction of the current flow in the coil. (a) The magnet is moved into the coil. (b) The magnetic-field lines of the bar magnet. (c) The magnet is moved out of the coil.

The current in the coil opposes this increase in the flux, as shown in Figure 40.7(a). The magnetic field generated by the induced current in the coil is indicated by the solid bold line. When the magnet is removed from the coil, shown in Figure 40.7(c), the flux within the coil decreases in the direction marked a. The induced current now flows in the opposite direction compared to the previous case, to oppose the decreasing flux, and increasing the flux in the direction marked a.

41

Magnetic Materials

The influence of a dielectric material on an electric field was discussed in detail in Chapter 37. Likewise, a material may also affect the magnetic field present in the material. Besides the magnetic field $\vec{H}$, a new electrical quantity $\vec{B}$, the magnetic-flux density, will be introduced, accounting for modifications to the field by a material. Such materials are known as magnetic materials.

A simplified version of the experiment from Reference [67], will be presented first that shows the need for a new quantity $\vec{B}$, besides $\vec{H}$. The complete version of the experiment will be presented in Chapter 49, after presenting the prerequisites. Different types of magnetic materials will be described in this chapter.

41.1 Ballistic-Galvanometer Experiment

The motivation for introducing a new electrical quantity $\vec{B}$, will be presented in this section. The early name of $\vec{B}$ is magnetic induction, but now known as magnetic-flux density, and the reason for their names will become apparent in future sections.

The primary coil, marked P, in Figure 41.1 is wound around a toroid, and is connected to a battery B via switch W. The experiment is carried out on a toroid made of iron and wood, as examples of a magnetic and a non-magnetic material, respectively. A secondary coil S with many turns is wound over the primary coil in the highlighted region.

S is connected to a ballistic galvanometer G, whose throw of the needle is a measure of the strength of the induced current. The reason for using a ballistic galvanometer, instead of a tangent galvanometer, will become clear after the complete version of the experiment is presented in Chapter 49.

A top view of the horizontal slice of the toroid is shown in Figure 41.2. The tightly-wound turns of the primary coil can be thought of as a collection of single-turn coils, as illustrated in Figure 41.2(a) by the dotted lines, and sparsely drawn to show the derivation of the magnetic-field pattern.

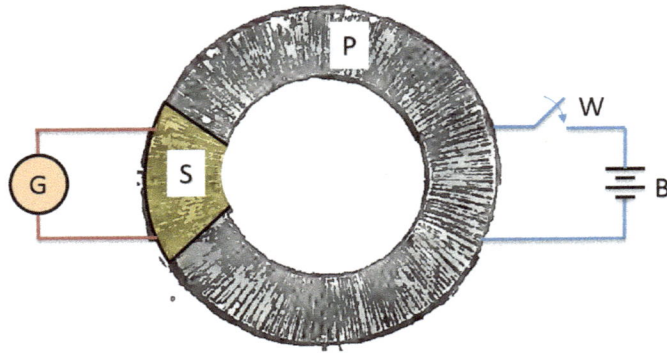

Figure 41.1: An experiment to show the effect of a magnetic material on an applied magnetic field.

The magnetic field at the center of each of the turns of the coil, using Figure 21.4(b), is shown by the arrows. The directions of the magnetic field have been determined using the right-hand rule in Section 21.2. From Figure 41.2(a), intuitively, it can be seen that in a tightly wound primary coil, the magnetic field within the toroid is shown by the dashed circle in Figure 41.2(b). Such a pattern may also be determined numerically, by calculating the field using Biot-Savart's law.

When the switch W is closed, the time-varying current in the primary coil sets up a time-varying magnetic field, which couples to the secondary coil, since S is wound over P in the shaded region in Figure 41.1. From Equation 40.2, the time-varying flux in the secondary coil induces a transient current in the secondary coil, which can be detected from the throw of the needle in the ballistic galvanometer.

If the toroid is made up of a magnetic material such as iron, the transient current in the sec-

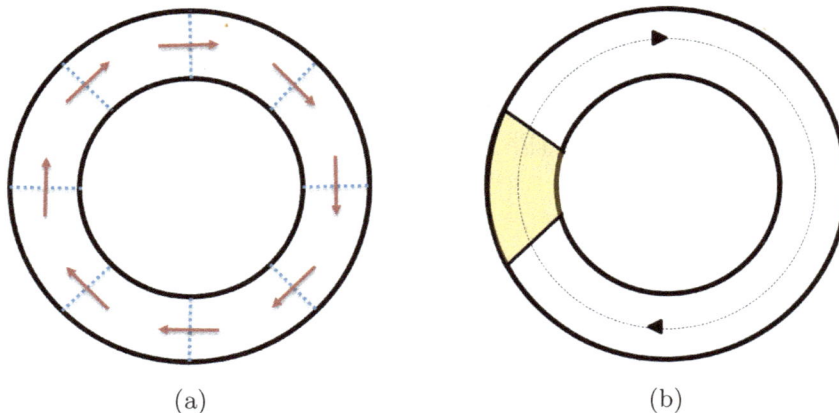

(a)   (b)

Figure 41.2: (a) The magnetic field at the center of single-turn coils. (b) The magnetic-field pattern in a tightly wound toroid.

284

ondary coil when W is closed, captured by the throw of the galvanometer's needle, is different from the case when the toroid is made up of a non-magnetic material such as wood. Multiple batteries may be connected in series to increase its strength, as explained earlier in Section 19.2. The same conclusion would be reached for different values of the voltage source connected to the primary. Also, the induced current with a non-magnetic toroid, would be the same as the absence of any material in the toroid, or when the toroid is air.

41.2   Magnetic-Flux Density $\vec{B}$

From the experiment in the previous section, using Equation 40.2, it can be concluded that the flux across the secondary coil must be different between a magnetic and a non-magnetic material. This shows that a magnetic material modifies the strength of the magnetic field.

The modified magnetic field at a point, by the material present at that point, is denoted as $\vec{B}$. Its early name is magnetic induction, since it is the field that generates an induced current caused by a time-varying magnetic flux. The field that would be present at a point, if the material at that point does not modify the applied magnetic field, is $\vec{H}$. The source of the applied magnetic field at a point can be from a current-carrying wire, as seen in the experiment described, and/or the magnetic field from a magnetized material near the point.

B •

A •

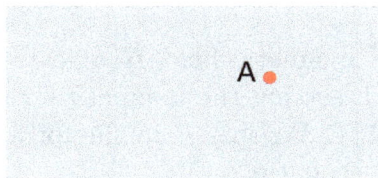

Figure 41.3: Two points A and B in different materials.

$\vec{B}$ is written as a scaled factor of $\vec{H}$,

$$\vec{B} = \mu_r \vec{H}, \qquad\qquad (41.1, \text{EMU})$$

where μ_r is called relative permeability. μ_r has no units in electromagnetic units, and only scales $\vec{H}$. Therefore, $[\vec{B}]$ and $[\vec{H}]$ have the same units in EMU.

Note that although $\vec{B}$ may be called the magnetic-flux density or magnetic induction, it is nevertheless a magnetic field. In electromagnetic units, $\vec{B}$ and $\vec{H}$ have the same units, and from Equation 14.2,

$$[\vec{B}] = \frac{gm^{1/2}}{cm^{1/2}\,s} = \text{gauss} \qquad\qquad (41.2, \text{EMU})$$

$$[\vec{H}] = \frac{gm^{1/2}}{cm^{1/2}\,s} = \text{oersted}. \qquad\qquad (41.3, \text{EMU})$$

285

However, the aliases gauss and oersted are used to differentiate between the two.

The value of μ_r relating $\vec{B}$ and $\vec{H}$ depends on the material present at a point. For example, if the relative permeability of the material at Point A in Figure 41.3 is μ_r^A,

$$\vec{B}(\text{Point A}) = \mu_r^A \, \vec{H}(\text{Point A}). \qquad (41.4, \text{EMU})$$

At Point B however, if the material is air, which is non-magnetic, $\mu_r = 1$, and therefore,

$$\vec{B}(\text{Point B}) = \vec{H}(\text{Point B}). \qquad (41.5, \text{EMU})$$

41.3 Magnetization Vector

An alternate to Equation 41.1 is to write $\vec{B}$ as a sum of two vectors, rather than scaling $\vec{H}$ by a factor of μ_r. Symmetric to Equation 37.1, $\vec{B}$ can be written as

$$\vec{B} = \vec{H} + 4\pi\vec{I}, \qquad (41.6, \text{EMU})$$

where $\vec{I}$ was known in the past as the intensity of magnetization vector, but now is called the magnetization vector. In the above equation, $\vec{H}$ is the magnetic field that would exist at a point, if the material at that point does not modify the field. The contribution of the material to the field at that point is $4\pi\vec{I}$, and the total field at the point $\vec{B}$, is the sum of the two, as written in the above equation.

Although there may be no applied magnetic field $\vec{H}$ at a point in a magnetic material, if the material is permanently magnetized, setting the magnetization vector $\vec{I}$ to a non-zero value, the magnetic field at the point is $\vec{B} = 4\pi\vec{I}$. Writing $\vec{B}$ in the form shown above has the advantage of being able to capture this type of a scenario.

41.4 Magnetic Susceptibility

If the magnetization vector $\vec{I}$ is written as

$$\vec{I} = \chi_m\vec{H}, \qquad (41.7, \text{EMU})$$

where χ_m is the magnetic susceptibility, Equation 41.6 becomes

$$\vec{B} = (1 + 4\pi\chi_m)\,\vec{H}. \qquad (41.8, \text{EMU})$$

Equating the above relation to Equation 41.1,

$$\mu_r = 1 + 4\pi\chi_m. \qquad (41.9)$$

The advantage of writing the relative permeability in terms of the susceptibility is that for values of μ_r that are approximately equal to 1, but not exactly equal to 1, such as 1.0000001 or 0.9999999, it is convenient to express the relative permeability in terms of the susceptibility.

41.5 Non-linear and Anisotropic Materials

In a non-linear material, the value of the permeability μ_r depends on the magnitude of the applied magnetic field $\vec{H}$ in the material. It will be shown by experiment in Chapter 49 that magnetic materials are highly non-linear. It will be shown that the value of μ_r depends on the history of the magnetic field $\vec{H}$ that has been applied in the material. This behavior is known as hysteresis.

In an isotropic material, the relative permeability is independent of the direction of $\vec{H}$. Similar to Equation 37.8, the relation between $\vec{B}$ and $\vec{H}$ in an isotropic material can be written as

$$\begin{pmatrix} B_x \\ B_y \\ B_z \end{pmatrix} = \begin{bmatrix} \mu_r & 0 & 0 \\ 0 & \mu_r & 0 \\ 0 & 0 & \mu_r \end{bmatrix} \begin{pmatrix} H_x \\ H_y \\ H_z \end{pmatrix}. \tag{41.10, EMU}$$

The above equation may be generalized so that the relative permeability has different values for different directions of $\vec{H}$,

$$\begin{pmatrix} B_x \\ B_y \\ B_z \end{pmatrix} = \begin{bmatrix} \mu_r^x & 0 & 0 \\ 0 & \mu_r^y & 0 \\ 0 & 0 & \mu_r^z \end{bmatrix} \begin{pmatrix} H_x \\ H_y \\ H_z \end{pmatrix}. \tag{41.11, EMU}$$

The permeability matrix written in the form above is known as the permeability tensor of rank 2. In the above equation, μ_r^x scales only the x-component of $\vec{H}$, H_x, to generate B_x, etc. Including off-diagonal elements in the permeability matrix, this restriction can be removed. There may exist a material, where H_x results in B_y. The permeability matrix enables the modeling of such materials.

In Equation 41.1, μ_r only scales the magnitude of the applied field $\vec{H}$, but keeping the direction of $\vec{B}$ the same as $\vec{H}$. Modeling the material with a permeability matrix gives the flexibility to scale the magnitude of $\vec{H}$, and modify its direction. The relation between $\vec{B}$ and $\vec{H}$ is

$$\vec{B} = \bar{\bar{\mu}}_r \vec{H}, \tag{41.12, EMU}$$

where $\bar{\bar{\mu}}_r$ is the permeability tensor.

Writing Equation 41.7 as a matrix, the susceptibility tensor $\bar{\bar{\chi}}_m$ is defined as

$$\vec{I} = \bar{\bar{\chi}}_m \vec{H}, \tag{41.13, EMU}$$

where $\vec{I}$ is the intensity of magnetization vector. Substituting the above expression in Equation 41.6,

$$\vec{B} = \vec{H} + 4\pi \bar{\bar{\chi}}_m \vec{H}$$
$$= (I + 4\pi \bar{\bar{\chi}}_m) \vec{H}, \tag{41.14, EMU}$$

where I is the 3×3 identity matrix. Equating the above equation to Equation 41.12,

$$\bar{\bar{\mu}}_r = I + 4\pi \bar{\bar{\chi}}_m. \tag{41.15}$$

41.6 Magnetization Vector $\vec{I}$ [Optional]

Equation 41.6 is similar to Equation 37.1,

$$\vec{D} = \vec{E} + 4\pi\vec{P} \tag{41.16}$$

$$\vec{B} = \vec{H} + 4\pi\vec{I}. \tag{41.17}$$

Likewise, Equation 41.1 and Equation 36.7 are similar,

$$\vec{B} = \mu_r\vec{H} \tag{41.18}$$

$$\vec{D} = \epsilon_r\vec{E}. \tag{41.19}$$

From these equations, it can be seen that $(\vec{B}, \vec{D})$, $(\vec{H}, \vec{E})$, and $(\vec{I}, \vec{P})$ are symmetrical. It was shown earlier that the polarization vector $\vec{P}$ can be defined as the average electric dipole-moment per unit volume. Analogous to the definition of $\vec{P}$, $\vec{I}$ is defined as the average magnetic dipole-moment per unit volume.

Similar to an electric dipole, as explained in Section 14.1, if a magnetic dipole made up of charges $\pm q_m$ is separated by distance ℓ, where $\vec{\ell}$ is the distance vector from $-q_m$ to $+q_m$, the magnetic dipole moment is

$$\vec{m} = q_m\vec{\ell}. \tag{41.20}$$

The average dipole moment per unit volume, similar to Equation 37.23, can be rigorously defined as

$$\vec{I} = \frac{\frac{1}{N}\sum_{i=1}^{N}\vec{m}_i}{\Delta V}, \tag{41.21}$$

where $\vec{m}_i$ is the ith dipole moment, and N is the number of dipoles in volume ΔV. Applying dimensional analysis in EMU in the above equation, the unit of $[\vec{I}]$ is the same as magnetic field $[\vec{H}]$ and magnetic-flux density $[\vec{B}]$, consistent with Equation 41.6. Of course, the above equation doesn't mean that one can count the dipoles, and compute the average dipole-moment value in a volume to calculate $\vec{I}$. The above definition provides a different interpretation of the magnetization vector.

The top view of a toroid is shown in Figure 41.4(a), and the applied magnetic field in the toroid created by the primary coil is $\vec{H}_a$. The toroid is discretized into tiny cuboids, a few of which are shown within the dotted rectangle. The magnetic material making up the toroid is modeled with positive and negative magnetic charges, similar to the positive and negative electric charges of a dielectric material in Figure 37.1. The applied magnetic field in the direction shown, exerts a force on the positive magnetic charge in the same direction as $\vec{H}_a$, and in the opposite direction on the negative magnetic charge, creating a net magnetic dipole moment, as shown in one of the volume elements in Figure 41.4(b).

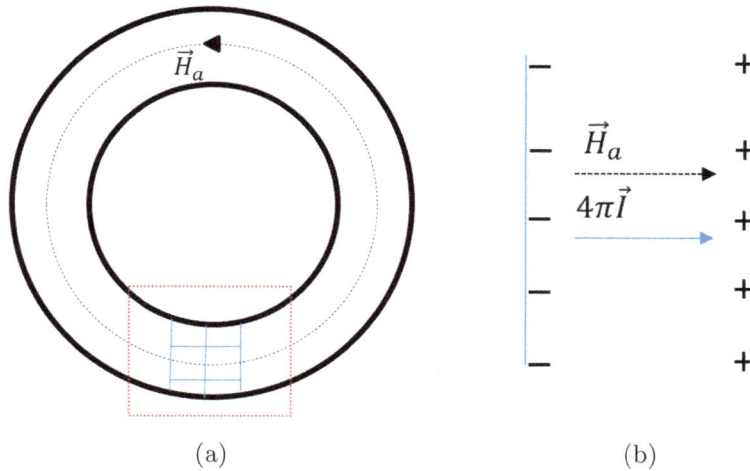

Figure 41.4: (a) A toroid discretized into volume elements. (b) The magnetic-flux density within a volume element.

The magnetic-flux density at the center of the discretized element in Figure 41.4(b) is

$$\vec{B} = \vec{H}_a + 4\pi\vec{I}, \qquad (41.22,\ \text{EMU})$$

where $\vec{H}_a$ is the applied magnetic field at the center of the volume element, and $\vec{I}$ is the average magnetic dipole moment per unit volume in the volume element. This model is a different perspective of $\vec{B}$.

41.7 Revisiting Ampere's Law to Verify the Formulation

A new electrical quantity $\vec{B}$, to account for the effect of a magnetic material on an applied magnetic field was introduced in this chapter. As stated earlier, $\vec{B}$ and $\vec{H}$ have the same electromagnetic units. Although they may be called by different names, nonetheless, $\vec{B}$ is also a magnetic field. Could it be that the path integral in Ampere's law in Equation 21.8, needs to be written as

$$\oint \vec{B} \cdot \vec{dl}, \qquad (41.23)$$

rather than

$$\oint \vec{H} \cdot \vec{dl} \ ? \qquad (41.24)$$

It will be shown in the remainder of this section that this is not the case. Ampere's law in Equation 21.8 does not require such a modification as stated above.

Two wires carrying current i are shown in Figure 41.5. Two cases are studied: In Case 1 in Figure 41.5(a), the material medium is air, but in Case 2 shown in Figure 41.5(b), the medium is a linear, isotropic magnetic material of relative permeability μ_r. The fields in Case 1 are denoted as $\vec{B}_1$ and $\vec{H}$, and in the second case, $\vec{B}_2$ and $\vec{H}$. Since the wires carry current i in both the cases, the magnetic field $\vec{H}$ generated by the currents are equal.

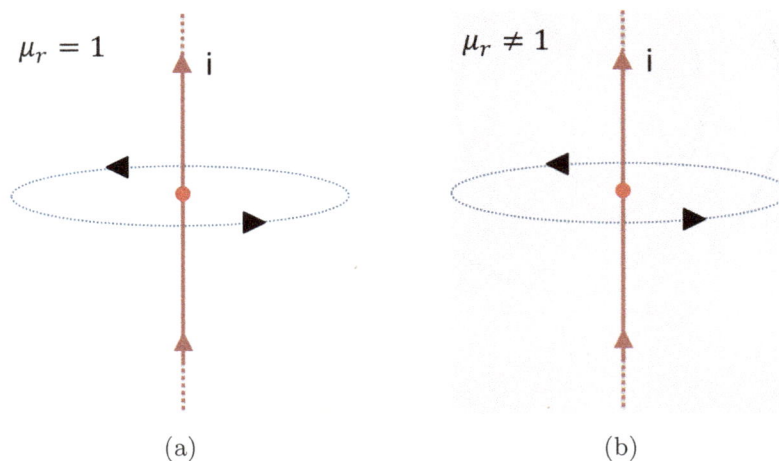

Figure 41.5: A current-carrying wire in two different media, (a) $\mu_r = 1$, and (b) $\mu_r \neq 1$.

In Case 1, since no magnetic material is present, using Equation 41.1,

$$\vec{B}_1 = \vec{H}, \qquad \text{(41.25, EMU)}$$

and in Case 2,

$$\vec{B}_2 = \mu_r \vec{H}. \qquad \text{(41.26, EMU)}$$

If the path integral of Ampere's law is formulated as Equation 41.23, the result of the path integral in Case 1 is different from that of Case 2, since $\vec{B}_1 \neq \vec{B}_2$, and the enclosed current by the Ampere loop in each of the cases are not equal. This clearly is incorrect, since equal currents flow in both the wires.

The $\vec{H}$ field, by definition, is the field without the material effect, and therefore, is related only to the current generating the field. The path integrals in Ampere's law, written as Equation 41.24, are the same in both the cases, and therefore, the current enclosed by the paths in the two cases are also equal, as it should be. This exercise shows that the path integral in Ampere's law, written as Equation 41.24, is the correct formulation.

42

Faraday's Law

The results of Faraday's experiments in Chapter 40, when translated to the language of mathematics is known as Faraday's law, which is the focus of this chapter. Part of this translation was presented earlier in Equation 40.2. It was shown in Chapter 41 that magnetic materials modify the strength of a magnetic field. Taking this into account, Faraday's law will be derived.

42.1 Revised Definition of Flux

The modified magnetic field at a point by a magnetic material is magnetic-flux density $\vec{B}$. Repeating Equation 41.1,

$$\vec{B} = \mu_r \vec{H}, \tag{42.1, EMU}$$

where $\vec{H}$ is the field at a point that is unmodified by the material at the point.

In the case shown in Figure 41.1, where the secondary coil is wound around a magnetic material, the magnetic field generated by the primary current that couples to the secondary coil is modified by the magnetic material. As noted in the experiment in Figure 41.1, the response of the ballistic galvanometer is different between the cases where the toroid is a magnetic and a non-magnetic material. The flux that couples to a loop, therefore, is written in general in terms of $\vec{B}$, rather than $\vec{H}$ in Equation 40.1,

$$\Phi = \int_A \vec{B} \cdot \hat{n}\, dA, \tag{42.2}$$

where A is the area of the loop, and $\hat{n}$ is the unit-normal vector to the area element dA. In the special case of a non-magnetic material $\vec{B} = \vec{H}$ in EMU, and the above equation reduces to Equation 40.1.

42.2 Faraday's Law

A loop of wire is shown in Figure 42.1(a). As seen in Faraday's experiments, a time-varying flux $\frac{\partial \Phi_s}{\partial t}$ across the loop area, generates a current flow i_{ind} in the wire. The subscript s stands for a single turn. The N-turn coil will be analyzed in the following chapter. Summarizing the previous conclusions, the current flows in a direction that opposes the changing magnetic flux, known as Lenz's law. The direction of current flow, for example, is shown by the arrows in the figure.

Figure 42.1(a) can be represented as a circuit, shown in Figure 42.1(b). The source of the induced current is captured using a voltage source V_{ind}, and R is the resistance of the wire. Measurement of the resistance of a wire will be discussed later, but the absolute value of the resistance is not required to be known at this point. V_{ind} will be defined in a moment. The polarity of V_{ind} is consistent with the clockwise direction of the current flow.

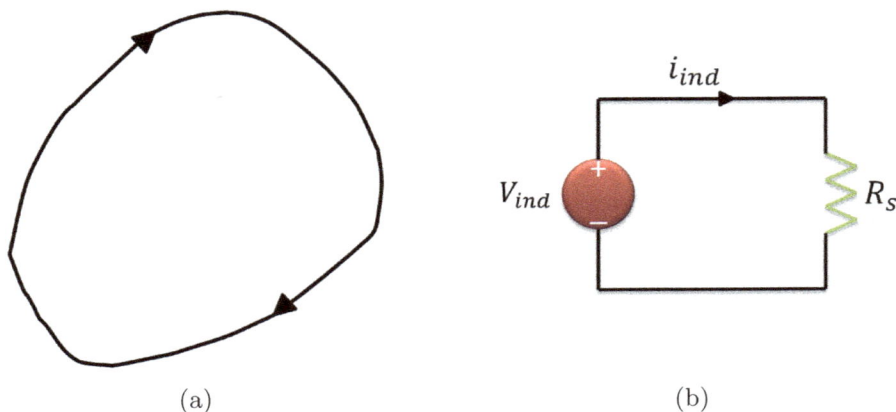

Figure 42.1: (a) A current loop. (b) The circuit representation of the current loop.

Repeating Equation 40.2,

$$i_{ind} \propto \frac{\partial \Phi_s}{\partial t}.$$

(42.3)

The induced current i_{ind} also depends on the wire properties, such as the material and its geometric dimensions, as noted in Ohm's law. If, for example, the loop of wire is made of a dielectric material, instead of a metal, then no current flow will be detected in the wire. From Equation 39.17,

$$i_{ind} \propto \frac{1}{R_s},$$

(42.4)

where R_s is the resistance of the loop of wire. It will be shown in Section 42.4, the value of R_s derived from Faraday's law, is the same result as the resistance of the loop of wire derived in Equation 39.18. From the above two equations,

$$i_{ind} \propto \frac{1}{R_s} \frac{\partial \Phi_s}{\partial t}.$$

(42.5)

This equation is of the form written in Ohm's law,

$$i_{ind} = \frac{V_{ind}}{R_s}.$$

(42.6)

From the above two equations, V_{ind} can be formulated as

$$V_{ind} \propto \frac{\partial \Phi_s}{\partial t}. \tag{42.7}$$

It will be shown in Chapter 46, the above equation will be used in the measurement of resistance. Any proportionality constant may be chosen in the above equation. The chosen proportionality constant will set the value of the wire resistance R_s, or the resistivity of the material in the wire. For simplicity, a proportionality constant of 1 is chosen, resulting in

$$V_{ind} = \frac{\partial \Phi_s}{\partial t}. \tag{42.8}$$

The above equation relates voltage and time-varying magnetic flux, which will be used to derive Faraday's law.

If a current flows in a wire, an electric field must exist in the wire. The electric field is in the same direction as the current flow, since the electric field exerts a force on the charges, and causing them to flow. By convention, it is the positive charges that are assumed to flow. In the example shown in Figure 42.1(a), the electric field is also in the clockwise direction, redrawn in Figure 42.2.

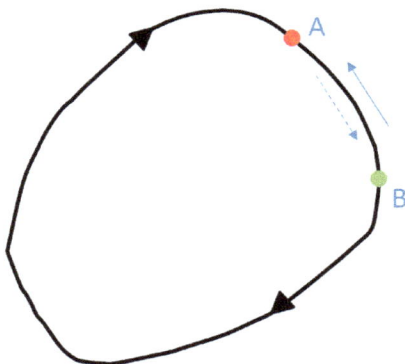

Figure 42.2: A current loop, and any two different points on the loop A and B.

Using the definition of voltage as the path integral of the electric field, Equation 42.8 may be written as

$$\int \vec{E} \cdot \vec{dl} = \frac{\partial \Phi_s}{\partial t}, \tag{42.9}$$

where $\vec{E}$ is the electric field that exists in the wire.

The path integral of the electric field may be evaluated between two points A and B, shown in Figure 42.2. There are two choices for the direction of the path integral: in the clockwise direction from A to B, or counter clockwise B to A. Likewise, there are two options for the direction of the unit-normal vector to calculate the magnetic flux Φ_s: into the page, or out of the page. From calculus, in any path integral of a vector field,

$$\int_A^B \vec{E} \cdot d\vec{l} = -\int_B^A \vec{E} \cdot d\vec{l}, \tag{42.10}$$

and in the case of a surface integral,

$$\int_A \vec{B} \cdot \hat{n}\, dA = -\int_A \vec{B} \cdot (-\hat{n})\, dA. \tag{42.11}$$

From the above equations, note that picking inconsistent directions for either of these two choices would make the signs differ between the left-hand side and the right-hand side of Equation 42.9. Therefore, picking the direction of the path integral would fix the unit-normal vector direction to calculate the flux, and conversely.

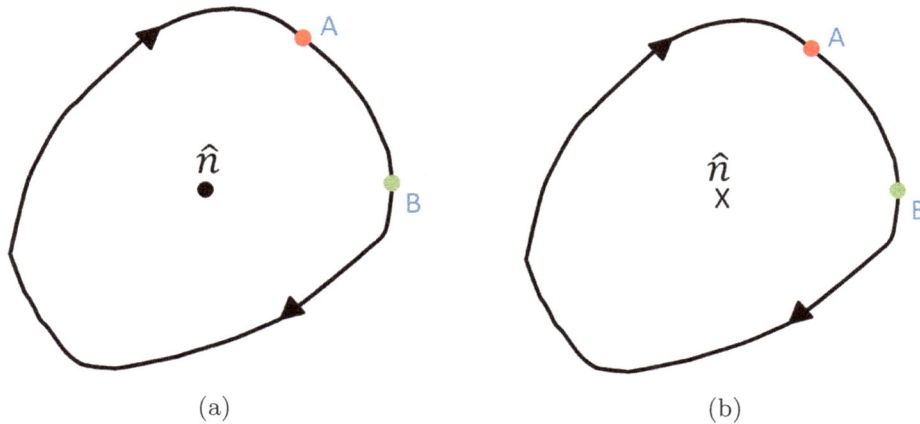

(a)   (b)

Figure 42.3: (a) The left-hand rule of Lenz's law corresponding to Equation 42.9. (b) The right-hand rule of Lenz's law corresponding to Equation 42.14.

Lets look at an example to illustrate this point. A wire is shown in Figure 42.3(a), and let the positive direction of the unit-normal vector $\hat{n}$ be out of the page, as marked by the solid circle. Lets say that the magnetic flux through the loop area is also increasing in the direction of $\hat{n}$, resulting in

$$\frac{\partial \Phi_s}{\partial t} > 0. \tag{42.12}$$

By Lenz's law, the induced current flow, and therefore, also the induced electric field, are in the clockwise direction to counteract the increasing magnetic flux. The path of integration must also be in the clockwise direction from A to B, as marked by the arrows, so that

$$\int_A^B \vec{E} \cdot d\vec{l} > 0, \tag{42.13}$$

294

and the signs of both sides of Equation 42.9 match.

From this example, a left-hand rule can be created, which states that if the thumb of the left hand points in the direction of $\hat{n}$, the direction chosen for the unit normal area vector to calculate the flux, then the fingers curl in the direction in which the path integral of the electric field needs to be evaluated. There are two possible directions for the path integral of the electric field, two for picking the direction to calculate the flux $\hat{n}$, and the magnetic flux may increase or decrease in the direction of $\hat{n}$, for a total of $2 \times 2 \times 2 = 8$ possible cases. It is left as an exercise for the reader to verify that the left-hand rule correctly captures the signs on both sides of Equation 42.9 on all these 8 possible scenarios.

The Kelvin-Stokes theorem will be used convert Faraday's law in the integral-equation form to the differential form, later in the chapter. The Kelvin-Stokes theorem uses the right-hand rule to relate the direction of the unit normal area vector to calculate the flux across an area, to the direction of the path integral. The reader is referred to a calculus textbook for more details on the Kelvin-Stokes theorem. The left-hand rule convention described earlier is inconsistent with the Kelvin-Stokes theorem, and the result of the inconsistency between the two is that the integral and the differential forms would differ by a negative sign.

To resolve the discrepancy between the integral and the differential forms, the left-hand rule proposed for Equation 42.9, can be converted into a right-hand rule by adding a negative sign to the equation,

$$\int \vec{E} \cdot \vec{dl} = -\frac{\partial \Phi_s}{\partial t}. \tag{42.14}$$

The right-hand rule for the above equation states that if the thumb of the right hand points in the direction chosen for the unit normal area vector of the loop area, the fingers curl in the direction in which the path integral of the electric field is to be evaluated. Conversely, if the fingers of the right hand curl in the direction chosen for the path integral of the electric field, the thumb points in the direction of the unit normal area vector to calculate the flux.

By the right-hand rule, if the path of integration is in the clockwise direction, as illustrated in Figure 42.3(b), the unit normal area vector must be pointing into the page, marked by $\times$. The unit normal area vector is in the opposite direction compared to the left-hand rule in Figure 42.3(a).

The question that would naturally arise is between which two points does the path integral on the left-hand side of Equation 42.14 needs to be integrated? The conservation of energy provides the remaining clue on how to complete Equation 42.14.

By the law of conservation of energy, energy is neither created nor destroyed, but converted to different forms. This is stated mathematically as

$$\Delta E_1 + \Delta E_2 + ... + \Delta E_N = 0, \tag{42.15}$$

where $\Delta E_1, \Delta E_2, ..., \Delta E_N$ are the exchanges in energies that occur in a system during any time interval. The change in energy ΔE_i during a time interval $[t_1, t_2]$ is

$$\Delta E_i = E_f - E_i, \tag{42.16}$$

where E_f and E_i are the final and the initial energies at the end, and at the beginning of the time interval, at t_2 and t_1. A positive ΔE_i means energy gain, and energy loss for a negative value.

A simple example to explain Equation 42.15 will now be presented. Lets say that an object falls from some height to ground in a certain time interval. If ΔE_1 represents the change in the object's gravitational potential energy, and ΔE_2 is the change in the object's kinetic energy, by the law of conservation of energy,

$$\Delta E_1 + \Delta E_2 = 0 \tag{42.17}$$

must be satisfied. As the object falls, its gravitational potential energy decreases and its kinetic energy increases. Therefore, ΔE_1 is a negative value, and ΔE_2 positive, and the above equation must be satisfied.

The current flowing in the wire loop ℓ would mean that a hypothesis can be made that an electric field exists in the wire in a *closed loop*, causing the charges to flow in the loop. Mathematically, this can be stated as

$$\oint_\ell \vec{E} \cdot \vec{dl} \neq 0. \tag{42.18}$$

Lets take a snapshot of the field pattern at some instant, and analyze it. As explained in Section 6.5, the work W done in moving a charge q in a closed loop, is given by the path integral

$$W = -q \oint \vec{E} \cdot \vec{dl}. \tag{42.19}$$

Similar to Section 6.5, the direction of the path integral is chosen such that the result of the above path integral is negative. This means that energy is gained in moving the charge q in a full loop, and the energy gain E_{gain} is

$$E_{gain} = q \oint_\ell \vec{E} \cdot \vec{dl}. \tag{42.20}$$

As reasoned in Section 6.5, if there is a net gain in energy when a charge traverses a closed loop and returns to its starting point, then energy must be lost by something else for the gain to occur. Equation 42.14 shows the way to justify the non-zero value of the above equation, unlike the electrostatic case, where the electric field is a conservative field.

Multiplying both sides of Equation 42.14 by the unit charge,

$$(\text{unit charge}) \int_\ell \vec{E} \cdot \vec{dl} = -(\text{unit charge}) \frac{\partial \Phi_s}{\partial t}. \tag{42.21}$$

From Table 25.1 and Table 26.1, it is left as an exercise for the reader to verify that the left-hand side of the above equation has the same unit as energy. From dimensional analysis, the right-hand side must also have the unit of energy. There is energy spent to create the time rate of change of magnetic flux, and the right-hand side of the above equation can be taken as the energy spent to account for E_{gain} in Equation 42.20.

If the above equation is modified and written as

$$q \oint_\ell \vec{E} \cdot \vec{dl} + (\text{unit charge}) \frac{\partial \Phi_s}{\partial t} = 0, \tag{42.22}$$

it can be interpreted as the law of conservation of energy, and to explain the energy gain in Equation 42.20, since its of the form written in Equation 42.15,

$$\Delta E_1 + \Delta E_2 = 0. \tag{42.23}$$

Cancelling the unit of charge in Equation 42.22,

$$q \oint_\ell \vec{E} \cdot \vec{dl} = -\frac{\partial \Phi_s}{\partial t}, \tag{42.24}$$

and now q is a real dimensionless positive number. It is only limited to a positive value to not affect the right-hand rule to relate the direction of the flux and the direction of the electric field.

If q is set to 1,

$$\oint_\ell \vec{E} \cdot \vec{dl} = -\frac{\partial \Phi}{\partial t}, \tag{42.25}$$

where Φ is the flux across a loop area, is the famous Faraday's law. It will be shown later in Chapter 57, however, any positive real number for q may be chosen.

Setting $q = 3$, for example, the energy gain in the excursion of a unit charge around the closed contour 3 times,

$$3 \oint \vec{E} \cdot \vec{dl}, \tag{42.26}$$

is equal to the energy expenditure of the time-varying magnetic flux at that instant. The above equation can also be viewed as the energy gain in moving a charge of 3 units around the loop once. There is some flexibility on how Faraday's law can be defined. Another perspective of Faraday's law is to view the left-hand side of Equation 42.25, as a measure of the energy of the electric field. The right-hand side is the energy spent to generate the energy of the electric field.

Lets look at a simple thought experiment that provides more insight into Faraday's law. A constant time-varying flux is used to generate an electric field in a closed loop. Lets look at two cases: in Case 1, a charged object is tossed into the electric field, and allowed to accelerate in the electric field, and in Case 2, there is no charged object accelerating in the electric field.

At a first glance, it may seem that Case 1 has additional energy in the form of the kinetic energy of the accelerating charged object, compared to Case 2. However, this is not the case. In Case 1, the charged object in motion is current that generates a magnetic field. From Lenz's law, this opposes the flux that is used to generate the electric field. By definition, current is the rate of flow of charges. The faster the charged object moves, the higher the current, and therefore, stronger the opposing flux. As a consequence, more work needs to be done to generate the flux needed to sustain the electric field. The gain in the kinetic energy of the charged object in motion, is balanced by the additional work to generate the flux. Therefore, there is no violation of energy conservation.

42.3   Voltage and Non-Conservative Electric Fields

From Faraday's law,

$$\oint \vec{E} \cdot \vec{dl} \neq 0, \tag{42.27}$$

and as explained in Chapter 6, if the closed path integral of a field is non-zero, then it means that it is a non-conservative field. In a non-conservative field, the path integral of the field between two points is not unique, and the result depends on the path in which the integration is done.

Voltage, which is the path integral of the electric field, is not well defined in a non-conservative field, since it does not have a unique definition. For example, in the loop of wire shown in Figure 42.4, there are two paths shown from A to B. Path 1 is shown in the clockwise direction

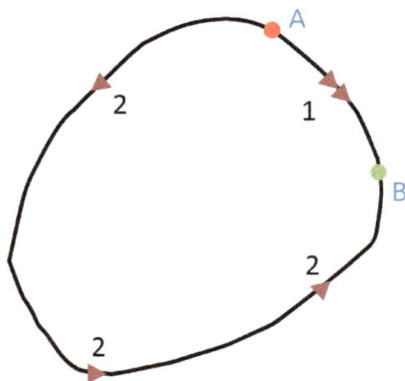

Figure 42.4: An example to show that voltage is not well defined in a non-conservative field.

marked by the double arrows, and Path 2 is in the counter-clockwise direction, marked by single arrows. If the electric field forms a closed loop, and for example, has the same value at all the points in the wire, then

$$\int_{Path\,1} \vec{E} \cdot \vec{dl} \neq \int_{Path\,2} \vec{E} \cdot \vec{dl}, \tag{42.28}$$

and the voltage between points A and B is not unique.

Since voltage is not well defined in a non-conservative field, Ohm's law, which is written in terms of voltage, is undefined as well. However, Ohm's law can be cast in a different form, repeating Equation 39.25,

$$\vec{J} = \sigma \vec{E}. \tag{42.29}$$

The above equation is without ambiguity, using electric field instead of voltage, and therefore, valid in a non-conservative field as well.

42.4   Circuit Representation of a Loop of Wire

The circuit representation of a wire loop, shown in Figure 42.1(b), can be derived from Faraday's law. A wire with resistivity $\rho = \frac{1}{\sigma}$ is shown in Figure 42.5. The loop of wire is drawn circular for simplicity, but can be of any shape. Faraday's law is applied on the dashed line marked ℓ, drawn at the center of the wire cross section,

$$\oint_\ell \vec{E} \cdot \vec{dl} = -\frac{\partial \Phi_s}{\partial t} \tag{42.30}$$

Substituting Equation 42.29 in Faraday's law,

$$\oint \frac{\vec{J}}{\sigma} \cdot \vec{dl} = -\frac{\partial \Phi_s}{\partial t}. \tag{42.31}$$

Faraday's law was interpreted as a conservation of energy equation earlier, where the energy of the time-varying magnetic flux is converted to the energy of the electric field. The above equation can be interpreted as a different form of the conservation of energy equation. The left-hand side is the energy dissipation resulting from the current flow in the wire, such as heat. The energy of the heat is converted from the energy of the time-varying magnetic flux.

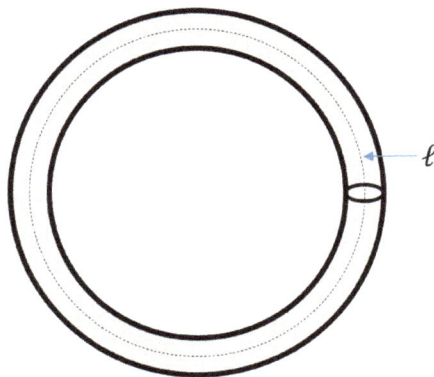

Figure 42.5: A loop of wire carrying uniform current.

The negative sign in the above equation will be ignored, and the direction of the current flow will be captured in the polarity of the induced voltage source. If the current flow is assumed to be in the direction parallel to $\vec{dl}$, the above equation simplifies to

$$\oint \frac{J}{\sigma} dl = \frac{\partial \Phi_s}{\partial t}. \tag{42.32}$$

At any time instant t, if J is assumed to be a uniform current density in all the segments making up the wire of uniform cross section A,

$$\frac{i}{A} \oint \frac{1}{\sigma} dl = \frac{\partial \Phi_s}{\partial t}, \tag{42.33}$$

where

$$J = \frac{i}{A}. \tag{42.34}$$

If the conductivity is a constant in all the wire segments,

$$i = \frac{\sigma A}{L} \frac{\partial \Phi_s}{\partial t}, \tag{42.35}$$

where L is the length of the loop of wire. Rewriting the above equation,

$$i = \frac{\frac{\partial \Phi_s}{\partial t}}{\rho \frac{L}{A}}. \tag{42.36}$$

ρ is not any arbitrary value, but the resistivity of the wire, which was stated at the beginning of the derivation. Equating the above equation to Equation 42.6,

$$V_{ind} = \frac{\partial \Phi_s}{\partial t} \tag{42.37}$$

and

$$R_s = \rho \frac{L}{A}. \tag{42.38}$$

In comparison of Equation 42.36 with Ohm's law, the current in the loop can be modeled using a resistor and a voltage source, shown in Figure 42.1(b). Their values are V_{ind} and R_s written above. This circuit representation will be extended to model an N-turn coil in the following chapter.

By definition of Ohm's law, and as shown earlier in Figure 19.1(a), by definition of voltage, a battery with voltage V_{ind} satisfies the electric-field path-integral relation in Equation 19.1. V_{ind} is a time-varying value, but must satisfy the path-integral relation in Equation 19.1 at any instant in time.

Using Kelvin-Stokes theorem, Faraday's law will be converted from the integral form to the differential form. Repeating Faraday's law in Equation 42.25, and using Equation 42.2,

$$\oint_\ell \vec{E} \cdot \vec{dl} = -\frac{\partial}{\partial t} \int_A \vec{B} \cdot \vec{dA}, \tag{42.39}$$

where the right-hand side is the rate of change of magnetic flux across a loop area A, and the left-hand side is the path integral of the electric field around the loop ℓ enclosing area A. The direction of the path of integration is related to the flux across the loop area by the right-hand rule stated earlier in the chapter.

Applying Kelvin-Stokes theorem, the path integral along the loop ℓ can be converted to a surface integral over the area A enclosing the loop,

$$\oint_\ell \vec{E} \cdot \vec{dl} = \int_A \left(\nabla \times \vec{E} \right) \cdot \vec{dA}, \tag{42.40}$$

where the direction of the path integral, and the direction of the unit normal area vector in the surface integral, are related by the right-hand rule, consistent with Faraday's law.

From the above equations,

$$\int_A \left(\nabla \times \vec{E} \right) \cdot \vec{dA} = -\frac{\partial}{\partial t} \int_A \vec{B} \cdot \vec{dA}. \tag{42.41}$$

Rearranging the above equation,

$$\int_A \left(\nabla \times \vec{E} + \frac{\partial \vec{B}}{\partial t} \right) \cdot \vec{dA} = 0. \tag{42.42}$$

For any value of $\vec{E}$ generated by any time varying $\vec{B}$, and any area A, the above equation is satisfied if the integrand is 0,

$$\nabla \times \vec{E} = -\frac{\partial \vec{B}}{\partial t}, \tag{42.43}$$

which is the differential form of Faraday's law.

43

Circuit Representation of a Coil with Time-Varying Magnetic Flux

The circuit model of a single loop of wire in the presence of a time-varying magnetic field was derived in Chapter 42. The circuit representation of a coil with many turns requires special attention. Faraday's law will be used to derive the model. A simple experiment that provides considerable insight into the behavior of an N-turn coil will be presented first, followed by the derivation of the circuit model.

43.1 N-turn Coil Experiment

The experiment in Reference [68] is presented here that provides considerable insight into the behavior of an N-turn coil. Three coils tightly wound with N, $2N$, and $4N$ turns are shown in Figure 43.1, and labeled a, b, and c. The load connected to these coils is the same in all the three cases, which is the resistance of the wire making up the galvanometer coil, as shown in Figure 27.2(a). The only difference between the three coils in this experiment is the number of turns.

When the magnet is moved in and out of the coil, the current flowing in the coil in each of the cases is proportional to the number of turns in the coil. If the current observed in the galvanometer in Coil a is current i, the currents in Coil b and Coil c are $2i$, and $4i$.

This result is a bit surprising. When more number of turns are present, the magnetic flux is greater. In Coil b, the flux is $2\times$ greater than Coil a, and $4\times$ in Coil c than Coil a. However from Ohm's law, when more number of turns are present, the resistance of the wire is also greater. One would expect the two to cancel each other, thereby resulting in approximately the same current flow in the coils a, b, and c. However, this is not the observed behavior of the current in the coils. A closer look at the electric field generated in the coil provides an understanding of this behavior.

43.2 Electric Field in an N-turn Coil

The difference in the electric field generated between a 1-turn coil and an N-turn coil will be analyzed in this section. The experiment presented earlier, clearly shows that the current induced

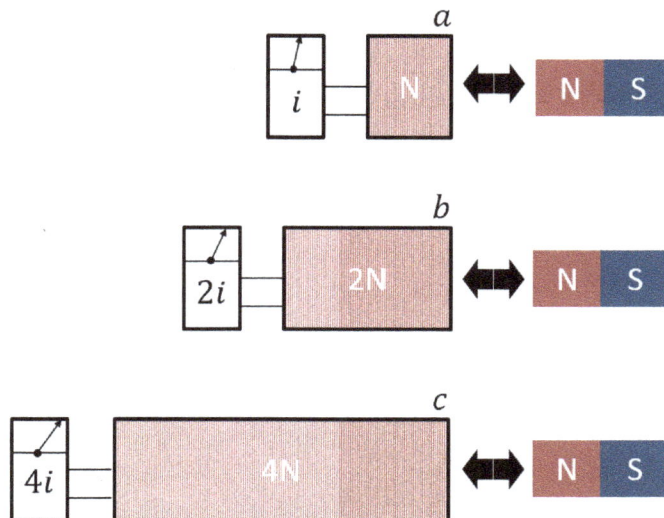

Figure 43.1: The induced current in coils made up of N, $2N$, and $4N$ turns.

in a coil is proportional to the number of turns in the coil. An N-turn coil has $N\times$ more current flowing in the coil, compared to a 1-turn coil, for the same load conditions.

Using Equation 42.34, assuming the current is uniformly distributed in the cross section of the wire, the volume current density is $N\times$ greater in the N-turn coil. From Equation 42.29, it can be concluded that the electric field in the N-turn coil is also $N\times$ greater than the 1-turn coil.

Although the total length of the wire is $N\times$ greater in the N-turn coil, its electric field is also $N\times$ greater. This behavior is so opposite to the electric field in a wire connected to a battery. Two wires made of identical materials, with lengths ℓ and 2ℓ connected to a battery, are shown in Figure 43.2(a) and Figure 43.2(b). A galvanometer connected to the circuit (not shown) would

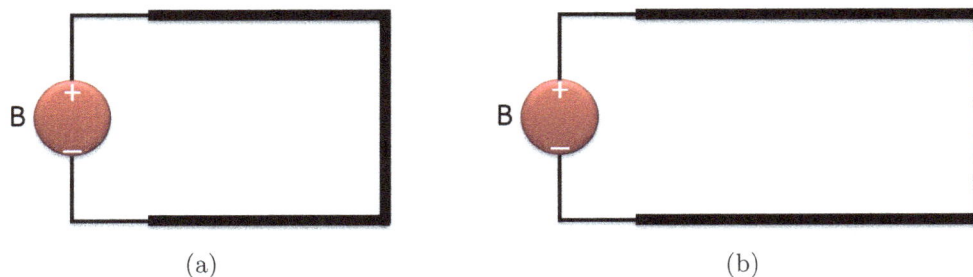

Figure 43.2: A wire of length (a) ℓ, and (b) 2ℓ, connected to a battery.

register half the current flowing in the case where the length is 2ℓ. Using Equation 42.29, the electric field in the wire with length 2ℓ must be half of what it is in the wire with length ℓ. In the experiment with the coils, however, although the N-turn coil wire length is $N\times$ more than the

1-turn coil, the electric field in the coil with N turns is $N\times$ greater.

The additional length of the N-turns of the coil, compared to a 1-turn coil, does not reduce the induced current. In effect, the N-turn coil behaves like a "1-turn super coil", with the flux, and by Faraday's law, also the induced current, $N\times$ greater compared to a 1-turn coil. This is illustrated in Figure 43.3. By Faraday's law, the path integral of the electric field in the super

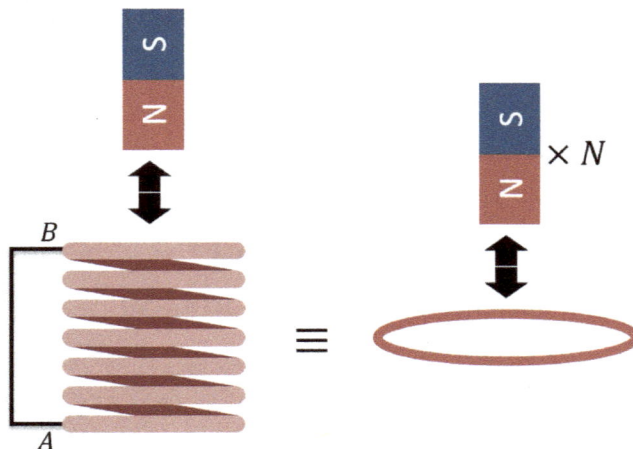

Figure 43.3: An N-turn coil behaves like a "1-turn super coil".

coil is

$$\oint \vec{E} \cdot \vec{dl} = -N\frac{\partial \Phi_s}{\partial t}. \tag{43.1}$$

The right-hand side is $N\times$ greater flux, and therefore, the electric field in the left-hand side of the equation is also $N\times$ greater than a 1-turn coil. From Equation 42.29, the current is also $N\times$ greater, caused by the $N\times$ greater electric field, consistent with the experiment results.

43.3 Circuit Model of the N-turn Coil

Following the steps in Section 42.4, the circuit model of an N-turn coil with its end points shorted together, can be derived from Equation 43.1. The resulting circuit model is shown in Figure 43.4(a). Note that when $N = 1$, the 1-turn circuit model is obtained.
The voltage source is

$$V_{ind} = N\frac{\partial \Phi_s}{\partial t}, \tag{43.2}$$

which is $N\times$ greater than a 1-turn coil. Intuitively, this is the expected behavior, since the N turns of the coil amplifies the flux $N\times$. The resistance R_s is the resistance of a 1-turn loop, and *not* all the N-turns. Repeating Equation 42.38,

$$R_s = \rho\frac{L}{A}, \tag{43.3}$$

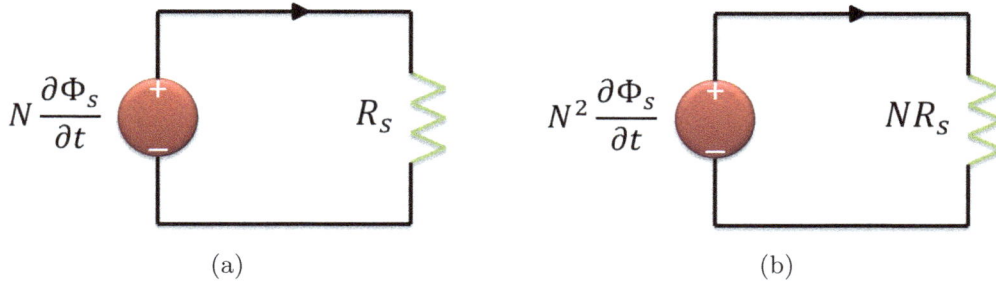

$$N\frac{\partial \Phi_s}{\partial t} \qquad R_s \qquad\qquad N^2\frac{\partial \Phi_s}{\partial t} \qquad NR_s$$

(a)   (b)

Figure 43.4: (a) The circuit model of an N-turn coil with its end points shorted together. (b) An incorrect circuit model of the N-turn coil.

where ρ is the resistivity of the wire, L is the length of a single turn loop, and A is the cross-section area of the wire. As explained in Section 43.2, the length of the coil does not impact the electric field in the typical way. Not including the resistance of the N-turns of the wire in the circuit model is a reflection of this anomalous behavior.

As derived in Equation 42.35, applying the same steps in the N-turn coil, the induced current i_{ind} is

$$i_{ind} = \frac{N}{R_s}\frac{\partial \Phi_s}{\partial t}, \qquad (43.4)$$

$N\times$ greater than the 1-turn loop, as it should be.

It would be very incorrect to model the shorted N-turn coil as shown in Figure 43.4(b). This model includes the resistance of all the N-turns of the coil. To obtain $N\times$ greater current, the voltage source is also scaled by N, resulting in $N^2\times$ greater rate of change of flux. The N-turns of the coil amplifies the flux $N\times$ greater than the flux of a single turn, but not $N^2\times$. Modeling the voltage source as $N^2\times$ the rate of change of flux, therefore, would be incorrect.

A coil connected to a load, for example, a wire of some length, whose resistance R_L affects the induced current, is shown in Figure 43.5(a). The circuit representation is shown in Figure 43.5(b). The measurement of resistance will be explained in Chapter 46. The circuit can always be verified by experiments, using a similar setup as Figure 43.1. By varying the values of R_L, but keeping the number of turns in the coil the same, it can be verified that the measured values of the induced current match with the circuit model.

To summarize, the circuit model of the N-turn coil is shown in Figure 43.6(a). The voltage source is written as

$$V_{ind} = \frac{\partial \Phi}{\partial t}, \qquad (43.5)$$

where

$$\Phi = N\Phi_s \qquad (43.6)$$

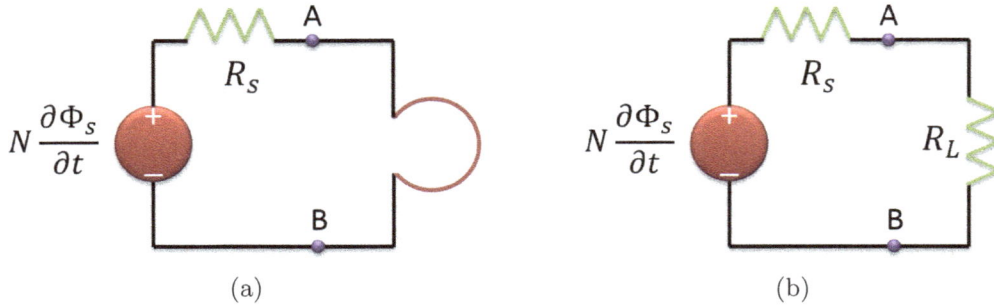

Figure 43.5: (a) A load, such as a wire of some length, connected to the end points of an N-turn coil. (b) The circuit model of Figure 43.5(a).

is the flux across all the N turns of the coil. R_s is the single-turn resistance. To simplify the model, R_s is often ignored, and the coil is represented as a voltage source, as shown in Figure 43.6(b). This model will be used in the future chapters.

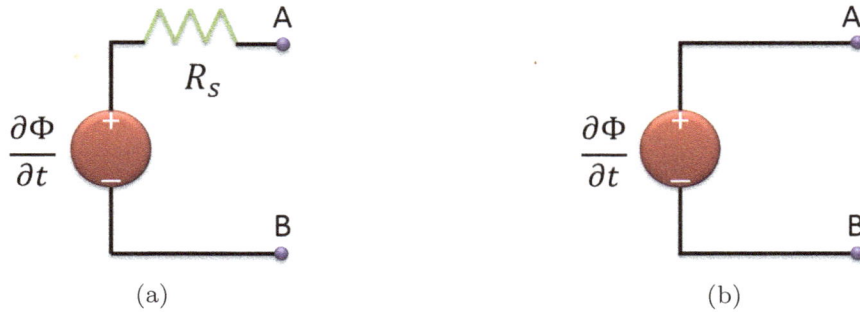

Figure 43.6: (a) The circuit model of an N-turn coil. (b) A simplified circuit model of the N-turn coil.

44

Self Inductance

A coil of wire marked S, also known as a solenoid, is shown in Figure 44.1(a). The solenoid is connected to a battery V, via switch W. When the switch W is closed, the current i in the wire increases from its initial value of 0 to its steady-state value in a short time duration. The time-varying current in the coil creates a time-varying flux within itself. This is known as self induction [69].

The time-varying flux across the turns of the solenoid, generates a voltage between the end-points of the coil, as seen by the model that represents this behavior in Figure 43.6(b). The induced voltage creates an induced current flow in the wire connected to it, such that it opposes the changing magnetic flux, known as Lenz's law.

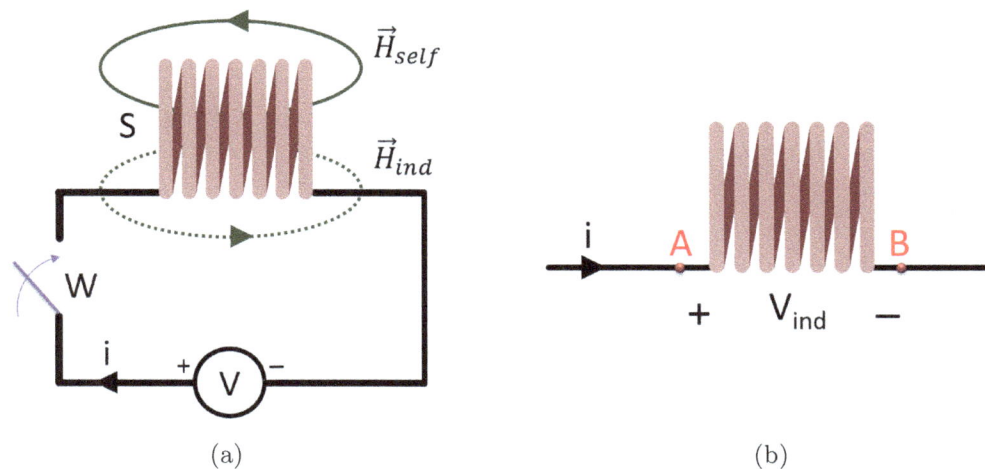

(a)   (b)

Figure 44.1: (a) Self inductance in a coil. (b) The induced voltage in a coil as a result of self inductance.

From Biot-Savart's law, the strength of a magnetic field generated by a current is proportional to the current. Therefore, the magnetic flux that is created in the solenoid Φ_{self} is also proportional to the current flow i in the solenoid,

$$\Phi_{self} \propto i. \tag{44.1}$$

The current i in the above equation is the net current in the solenoid, which is the superposition of the current that generates the flux, and the current generated by the solenoid in response to the changing flux.

The self inductance L relates both the sides of the above equation, and is defined as

$$L = \frac{\Phi_{self}}{i}. \tag{44.2}$$

In Section 43.3, from Faraday's law, the induced voltage across a solenoid was derived as the time rate of change of flux. Substituting Φ_{self} in the above relation in Equation 43.5,

$$V_{ind} = L\frac{di}{dt}. \tag{44.3}$$

The induced voltage across the solenoid is illustrated in Figure 44.1(b). This model will be revisited later in the chapter to discuss the polarity of the induced voltage across the coil.

An approximate value of the self inductance L of a solenoid will be derived in the next section. L depends only on the geometry of the coil, is a positive number, and non-time varying. L can therefore be moved out of the time derivative in the above equation. The polarity of the induced voltage will be addressed in Section 44.2.

44.1   Self Inductance of a Solenoid Using Ampere's Law

It will be assumed that the current in a solenoid flows in a wire of infinitesimal diameter, or a line, which is known as a filament current. From Ampere's law, the magnetic field $\vec{H}$, and current i_{enc} enclosed by a closed contour C, satisfy the relation,

$$\oint_C \vec{H} \cdot \vec{dl} = 4\pi\, i_{enc}, \tag{44.4}$$

Applying the above law to a solenoid of length ℓ in Figure 44.2, the self-inductance formula will be derived. The rectangular Ampere loop C is drawn as shown in the figure that encloses the N turns of the solenoid, and whose lengths are marked 1-4. The magnetic- field pattern generated by the solenoid is shown in the figure. The pattern can be observed using iron filings, as described in Section 21.1, or verified by numerical solution using Biot-Savart's law.

In the magnetic-field pattern of the solenoid, it will be assumed that the field within the solenoid is uniform, and the field at the edges of the solenoid will be ignored. It will also be assumed that the edge of the Ampere loop marked 1, is far away from the solenoid, and its contribution to the path integral is 0. The only edge that contributes to the path integral is 3 that lies within the solenoid.

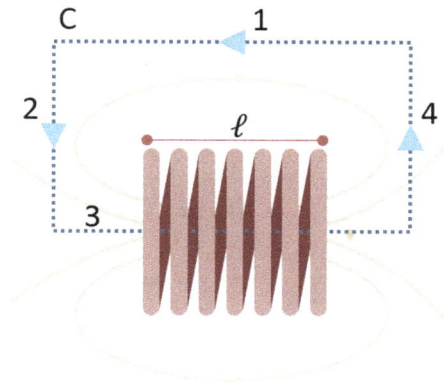

Figure 44.2: The magnetic-field pattern of a current-carrying coil.

If $\vec{H}$ is assumed to be uniform within the solenoid, and its pattern as shown, the path integral evaluated along the edge marked 3, in the direction shown by the arrows is

$$\oint \vec{H} \cdot \vec{dl} = H\ell, \tag{44.5}$$

where H is the magnitude of the uniform magnetic field along the path integral, and ℓ is the length of the solenoid, which is the same as the length of the edge 3. For the chosen direction of the path of integration, by the right-hand rule of Section 21.2, the positive direction of current flow is out of the page. For the magnetic-field direction shown in the figure, by the right-hand rule, Ni is the current flowing out of the page in the N-turns of the solenoid. Using Equation 44.4,

$$H\ell = 4\pi Ni, \tag{44.6}$$

Rearranging the above equation, and solving for H,

$$H = 4\pi \frac{N}{\ell} i. \tag{44.7}$$

The magnetic flux is given by Equation 42.2. In the special case where no magnetic material is present in the solenoid,

$$\vec{B} = \vec{H}. \tag{44.8, EMU}$$

If A is the area of a single-turn loop, the magnetic flux across the N turns Φ_{self} is $N\times$ greater,

$$\Phi_{self} = HNA$$
$$= 4\pi A \frac{N^2}{\ell} i. \tag{44.9, EMU}$$

The above equation may also be written as

$$\Phi_{self} = 4\pi An^2 \ell i, \tag{44.10, EMU}$$

where $n = \frac{N}{\ell}$ is the number of turns per unit length in the solenoid. Applying the definition of self inductance in Equation 44.2,

$$L = \frac{\Phi_{self}}{i}$$
$$= 4\pi A n^2 \ell. \qquad (44.11, \text{EMU})$$

The self inductance of a solenoid is a function of the geometry of the coil, and is a positive value. A more accurate formula for the self inductance of a solenoid will be derived in Chapter 45.

Applying dimensional analysis on the definition of inductance in Equation 44.2, the unit of inductance $[L]$ is

$$[L] = cm$$
$$= abhenry, \qquad (44.12, \text{EMU})$$

and the alias abhenry is the unit of inductance in EMU.

44.2 Voltage Polarity Convention Used in Self Inductance

Using the self-inductance formula derived in the previous section, the induced voltage across a solenoid can be calculated using Equation 44.3. What is the polarity of the induced voltage, or which side of the solenoid is at the higher potential? Lenz's law provides the answer to this question. Lenz's law states that the induced current is always in a direction that opposes the changing magnetic flux across the solenoid.

A solenoid is shown in Figure 44.3(a). Once the direction of current flow is chosen, the convention followed is that the positive node of the induced voltage across the solenoid is the node where the current enters the solenoid. This is illustrated in the figure.

To show that the convention is in agreement with Lenz's law, the different possibilities of the rate of change of current will be analyzed. There are only 3 possibilities, as illustrated in Figure 44.3(b): $\frac{di}{dt} > 0$, $\frac{di}{dt} = 0$, or $\frac{di}{dt} < 0$. The direction chosen for current flow is shown by the arrow marked i.

Suppose $\frac{di}{dt} > 0$, as shown in Case I. Using Equation 44.3, the induced voltage is also positive. The solenoid is modeled as a voltage source in the figure. Since the induced voltage is positive, the positive and negative terminals of the voltage source are at the left side and the right side of the solenoid, according to the convention in Figure 44.3(a). The increasing current generates an increasing flux in the solenoid. The induced current flows out of the positive terminal of the voltage source (see Section 19.3) to oppose the increasing current, and therefore, is in agreement with Lenz's law.

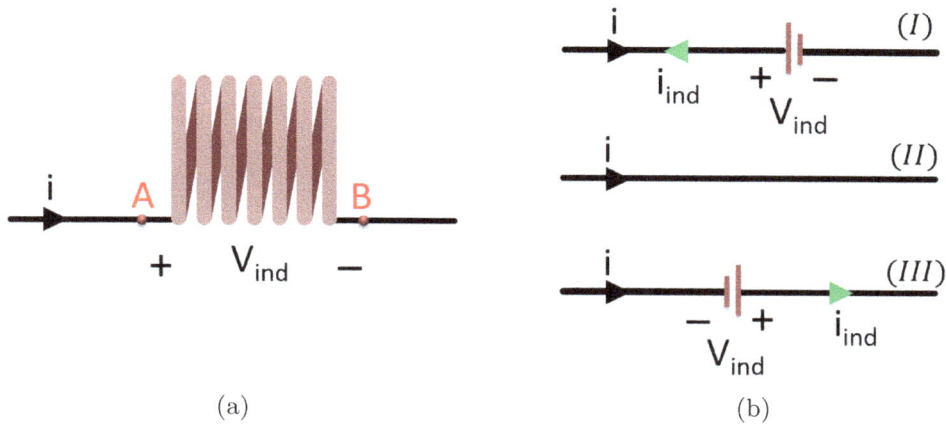

Figure 44.3: (a) The convention followed to relate the induced voltage, and the current in the coil. (b) The three possible cases, $\frac{di}{dt} > 0$, $\frac{di}{dt} = 0$, and $\frac{di}{dt} < 0$, that may occur for the rate of change of current flow in the coil.

In Case II, $\frac{di}{dt} = 0$. There is neither an increase nor decrease in the current, and the magnetic flux. Therefore, $V_{ind} = 0$, and the voltage source is absent in the figure. This is also in agreement with Lenz's law, since there is no opposing current required to fight the changing flux.

In Case III, $\frac{di}{dt} < 0$. V_A and V_B are the voltages at the left-side terminal and the right-side terminal of the induced voltage source, as shown in Figure 44.3(a). $V_{ind} = V_A - V_B$ is negative, which means that V_B is at a higher potential than V_A. Current flows out of the terminal of a voltage source that is at a higher potential. The direction of i_{ind} is marked in Figure 44.3(b), and adds to i. i is decreasing, since $\frac{di}{dt} < 0$. i_{ind} counteracts the decreasing magnetic flux caused by the decreasing current i, by adding to it. This too, therefore, is in agreement with Lenz's law, thereby verifying the convention in Figure 44.3(a).

45

Mutual Inductance

In the previous chapter, the self inductance was introduced, which is the result of a time-varying magnetic flux across a solenoid, caused by a time-varying current flowing in the same solenoid. It is also possible that a time-varying magnetic flux on a coil, is caused by a time-varying current on a different coil that is in close proximity. This is called mutual inductance.

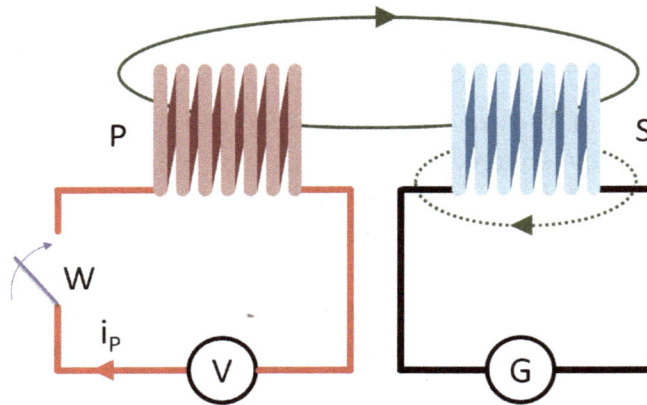

Figure 45.1: The coupling of magnetic field between two coils in proximity.

An example is shown in Figure 45.1. Two coils, P for primary and S for secondary, are in close proximity. When the switch W is closed in the primary circuit, a time-varying current flows in the circuit until it reaches the steady state in a short time duration. The time-varying magnetic field caused by the primary current, couples to the secondary coil. This creates a time-varying flux across the secondary coil, and generates an induced voltage across the secondary coil.

From Biot-Savart's law, the magnetic field coupling to the secondary coil is proportional to the primary current. This also means that

$$\Phi_S \propto i_P, \tag{45.1}$$

where Φ_S is the the magnetic flux across the secondary coil, and i_P is the current in the primary coil. Mutual inductance M is defined as

$$M = \frac{\Phi_S}{i_P}, \tag{45.2}$$

similar to the definition of self inductance. Using Equation 43.5, and from the above definition of M, the induced voltage across the secondary coil is

$$V_{ind} = M\frac{di_P}{dt}. \tag{45.3}$$

Similar to self inductance, a filament current approximation for the current will be assumed, and the mutual inductance value, which will be derived in Section 45.1–Section 45.3, depends only on the geometry of the coils.

Accurate formulae for the self inductance of a solenoid, and the mutual inductance between coils, will be derived in this chapter. This is needed for resistance measurement, as discussed in Chapter 46. The polarity of the induced voltage in the secondary coil, unlike the case in self inductance, depends on how the primary and the secondary coils are wound. The dot convention, explained later in the chapter, is used to indicate the polarity of the induced voltage in the secondary coil.

45.1 Neumann's Formula to Calculate Mutual Inductance [Optional]

A generic formula to calculate the mutual inductance between two current loops will be presented in this section. This result will be used in the next section to calculate the mutual inductance between two coils.

Two current loops, Loop P for primary and Loop S for secondary, are shown in Figure 45.2. Assume that the current i in Loop P is in the direction marked in the figure. The filament current approximation will be assumed. The magnetic field generated by the current in Loop P generates a magnetic flux in Loop S. This flux will be calculated using Biot-Savart's law, and Equation 45.2 will be used to formulate the mutual-inductance equation.

To calculate the flux across the area of Loop S, assume that its area is discretized into tiny patches, one of which is illustrated, and whose center has been marked Point A. Biot-Savart's law will be used to solve for the magnetic field at the center of each of the infinitesimal patch areas, from which the flux will be calculated by integration.

Using a similar setup of the coordinate system in Chapter 30, the primary loop carrying the current is placed in the primed coordinate system, and the point on the secondary loop area at which the magnetic field is calculated is in the unprimed coordinate system. The current source region is a function of the primed variables (x', y', z'), and the point at which the magnetic field is calculated is at (x, y, z), which is differentiated from the source region using unprimed variables.

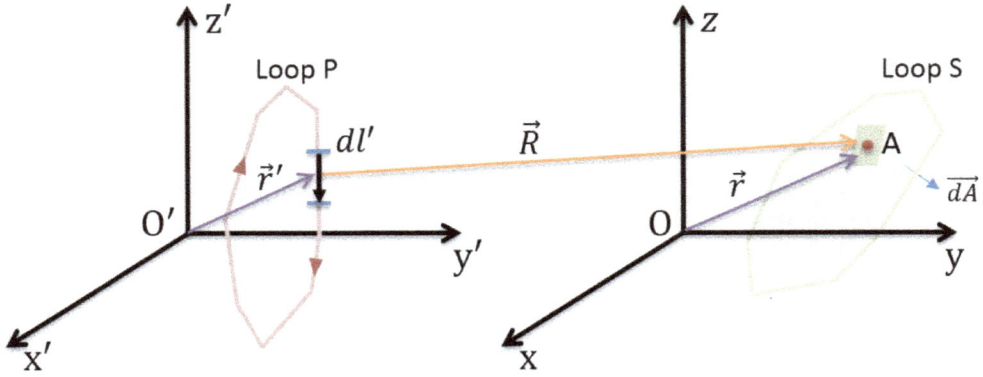

Figure 45.2: The calculation of mutual inductance between two loops.

The origins O and O' overlap, but shown separated.

The ∇ operator in curl $\nabla\times$, and gradient ∇ operations, is defined on the unprimed variables, and ∇' on the primed variables, as written in Equation 30.8 and Equation 30.9. When applying a primed operation, the unprimed variables are treated as constants, and vice versa. The primed operator operates on the source region, while keeping the point at which the field is calculated constant, and the unprimed operator operates on the point at which the field is calculated, while keeping the source region constant.

Φ_S, the flux across the secondary loop S is

$$\Phi_S = \int_S \vec{B} \cdot d\vec{A}, \tag{45.4}$$

where $\vec{B}$ is the magnetic-flux density generated by the current in the primary loop, and $d\vec{A}$ is the area normal vector to loop S. There are two choices for the direction of the area normal vector of Loop S: the direction marked $d\vec{A}$ in the figure, or opposite to that. Either of these directions may be chosen. The sign in the calculations will be disregarded, and the value of mutual inductance will always be a non-negative value.

Assuming a non-magnetic medium, the magnetic-flux density $\vec{B}$ is the same as the magnetic-field strength $\vec{H}$,

$$\vec{B} = \vec{H}. \tag{45.5, EMU}$$

To apply Biot-Savart's law, the primary loop is discretized into infinitesimal current elements, one of which is marked dl'. $\vec{R}$ is the vector from the midpoint of this infinitesimal current element to Point A, marked in the figure, where the magnetic field is calculated. If $\vec{r}' = (x', y', z')$ is the midpoint of the current element, and A has the coordinates $\vec{r} = (x, y, z)$,

$$\vec{R} = \vec{r} - \vec{r}', \tag{45.6}$$

and the magnitude of the vector $\vec{R}$ is written in Equation 30.17, and denoted as R. The unit vector of $\vec{R}$ is

$$\hat{R} = \frac{\vec{R}}{R}. \tag{45.7}$$

Substituting Biot-Savart's law in Equation 45.4, and using the above definitions,

$$\Phi_S = \int_S \left(\oint_p \frac{i \, \vec{dl}' \times \hat{R}}{R^2} \right) \cdot \vec{dA}, \tag{45.8, EMU}$$

The lowercase p represents the path integral over Loop P, and the uppercase S denotes integration over the area of Loop S. The lowercase and the uppercase letters are used to differentiate between a path integral and an area integral. Using the identity in Equation 30.31, the above expression is written as

$$\Phi_S = \int_S \left(\oint_p -i \, \vec{dl}' \times \nabla \left(\frac{1}{R} \right) \right) \cdot \vec{dA}. \tag{45.9, EMU}$$

Using the expressions

$$f = \frac{1}{R} \tag{45.10}$$

$$\vec{g} = \vec{dl}', \tag{45.11}$$

on the identity in Equation 30.33, repeated here,

$$\nabla \times (f\vec{g}) = f\nabla \times \vec{g} - \vec{g} \times \nabla f, \tag{45.12}$$

Equation 45.9 can be written as

$$\Phi_S = i \int_S \left(\oint_p \nabla \times \left(\frac{\vec{dl}'}{R} \right) \right) \cdot \vec{dA} - i \int_S \left(\oint_p \frac{1}{R} \nabla \times \vec{dl}' \right) \cdot \vec{dA}. \tag{45.13, EMU}$$

The second term in the above equation reduces to 0, since ∇, an unprimed operator, operates on the primed variable $\vec{dl}'$, which is treated as a constant. If the curl operation on an infinitesimal element $\vec{dl}'$ is confusing, one could think of the integral as a Riemann sum, and $\vec{dl}'$ as a discretized tiny vector element $\vec{\Delta l}'$. The operand of the curl operation is a tiny vector element, rather than an infinitesimal element. The conclusion is the same. The above equation, therefore, reduces to

$$\Phi_S = i \int_S \left(\nabla \times \oint_p \frac{\vec{dl}'}{R} \right) \cdot \vec{dA}, \tag{45.14, EMU}$$

where the $\nabla \times$ operator has been moved out of the integral. This is possible because the integration is with respect to the primed variables, but the $\nabla \times$ operator is with respect to the unprimed variables. Using simple integrals and 1D operators, it is left as an exercise for the reader to verify this.

Applying the Kelvin-Stokes theorem to the above expression,

$$\Phi_S = i \oint_s \oint_p \frac{\vec{dl'} \cdot \vec{dr}}{R}. \qquad \text{(45.15, EMU)}$$

This is a very interesting result. The magnetic flux is first written as an area integral, but using Kelvin-Stokes theorem, the area integral has been transformed into nested path integrals. The lower-case subscripts, p and s, represent the path integrals over the primary and the secondary loops.

In Kelvin-Stokes theorem, the direction of the path integral over the secondary loop is given by the right-hand rule: if the thumb of the right hand points in the direction of the area normal vector, the fingers curl in the direction of the path integral. If the path integral is evaluated in the reverse direction, the sign of the result would be opposite. However, the sign is disregarded in the calculation, and M is always non-negative, as written in the above equation. The polarity of the induced voltage in Equation 45.3 is indicated by the dot convention described later.

Reverting to a single unprimed coordinate system, and applying the definition of mutual inductance, the above integral is written with unprimed variables as

$$M = \oint_s \oint_p \frac{\vec{dl} \cdot \vec{dr}}{R}, \qquad \text{(45.16, EMU)}$$

where R is the distance between the infinitesimal segments $\vec{dl}$ and $\vec{dr}$. This is the Neumann's formula, and will be used in the following sections to derive an expression for the mutual inductance between two coils. From the above equation, it can be observed that mutual inductance is independent of which loop is chosen as the primary or the secondary. The value of the mutual inductance would be the same.

45.2 Mutual Inductance of Two Coaxial Circular Wires

Two circular wires L_1 and L_2, of radii r_1 and r_2, are shown in Figure 45.3. The loop areas are parallel to the xy-plane, and the line joining their centers is along the z-axis. The wires are said to be in a coaxial arrangement. The wires are separated by length ℓ, and are equidistant at length $\frac{\ell}{2}$ from the xy-plane. The mutual inductance between the two loops M_s, subscript s for single-turn loops, will be calculated using Neumann's formula,

$$M_s = \oint_{L_1} \oint_{L_2} \frac{\vec{dL_1} \cdot \vec{dL_2}}{R}, \qquad \text{(45.17, EMU)}$$

where $\vec{dL_1}$ and $\vec{dL_2}$ are the differential elements of the loops L_1 and L_2, and R is the distance between them [70]–[73]. This result will be extended in the next section to calculate the mutual inductance between two coils.

316

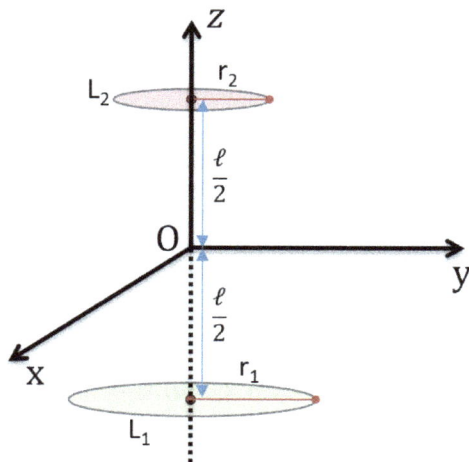

Figure 45.3: Two loops in a coaxial arrangement.

The loops L_1 and L_2 are parametrized using the parameters ϕ and θ,

$$\vec{L}_1 = r_1 \left(\cos \phi \, \hat{x} + \sin \phi \, \hat{y} \right) - \frac{\ell}{2} \, \hat{z}, \quad 0 \le \phi < 2\pi \tag{45.18}$$

$$\vec{L}_2 = r_2 \left(\cos \theta \, \hat{x} + \sin \theta \, \hat{y} \right) + \frac{\ell}{2} \, \hat{z}, \quad 0 \le \theta < 2\pi. \tag{45.19}$$

As explained in Section 3.3, angles will be measured in radians, rather than in degrees. Differentiating the above expressions with respect to ϕ and θ,

$$d\vec{L}_1 = r_1 \left(-\sin \phi \, \hat{x} + \cos \phi \, \hat{y} \right) d\phi \tag{45.20}$$

$$d\vec{L}_2 = r_2 \left(-\sin \theta \, \hat{x} + \cos \theta \, \hat{y} \right) d\theta. \tag{45.21}$$

Taking the dot product of the above equations and simplifying,

$$d\vec{L}_1 \cdot d\vec{L}_2 = r_1 r_2 \cos \left(\phi - \theta \right) d\phi \, d\theta. \tag{45.22}$$

The distance between the differential elements is

$$R = \left| \vec{L}_1 - \vec{L}_2 \right|$$
$$= \sqrt{r_1^2 + r_2^2 + \ell^2 - 2 r_1 r_2 \cos \left(\phi - \theta \right)}. \tag{45.23}$$

Substituting the above equations in Equation 45.17,

$$M_s = \int_0^{2\pi} \int_0^{2\pi} \frac{r_1 r_2 \cos \left(\phi - \theta \right) d\phi \, d\theta}{\sqrt{r_1^2 + r_2^2 + \ell^2 - 2 r_1 r_2 \cos \left(\phi - \theta \right)}}. \tag{45.24, EMU}$$

Let

$$\gamma = \phi - \theta. \tag{45.25}$$

θ is a constant within the ϕ integral in Equation 45.24. Differentiating the above equation with respect to ϕ, and treating θ as a constant,

$$d\gamma = d\phi. \tag{45.26}$$

From the above equations,

$$M_s = \int_0^{2\pi} \int_{-\theta}^{2\pi-\theta} \frac{r_1 r_2 \cos\gamma \, d\gamma d\theta}{\sqrt{r_1^2 + r_2^2 + \ell^2 - 2r_1 r_2 \cos\gamma}}. \tag{45.27, EMU}$$

The integrand is a periodic function with period 2π, and integrated over an interval of 2π radians. From calculus, the periodic function may be integrated over *any* interval of length 2π radians. Integrating the γ integral over $[0, 2\pi]$, and evaluating the outer θ integral,

$$M_s = 2\pi \int_0^{2\pi} \frac{r_1 r_2 \cos\gamma \, d\gamma}{\sqrt{r_1^2 + r_2^2 + \ell^2 - 2r_1 r_2 \cos\gamma}}. \tag{45.28, EMU}$$

Only the solution to the above integral will be provided, and the reader is referred to other resources for more details on solving the integral. The closed-form solution to the above integral is written in terms of the complete elliptic integral of the first kind $K(k)$ and the second kind $E(k)$ [56],

$$M_s = 4\pi \sqrt{r_1 r_2} \left[\left(k - \frac{2}{k} \right) K + \frac{2}{k} E \right],$$

$$k = \frac{2\sqrt{r_1 r_2}}{\sqrt{(r_1 + r_2)^2 + \ell^2}}. \tag{45.29, EMU}$$

Although not explicitly stated in the above equation, K and E are evaluated at k, known as the modulus. Numerical tools or integral tables can be used to evaluate K and E at m, where $m = k^2$.

45.3   Mutual Inductance Between Two Solenoids

The result for the mutual inductance between two single-turn loops, derived in Equation 45.29, can be extended to approximate the mutual inductance between two solenoids in a coax arrangement. N_1 and N_2 are the number of turns in Loop 1 and Loop 2. The thickness of the coils will be assumed to be negligible.

Neumann's formula and mutual inductance do not depend on which coil is chosen as the primary and the secondary. Coil 2 in Figure 45.4 will be chosen to be the secondary coil, and Coil 1 the primary. Reversing this assumption does not change the outcome of the mutual-inductance formula that will be derived.

Φ_s is the flux across a single-turn secondary loop caused by a single-turn primary, separated

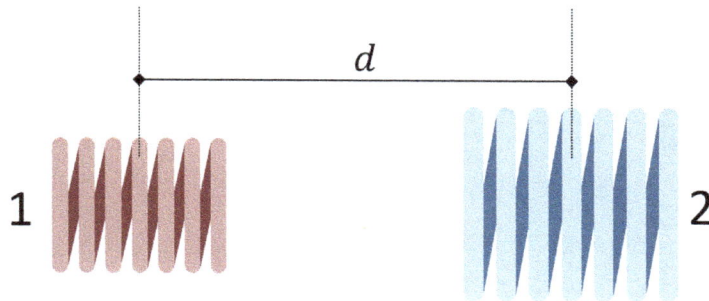

Figure 45.4: Two solenoids in a coaxial arrangement.

by distance d between the midpoint of the coils, as shown in the figure. M_s is the mutual inductance between the two single-turn loops in Equation 45.29. Using Equation 43.6, the flux across the secondary coil is $N_2\times$ greater than the flux across a single-turn secondary coil. In addition, by Biot-Savart's law, the N_1 turns of the primary loop will amplify the magnetic field $N_1\times$, and therefore, increasing the flux coupled to the secondary coil $N_1\times$. Taking both of these factors into account, the flux across the secondary is $N_1 N_2\times$ greater.

Applying the mutual-inductance definition, where i is the current in the primary,

$$M = N_1 N_2 \left(\frac{\Phi_s}{i} \right)$$
$$= N_1 N_2 \, M_s. \tag{45.30}$$

The mutual inductance between the two single-turn coils is scaled by $N_1 N_2$. A more accurate calculation of the mutual inductance between two coils is presented in Section 45.5. This is important, because it will be used in the measurement of resistance, as discussed in Chapter 46.

45.4 Dot Convention

The voltage induced in the secondary coil is given by Equation 45.3. The polarity of the induced voltage is addressed in this section. Unlike the polarity of the induced voltage in self inductance, the polarity of the induced voltage in mutual inductance depends on how the coils are wound. This is illustrated by an example.

Two coils A and B are shown in Figure 45.5, and placed in close proximity so that the fields couple. The field generated by Coil A is shown by the solid line in the direction indicated by the arrow. By Lenz's law, the polarity of the induced voltage in Coil B, and the current flowing in it, is such that it opposes the field generated by A, and is in the direction shown by the dashed line.

A is wound the same way in both the cases shown in the figure, but not B. Using the right-hand rule, the direction of the curent flow in B has been marked, which generates the magnetic field shown by the dashed line. The current in Coil B enters the coil at the top in Figure 45.5(a), but leaves the coil in Figure 45.5(b). If the coil is modeled as a voltage source, whose value is given by Equation 45.3, the side of the coil where the current exits the coil is the positive node of the voltage source, since current flows from the positive node to the negative node of the voltage source. The polarity of the voltage source has been marked in each of the figures, and they are opposite.

Since the voltage polarity depends on how the coils are wound, the dot convention is used to indicate the voltage polarity in mutual inductance. The schematic of the primary-secondary coil is shown in Figure 45.6. The positive direction of the currents in the coils may be chosen arbitrarily, for example, as the ones marked in the figure. Dots are placed such that current entering a dotted node of the coil, such as i_1, generates a voltage that is at a higher potential at the dotted node of the other coil, such as V_2. If the current leaves a dotted node, such as i_2, the node of the other coil with no dot is at a higher potential, such as V_1.

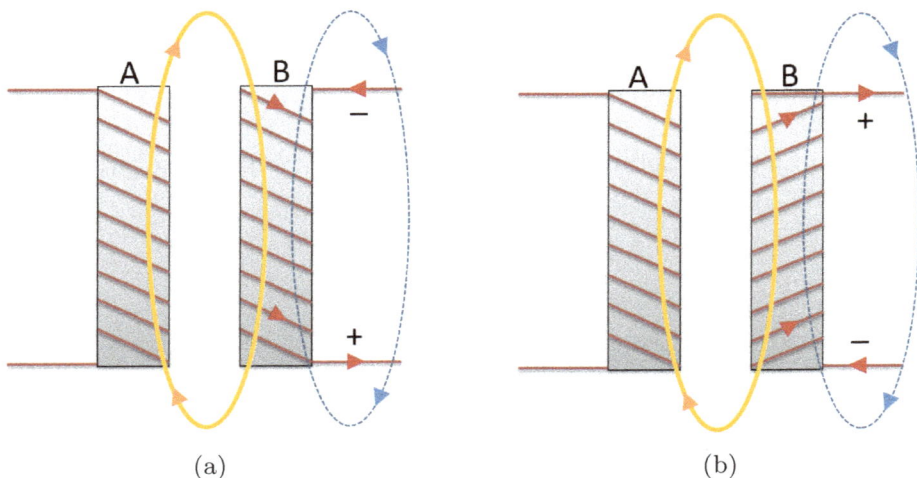

(a)   (b)

Figure 45.5: (a) The polarity of the induced voltage in Coil B is opposite to that of (b), although Coil A is wound the same way in both cases.

In the example shown in Figure 45.6,

$$V_1 = M\frac{di_2}{dt} \tag{45.31}$$

$$V_2 = M\frac{di_1}{dt}. \tag{45.32}$$

In addition to mutual inductance, the self inductance of the coils must also be taken into account. Superposition will be assumed to be valid for the voltages generated across the coils due to self and mutual inductances. By the voltage-polarity convention for self inductance, and for the current

320

directions assumed in Coil 1 and Coil 2, the above equations are modified to be

$$V_1 = -L_1 \frac{di_1}{dt} + M \frac{di_2}{dt} \qquad (45.33)$$

$$V_2 = -L_2 \frac{di_2}{dt} + M \frac{di_1}{dt}, \qquad (45.34)$$

where L_1 and L_2 are the self inductances of Coil 1 and Coil 2.

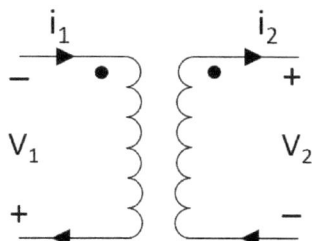

Figure 45.6: A circuit schematic to illustrate the dot convention that is used to set the polarity of the induced voltage resulting from the mutual inductance between two coils.

A time-varying current in Coil 1 induces a current in Coil 2. The induced current in Coil 2 generates a time-varying flux in Coil 1, and in turn induces a current in Coil 1, and so on. The induced voltages are written in terms of both the coil currents i_1 and i_2. The solution to the differential equations, therefore accounts for such "feedback" effects.

45.5 Accurate Calculations of Self and Mutual Inductances of Coils [Optional]

To determine the resistance of a wire, it will be explained in Chapter 46, an accurate value of the mutual inductance between two coils is required. Accurate calculations of self and mutual inductances will be presented in this section, using the mutual-inductance formula of single-turn coils in Equation 45.29. This is a classic formula known as the Lorenz current-sheet formula, and the same result may be derived in several ways [74][75][76], as noted in [75]. However, the method in Reference [76] will be used. The integral will be setup, and the final solution provided. But the step-by-step method to solve the integral is well documented in the reference, and will not be presented here.

Two coils 1 and 2 of lengths $2\ell_1$ and $2\ell_2$, and radii r_1 and r_2 are shown in Figure 45.7. The length between the midpoint of the coils is d, marked in the figure. x_1 and x_2 are variables swept over the lengths of Coil 1 and Coil 2, respectively. x_1 varies over $[-\ell_1, +\ell_1]$, and x_2 over $[-\ell_2, +\ell_2]$, as shown in the figure. The N_1 and N_2 turns of the coils will be viewed as N_1 and N_2 single-turn loops. If x_1 and x_2 are the locations of two single-turn loops, the distance between them is $d + x_2 - x_1$. The mutual inductance between the single-turn coils can be calculated using Equation 45.29.

Taking the thickness of the coils into account, the average value of the mutual inductance between the coils may be calculated as

$$M_{avg} = N_1 N_2 \left[\left(\frac{1}{2\ell_1} \right) \left(\frac{1}{2\ell_2} \right) \int_{-\ell_2}^{+\ell_2} \int_{-\ell_1}^{+\ell_1} M(x_1, x_2) \, dx_1 \, dx_2 \right], \qquad (45.35, \text{EMU})$$

where $M(x_1, x_2)$ is the mutual inductance between two single-turn coils separated by $d + x_2 - x_1$, and N_1, N_2 are the number of turns in Coil 1 and Coil 2. The term within the square brackets is an average value of the mutual inductance of two single-turn coils over the lengths of the coils. Multiplying this value by $N_1 N_2$ is the mutual inductance between the coils, similar to how Equation 45.30 was calculated.

Once the mutual inductance between the coils is formulated accurately, if $r_1 = r_2$, and if $d = 0$, the two coils will merge into a single coil. In this case, the ratio of the flux across Coil 2, and the current in Coil 1, is the same as the definition of self inductance. By setting $r_1 = r_2$ and $d = 0$ in the mutual inductance relation between the two coils, the self inductance of the solenoid can be obtained. This is the idea behind calculating self inductance using mutual inductance in Cohen's derivation [76].

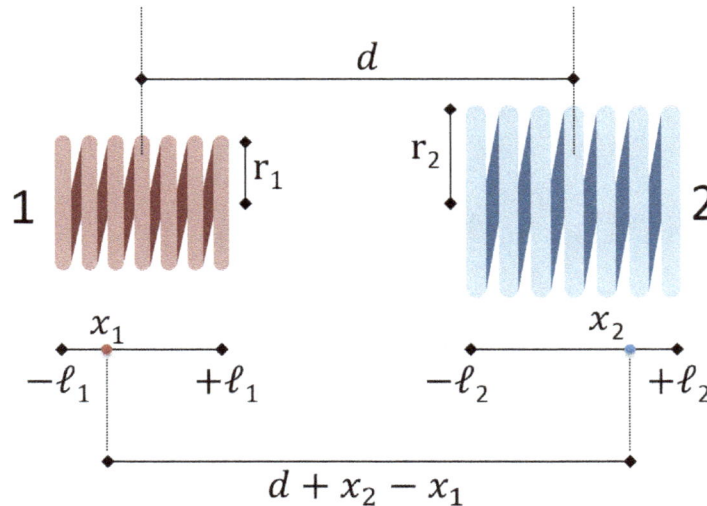

Figure 45.7: The calculation of mutual inductance between two solenoids in a coaxial arrangement.

The integral in Equation 45.28 is substituted in the above equation for $M(x_1, x_2)$ with $\ell = d + x_2 - x_1$, and the closed-form solution to the mutual inductance is presented in detail in [76]. Skipping a couple of solution steps in solving the integral, setting $r_1 = r_2 = r$, $\ell_1 = \ell_2 = \ell$, and $d = 0$ in the solution, the self-inductance L of the solenoid is

$$L = 4\pi \left(\frac{N}{2\ell} \right)^2 \left\{ \frac{c^4 + 4r^2 c^2}{3\sqrt{4r^2 + c^2}} K + \frac{4r^2 - c^2}{3} \sqrt{4r^2 + c^2} E - \frac{8r^3}{3} \right\},$$

$$c = 2\ell. \qquad (45.36, \text{EMU})$$

K and E are the complete elliptic integral of the first and the second kind at k,

$$k = \frac{2r}{\sqrt{4r^2 + c^2}}.$$ (45.37)

The self inductance of a solenoid was calculated using Ampere's law in Section 44.1. However, the magnetic field was approximated to be uniform within the coil. The calculation of self inductance using the mutual-inductance formula in Equation 45.29 is a more accurate solution.

The closed-form solution shown above is compared with numerical integration of Equation 45.35 in Figure 45.8 and Figure 45.9, along with the self inductance of the solenoid using Ampere's law. The circle markers A is the self inductance calculated by numerical integration of Equation 45.35, B is the solid line computed using Ampere's law in Equation 44.11, and C is the dashed line calculated using the closed-form solution given above. There is an excellent correlation between A and C, verifying the closed-form solution of self inductance.

In Figure 45.8, the number of turns per unit length of the solenoid is kept constant at 15 turns/cm, and the length of the solenoid is swept between 2 cm – 20 cm. The x-axis is the length of the solenoid, and the y-axis is the inductance in *abhenry*. The difference in the results, between Ampere's law and the Lorenz current-sheet formula, is 4% for the test case analyzed.

In Figure 45.9, the length of the solenoid is fixed at 2 cm, and the number of turns is swept from 30 turns to 75 turns. The results of self inductance calculated using Ampere's law is 45% higher than those of the Lorenz current-sheet formula.

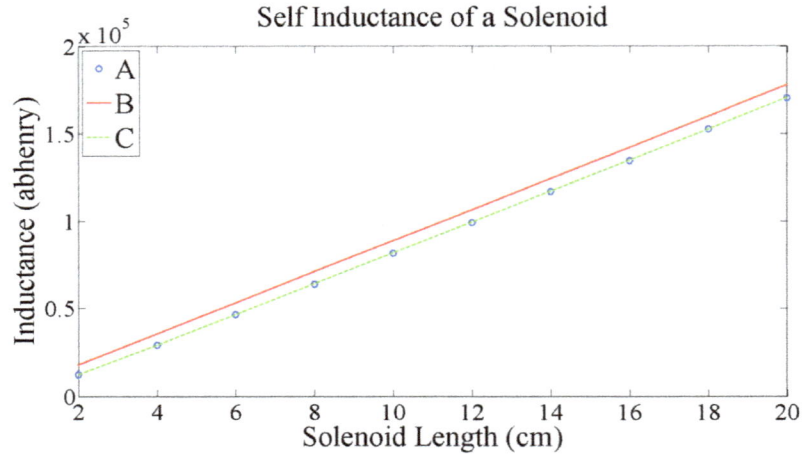

Figure 45.8: The variation of self inductance as a function of the length of a solenoid, while keeping the number of turns per unit length constant at 15 turns/cm. The self inductance is calculated by (A) numerical integration of Equation 45.35, (B) Ampere's law in Equation 44.11, and (C) closed-form solution in Equation 45.36.

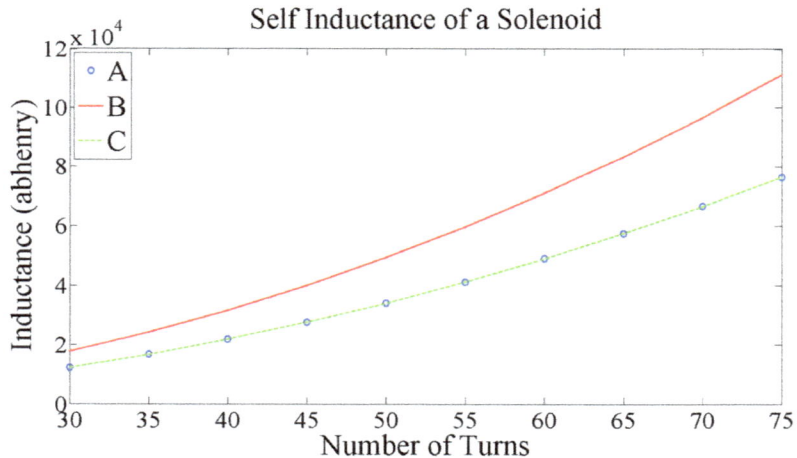

Figure 45.9: The variation of self inductance as a function of number of turns, while keeping the length of the solenoid fixed at 2 cm. The self inductance is calculated by (A) numerical integration of Equation 45.35, (B) Ampere's law in Equation 44.11, and (C) closed-form solution in Equation 45.36.

46

Absolute Measurement of
Resistance in Electromagnetic Units

Kirchoff was the first to measure resistance in the year 1849 [56]. Kirchoff is well known for the voltage and current laws in circuits that have been named after him, and covered in Chapter 47. A variation of his experiment from Reference [58] will be presented in this chapter, but the idea is the same. The measurement of resistance in electromagnetic units led to the development of resistance coils, and the resistance box is the precursor to the present day resistor, and will be described later in the chapter.

46.1   Resistance Measurement Technique

The experiment setup to measure resistance R is shown in Figure 46.1. It consists of two coils placed in close proximity for the magnetic field to couple. G_1 is a galvanometer to measure the

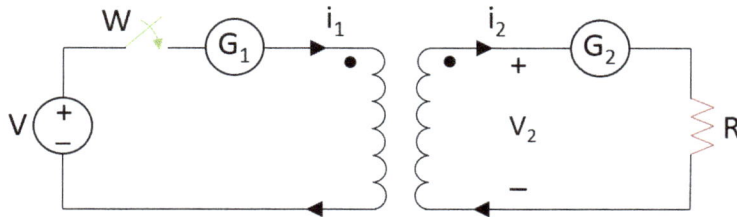

Figure 46.1: The schematic of the experiment setup to measure resistance in electromagnetic units.

steady-state current in the primary circuit, after the switch W is closed. When the switch W is closed, a time-varying current exists for a short duration in the primary, until the steady-state current is reached. This creates a time-varying magnetic-field flux across the secondary coil, and from Faraday's law, an induced current in the secondary circuit, which decays to 0 after the steady state is reached in the primary. G_2 is a ballistic galvanometer to measure the total charge flowing in the secondary circuit. R is the total resistance of the coil to be measured, including the resistance of the Galvanometer coil G_2. The secondary coil is modeled as a voltage source, as shown in Figure 43.6(b).

When the switch W is closed at time $t = 0$, the primary current can be modeled approximately as shown in Figure 46.2, as a step waveform with a short rise time. The induced voltage in the secondary circuit caused by mutual inductance M is

$$M\frac{di_1}{dt},$$ (46.1)

This waveform is also plotted in the figure. The induced voltage creates a burst of current in the secondary coil. The burst of current exerts a force on the magnetic needle of the ballistic galvanometer G_2, causing it to deflect, and oscillate from the momentum imparted to the needle, before coming to rest. The charge contained in the burst of current can be determined using the ballistic galvanometer, as explained earlier.

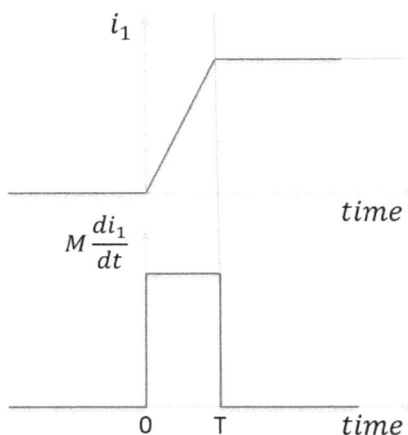

Figure 46.2: A plot of the current step in the primary coil, and the plot of the induced voltage in the secondary coil.

The induced voltage V_2 in the secondary circuit, using the dot convention,

$$V_2 = M\frac{di_1}{dt} - L\frac{di_2}{dt},$$ (46.2)

where i_1 and i_2 are the primary and the secondary currents. Substituting Ohm's law,

$$V_2 = i_2 R,$$ (46.3)

in the above equation,

$$i_2 R = M\frac{di_1}{dt} - L\frac{di_2}{dt}.$$ (46.4)

Let T be the duration over which the transient current i_2 exists in the secondary circuit. Integrating the above equation over $[0, T]$,

$$R \int_0^T i_2 \, dt = M \int_0^T \frac{di_1}{dt} dt - L \int_0^T \frac{di_2}{dt} dt$$

$$= M \int_{i_{1i}}^{i_{1f}} di_1 - L \int_{i_{2i}}^{i_{2f}} di_2. \tag{46.5}$$

One can get a better understanding of the right-hand side of the above equation by writing the integral as a Riemann sum, as written in Equation 31.10, and is left as an exercise for the reader.

i_{1i} and i_{1f} are the initial and the final steady-state currents in the primary circuit, and i_{2i} and i_{2f} are the initial and final currents in the secondary circuit,

$$i_{1i} = 0 \tag{46.6}$$
$$i_{1f} = i_s \tag{46.7}$$
$$i_{2i} = 0 \tag{46.8}$$
$$i_{2f} = 0, \tag{46.9}$$

where i_s is the steady-state current in the primary circuit. Since the initial and final currents are 0 in the secondary circuit, Equation 46.5 can be simplified as

$$R = M \frac{i_s}{Q_2}, \tag{46.10}$$

where,

$$\int_0^T i_2 \, dt = Q_2, \tag{46.11}$$

and Q_2 is the charge contained in the transient secondary current i_2. Q_2 can be calculated using the measurements made from the throw of the ballistic galvanometer, as discussed in detail in Chapter 31. i_s is measured using the galvanometer G_1, and the mutual inductance M can be calculated from the geometry of the coils. R can therefore, be measured using Equation 46.10. Once R has been determined, the resistivity of the material ρ, and the conductivity σ, can be calculated using Equation 39.18.

Substituting Equation 27.13 and Equation 31.30 in Equation 46.10, and simplifying,

$$R = \frac{M}{2} \frac{G_2}{G_1} \frac{\pi \tan \theta_1}{T_{1/2} \sin \left(\frac{\theta_2}{2} \right)}, \tag{46.12}$$

where G_1 and G_2 are the galvanometer constants, θ_1 and θ_2 are the angle measurements of the magnetic needles of the galvanometers, and $T_{1/2}$ is half the period of oscillations of the ballistic galvanometer G_2.

Note that the terrestrial magnetic-field strength H cancels in the above equation. In Chapter 56, the measurement of the terrestrial magnetic field using Gauss's magnetometer will be superseded by a different technique. Even when the value of H is revised, the measurement result of R using the technique presented does not change, since H cancels in the above equation. Other techniques to measure R do exist, such as the ones in References [77]-[80], which serve as a confirmation of the resistor values, but these are not covered in this book.

Although the self inductance of the ballistic galvanometer coil has not been explicitly included in Equation 46.5, L in Equation 46.2 can be thought of as also including the inductance of the galvanometer coil. After Kirchoff's circuit laws have been presented in the following chapter, it can be shown from the circuit equations of the secondary circuit, the term with the self inductance of the galvanometer coil also reduces to 0, similar to the self-inductance term of the secondary coil. The resistance coil can be wound twice so that it loops in the forward and reverse directions to cancel the magnetic flux on itself, as explained in the following section, and resulting in negligible self inductance. The resistance coil will be referred to as a resistor.

According to dimensional analysis applied to Equation 46.10,

$$[R] = \frac{abhenry \cdot abampere}{abampere \cdot s}$$
$$= \frac{cm}{s}, \tag{46.13, EMU}$$

the unit of resistance in electromagnetic units has the same unit as velocity! To avoid this meaningless unit, a new base unit for current in the SI system will be introduced.

46.2 Resistance Box

A resistance box is a tunable resistor and is shown in Figure 46.3. It consists of a set of resistance coils 1–4, and plugs P_1–P_3, attached to a metal bar B, as shown in the figure. Note that the resistance coils are double wound so that the current in the coil flows in opposite directions. Consequently, there is no self-induced magnetic flux, and therefore the self inductance of the coil is 0. The two terminals of the resistor are T_1 and T_2.

From Ohm's law, less current flows across a resistor with a higher resistance. When a plug is put in place, such as P_2, most of the current flows across the metal bar between a and b, and very little through the resistance coil. The resistance between a and b through the bar is small, since the length of the bar between a and b is tiny compared to the coil, and has a much larger cross-section area than the resistance coil. The preferred path for the current flow is through the bar, which can be empirically verified. This can also be proven mathematically using Kirchoff's circuit laws presented in the following chapter. Inserting a plug decreases the resistance between the terminals of the resistance box.

When a plug is removed, the current is forced through the resistance coil, thereby increasing

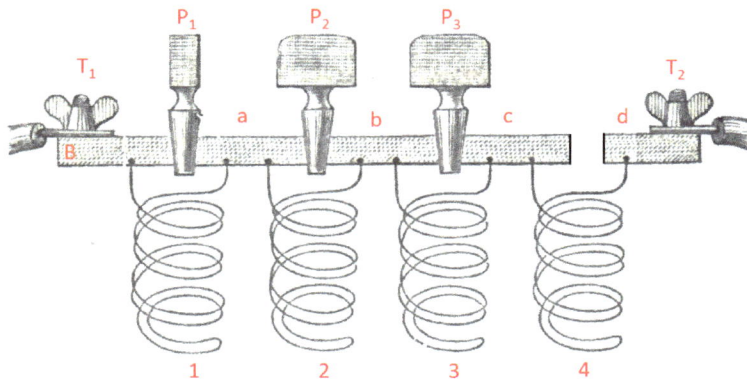

Figure 46.3: A resistance box [27].

the resistance between the terminals T_1 and T_2. By inserting a plug or removing it, the resistance between the terminals T_1 and T_2 can be varied.

47

Kirchoff's Circuit Laws on Voltage and Current

Kirchoff's voltage law (KVL) and Kirchoff's current law (KCL) are used to setup circuit equations, and solving for branch currents in circuits, and voltages across components such as resistors, inductors, and capacitors, etc. The focus of this chapter will be on the derivation and proof of Kirchoff's voltage and current laws. The reader is referred to other books for more details on setting up and solving circuit equations.

47.1 Kirchoff's Voltage Law

Kirchoff's voltage law will be derived using the simple circuit shown in Figure 47.1, but the result can be generalized to any circuit. Three resistance coils C_1, C_2, and C_3 are connected in series with a voltage source as shown. The resistance values are different between the coils, which is illustrated by the coils of different thicknesses.

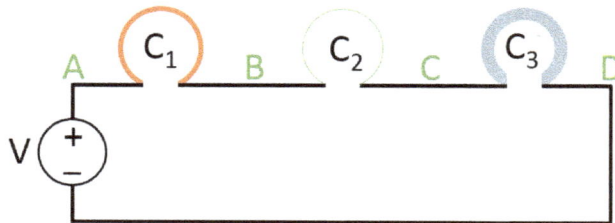

Figure 47.1: Three resistance coils connected to a voltage source.

The resistance of a coil of wire can be measured in absolute measure, as outlined in Chapter 46, from which the resistivity can be determined. With resistivity known, the resistance of a wire of a certain length, can be computed using Equation 39.18. This value can be used to check if the resistance of a wire is negligible or not. The wires connecting the components, marked A, B, C, and D, are treated as ideal wires with 0 resistance. If the resistance cannot be ignored, a resistor component can be added to represent the wire resistance explicitly, and the wires can then be treated as being ideal.

A wire with little resistance or an ideal wire, and two points marked 1 and 2 on the wire, is shown in Figure 47.2. If current i is flowing in the wire, using Ohm's law, the voltage difference V_{12} between 1 and 2 is

$$V_{12} = iR_w, \tag{47.1}$$

where R_w is the resistance of the wire between points 1 and 2. If R_w is a very small value, or 0 in the ideal case,

$$V_{12} \approx 0, \tag{47.2}$$

and therefore, the voltages at points 1 and 2 are equal, $V_1 = V_2$. Since voltage is the path integral of the electric field, the electric field between Points 1 and 2 is also 0.

Figure 47.2: A wire, and two points 1 and 2.

With the battery connected to the coils, current flows in the wire, which can be measured using a galvanometer. The current flowing through each of the coils is the same, and can be empirically verified. Using Ohm's law,

$$\Delta V_n = iR_n, \tag{47.3}$$

is the voltage across a coil of resistance R_n, and current i flowing in the coil .

Lets do a thought experiment on the circuit. Imaginary capacitors with small capacitances are connected across the components, as shown in Figure 47.3(a). The voltage difference across a component, as written in Equation 47.3, charges the capacitor, resulting in an electric field across the capacitor plates as shown. The capacitors and the wires connecting their terminals are isolated from the rest of the circuit, and shown in Figure 47.3(b). This circuit is in electrostatic equilibrium, since there are no moving charges.

It was proven in Chapter 5, the path integral of an electric field over a closed loop in the electrostatic case is 0,

$$\oint \vec{E} \cdot \vec{dl} = 0. \tag{47.4}$$

But voltage is the path integral of the electric field, defined in Equation 10.6. The above equation can be written in terms of voltage as described next.

A closed loop path is shown in Figure 47.4. The path is broken into N segments. The ith segment has endpoints marked a and b. The voltage between the two endpoints is

$$\Delta V_i = V_a - V_b$$

$$= \int_a^b \vec{E} \cdot \vec{dl}. \tag{47.5}$$

(a)

(b)

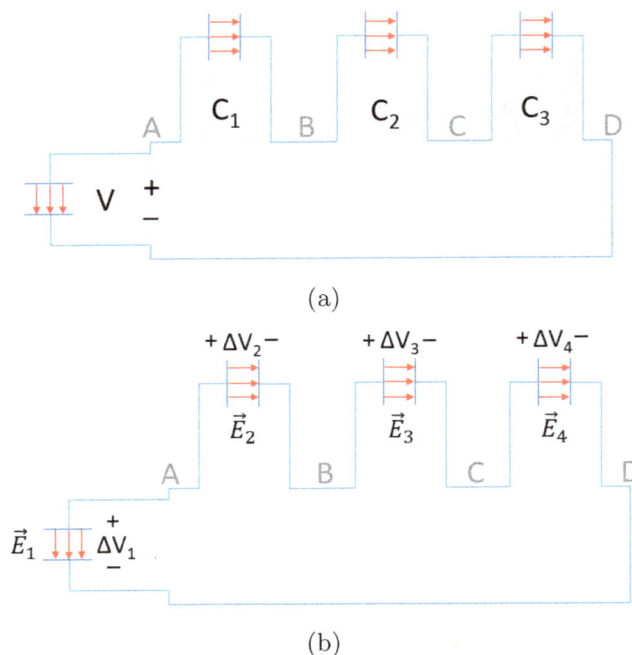

Figure 47.3: (a) Imaginary capacitors connected across the components in the circuit. (b) The imaginary capacitors without the circuit components is an electrostatic condition.

If ΔV_1, ΔV_2, ..., ΔV_N are the voltage changes across the segments in the closed loop, Equation 47.4 is rewritten as

$$\sum_{i=1}^{N} \Delta V_i = 0. \tag{47.6}$$

The above equation is Kirchoff's voltage law, which states that the sum of the voltage changes in a closed loop is 0. The directions of the path integrals of the electric field defining ΔV_1, ΔV_2, ..., ΔV_N, must all be in the same direction, either clockwise or counter clockwise, to be equivalent to Equation 47.4.

Applying KVL to the circuit in Figure 47.3(b), starting at Node A, moving within the wires, and returning back to Node A, for example, in a clockwise loop,

$$\Delta V_2 + \Delta V_3 + \Delta V_4 - \Delta V_1 = 0. \tag{47.7}$$

Note that there is a change in the voltage level only when traversing across the capacitor plates. In the electrostatic case, a metal wire is an equipotential volume, and no electric field is present within the wire, and no change in voltage exists during the traversal within the wires or within any metal.

Using Ohm's law in Equation 47.3, the above equation is written as

$$iR_2 + iR_3 + iR_4 - V = 0, \tag{47.8}$$

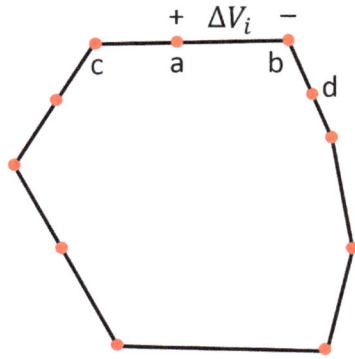

Figure 47.4: A closed path with voltage differences between the nodes. A closed path across the capacitors in Figure 47.3(b) can be represented in such a manner.

where $V = \Delta V_1$ is the voltage of the battery source. Measurement of the voltage of a battery will be presented in a later chapter. The above equation can be solved for the current, and from which the voltages across the resistor components can be calculated. It is customary to define one of the nodes, the reference node, as $0\,V$. The voltages at all other nodes are calculated with respect to the reference node. The reader is referred to other books for more details and examples.

47.2   Using KVL on Circuits With Time-Varying Voltages and Currents

Suppose a circuit consists of components connected together, and each component is described by a voltage-current behavior. For example, in the case of an inductor, the voltage-current relation is given in Equation 44.3. If time-varying sources, voltages, and currents are present in the circuit, it will be shown in this section that Kirchoff's voltage law is still applicable. It will be assumed that there is no propagation delay of voltages. Any change in the voltage of the source is reflected at any node in the circuit instantaneously.

The time-varying values may be viewed as a series of snapshots in discrete time intervals. At every moment, performing the thought experiment of attaching capacitors across the components, and isolating them like Figure 47.3(b), one can prove that KVL is valid at every time instant. Therefore, KVL also holds true in circuits with time-varying voltages and currents, and for any set of components making up the circuit.

How can KVL be valid when Faraday's law states that

$$\oint \vec{E} \cdot \vec{dl} = -\frac{\partial \Phi}{\partial t}$$
$$\neq 0?$$
(47.9)

The above equation is inconsistent with Equation 47.4, from which KVL was derived. This question is addressed next.

A solenoid S is connected to a resistance coil C in Figure 47.5(a). A time-varying flux in the solenoid induces a current flow in the wire to oppose the changing flux. In the derivation of Faraday's law, it was hypothesized that an electric field in a closed loop must exist in the wire for the current to flow. Clearly, the path integral of the electric field is non-zero, and voltage is not even defined. However, the circuit representation of the solenoid-resistance coil enables the use of KVL.

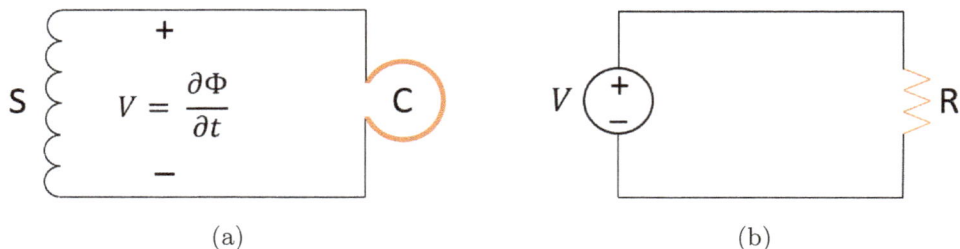

Figure 47.5: (a) A solenoid connected to a resistance coil. (b) The circuit schematic of (a).

The circuit schematic of Figure 47.5(a) is shown in Figure 47.5(b). The induced voltage across the solenoid is modeled as a voltage source, as explained in Chapter 43. The wires connecting the solenoid and the resistance coil is approximated to be ideal. The thought experiment of attaching imaginary capacitors to the circuit, can now be used to show that KVL is applicable. Modeling the solenoid as a voltage source enables the use of KVL. In summary, KVL is also applicable when time-varying voltages and currents are present in a circuit.

47.3   Kirchoff's Current Law

Lets start with magnetostatics first, and Kirchoff's current law (KCL) will be derived assuming that no time-varying currents are present in the circuit. In the next step, this law will be extended to show that it is also valid for time-varying currents.

Kirchoff's current law is the same as the current-continuity equation derived in Equation 29.12, but cast in a different form. The cross section of a wire with branches b_1, b_2, and b_3 is shown in Figure 47.6(a), and the circuit schematic in Figure 47.6(b). A closed surface S with areas S_1, S_2, and S_3 in the branches b_1, b_2, and b_3 has been drawn. Let the volume-current density in the wire be denoted as $\vec{J}$. The outward unit-normal vectors to the surface areas are denoted as $\hat{n}_1$, $\hat{n}_2$, and $\hat{n}_3$. Using Equation 28.13,

$$\oint_S \vec{J} \cdot \hat{n} \, dA = \int_{S_1} \vec{J} \cdot \hat{n}_1 \, dA + \int_{S_2} \vec{J} \cdot \hat{n}_2 \, dA + \int_{S_3} \vec{J} \cdot \hat{n}_3 \, dA \tag{47.10}$$

$$= i_1 + i_2 + i_3, \tag{47.11}$$

where positive values of i_1, i_2, and i_3 are the currents flowing *out* of the surfaces S_1, S_2, and S_3

for the chosen values of the unit-normal vectors. Note that

$$-i_i = -\int_{S_i} \vec{J} \cdot \hat{n}_i \, dA \qquad (47.12)$$

is the current flowing *into* the surface S_i, and $\hat{n}_i$ is the outward unit-normal vector to S_i.

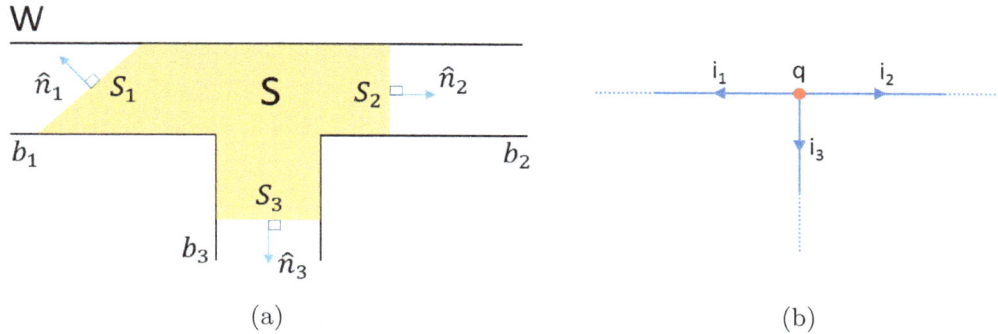

(a)   (b)

Figure 47.6: (a) A closed surface S at a wire junction. (b) The circuit schematic of (a).

Applying the current-continuity equation for magnetostatics derived earlier in Equation 29.12,

$$\oint_S \vec{J} \cdot \hat{n} \, dA = 0. \qquad (47.13)$$

Equation 47.11 reduces to

$$i_1 + i_2 + i_3 = 0. \qquad (47.14)$$

Although the above equation was derived for 3 branch currents, it can be generalized to any number of branches N splitting from a node,

$$\sum_{i=1}^{N} i_i = 0, \qquad (47.15)$$

where i_i is the current in the ith branch flowing *out* of a node. Both sides of the above equation may be multiplied by -1, and in this case $-i_i$ is the current flowing *into* a node. The above equation is Kirchoff's current law, which states that the sum of all the currents flowing out of (or into) a node is 0.

For the above equation to be true, some of the current terms in the branches must be negative, and some positive. For the chosen direction of the unit- normal vectors defined out of a surface, a positive value of the branch current means that current is flowing out of the surface, and a negative value is current flowing into the surface. The terms in the above equation may be rearranged so that the positive current terms are all on one side, and the negative current terms, are written as positive current terms on the other side of the equation. This equation lends itself to a different interpretation of KCL: the sum of the currents flowing into a node is equal to the sum of the currents flowing out of the node.

It will be shown next that KCL also holds true for time-varying currents. To get an idea of when KCL will not be valid, lets revisit the same example in Figure 47.6(a), but with time-varying currents. Suppose the surfaces S_1, S_2 and S_3 are each extended along the wire so that the surface S is made very large. S is large enough that if the current density at S_1 changes, there is a time lag for the current density at S_2 and S_3 to change.

As a result, the charge contained within the surface will vary with time. For example, if the current flowing into S_1 increases, the charge contained within the volume S would also increase, since there is a time lag for the current to flow out of S_2 and S_3. During this time, repeating Equation 29.3,

$$\oint_S \vec{J} \cdot d\vec{A} = -\frac{\partial Q}{\partial t}$$
$$\neq 0, \tag{47.16}$$

and KCL is no longer applicable.

It will be assumed that there is no propagation delay for the currents, although they may be time varying. Any change in the source at time t, is reflected in the voltages and the currents in the circuit instantaneously. At any time instant t, the voltages and currents in the circuit can be viewed as non-time varying, or in a magnetostatic state. Therefore, KCL is valid at any instant in time, and is applicable to time-varying currents as well.

47.4   Circuit Theory

Using KCL and KVL, circuits may be simplified without changing its voltage and current behavior. For example, resistors connected in series, as shown in Figure 47.1, may be replaced by a single resistor, whose value is the sum of the resistors in series. The reader is referred to other books on circuit theory for more information.

48

Measurement of the Voltage of a Battery in Electromagnetic Units

The absolute measurement of resistance in EMU, enables quantifying the voltage of a battery in EMU. This technique is presented in this chapter.

Many of the equations presented until now have the term voltage in them. For example, the definition of voltage as the path integral of an electric field, the capacitance relation in Equation 35.7, and Ohm's law. It may seem that there are multiple definitions of the same electrical quantity, and may be inconsistent. This is not the case, and the definitions in ESU and EMU presented until now, will be summarized in the form of a flowchart in Chapter 58 and Chapter 59. From this it will become clear that there are no circular definitions or multiple definitions of the same electrical quantity. It will also become clear, why some of the equations are common to both ESU and EMU, while others are specific to ESU or EMU.

48.1 Measurement of the Voltage of a Battery

Clark's experiment [81] differs slightly from the technique presented here. In Clark's experiment, the setup is such that no current is drawn from the battery whose voltage is to be measured. The advantages of Clark's experiment setup are (1) the strength of the battery is not diminished in the process, (2) the internal resistance of the battery does not affect the results of the measurement.

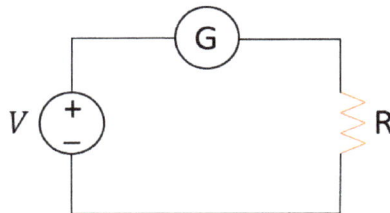

Figure 48.1: The experiment setup to determine the voltage of the battery in electromagnetic units.

The reader is referred to Clark's original paper in Reference [81] for more details. In this chapter, the original experiment has been slightly modified for simplicity.

A battery of voltage V that is to be measured, is connected to a resistor of known resistance R in EMU, shown in Figure 48.1. The current flowing in the circuit i in EMU, is measured using a galvanometer G. Using Ohm's law, the voltage of the battery is

$$V = iR. \tag{48.1}$$

For higher accuracy, if the resistance of the coil of wire in the galvanometer is significant compared to R, then this value must be added to the resistor R.

49

B vs H Measurement Using the Ballistic-Galvanometer Method

A simplified version of the ballistic-galvanometer method to introduce the magnetic-flux density $\vec{B}$, was presented in Chapter 41. The complete version of the experiment is now ready to be presented [67]. The calculation of the magnetic field within a toroid, or doughnut shape, is presented first, followed by the experiment, typical results of a magnetic material, and its properties.

49.1   Magnetic Field in a Toroid

The cross section of a toroid is shown in Figure 49.1. Although not shown in the figure, a wire carrying current i is tightly wound around the toroid with N_{pri} turns, *pri* stands for the *pri*mary circuit, marked P in Figure 49.2. The magnetic field within the toroid will be solved using Ampere's law.

As discussed in Section 41.1, the magnetic field generated by the primary current within the toroid is circular, as shown by the dotted line of radius r in the figure. The magnetic-field strength $H = |\vec{H}|$ will be assumed to be uniform within the toroid. Assume that the primary current direction is such that the magnetic field generated is in the clockwise direction marked by the arrows.

The dotted line at radius r is also the Ampere loop. Although r changes over the cross section of the toroid, the midpoint of the cross section will be used for the value of r. By Ampere's law,

$$\oint \vec{H} \cdot \vec{dl} = 4\pi i_{enc},\tag{49.1}$$

where i_{enc} is the current enclosed by the Ampere loop, and the path integral is evaluated along the dotted line. If H is assumed to be uniform, the above equation simplifies to

$$H(2\pi r) = 4\pi N_{pri} i,\tag{49.2}$$

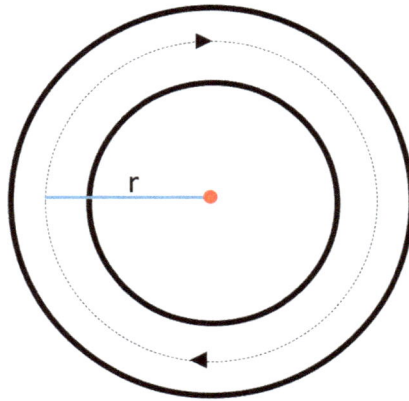

Figure 49.1: The cross section of a toroid with a wire (not shown) tightly wound around it.

where $N_{pri}i$ is the current enclosed by the Ampere loop. Since

$$N'_{pri} = \frac{N_{pri}}{2\pi r}$$ (49.3)

is the number of turns per unit length, Equation 49.2 is often written as

$$H = 4\pi i N'_{pri}.$$ (49.4)

From the above equation, if the current changes by Δi, the magnetic field changes by

$$\Delta H = 4\pi N'_{pri}\Delta i.$$ (49.5)

This result will be used in the calculations presented in the following section.

49.2   Ballistic-Galvanometer Method

The setup of the experiment to characterize the magnetic-flux density B vs magnetic-field strength H behavior of a magnetic material is shown in Figure 49.2. The material to be characterized is in the shape of a toroid with a tightly wound wire around its entire length, and marked P for the primary coil. The secondary coil S is wound over the primary in a smaller section, shown shaded.

A battery B is connected to a galvanometer G_1 to measure the steady-state current in the primary coil. R is a resistance box, and K are simple switches that can be opened or closed, which are used to reverse the current flow in the primary coil. K is used to set the current in the path $\{a, b\}, \{c, d\}$ or in the opposite direction $\{a, c\}, \{b, d\}$.

When the plug in the resistance box is removed or inserted, the current in the primary coil changes quickly by Δi. This quick change in the current, creates a time-varying magnetic flux in the secondary coil that induces a burst of transient current in the secondary coil. G_2 is a ballistic galvanometer that will be used to calculate the charge contained in the burst of current in the

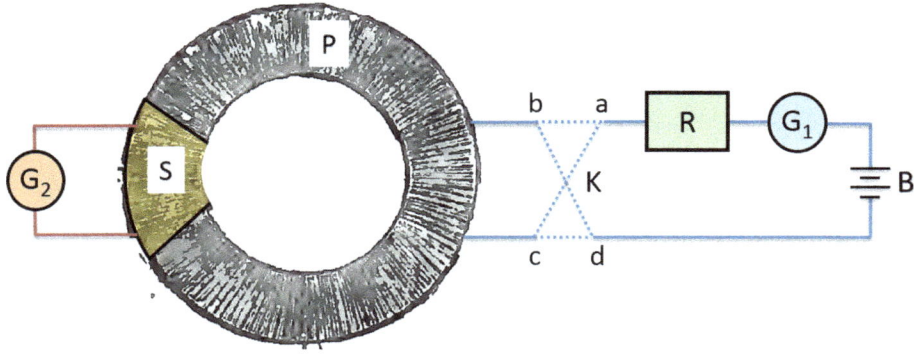

Figure 49.2: The setup of the ballistic galvanometer method [67].

secondary coil. The material can be characterized using this information as explained next.

Using the same exact setup in Figure 49.2, two experiments are done: in the first experiment, the toroid is a non-magnetic material such as wood, and in the second, the toroid is the magnetic material that is to be characterized. A change in the current in the primary coil Δi, creates a change in the magnetic field ΔH within the toroid. A magnetic material alters the strength of the magnetic field, as discussed in detail in Chapter 41. In the case of the toroid made of magnetic material, ΔH is altered by the magnetic material to ΔB.

The primary-coil current varying from 0 to i_{max}, for example, is varied in small steps, using the plugs of the resistance box. The sudden change in the current in the primary coil, generates an induced current in the secondary coil. For the primary coil current steps $\Delta i \in \{\Delta i_1, \Delta i_2, \dots , \Delta i_n\}$, the throw of the ballistic galvanometer G_2, $\theta \in \{\theta_1, \theta_2, \dots, \theta_n\}$ in the non-magnetic toroid, and $\theta' \in \{\theta'_1, \theta'_2, \dots, \theta'_n\}$ in the magnetic toroid are noted down. For the current steps in the primary coil, the change in the magnetic field in the non-magnetic toroid are $\Delta H \in \{\Delta H_1, \Delta H_2, \dots, \Delta H_n\}$, and $\Delta B \in \{\Delta B_1, \Delta B_2, \dots, \Delta B_n\}$ in the toroid made of the magnetic material.

In the non-magnetic material, ΔH can be calculated from Δi using Equation 49.5. Using θ and θ', ΔB in the magnetic toroid caused by Δi can be calculated as explained next.

The circuit schematic of the experiment is shown in Figure 49.3. V_s is the voltage across the secondary coil, and R_L is the load resistance. In this case, the load resistance is the resistance of the coil in the galvanometer G_2, but it is not necessary to know the value of R_L. This will become apparent shortly. The induced current in the secondary coil i_s can be formulated as

$$i_s = \frac{V_s}{R_L}$$
$$= \frac{1}{R_L} \left(\frac{d\Phi_m}{dt} - L \frac{di_s}{dt} \right), \tag{49.6}$$

where Φ_m is the mutual coupling of the magnetic flux in the secondary coil caused by the primary

341

current step, and L is the self inductance of the secondary coil. Integrating the above equation

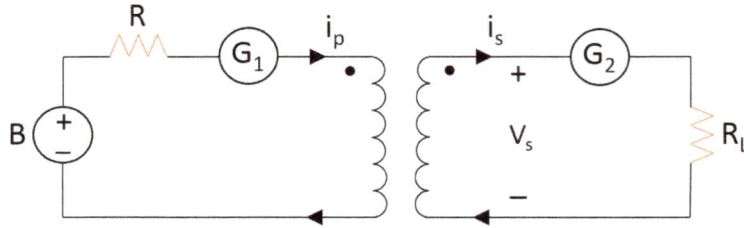

Figure 49.3: The circuit schematic of Figure 49.2.

over the duration of the burst of the induced current in the secondary coil $[0, T]$,

$$\int_0^T i_s \, dt = \frac{1}{R_L} \left(\int_0^T \frac{d\Phi_m}{dt} dt - L \int_0^T \frac{di_s}{dt} dt \right). \tag{49.7}$$

The term with the self inductance L reduces to 0, since the secondary current i_s is 0 at the beginning, and at the end of the time interval. Simplifying the above equation,

$$Q = \int_0^T i_s \, dt$$
$$= \frac{N_{sec}A}{R_L} \int_0^T \frac{dB}{dt} dt, \tag{49.8}$$

where Q is the charge contained in the burst of current induced in the secondary coil, $\Phi_m = BN_{sec}A$ is the flux across the secondary coil with N_{sec} turns, subscript sec stands for $secondary$, and A is the loop area of the secondary coil. Simplifying the above equation,

$$Q = \frac{N_{sec}A \, \Delta B}{R_L}, \tag{49.9}$$

where ΔB is the change in the value of the magnetic-flux density in the toroid, between the beginning and the end of the interval $[0, T]$. In the case where the toroid is made of a magnetic material, using Equation 31.77, Equation 49.9 can be written as

$$\Delta B \propto \theta'. \tag{49.10}$$

In the case of the non-magnetic toroid, if the experiment setup is identical to the magnetic toroid,

$$\Delta H \propto \theta. \tag{49.11, EMU}$$

θ' and θ are the throw of the ballistic galvanometer G_2 in the experiments using a magnetic, and a non-magnetic toroid. Dividing Equation 49.10 by Equation 49.11,

$$\Delta B = \Delta H \left(\frac{\theta'}{\theta} \right). \tag{49.12}$$

Substituting Equation 49.5 in the above expression,

$$\Delta B = 4\pi N'_{pri} \, \Delta i \left(\frac{\theta'}{\theta}\right). \qquad (49.13, \text{ EMU})$$

To summarize, the current steps Δi are generated by removing/inserting the plugs in the resistance box, and measured using the galvanometer G_1. From the observations θ' and θ made on the galvanometer G_2, using the above equation, the changes in ΔB can be calculated. ΔH can be calculated from the current steps using Equation 49.5. B vs H can then be plotted from this result. A typical plot of the behavior of a magnetic material will be discussed in the next section.

49.3   Hysteresis Plot

A typical plot of B vs H of a magnetic material is sketched in Figure 49.4. The x-axis is the applied magnetic field H, and the y-axis is the magnetic field B that includes the effect of the magnetic material. The plot is separated into three segments, ab, bc, and cb. To begin with, the magnetic toroid is fully demagnetized, and the process of demagnetization will be discussed in Section 49.4.

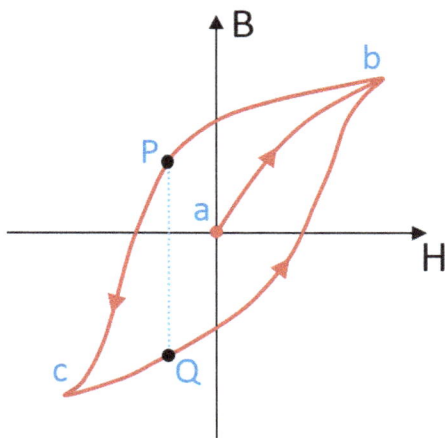

Figure 49.4: A plot of the hysteresis of a magnetic material.

The arrows on the plot is the order in which H is applied. Positive and negative values of H are magnetic fields in opposite directions. The current steps Δi to characterize the magnetic material is shown in the number lines in Figure 49.5. Each square and its neighbor, in the direction shown by the arrow, represents the change in current in the primary coil by inserting or removing a plug in the resistance box. The plot from a to b is generated using the currents steps from 0 to $+i_{max}$, shown in the number line marked 1, b to c is generated by the current steps from $+i_{max}$ to $-i_{max}$, marked 2, and finally c to b by varying from $-i_{max}$ to $+i_{max}$. The current direction in the primary coil is reversed using the switches marked K in Figure 49.2.

A characteristic of the plot is that the value of B for an applied H depends on the past history of H that has been applied in the material. Such a plot is known as a hysteresis plot. For

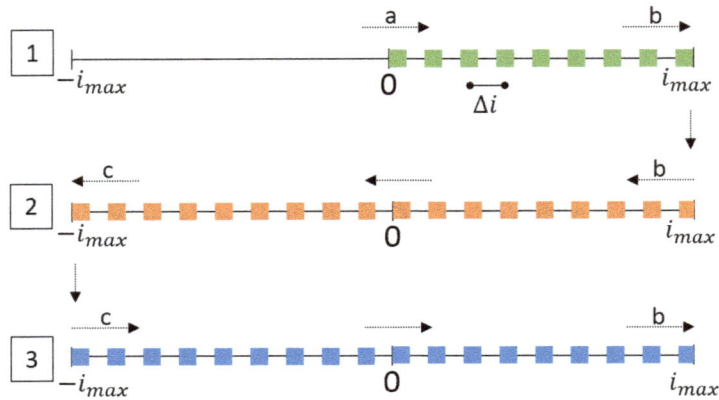

Figure 49.5: The current steps in the primary coil to generate the B vs H plot of a magnetic material.

example, for the same applied value of H, the value of B may be P or Q, as marked in the figure, depending on the history of H that has been applied. Along the path from b to c, B takes on the value P, but Q along the path c to b.

As shown in the figure, the hysteresis plot is highly non-linear. A linear behavior would be a straight line through the origin. The hysteresis plot, however, is far from being linear. Using Equation 41.1, the ratio of B and H is the permeability μ_r,

$$\mu_r = \frac{B}{H}. \tag{49.14, EMU}$$

Measurement data using the ballistic-galvanometer method in Reference [67] is plotted in Figure 49.6. The current steps in the primary coil are increased starting at 0 to a final value of $+i_{max}$ to generate the curve labeled 1. Following this, the current is decreased from $+i_{max}$ to 0 to generate 2, and finally the current is increased again from 0 to $+i_{max}$ to generate 3. The hysteresis behavior can be clearly seen in the plot. Since the current steps are varied different compared to Figure 49.4, the plot looks different from the typical curve.

49.4 Demagnetization by Current Reversals

To ensure the toroid is not magnetized to begin with, so that the characterization of B vs H may be done accurately, a series of reversing currents are applied to the toroid that decrease in magnitude each time. Adjusting the plugs in the resistance box, a large current is first applied. Following this, the switches K are flipped to route the current in the opposite direction in the toroid. This process is repeated, and each time, the strength of the current is decreased by removing a plug from the resistance box, until the current becomes 0.

In the experiment technique in Reference [67], Ewing uses a liquid rheostat, which is a type of tunable resistor, and a commutator to reverse the current direction, to make this process more

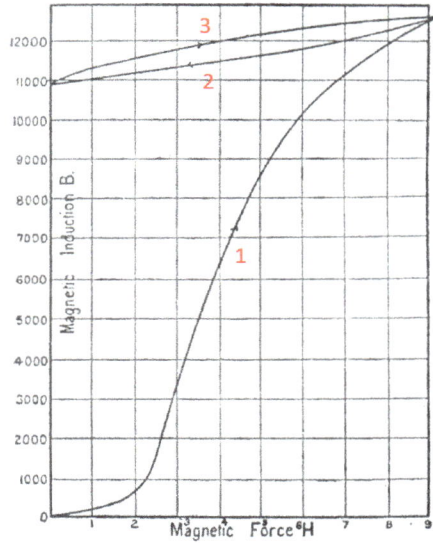

Figure 49.6: An example of a B vs H plot from Reference [67].

efficient. The setup presented here differs from Reference [67] for simplicity, but the idea is the same. The toroid can be verified to be demagnetized by checking to see if it has any effect in altering the direction of a magnetic needle.

345

50

Gauss's Law for Time-Varying Charge Density and Time-Varying Fields

Gauss's law was derived for stationary charges in Chapter 36. Gauss's law, however, is a much more general law, and a much stronger description of the behavior of fields. It is also valid for time-varying fields and time-varying charge density [82][83]. In addition, it is also valid in a non-uniform dielectric medium. It is always valid, no matter what! The conclusions will be stated without a proof for now, but the logical proof will be presented in the next chapter. Gauss's law for time-varying fields is often neglected, and to signify its importance, this chapter is wholly dedicated for this discussion.

Lets look at an example to illustrate the point. The cross section of a closed 3D surface S is shown in Figure 50.1, in any type of dielectric medium, uniform or non-uniform. Lets say that a charge q is stationary at Point A. By Gauss's law,

$$\oint_S \vec{D} \cdot d\vec{A} = 4\pi q, \tag{50.1}$$

where $q_{enc} = q$ is the charge enclosed within the surface S. In the differential form, repeating Equation 36.16, the above equation is written as

$$\nabla \cdot \vec{D} = 4\pi \rho, \tag{50.2}$$

where ρ is the volume charge density at the point where the divergence is calculated.

The charge moves from A to B, for example, in the path shown by the arrow. It may be moved at a constant speed, or the charge may accelerate/decelerate during the transit. Since a moving charge constitutes a current, this will modify the electric and magnetic fields in space and time. However, it will be stated without a proof for now, Gauss's law must always be satisfied at any time t. The logical proof will be presented in the next chapter.

It is possible to formulate the electric field generated by charges in motion. This is a general

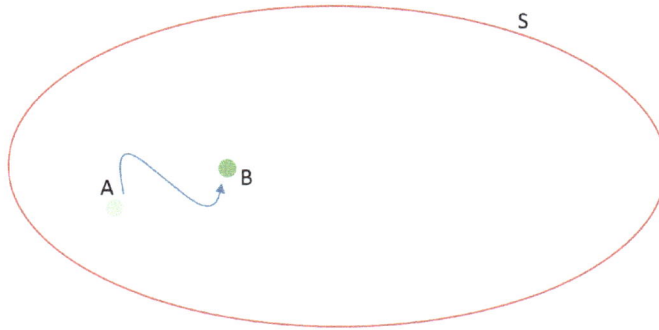

Figure 50.1: Although a point charge is moving from A to B, Gauss's law always holds true at any time instant.

form of Coulomb's law, known as time dependent generalization of Coulomb's law [84]. It is different from the electric field calculated using Coulomb's law in Equation 4.1, which is valid only for stationary charges. The derivation of the time-dependent generalization of Coulomb's law is beyond the scope of this book.

To emphasize that Gauss's law is also valid for time-varying fields and time-varying charge density, Gauss's law is written as

$$\nabla \cdot \vec{D}(\vec{r}, t) = 4\pi \rho(\vec{r}, t), \tag{50.3}$$

where $\vec{D}(\vec{r}, t)$ is the electric-flux density at the point $\vec{r} = (x, y, z)$ at time t, and $\rho(\vec{r}, t)$ is the volume-charge density at the point $\vec{r}$ at time t. Although wave propagation is not covered in this book, it will be stated that in the above equation, since $\vec{D}$ and ρ are related at the same point, there is no propagation delay between any change in ρ at a point, and its effect on $\vec{D}$ at the same point.

The above equation can be converted into the integral form, and shown that the integral form must also be valid for time-varying fields and time-varying charge density. Lets take a snapshot of the field $\vec{D}(t)$ at some instant t. By Gauss's divergence theorem, given a closed surface S,

$$\oint_S \vec{D}(t) \cdot d\vec{A} = \int_V \nabla \cdot \vec{D}(t)\, dV, \tag{50.4}$$

where the right-hand side is the volume integral in the space enclosed by the surface S. If Equation 50.3 holds true, then substituting this equation in the above relation,

$$\oint_S \vec{D}(t) \cdot d\vec{A} = 4\pi \int_V \rho(t)\, dV. \tag{50.5}$$

But the right-hand side of the above equation is the charge enclosed within the surface S at time t, and therefore,

$$\oint_S \vec{D}(t) \cdot d\vec{A} = 4\pi q_{enc}(t), \tag{50.6}$$

where $q_{enc}(t)$ is the charge enclosed within the surface at time t. Assuming Equation 50.3 is true, the integral form of Gauss's law is also valid for time-varying fields, and time-varying charge density.

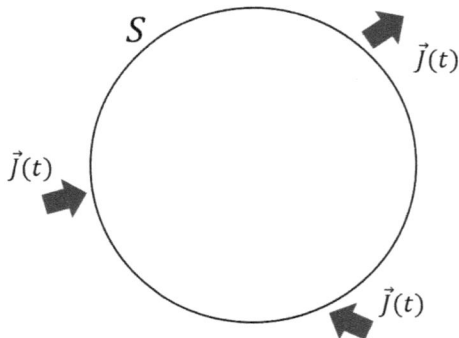

Figure 50.2: A time-varying charge density within the surface S caused by a current flow.

Another example of a time-varying charge density is shown in Figure 50.2. A closed surface S is shown in the figure. A time-varying current density $\vec{J}(t)$ may increase or decrease the charge contained in the volume. In this case, the time-varying charge density is caused by current flow. Although a time-varying field is present, the most perplexing aspect of Gauss's law is that Equation 50.6 must always be satisfied at any time t. The logical proof will be presented in the following chapter.

51

Displacement Current

Maxwell included the displacement current term in Ampere's law, so that it is also valid for time-varying fields and sources. This will be derived in this chapter [85]. The logical proof that Gauss's law must always be satisfied, including when time-varying fields and time-varying charge density are present, as well as in a non-uniform dielectric medium, will also be presented.

51.1 Ampere's Law for Time-Varying Fields and Currents

Faraday's law was formulated as

$$\nabla \times \vec{E} = -\frac{\partial \vec{B}}{\partial t},\tag{51.1}$$

relating magnetic-flux density $\vec{B}$, and electric field $\vec{E}$ at any point. The above equation states that a changing magnetic field, results in the generation of an electric field. By symmetry, a hypothesis will be made that the converse is also true: a time-varying electric field generates a magnetic field. Ampere's law will be modified to write the hypothesis as an equation.

Ampere's law was formulated in the case of a steady-state current density $\vec{J}$, and magnetic field $\vec{H}$ as

$$\nabla \times \vec{H} = 4\pi \vec{J}.\tag{51.2}$$

The left-hand sides of Equation 51.1 and Equation 51.2 have the same form. The right-hand side of Faraday's law in Equation 51.1 can be used to rewrite Ampere's law as

$$\nabla \times \vec{H} = 4\pi \vec{J} + \frac{\partial \vec{D}}{\partial t}.\tag{51.3}$$

Although Ampere's law was derived for the magnetostatic case, the above equation will be assumed to be valid for time-varying currents and time-varying magnetic fields. This will be proved next. Breaking symmetry with Faraday's law, it will become clear in a moment, it would be incorrect to include the negative sign in the time derivative.

349

Applying the divergence operation on both sides of the above equation,

$$\nabla \cdot \left(\nabla \times \vec{H} \right) = 4\pi \left(\nabla \cdot \vec{J} \right) + \frac{\partial}{\partial t} \left(\nabla \cdot \vec{D} \right). \tag{51.4}$$

From calculus, the divergence of curl operation $\nabla \cdot \nabla \times$ on a vector field is 0, and the left-hand side of the above equation reduces to 0,

$$0 = 4\pi \left(\nabla \cdot \vec{J} \right) + \frac{\partial}{\partial t} \left(\nabla \cdot \vec{D} \right). \tag{51.5}$$

In Equation 51.4, what if $\vec{H}$ is not defined at a point? $\nabla \cdot \left(\nabla \times \vec{H} \right)$ will be undefined. For example, in steady-state line currents, $\vec{H}$ is singular at the location of the line current, as written in Equation 23.19. The answer is that a line current is only an approximation. In reality, the wire has a cross section, and a well defined volume current density $\vec{J}$. $\nabla \times \vec{H}$ at a point is the volume current density $\vec{J}$ at that point, and there is no singularity at the location of the current flow.

Substituting Gauss's law in Equation 50.3 in the above equation,

$$0 = 4\pi \left[\nabla \cdot \vec{J} + \frac{\partial \rho_e}{\partial t} \right], \tag{51.6}$$

where ρ_e is the volume electric charge density. Simplifying the above result,

$$\nabla \cdot \vec{J} = -\frac{\partial \rho_e}{\partial t}, \tag{51.7}$$

which is the current-continuity equation that was derived earlier in Chapter 29. The derivation of the current-continuity equation from Ampere's law, is taken to be a logical proof of Ampere's law for time-varying sources and fields.

Since the current-continuity equation must always be satisfied, the above derivation also shows that Gauss's law must be applicable to time-varying fields and sources, as well as in any dielectric medium, uniform or non-uniform. As explained in Chapter 50, the integral form of Gauss's law must also be valid for time-varying fields and sources, and in any type of dielectric medium.

Factoring 4π from the right-hand side in Equation 51.3,

$$\nabla \times \vec{H} = 4\pi \left[\vec{J} + \frac{1}{4\pi} \frac{\partial \vec{D}}{\partial t} \right]. \tag{51.8}$$

Equating the above equation to Ampere's law for a steady-state current, the sum within the square brackets is an equivalent current density.

$\vec{J}_D$ is called the displacement-current density,

$$\vec{J}_D = \frac{1}{4\pi} \frac{\partial \vec{D}}{\partial t}. \tag{51.9}$$

Since $\vec{J}_D$ is added to the conduction-current density term $\vec{J}$, by dimensional analysis, they must have the same unit.

Repeating Faraday's law and Ampere's law,

$$\nabla \times \vec{E} = -\frac{\partial \vec{B}}{\partial t} \tag{51.10}$$

$$\nabla \times \vec{H} = 4\pi \vec{J} + \frac{\partial \vec{D}}{\partial t}, \tag{51.11}$$

there is still an asymmetry between the two equations. As noted in Chapter 42, it may be possible to formulate Faraday's law without the negative sign in front of the time derivative. Therefore, the asymmetry in the signs of the partial time derivatives can be ignored. There is an electric-current density term $\vec{J}$ in Ampere's law, but no symmetric term in Faraday's law. This will be addressed in Chapter 52.

51.2   Integral Form of Ampere's Law

The integral form of Ampere's law will be derived using the Kelvin-Stokes theorem [86][87]. The reader is referred to other resources for more details on the Kelvin-Stokes theorem. Kelvin-Stokes theorem may be applied on a 3D surface. However, the simplest surface is a planar surface, which will be used to derive the integral form of Ampere's law.

A planar surface S with boundary C is shown in Figure 51.1. The normal vector to the planar surface is $\hat{n}$. Integrating the differential form of Ampere's law over the surface S,

$$\int_S \left(\nabla \times \vec{H} \right) \cdot d\vec{A} = 4\pi \left[\int_S \vec{J} \cdot d\vec{A} + \frac{1}{4\pi} \frac{\partial}{\partial t} \int_S \vec{D} \cdot d\vec{A} \right]. \tag{51.12}$$

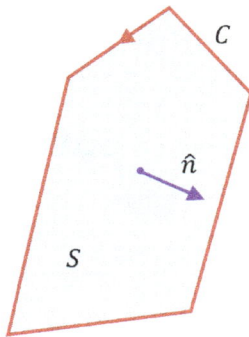

Figure 51.1: A planar surface S to derive the integral form of Ampere's law.

Applying the Kelvin-Stokes theorem on the left-hand side of the above equation, the surface integral is transformed into a line integral of magnetic field $\vec{H}$ over its boundary C,

$$\oint_C \vec{H} \cdot d\vec{l} = 4\pi \left[i_{enc} + \frac{1}{4\pi} \frac{\partial}{\partial t} \int_S \vec{D} \cdot d\vec{A} \right], \tag{51.13}$$

where by definition of $\vec{J}$,

$$\int_S \vec{J} \cdot d\vec{A} = i_{enc}, \tag{51.14}$$

is the current that flows across the surface S, into the region that $\hat{n}$ points towards. In Kelvin-Stokes theorem, the direction of the path of integration must conform to the right-hand rule: if the fingers of the right hand curl in the direction of the path integral, the thumb points in the direction of the surface area unit-normal vector. This is consistent with the third right-hand rule for Ampere's law in Chapter 21.

Equation 51.13 may be written as

$$\oint_C \vec{H} \cdot d\vec{l} = 4\pi \left[i_{enc} + i_d \right], \tag{51.15}$$

where i_d is the displacement current,

$$i_d = \frac{1}{4\pi} \frac{\partial}{\partial t} \int_S \vec{D} \cdot d\vec{A}. \tag{51.16}$$

By dimensional analysis, i_d must have the same unit as current, since it is added to the conduction-current term i_{enc}. The sum of i_{enc} and i_d may be viewed as an equivalent "current" flow.

51.3   Current-Voltage Relation in a Capacitor

The equation relating the current flowing into a capacitor, and the resulting voltage that forms across the capacitor at some time instant t, will be derived in this section. Repeating the capacitance relation,

$$Q = CV. \tag{51.17}$$

From the above equation, given a change in the voltage ΔV between the capacitor plates, the charge stored changes by ΔQ,

$$\Delta Q = C \Delta V. \tag{51.18}$$

If current i_c flows into the capacitor for duration Δt, the change in the charge stored in the capacitor plates is

$$\Delta Q = i_c \Delta t. \tag{51.19}$$

From the above two equations,

$$i_c \Delta t = C \Delta V. \tag{51.20}$$

If Δt is infinitesimal,

$$i_c = C \frac{dV}{dt}. \tag{51.21}$$

To ensure that the left-hand side and the right-hand side of the equation have the same sign, a convention is followed: for a chosen positive direction of the branch current, as shown in Figure 51.2, the terminal of the capacitor where the current flows into, has the positive polarity for voltage calculations.

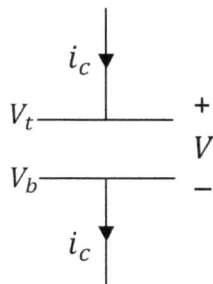

Figure 51.2: The convention chosen for the polarity of the voltage across a capacitor, and the direction of current flow.

For example, lets say that current i_c flowing in the direction shown in the schematic in Figure 51.2 is positive. The charge stored on the plate where the current flows into increases in positive charges, and the opposite plate more negatively charged. If the voltage between the capacitor plates is defined as

$$V = V_t - V_b, \tag{51.22}$$

as marked in the figure, V increases in the short time interval dt, and $\frac{dV}{dt}$ is a positive number. The signs between the left-hand side and the right-hand side in Equation 51.21 match. If, however, V is defined as $V_b - V_t$, the voltage becomes more negative after time dt for the current flowing in the direction shown, and there is a mismatch in the signs. The convention described prevents the mismatch in the signs in Equation 51.21.

51.4 Displacement Current in a Capacitor

It will be shown that the displacement current across an ideal parallel-plate capacitor dielectric, is equal to the conduction current flowing into the capacitor. An ideal capacitor will be assumed in this exercise, without any fringing fields. This is an interesting exercise, which shows that the displacement current acts like a current flowing across a capacitor dielectric, resulting in a continuous "current" flow. Since the displacement current is not a real current flow, there is no violation of a physical law, even if the displacement current does not enforce the current continuity.

A parallel-plate capacitor is shown in Figure 51.3(a). The top and bottom metal plates have been marked t and b, and for example, charged positive and negative at some time instant t, as a result of current flow. An ideal capacitor is assumed, where the electric field is uniform, and without fringing fields, shown by the arrows in the figure.

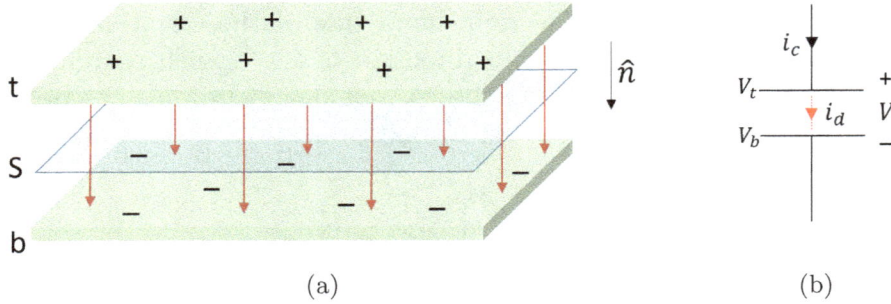

Figure 51.3: (a) A parallel-plate capacitor. (b) The circuit schematic of the capacitor.

The schematic of the capacitor is shown in Figure 51.3(b). The current flow into the capacitor is chosen, for example, in the direction shown, marked i_c. By the convention described in the previous section, the voltage across the capacitor plates V is marked in the schematic, for the relation

$$i_c = C \frac{dV}{dt} \tag{51.23}$$

to be correct.

In the region between the capacitor plates, there is no conduction current. The displacement current i_d, using Equation 51.16, is

$$i_d = \frac{1}{4\pi} \frac{\partial}{\partial t} \int_S \vec{D} \cdot d\vec{A}, \tag{51.24}$$

where the surface integral is calculated over the planar surface S marked in Figure 51.3(a).

In this example, $\vec{J}_D$ is either parallel or anti-parallel to $d\vec{A}$. As explained in Section 28.3, in this special case, $\vec{J}_D \cdot d\vec{A}$ is the current flowing in the direction of $d\vec{A}$ across dA. The unit normal vector to S, marked $\hat{n}$, is chosen in the direction from the top plate to the bottom plate, so that

$$i_d = \int_S \vec{J}_D \cdot d\vec{A}, \tag{51.25}$$

is the displacement current flowing in the direction of i_c to complete the loop.

If a uniform electric field is assumed, $\vec{D}$ is perpendicular to S, and has the same value at all points on S. The above integral simplifies to

$$i_d = \frac{A}{4\pi} \frac{dD}{dt}, \tag{51.26}$$

where D is the magnitude of the electric-flux density, and A is the area of the metal plate of the capacitor. If i_c is a positive value, positive charges are flowing into the top plate, and removed from the bottom plate, resulting in a positive value for $\frac{dD}{dt}$ and i_d. i_c and i_d are both positive, and flowing in the same direction.

The cross section of the capacitor is shown in Figure 51.4. Applying Gauss's law on the Gaussian surface shown by the dotted line, D can be related to Q stored in the top plate at some time instant t as

$$DA = 4\pi Q. \tag{51.27}$$

From the above two equations,

$$i_d = \frac{dQ}{dt}. \tag{51.28}$$

The charge stored in the top plate Q, relates to the voltage V across the capacitor plates, as shown in the schematic in Figure 51.3(b), by the capacitance-voltage relation,

$$Q = CV. \tag{51.29}$$

From the above equations,

$$i_d = C\frac{dV}{dt}, \tag{51.30}$$

and is equal to the conduction current in Equation 51.21.

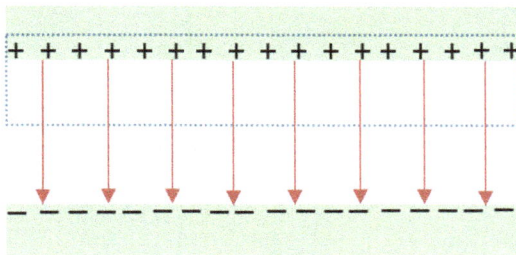

Figure 51.4: The cross section of the parallel-plate capacitor.

The displacement current i_d completes the current flow in the loop i_c, in the dielectric region between the capacitor plates, as shown in Figure 51.3(b). Considering all the possible permutation combinations, it is left as an exercise for the reader to verify that regardless of what direction is chosen for the area normal vector to S, the positive direction of the capacitor current i_c, or if the top/bottom plates are charged positive/negative or negative/positive, the displacement current completes the current flow in the loop.

52

Divergence-Free Condition of Magnetic-Flux Density $\vec{B}$

The divergence-free condition will be derived from Faraday's law. An important consequence of the formulation of the divergence-free equation is the boundary condition of $\vec{B}$ at material interfaces. It will explained in this chapter, this mathematical condition will ensure that a magnetized material, not only has a stronger magnetic field within the material, but also in the region surrounding the material. The conformity of Biot-Savart's law to the divergence-free condition will be proven last.

52.1 Derivation of the Divergence-free Condition of $\vec{B}$ from Faraday's Law

Applying the divergence operation $\nabla\cdot$ on both sides of Faraday's law,

$$\nabla \cdot \left(\nabla \times \vec{E} \right) = -\frac{\partial}{\partial t} \left(\nabla \cdot \vec{B} \right). \tag{52.1}$$

From calculus, $\nabla \cdot \nabla \times$ of a vector field is 0,

$$0 = -\frac{\partial}{\partial t} \left(\nabla \cdot \vec{B} \right). \tag{52.2}$$

The above equation means that $\nabla \cdot \vec{B}$ may have a spatial variation, but must be a non time-varying value at any point, although $\vec{B}(t)$ may be a time-varying field.

Using a logical proof by contradiction, it will be shown that the only value for $\nabla \cdot \vec{B}$ is 0. Lets assume that at some Point A, the source of $\vec{B}$ is present at or near Point A resulting in

$$\nabla \cdot \vec{B} \neq 0. \tag{52.3}$$

If these sources, such as current, magnets, etc., resulting in the above non-zero value at Point A, are moved far away, so far away that $\vec{B}$ decays to 0 at Point A. The above result at Point A changes to

$$\nabla \cdot \vec{B} = 0, \tag{52.4}$$

since $\vec{B}$ decays to 0. This contradicts the condition in Equation 52.2 that $\nabla \cdot \vec{B}$ cannot vary in time.

This shows that the only possible value that $\nabla \cdot \vec{B}$ can be, which satisfies the condition that $\nabla \cdot \vec{B}$ cannot be a time-varying value, is

$$\nabla \cdot \vec{B} = 0, \tag{52.5}$$

at any point. This equation is known as the divergence-free condition of $\vec{B}$, and must always be satisfied at any point.

52.2 Magnetic-Field Lines and the Divergence-Free Condition

A consequence of Equation 52.5 is that $\vec{B}$ and $\vec{H}$ form closed loops. This is easier to see from the integral form of the divergence-free condition. Gauss's divergence theorem can be used to obtain the integral form of Equation 52.5.

Integrating Equation 52.5 over a volume V,

$$\int_V \nabla \cdot \vec{B} \, dV = \int_V 0 \, dV. \tag{52.6}$$

The right-hand side is 0. Applying Gauss's divergence theorem, the integral form of the divergence-free condition is

$$\oint_A \vec{B} \cdot d\vec{A} = 0, \tag{52.7}$$

where A is a closed 3D surface over which the flux is calculated, and $d\vec{A}$ is the differential surface-area vector pointing outward to the surface. In words, from calculus, the above equation states that the flux "entering" a closed surface is equal to the flux "leaving" the surface. This is only possible if the field forms closed loops. This is also true for magnetic field $\vec{H}$. Substituting Equation 41.1 in the above equation,

$$\oint_A \mu_r \vec{H} \cdot d\vec{A} = 0. \tag{52.8, EMU}$$

In general, if a magnetic material is non-uniform, μ_r cannot be moved out of the integral in the above equation. Similar to Equation 52.7, the above equation is satisfied only if $\vec{H}$ forms closed loops. The field pattern around a magnet is shown in Figure 52.1(a). If the field does not form a closed loop, as shown in Figure 52.1(b), there is only flux entering the closed surface A, and the above equation is not satisfied.

52.3 $\vec{B}$ at the Boundary of Different Material Interfaces

A magnetic material modifies the applied magnetic field within the material, as well as in the region surrounding the material. A material may temporarily become magnetized until the applied field

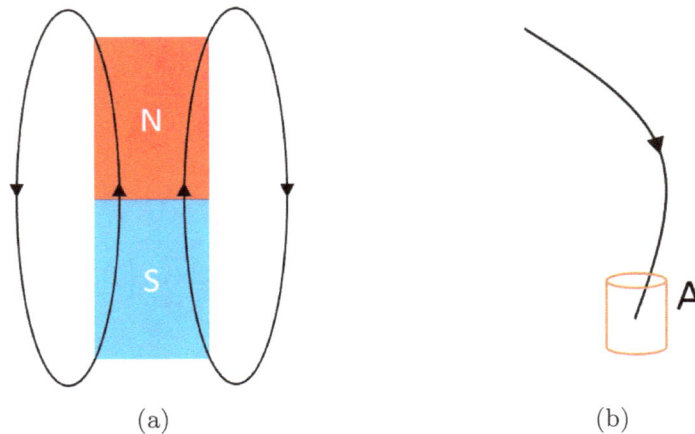

(a)   (b)

Figure 52.1: (a) The magnetic-field lines in a magnet. (b) An example to show that magnetic-field lines must form closed loops.

exists, or permanently retaining its magnetization, even when the applied field is no longer present. Within the material, the modified field in the magnetic material is captured in the relation,

$$\vec{B} = \mu_r \vec{H}, \tag{52.9, EMU}$$

or equivalently,

$$\vec{B} = \vec{H} + 4\pi \vec{I}, \tag{52.10, EMU}$$

where μ_r is the relative permeability of the magnetic material, and $\vec{I}$ is the magnetization vector. It will be proven in this section, the divergence-free condition captures the modified magnetic field surrounding the magnetic material.

The magnetization of a material, such as iron, can be easily demonstrated using an iron nail with a wire wound around it [88]. The current transmitted in the wire with a battery magnetizes the iron nail, which can then attract iron paper clips, similar to a magnet.

The cuboid in Figure 52.2(a) is a non-magnetic material, and magnetic in Figure 52.2(b). A wire is wound around it, and connected to a battery V. In the case of the magnetic material, the stronger field caused by the magnetization of the magnetic material, within and around, is shown by the thicker field line, compared to the non-magnetic material. The field outside the material is differentiated from the field within by the dotted line.

The interface between the magnetic material and air is shown in Figure 52.3. A closed surface A straddling the interface boundary is also shown. The flux entering A must equal the flux leaving A, as noted earlier. Equation 52.5 "carries forward" the stronger field in a magnetic material to its surrounding region. Using Gauss's magnetometer in Chapter 16, the stronger field can be measured and the hysteresis of the magnetic material can be observed [67].

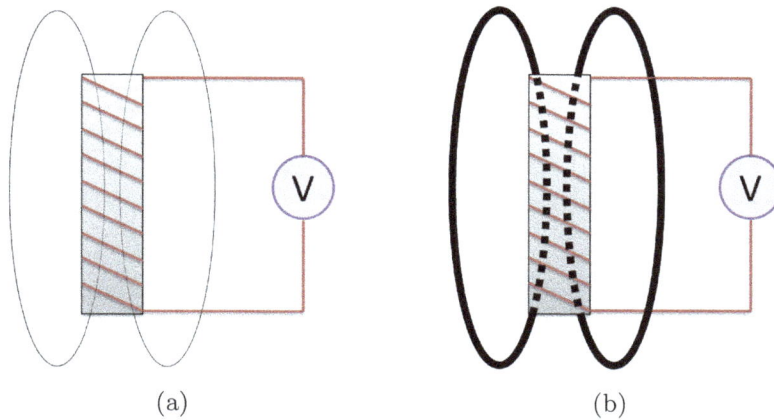

Figure 52.2: The magnetic-field pattern around (a) a non-magnetic, (b) a magnetic material, resulting from an applied magnetic field.

52.4 $\nabla \cdot \vec{B}$ vs $\nabla \cdot \vec{H}$

In general,

$$\nabla \cdot \vec{B} = 0 \nRightarrow \nabla \cdot \vec{H} = 0. \tag{52.11}$$

Applying the divergence operation on both sides of Equation 52.9,

$$\nabla \cdot \vec{B} = \nabla \cdot \left(\mu_r \vec{H} \right). \tag{52.12, EMU}$$

From calculus, if μ_r is a constant, it can be factored out of the divergence operator,

$$\nabla \cdot \vec{B} = \mu_r \nabla \cdot \vec{H}, \tag{52.13, EMU}$$

showing that

$$\nabla \cdot \vec{B} = 0 \Rightarrow \nabla \cdot \vec{H} = 0 \tag{52.14}$$

is true only in the special case where μ_r is a constant.

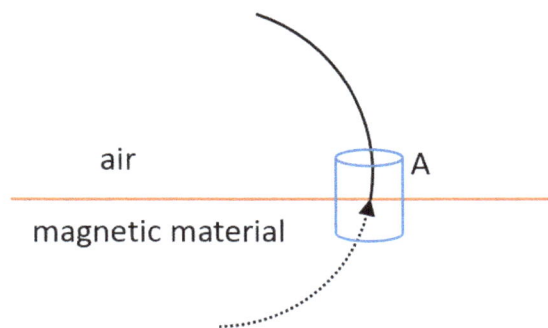

Figure 52.3: The material boundary between a magnetic and a non-magnetic material.

359

52.5 $\nabla \cdot \vec{B} = 0$ and Biot-Savart's Law [Optional]

It will be shown that Biot-Savart's law conforms to the divergence-free condition. It will be proven that Biot-Savart's law may be used to solve for the magnetic field in cases where a material of uniform relative permeability μ_r is present. After applying Biot-Savart's law to solve for $\vec{H}$, $\vec{B}$ can be obtained from

$$\vec{B} = \mu_r \vec{H}, \tag{52.15, EMU}$$

assuming a uniform material of permeability μ_r is present everywhere. The above equation captures the effect of the material medium. For a valid solution, $\vec{B}$ must also satisfy the divergence-free condition. This will be shown to be true in the remainder of the section.

Using Equation 30.55 that was derived earlier from Biot-Savart's law, the magnetic-flux density at some point given by the position vector $\vec{r}$ is

$$\vec{B}\,(\vec{r}) = \mu_r \left[\nabla \times \int_{V'} \left(\frac{\vec{J}}{|\vec{r} - \vec{r}\,'|} \right) dV' \right], \tag{52.16}$$

where the volume integration is carried out over V', the region where the volume current density $\vec{J}(\vec{r}\,')$ is present. Applying the divergence operation $\nabla\cdot$ on both the sides of the above equation,

$$\nabla \cdot \vec{B}\,(\vec{r}) = \mu_r \nabla \cdot \left[\nabla \times \int_{V'} \left(\frac{\vec{J}}{|\vec{r} - \vec{r}\,'|} \right) dV' \right], \tag{52.17}$$

where μ_r is moved out of the $\nabla\cdot$ operation, since its a constant. From calculus, the divergence of curl of a vector function is 0, and the right-hand side of the above equation reduces to 0,

$$\nabla \cdot \vec{B}\,(\vec{r}) = 0. \tag{52.18}$$

Therefore, Biot-Savart's law applied in a medium with uniform permeability, satisfies the divergence-free condition of $\vec{B}$. This shows that in the special case of a uniform medium with constant μ_r, $\vec{H}$ may be solved using Biot-Savart's law, and $\vec{B}$ can be calculated from Equation 52.15. The solution satisfies the effect of the material medium in Equation 52.15, and also meets the divergence-free condition, thereby confirming that it is the correct solution.

52.6   Biot-Savart's Law and Non-Uniform Magnetic Materials

The use of Biot-Savart's law is limited to cases with uniform magnetic materials. It will be shown by an example that Biot-Savart's law may not satisfy the divergence-free condition at the boundary of different material interfaces, thereby limiting its use to a medium with uniform permeability.

The interface between two regions with different permeability μ_1 and μ_2, is shown in Figure 52.4(a). Given a current distribution, not shown in the figure, Biot-Savart's law can be used to calculate

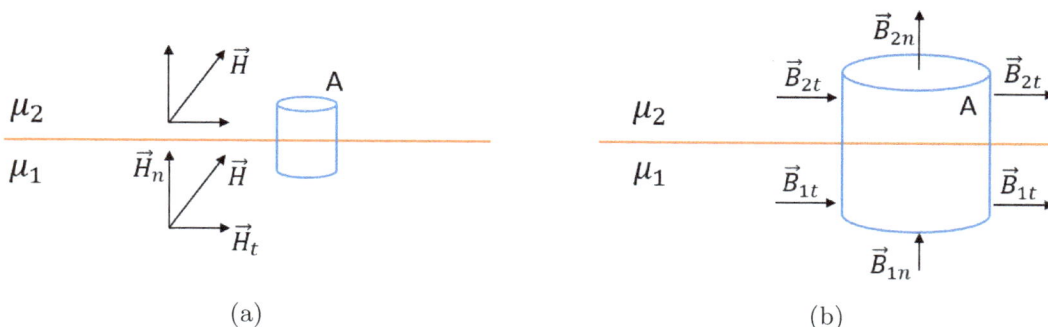

Figure 52.4: (a) The material boundary between two different materials. (b) A zoom of the closed surface A, and the tangential/normal components of $\vec{B}$.

the magnetic field. A closed cylindrical surface A across the interface is shown. A is small enough that the magnetic field $\vec{H}$, calculated from Biot-Savart's law, is uniform over the cylinder surface.

The magnetic flux across the surface of A in Figure 52.4(a) will be calculated, and shown that it does not satisfy Equation 52.7.

The zoom of A in Figure 52.4(a) is shown in Figure 52.4(b). $\vec{H}$ in each of the regions can be resolved into a component normal to the material boundary, and a component tangential to it. The tangential component of $\vec{H}$, $\vec{H}_t$, is related to $\vec{B}_{1t}$ and $\vec{B}_{2t}$, in the μ_1 and μ_2 regions, as

$$\vec{B}_{1t} = \mu_1 \vec{H}_t, \tag{52.19, EMU}$$
$$\vec{B}_{2t} = \mu_2 \vec{H}_t. \tag{52.20, EMU}$$

Likewise, the normal components are related as

$$\vec{B}_{1n} = \mu_1 \vec{H}_n, \tag{52.21, EMU}$$
$$\vec{B}_{2n} = \mu_2 \vec{H}_n. \tag{52.22, EMU}$$

As shown in the figure, the flux caused by the tangential components, $\vec{B}_{1t}$ and $\vec{B}_{2t}$, into the surface is equal to the flux out of the surface. Their contribution to the net flux is 0.

If $\mu_1 \neq \mu_2$, then $\vec{B}_{1n} \neq \vec{B}_{2n}$. Therefore, the flux due to the normal components of $\vec{B}$ do not cancel each other, resulting in

$$\oint_A \vec{B} \cdot d\vec{A} \neq 0. \tag{52.23}$$

This violates the divergence-free condition of $\vec{B}$. This exercise shows that using Biot-Savart's law in a non-uniform magnetic material medium to calculate the magnetic field, may result in an incorrect solution.

Numerical techniques in magnetostatics use Ampere's law,

$$\nabla \times \vec{H} = \vec{J},$$

$$(52.24)$$

together with the divergence-free condition of $\vec{B}$,

$$\nabla \cdot \vec{B} = 0,$$

$$(52.25)$$

to solve for the fields [89]. However, numerical techniques are not covered in this book.

52.7 Faraday's Law Including Magnetic-Current Density

The irony of electromagnetic theory is that although the existence of magnetic charges was hypothesized and was used to advance the theory, once developed, the definitions of electrical quantities will be formulated without the use of magnetic charges. The magnetic-flux density $\vec{B}$ will be defined in the following chapter without the use of magnetic charges, which is the present definition of $\vec{B}$. From the revised definition of $\vec{B}$, $\vec{H}$ will also be defined.

Naturally occurring magnetic charges have not been detected until now. A magnet, such as the magnetic needle in a compass in Figure 14.1(a), has been modeled using magnetic charges. However, this is only a model. If they do exist, magnetic charges seem to exist in complementary pairs, such as the positive and negative magnetic charges in a magnet.

In Figure 52.5, a wire carrying a steady current of electric charges in the direction shown, creates a magnetic field on a horizontal plane P. Likewise, if a steady magnetic current was to flow in the wire, assuming symmetric behavior between electric and magnetic charges, an electric field will be generated. A way by which the electric field generated by the magnetic current can be detected using an electric compass, was discussed in Chapter 12.

Alternately, if a wire w_P is to be placed on the plane P, centered on the magnetic current-carrying wire and forming a loop, as shown by the dotted line in the figure, the generated electric field will create a current flow in w_P. However, neither magnetic currents, nor induced electric fields from magnetic currents, have been observed until now.

It was noted in Chapter 51 that there still exists an asymmetry between Faraday's law and Ampere's law, even after including the displacement current term in Ampere's law:

$$\nabla \times \vec{E} = -\frac{\partial \vec{B}}{\partial t}$$

$$(52.26)$$

$$\nabla \times \vec{H} = 4\pi \vec{J} + \frac{\partial \vec{D}}{\partial t}.$$

$$(52.27)$$

In Ampere's law, there exists an electric current density term $\vec{J}$, but no such term in Faraday's law. As seen earlier, current flow creates a magnetic field, and this effect is captured in Ampere's

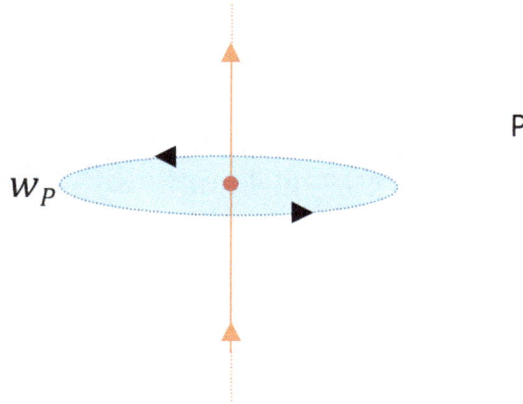

Figure 52.5: The magnetic field generated around an electric current-carrying wire. By symmetry, a magnetic current generates an electric field.

law. If symmetry is assumed, magnetic current flow must create an electric field. This result can be captured by modifying Faraday's law to look like Ampere's law,

$$\nabla \times \vec{E} = -4\pi\vec{K} - \frac{\partial \vec{B}}{\partial t}, \tag{52.28}$$

where $\vec{K}$ is magnetic current density, and defined similar to electric current density $\vec{J}$, except that it is the flow of magnetic charges, instead of electric charges. The reason for including the negative sign with $\vec{K}$ is to maintain symmetry with the electric current continuity equation, as explained next.

As done before in Chapter 51 to derive the electric current continuity equation from Ampere's law, applying the divergence operation $\nabla\cdot$ on both sides of the above equation,

$$\nabla \cdot \left(\nabla \times \vec{E}\right) = -4\pi \left(\nabla \cdot \vec{K}\right) - \frac{\partial}{\partial t}\left(\nabla \cdot \vec{B}\right). \tag{52.29}$$

In applying $\nabla \cdot \left(\nabla \times \vec{E}\right)$, as mentioned in the note on Equation 51.4, $\vec{K}$ flows across a cross section of non-zero area, rather than non-physical line currents, to avoid any singularity where the $\vec{E}$ field may be undefined. From calculus, the divergence of curl operation $\nabla\cdot\nabla\times$ on a vector field is 0, and multiplying the resulting equation by -1 to remove the negative sign,

$$0 = 4\pi \left(\nabla \cdot \vec{K}\right) + \frac{\partial}{\partial t}\left(\nabla \cdot \vec{B}\right). \tag{52.30}$$

This is the same form as Equation 51.5. The magnetic current in Equation 52.28 is written with a negative sign to maintain symmetry with the electric current continuity equation.

To complete the magnetic current continuity equation, following the derivation of Equation 51.6, it would be incorrect to write

$$\nabla \cdot \vec{B} = \rho_m, \tag{52.31}$$

363

or in the integral form,

$$\oint \vec{B} \cdot d\vec{A} = q_m, \tag{52.32}$$

similar to Gauss's law, where ρ_m is the magnetic charge density, and q_m is the enclosed magnetic charge. The above equation is incorrect, even if magnetic charges were to exist.

Here's an example to show that the above equation is incorrect. Two "magnetic parallel-plate capacitors" are shown in Figure 52.6. The magnetic material sandwiched between the capacitor plates are different in the two cases. The magnetic material in Figure 52.6(a) has permeability μ_1, and μ_2 in Figure 52.6(b).

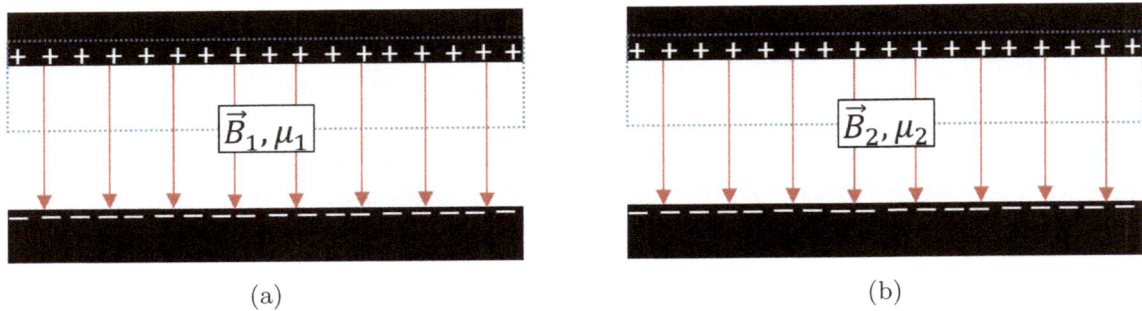

(a) (b)

Figure 52.6: A magnetic parallel-plate capacitor with permeability (a) μ_1, and (b) μ_2.

The magnetic charge density in the capacitor plates are identical in both the cases. Therefore, the applied magnetic field $\vec{H}$ in the magnetic material of the capacitors, caused by the magnetic charges, are equal in both the cases. $\vec{B}$ by definition, is the field that includes the effect of a magnetic material on the applied field $\vec{H}$. $\vec{B}$ are unequal in the two cases, since the magnetic materials have different permeability values, and marked $\vec{B}_1$ and $\vec{B}_2$.

Equation 52.32 is applied on a Gaussian surface with a rectangular cross-section, shown by the dotted rectangle. Since the values of $\vec{B}$ are different, two different values of the enclosed charge q_m are obtained. This is incorrect, since both the capacitors have the same magnetic charge density. This shows that the formulation of Equation 52.32 is incorrect.

If magnetic charges do exist, however, it may be possible that

$$\oint \vec{B} \cdot d\vec{A} \neq 0, \tag{52.33}$$

such as in the example presented. For this reason, the divergence-free condition in Equation 52.5, or the integral form in Equation 52.7, implicitly mean the absence of magnetic charges.

53

Lorentz Force Law and the New Definition of Magnetic-Flux Density $\vec{B}$

The Lorentz force law will be derived from Faraday's law. The derivation from Reference [90] will be presented here. This proof will be extended in the following chapter to show that the Lorentz force law is also applicable for a time-varying magnetic-flux density $\vec{B}$. This law is named in honor of the Dutch physicist Hendrik Antoon Lorentz (1853-1928). It describes the force exerted on a moving charge in a magnetic field. From this law, $\vec{B}$ can be defined without the need of magnetic charges, which is the current definition.

The Lorentz force equation will be used to derive the force between current-carrying wires in Chapter 55. The force between current-carrying wires can be measured using the Kelvin-Ampere current balance. This provides a way by which the Lorentz force law, as well as Faraday's law, from which the Lorentz law is derived, can be verified to be correct.

53.1   Discrete Version of Faraday's Law

Repeating the integral form of Faraday's law,

$$\oint \vec{E} \cdot \vec{dl} = -\frac{\partial \Phi}{\partial t}$$
$$= -\frac{\partial}{\partial t} \int \vec{B} \cdot \vec{dA}. \tag{53.1}$$

The direction of the path integral and the direction of the surface area vector are chosen according to the right-hand rule, as discussed earlier.

An example of a time-varying flux waveform is shown in Figure 53.1. The slope of the tangent $\frac{\partial \Phi}{\partial t}$ at any point, is related to the path integral of the electric field, as defined by Faraday's law. The slope of the tangent at Point A marked in the figure, for example, can be approximated in the discrete domain as

$$\frac{\partial \Phi}{\partial t} \approx \frac{\Delta \Phi}{\Delta t}, \tag{53.2}$$

where $\Delta\Phi$ is the change in flux during the time Δt,

$$\Delta\Phi = \Phi(t + \Delta t) - \Phi(t). \tag{53.3}$$

As Δt becomes smaller and smaller, the above approximation gets more and more accurate. In the limit as $\Delta t \to 0$, Equation 53.2 is exactly equal. In the derivation of the Lorentz law, Faraday's law will be approximated in the discrete domain, and the exact expression will be derived in the limit as $\Delta t \to 0$.

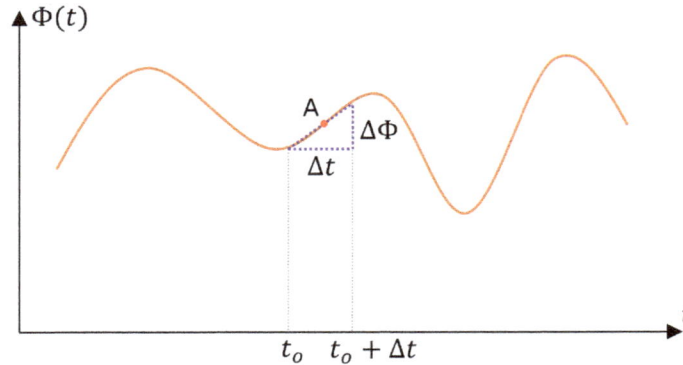

Figure 53.1: An example of a time-varying magnetic-flux waveform.

A loop at time t is divided into tiny segments, as shown in Figure 53.2. The ith segment is represented as a vector $\vec{\Delta l_i}$, which is along the same direction in which the path integral of the electric field is evaluated, and its magnitude is the length of the segment. In the example shown, the path of integration is assumed to be clockwise. The path integral of the electric field in

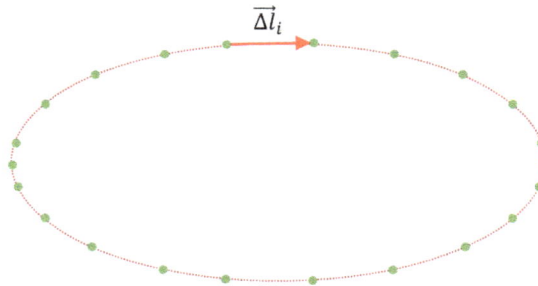

Figure 53.2: A loop is divided into N tiny segments.

Faraday's law is written as the Riemann sum

$$\sum_{i=1}^{N} \vec{E}_i \cdot \vec{\Delta l_i}, \tag{53.4}$$

where the loop is divided into N segments, and $\vec{E}_i$ is the electric field in the ith segment $\vec{\Delta l_i}$, and will be approximated to be uniform in the interval $[t, t+\Delta t]$. Equating the above discretized path integral, to the rate of change of flux, Faraday's law in the discrete domain is

$$\sum_{i=1}^{N} \vec{E}_i \cdot \vec{\Delta l}_i = -\frac{\Delta \Phi}{\Delta t}. \tag{53.5}$$

In the limit as $N \to \infty$, and $\Delta t \to 0$, the continuous form of Faraday's law can be obtained from the above equation.

53.2 A Thought Experiment to Derive the Lorentz Force Law

Typically in experiments related to Faraday's law, a time-varying flux is caused by a time-varying magnetic-flux density $\vec{B}$. However, a time-varying flux may also be generated when $\vec{B}$ is a constant, and the loop area changes over time. In general, both a loop area and/or $\vec{B}$ may vary over time to generate the time-varying flux. In this chapter, it will be assumed that $\vec{B}$ stays constant, and only the loop area is time varying. The derivation will be extended in the following chapter, to show that the Lorentz equation is also applicable in the case of a time-varying $\vec{B}$.

A $\vec{B}$ field is shown in Figure 53.3 by the dots, and for simplicity, lets assume that $\vec{B}$ may vary spatially, but is not time varying. A wire loop marked A_1, and dotted, is shown in the figure. It can be of any shape, but drawn circular. Suppose the loop A_1 changes shape, as shown by the arrows, into loop A_2, for example. Both the loops lie on the plane of the page.

A_1 is discretized into segments, as shown in Figure 53.2. It will be assumed that the perimeter of A_1 and the new shape A_2 are the same, and only the loop area changes. By making this assumption, each of the segments on A_1 translates to its new location on A_2. It will be proven later in the chapter, the derivation is still applicable if the wire stretches or shrinks, and the perimeter changes over time. By Faraday's law, the time-varying loop area creates a changing flux, and generates an electric field in the wire. This experiment will be used to derive the Lorentz force equation.

The flux across each of the areas are denoted as Φ_a through Φ_e. A_1 is the area at some time t_o, and changes to A_2 after time $t_o + \Delta t$. The flux across A_1 at time t_o is

$$\Phi(t_o) = \Phi_c + \Phi_e + \Phi_d. \tag{53.6}$$

Area A_1 changes to A_2 after Δt. The flux across A_2 at time $t_o + \Delta t$ is

$$\Phi(t_o + \Delta t) = \Phi_a + \Phi_e + \Phi_b. \tag{53.7}$$

From the above two equations,

$$\begin{aligned} \Delta \Phi &= \Phi(t_o + \Delta t) - \Phi(t_o) \\ &= \Phi_a + \Phi_b - \Phi_c - \Phi_d. \end{aligned} \tag{53.8}$$

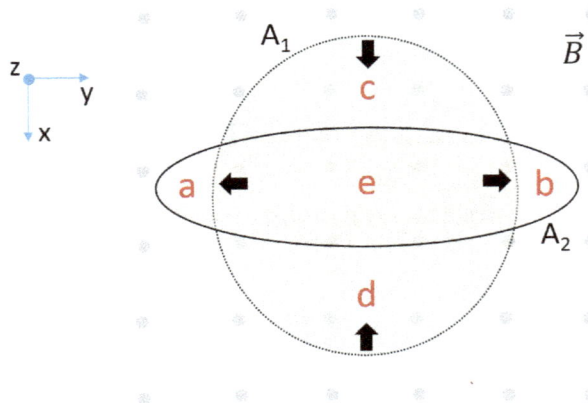

Figure 53.3: A loop changes shape from A_1 to A_2 during time Δt. Although the area enclosed by the loop changes, for simplicity, it will be assumed that the perimeter of A_1 and A_2 are equal.

Note that the flux across the area common to both A_1 and A_2, Φ_e, cancels in the above equation. If $\vec{B}$ is a constant, only the flux across the area that changes during time Δt is needed to calculate $\Delta \Phi$. This observation will be used to derive an expression of Faraday's law.

53.3 Calculating the Change in Area

In the thought experiment, a loop shape is continually and gradually changing over time in the presence of a constant magnetic field. The loop area also changes over time, and this creates a time-varying magnetic flux. It was shown in the previous section that only the flux across the area that changes after time Δt is needed to calculate $\Delta \Phi$. Given a loop at any time t, an expression to calculate the change in the loop area after time Δt will be derived in this section. From this result, the change in the flux can be calculated to write an equation of Faraday's law.

A loop is divided into N segments, similar to Figure 53.2. Each of the segments can be viewed as having a velocity at time t, written as the set $\{\vec{v}_1, \vec{v}_2, ..., \vec{v}_i, ..., \vec{v}_N\}$ for the N segments that captures this change in the shape of the loop. N is chosen large enough that the velocity can be considered to be uniform within a segment. Each of the segments translates to its new location described by its velocity vector. If Δt is assumed to be very small, the velocity vector can be assumed to be constant during that time interval.

This idea is illustrated by the example shown in Figure 53.4(a). A portion of a loop ℓ is shown in the figure. The shaded region R represents the area that the loop encloses. The loop changes over time, and ℓ becomes the dotted line ℓ' after time Δt. For example, two of the segments in the loop $\vec{\Delta l}_i$ and $\vec{\Delta l}_j$, move by the displacement vectors

$$\vec{\Delta r}_i = \vec{v}_i \Delta t \qquad (53.9)$$

$$\vec{\Delta r}_j = \vec{v}_j \Delta t, \qquad (53.10)$$

368

where $\vec{v}_i$ and $\vec{v}_j$ represent the velocity vectors of segments $\vec{\Delta l}_i$ and $\vec{\Delta l}_j$. The displacement of $\vec{\Delta r}_i$ increases the area of the enclosing loop, while $\vec{\Delta r}_j$ results in a decrease in the enclosed area.

The change in the area associated with the displacement of $\vec{\Delta r}_i$ has the shape of a parallelogram, as shown in Figure 53.4(b). In general, $\vec{\Delta l}_i$ may translate by $\vec{\Delta r}_i$ at any angle θ, and the resulting shape is a parallelogram.

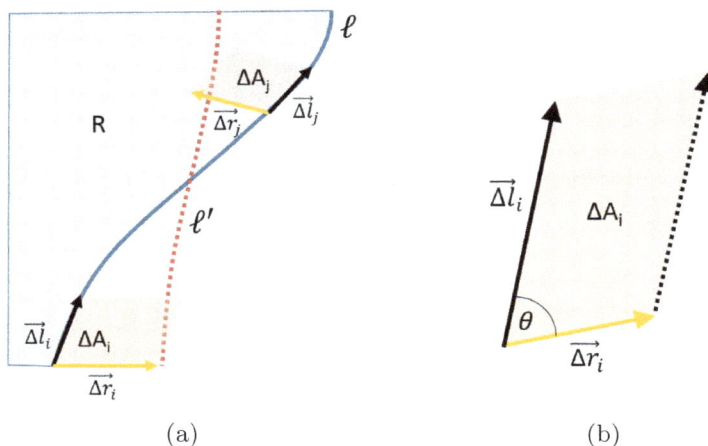

(a)   (b)

Figure 53.4: (a) A part of a loop perimeter ℓ, and a part of the perimeter ℓ' of the new shape, after the shape changes in time Δt. (b) The change in the area ΔA_i after $\vec{\Delta l}_i$ translates by $\vec{\Delta r}_i$.

The area of the parallelogram is

$$\begin{aligned} \Delta A_i &= |\vec{\Delta l}_i||\vec{v}_i \Delta t| \sin \theta \\ &= |\vec{\Delta l}_i \times \vec{v}_i \Delta t| \\ &= |\vec{v}_i \Delta t \times \vec{\Delta l}_i|. \end{aligned} \tag{53.11}$$

where $|\vec{\Delta l}_i|$ and $|\vec{v}_i \Delta t|$ are the magnitudes of the vectors. The area can be compactly represented as a cross product written above.

There is a change in the magnetic flux $\Delta \Phi_i$ that corresponds to the change in the area of each of the segments ΔA_i, after a short time Δt. The total change in the flux across the loop area after time Δt, $\Delta \Phi$, can be calculated by properly accounting for all the flux changes caused by each of the segments. This will be used to write an equation of Faraday's law in the following sections.

The result of the cross-product operation of two vector operands is a vector, and is perpendicular to the plane containing the operands. The operands $\vec{\Delta l}_i$ and $\vec{v}_i \Delta t$ lie on the plane of the loop area, and their cross-product area vector is perpendicular to the parallelogram area. The direction of the cross-product vector is given by the right-hand rule. The reader is referred to an algebra textbook for more details.

From algebra,

$$\vec{\Delta l}_i \times \vec{v}_i \Delta t, \tag{53.12}$$

and

$$\vec{v}_i \Delta t \times \vec{\Delta l}_i, \tag{53.13}$$

although they have the same magnitude, are in the opposite directions. But only one of them is correct, and in agreement with Lenz's law, discussed next.

53.4 Parallelogram Area and Lenz's Law

Although there are two possibilities to pick the parallelogram area swept by a segment, it will be justified in this section that

$$\vec{\Delta A}_i = \vec{v}_i \Delta t \times \vec{\Delta l}_i, \tag{53.14}$$

is the correct formulation of the parallelogram area $\vec{\Delta A}_i$, swept by a segment $\vec{\Delta l}_i$ with velocity $\vec{v}_i$. This will be shown in a brute-force way by looking at all the eight possible scenarios that may arise, and showing that in each of the cases, only the above equation captures the area element that conforms to Lenz's law.

The eight possible cases that may arise are shown in Figure 53.5. Each of the subfigures shows the translation of two segments: one whose translation increases the loop area, and the other that decreases the loop area. In Figure 53.5(a), the translation of $\vec{\Delta l}_i$ by $\vec{v}_i \Delta t$ increases the loop area, and the translation of the segment $\vec{\Delta l}_k$ by $\vec{v}_k \Delta t$ decreases the loop area. Only Figure 53.5(a) has been labeled. The remaining figures follow the same naming convention, and the labels have been omitted to avoid repetition.

There are two choices for picking the direction of $\vec{\Delta l}_i$: clockwise or counter clockwise. The counter-clockwise case is shown in Figure 53.5(a) and Figure 53.5(b), and the clockwise in Figure 53.5(c) and Figure 53.5(d).

The area swept by the translation of a segment $\vec{\Delta l}$ is shown by the shaded parallelogram. $\vec{\Delta l}$ and $\vec{v}\Delta t$ are assumed to be small enough that the magnetic-flux density $\vec{B}$ within a parallelogram area can be assumed to be uniform. There are two possible choices for the direction of $\vec{B}$ that is perpendicular to the loop area, that contributes to the flux: into the page or out of the page. The arrow convention will be used to denote the direction of $\vec{B}$ perpendicular to the loop area: '×' for the direction into the page, and '•' for the direction out of the page. In Figure 53.5(b) and Figure 53.5(d), $\vec{B}$ is into the page as marked, and out of the page for the other cases.

To summarize, the translation of a segment may increase or decrease the loop area, there are two choices for choosing the direction of $\vec{\Delta l}_i$, and there are two choices for the direction of $\vec{B}$ perpendicular to the loop area. These result in $2 \times 2 \times 2 = 8$ cases, illustrated in the figure.

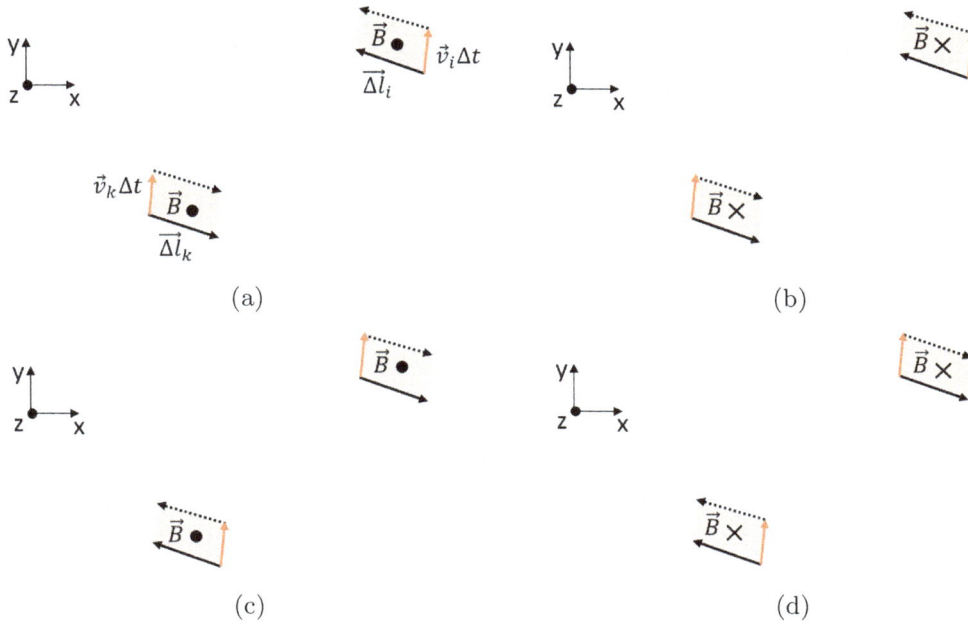

Figure 53.5: The different possibilities in the translation of a segment $\vec{\Delta \ell_i}$: the area may increase or decrease, $\vec{\Delta \ell_i}$ may be clockwise or counter clockwise, and $\vec{B}$ may be into the page or out of the page.

Figure 53.5(a) will be analyzed in detail. Similar explanation applies to the other cases, and it is left as an exercise for the reader to verify the remaining ones. Using Equation 53.14, the area increase caused by the segment $\vec{\Delta l_i}$ is the cross product

$$\vec{\Delta A_i} = \vec{v}_i \Delta t \times \vec{\Delta l_i}. \tag{53.15}$$

From algebra, the direction of the cross product is calculated using the right-hand rule. In this case, the resulting vector $\vec{\Delta A_i}$, like $\vec{B}$, is also pointing out of the page, in the $+\hat{z}$ direction. Using the above equation, lets verify that the change in the flux $\Delta \Phi_i$ caused by the translation of $\vec{\Delta l_i}$, can be calculated as

$$\Delta \Phi_i = \vec{B} \cdot \vec{\Delta A_i}. \tag{53.16}$$

Dividing both sides of the above equation by Δt, the rate of change of flux is

$$\frac{\Delta \Phi_i}{\Delta t} = \frac{\vec{B} \cdot \vec{\Delta A_i}}{\Delta t}. \tag{53.17}$$

Since $\vec{\Delta A_i}$ and $\vec{B}$ are in the $\hat{z}$ direction, the result of $\Delta \Phi_i$ is a positive number. Δt is a positive number, and therefore, so is the rate of change of flux $\frac{\Delta \Phi_i}{\Delta t}$.

If this is the only change in the flux across the loop area, by Faraday's law,

$$\oint \vec{E} \cdot \vec{dl} = -\frac{\Delta \Phi_i}{\Delta t}, \tag{53.18}$$

and the right-hand side of the above equation is a negative number. If the path integral of $\vec{E} \cdot \vec{dl}$ is negative, the electric field $\vec{E}$ in the wire is anti parallel to $\vec{\Delta l}_i$. The generated current in the wire, flowing in the direction of the electric field, decreases the flux in the loop area, and counter acts the increase in the flux caused by the translation of the segment. This is Lenz's law, and it proves that the formulation of the equations is correct.

If however, the cross product in Equation 53.12 is used in the calculation, the signs in the equations would not be in agreement with Lenz's law. Note that Equation 53.14 is correct on all the cases shown in the figure, regardless of whether the change in the flux is positive or negative. It is left as an exercise for the reader to verify this for the remaining cases shown in the figure. This shows that the parallelogram area in Equation 53.14, and the change in flux caused by the translation of a segment in Equation 53.16, are the correct formulation for use in Faraday's law.

53.5   Reduction in the Error of the Change in Area Calculation With Increasing N

The improvement in the accuracy of the change in area calculation, as the number of discretized loop segments N increases, will be illustrated by a simple example in Figure 53.6. ℓ shown by the dotted line, changes to ℓ' after time Δt. Segment marked by the endpoints 1-2 in Figure 53.6(a), translates to Segment 3-5. In Equation 53.14, however, the change in the area of each of the segments is treated as a parallelogram. Lets assume that the velocity of Segment 1-2 is such that it translates to Segment 3-4 after time Δt.

The parallelogram area with the vertices {1, 3, 4, 2}, does not quite exactly capture the area of the quadrilateral with vertices {1, 3, 5, 2}. The error in the change in area calculation is shown by the shaded triangle with vertices {3, 5, 4}.

If ℓ is divided into finer segments, it will be shown that the error in the area calculation decreases. Segment 1-2 is divided into finer segments, Segment 6-7 and Segment 7-8, shown in Figure 53.6(b). Segment 6-7 translates to 9-12, and Segment 7-8 translates to 12-13, as ℓ changes to ℓ' after time Δt. The change in the area is approximated with the parallelogram areas formed by the vertices {6, 9, 10, 7}, and {7, 12, 11, 8}. The error in the change in area with such a representation is shown by the shaded triangles p and q. The error has decreased by the area marked r, compared to the discretization in Figure 53.6(a). If the segments are made more and more finer, the calculation becomes more and more accurate, as this example shows. In the limit, the results are exact. The segments joining the nodes {9, 10, 12, 11, 13} resembles a "staircase", and is said to be a staircase approximation to ℓ'.

As N increases, the velocity of each of the segments $\vec{v}_i$ is captured more accurately. The piecewise linear representation of the curvature of the wire becomes more accurate, when more segments are added to the discretization of the loop contour. As Δt becomes smaller, the velocity of the segment is more uniform in the time interval. Therefore, the area swept by the parallelogram will

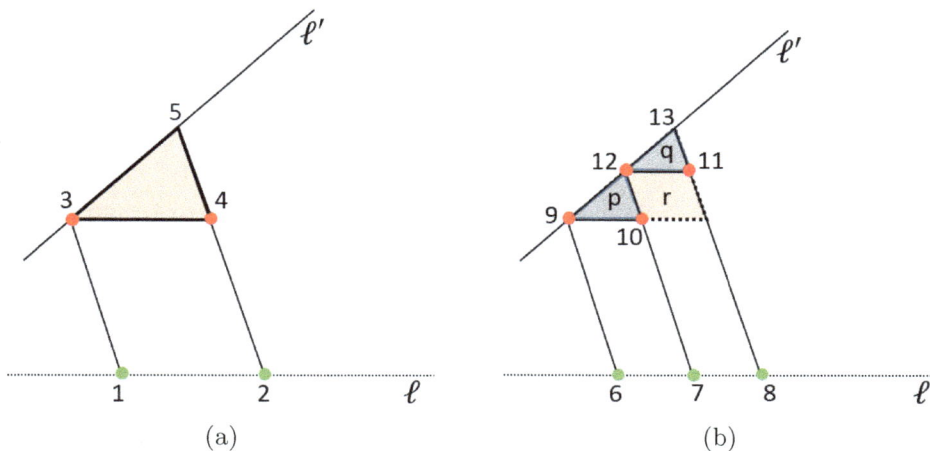

Figure 53.6: An example to show that the change in area calculation becomes more accurate, when the perimeter is discretized into finer segments, as shown in (b), compared to (a).

be more accurate. In the limit as $N \to \infty$, and as $\Delta t \to 0$, the term

$$\sum_{i=1}^{N} \vec{v}_i \Delta t \times \vec{\Delta l}_i \tag{53.19}$$

converges to the exact value of the change in the loop area that occurs after time Δt.

53.6 Lorentz Force Law

From the exercise in Section 53.4, repeating Equation 53.16 and Equation 53.14,

$$\Delta \Phi_i = \vec{B} \cdot \vec{\Delta A}_i \tag{53.20}$$

$$\vec{\Delta A}_i = \vec{v}_i \Delta t \times \vec{\Delta l}_i, \tag{53.21}$$

where $\Delta \Phi_i$ captures the change in the flux across a loop area, caused by the translation of a segment $\vec{\Delta l}_i$.

Adding the contribution of all the segments to the change in flux, the total change in flux is

$$\Delta \Phi = \sum_{i=1}^{N} \vec{B} \cdot \vec{\Delta A}_i. \tag{53.22}$$

Substituting this result in Faraday's law in the discrete domain in Equation 53.5,

$$\sum_{i=1}^{N} \vec{E}_i \cdot \vec{\Delta l}_i = -\frac{1}{\Delta t} \sum_{i=1}^{N} \vec{B} \cdot \vec{\Delta A}_i. \tag{53.23}$$

373

Substituting the expression for $\vec{\Delta A_i}$, the Δt term cancels,

$$\sum_{i=1}^{N} \vec{E}_i \cdot \vec{\Delta l_i} = -\sum_{i=1}^{N} \vec{B} \cdot \left(\vec{v}_i \times \vec{\Delta l_i} \right). \tag{53.24}$$

As $N \to \infty$ and $\Delta t \to 0$, the above Riemann sum becomes the definite integral,

$$\oint \vec{E} \cdot \vec{dl} = -\oint \vec{B} \cdot \left(\vec{v} \times \vec{dl} \right). \tag{53.25}$$

From algebra, given three vectors $\vec{a}$, $\vec{b}$, and $\vec{c}$, an identity of the scalar triple product is

$$\vec{a} \cdot \left(\vec{b} \times \vec{c} \right) = -\left(\vec{b} \times \vec{a} \right) \cdot \vec{c}. \tag{53.26}$$

Applying the above identity in Equation 53.25,

$$\oint \vec{E} \cdot \vec{dl} = \oint \left(\vec{v} \times \vec{B} \right) \cdot \vec{dl} \tag{53.27}$$

Rearranging the above equation,

$$\oint \left(\vec{E} \cdot \vec{dl} - \left(\vec{v} \times \vec{B} \right) \right) \cdot \vec{dl} = 0. \tag{53.28}$$

If the above equation is satisfied for any loop, and $\vec{E}$ generated by any change in the loop area,

$$\vec{E} = \vec{v} \times \vec{B}, \tag{53.29}$$

is a guaranteed solution. In the context of the derivation presented, Equation 53.29 can be interpreted as the electric field generated in a wire segment moving at velocity $\vec{v}$, in a magnetic field $\vec{B}$.

It is not a requirement that a wire needs to be a *closed* loop for a changing magnetic flux to generate an electric field. From Faraday's experiments, if a magnet is moved in and out of the loop area shown in Figure 53.7(a), an electric field is generated in the wire, which generates a current flow. If the same experiment is done using only half the loop, as shown by the solid line in Figure 53.7(b), no current flows in the wire, since its only half the wire. However, a changing magnetic flux creates an electric field, and therefore, an electric field must be generated in the wire segment, although no current flows.

It makes physical sense that it is not the presence of the closed loop of wire that generates the electric field, rather it is the changing magnetic flux. From this observation, it can be concluded that any wire segment, regardless of whether it is part of a closed loop or not, moving at velocity $\vec{v}$ through a magnetic field must encounter the influence of the electric field generated.

Furthermore, it is immaterial what material a wire loop is made of. Regardless of whether it is a

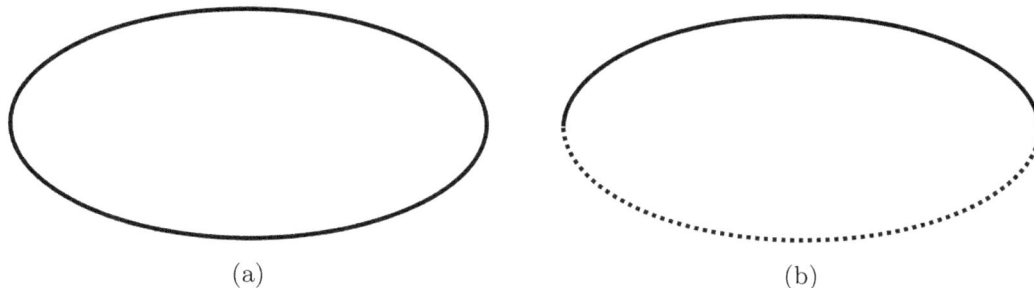

Figure 53.7: (a) A loop of wire. (b) Half the wire loop of Figure 53.7(a).

copper wire or a wire made of a dielectric material, an electric field is generated in the wire with changing magnetic flux. However, current flow occurs only in the copper wire. Equation 53.29, derived from Faraday's law, must therefore be true for any object moving in a magnetic field, rather than just a metal wire segment. It must also hold for an electric charge moving in a magnetic field.

The generated electric field on a moving electric charge in a magnetic field, modifies the charge trajectory by exerting a force on the charge. This force experienced by the moving electric charge, can be written as an equation by substituting Equation 5.2 in Equation 53.29,

$$\vec{F} = q\vec{v} \times \vec{B}. \tag{53.30}$$

This equation can be verified using the Kelvin-Ampere current balance, as discussed in Chapter 56. Note that by definition of the cross product, the force on the electric charge is orthogonal to the plane containing $\vec{v}$ and $\vec{B}$ vectors. Another way to state this is that the force on a moving charge is perpendicular to both $\vec{B}$ at the point where the charge is present at time t, and its velocity $\vec{v}$ at time t.

Using superposition of forces, combining this equation with the force experienced by the charge in an electric field $\vec{E}$,

$$\vec{F} = q\vec{E} + q\vec{v} \times \vec{B}. \tag{53.31}$$

This is known as the Lorentz force equation. It describes the force on an electric charge, caused by an electric field and a magnetic field.

Although the thought experiment assumes the perimeter of the loop does not change over time, the derivation is still applicable if the perimeter does change. The perimeter will change if the wire stretches or shrinks as the loop changes shape. If $P(t + \Delta t)$ is the perimeter at time $t + \Delta t$, and $P(t)$ is the perimeter at time t, as $\Delta t \to 0$, which is the assumption made in the derivation, $P(t + \Delta t) \to P(t)$. As $\Delta t \to 0$, the results become and more and more closer to the assumption made in the derivation that the perimeter does not change. In the limit, the conclusions will be the same as the case derived.

53.7   New Definition of Magnetic-Flux Density $\vec{B}$

The Lorentz force law provides an alternate definition of the magnetic-flux density $\vec{B}$, without the need of magnetic charges: if a charge q moving with velocity $\vec{v}$, experiences a force $\vec{F}$,

$$\vec{F} = q\vec{v} \times \vec{B}, \tag{53.32}$$

then $\vec{B}$ is the value of the magnetic-flux density at that point. This is the present definition of $\vec{B}$.

54

Lorentz Force Law and Time-Varying $\vec{B}$ [Optional]

For simplicity, it was assumed in the Lorentz law derivation in Chapter 53 that $\vec{B}$ is not time varying. However, the derivation can be modified a little, to show that it is also applicable for a time-varying $\vec{B}$.

In general, the change in flux across a loop may be created by a time-varying field and a time-varying area. Lets revisit the example in Figure 53.3, redrawn in Figure 54.1, to include both the time-varying flux causing mechanisms.

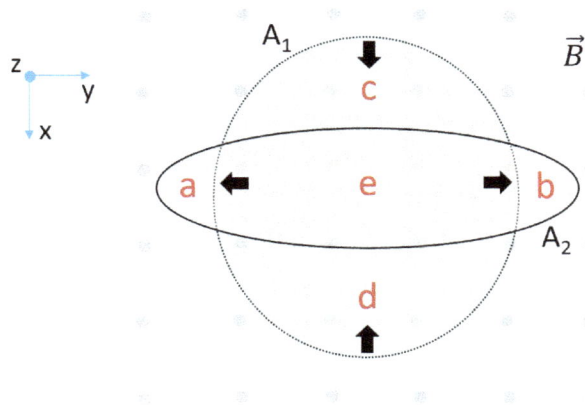

Figure 54.1: A loop changes shape from A_1 to A_2 in time Δt.

To simplify the notation, the magnetic-flux density at time t, $\vec{B}(t)$, will be referred to as simply $\vec{B}$. Since $\vec{B}$ is time varying, $\vec{B}(t+\Delta t)$ will be referred to as $\vec{B}'$. For brevity, the area integral over the area A

$$\int_A \vec{B} \cdot d\vec{A},\tag{54.1}$$

will be written as

$$\vec{B} \cdot \vec{A}.\tag{54.2}$$

As before, A_1 is the loop at time t. The loop changes to A_2 at $t + \Delta t$. The flux at time t is

$$\Phi(t) = \vec{B} \cdot \vec{A}_c + \vec{B} \cdot \vec{A}_e + \vec{B} \cdot \vec{A}_d, \tag{54.3}$$

where A_c, A_e, and A_d are the areas making up A_1. The flux at time $t + \Delta t$ is

$$\Phi(t + \Delta t) = \vec{B}' \cdot \vec{A}_a + \vec{B}' \cdot \vec{A}_e + \vec{B}' \cdot \vec{A}_b, \tag{54.4}$$

where A_a, A_e and A_b are the areas making up A_2. The change in the flux after Δt is

$$\begin{aligned}
\Delta \Phi &= \Phi(t + \Delta t) - \Phi(t) \\
&= \left(\vec{B}' \cdot \vec{A}_a + \vec{B}' \cdot \vec{A}_b - \vec{B} \cdot \vec{A}_c - \vec{B} \cdot \vec{A}_d \right) + \\
&\quad \left(\vec{B}' \cdot \vec{A}_e - \vec{B} \cdot \vec{A}_e \right).
\end{aligned} \tag{54.5}$$

The first term in the parenthesis,

$$\Delta \Phi_{\overline{A_1 \cap A_2}} = \vec{B}' \cdot \vec{A}_a + \vec{B}' \cdot \vec{A}_b - \vec{B} \cdot \vec{A}_c - \vec{B} \cdot \vec{A}_d, \tag{54.6}$$

is the change in the flux caused by time-varying area. Using set notation, the area that does not change during Δt is the area common to A_1 and A_2, denoted by the intersection $A_1 \cap A_2$. In this example, $A_1 \cap A_2$ is A_e. The complement of $A_1 \cap A_2$, denoted as $\overline{A_1 \cap A_2}$, is the area that changes during Δt, and denoted in the subscript in the above equation. The areas corresponding to $\overline{A_1 \cap A_2}$ are $\{A_a, A_b, A_c, A_d\}$.

Note that the above equations have terms with both $\vec{B}$ and $\vec{B}'$, unlike the derivation in Chapter 53. In the calculation of $\Delta \Phi_{\overline{A_1 \cap A_2}}$, $\vec{B}'$ appears in the calculation of the flux across the areas that increase after time Δt, which are $\{A_a, A_b\}$. $\vec{B}$, however, is used in the calculation of the flux across the areas that decrease after time Δt, which are $\{A_c, A_d\}$. This observation will be used later in the calculations.

The second term in Equation 54.5

$$\Delta \Phi_{A_1 \cap A_2} = \vec{B}' \cdot \vec{A}_e - \vec{B} \cdot \vec{A}_e \tag{54.7}$$

is the change in the flux across area A_e, the area that does not change after Δt.

From this exercise, it can be seen that a change in flux can be written as the superposition of the change in flux across the area that changes, and the change in flux across the area that remains the same,

$$\Delta \Phi = \Delta \Phi_{\overline{A_1 \cap A_2}} + \Delta \Phi_{A_1 \cap A_2}. \tag{54.8}$$

In the special case where no change in area occurs, and the change in the flux is caused only by the time-varying magnetic field, $\Delta \Phi_{\overline{A_1 \cap A_2}}$ is 0, since this is the term that represents the change in the flux across the area that changes. In the special case where $\vec{B}$ is not time varying, $\Delta\Phi_{A_1 \cap A_2}$ reduces to 0, since this is the term that represents the change in the flux across the area that remains the same. The latter case is the same as what was derived before in Equation 53.8. The time-varying magnetic flux gives rise to additional terms in the calculation of $\Delta\Phi$, represented by $\Delta\Phi_{A_1 \cap A_2}$, which was not seen before in Chapter 53. Also, unlike the static magnetic-flux density case, both $\vec{B}$ and $\vec{B}'$ are used in the calculations.

Dividing the above equation by Δt, the rate of change of flux is

$$\frac{\Delta\Phi}{\Delta t} = \frac{\Delta\Phi_{\overline{A_1 \cap A_2}}}{\Delta t} + \frac{\Delta\Phi_{A_1 \cap A_2}}{\Delta t}. \tag{54.9}$$

Likewise, superposition allows the path integral of the electric field in Faraday's law to be separated into

$$\oint \vec{E} \cdot \vec{dl} = \oint \vec{E}_{\overline{A_1 \cap A_2}} \cdot \vec{dl} + \oint \vec{E}_{A_1 \cap A_2} \cdot \vec{dl}, \tag{54.10}$$

where $\vec{E}_{A_1 \cap A_2}$ is the electric field generated by the changing flux $\Delta\Phi_{A_1 \cap A_2}$, and $\vec{E}_{\overline{A_1 \cap A_2}}$ is caused by $\Delta\Phi_{\overline{A_1 \cap A_2}}$. In the derivation of the Lorentz equation, only the terms related to the time-varying area, $\overline{A_1 \cap A_2}$, is of interest. The terms related to $A_1 \cap A_2$ will not be used in the remainder of this chapter.

As noted in Equation 54.6, both $\vec{B}$ and $\vec{B}'$ are present in the $\Delta\Phi_{\overline{A_1 \cap A_2}}$ calculation. Equation 53.22 for $\Delta\Phi$ can be divided into two sums,

$$\frac{\Delta\Phi_{\overline{A_1 \cap A_2}}}{\Delta t} = \frac{1}{\Delta t} \sum_p \vec{B} \cdot \vec{\Delta A}_p + \frac{1}{\Delta t} \sum_q \vec{B}' \cdot \vec{\Delta A}_q, \tag{54.11}$$

where the first summation accounts for the terms with $\vec{B}$, and the second, $\vec{B}'$. The values of the indices of summation p and q vary over the discretized segments making up the loop A_1. p operates on $\vec{B}$, and as noted earlier, varies over the segments that decrease the loop area after time Δt. q operates on $\vec{B}'$ and varies over the segments that increase the loop area. Substituting Equation 53.14 for $\vec{\Delta A}_p$ and $\vec{\Delta A}_q$ in the above equation, the Δt term cancels,

$$\frac{\Delta\Phi_{\overline{A_1 \cap A_2}}}{\Delta t} = \sum_p \vec{B} \cdot \left(\vec{v}_p \times \vec{\Delta l}_p \right) + \sum_q \vec{B}' \cdot \left(\vec{v}_q \times \vec{\Delta l}_q \right). \tag{54.12}$$

As $\Delta t \to 0$, $\vec{B}' \to \vec{B}$, and the above equation simplifies to

$$\frac{\Delta\Phi_{\overline{A_1 \cap A_2}}}{\Delta t} = \sum_p \vec{B} \cdot \left(\vec{v}_p \times \vec{\Delta l}_p \right) + \sum_q \vec{B} \cdot \left(\vec{v}_q \times \vec{\Delta l}_q \right), \tag{54.13}$$

without the need of both $\vec{B}$ and $\vec{B}'$. Since there is no need to distinguish the areas that decrease and increase the loop area, hence the need for $\vec{B}$ and $\vec{B}'$, the right-hand side of the equation can be written as

$$\frac{\Delta\Phi_{\overline{A_1 \cap A_2}}}{\Delta t} = \sum_{i=1}^{N} \vec{B} \cdot \left(\vec{v}_i \times \vec{\Delta l}_i \right), \tag{54.14}$$

where N is the number of discretized segments making up the loop A_1. Using the above equation, the discretized version of Faraday's law in Equation 53.5 can be written as

$$\sum_{i=1}^{N} \vec{E}_i \cdot \vec{\Delta l}_i = -\sum_{i=1}^{N} \vec{B} \cdot \left(\vec{v}_i \times \vec{\Delta l}_i \right),$$ (54.15)

which is the same as Equation 53.24. Following the same steps outlined in Section 53.6, it follows that the Lorentz equation is also valid for a time-varying magnetic-flux density $\vec{B}$.

Repeating the Lorentz force equation,

$$\vec{F} = q\vec{E} + q\vec{v} \times \vec{B},$$ (54.16)

a charge q with velocity $\vec{v}(t)$ at point $\vec{r}$ and time t, moving in the field $\vec{B}(\vec{r},t)$ and $\vec{E}(\vec{r},t)$, experiences the force in the above equation.

55

Force Between Current-Carrying Wires

Ampere studied the force between current-carrying wires. He was able to quantify the force between current elements [91], although, the form of the equation that he developed is seldom used. The Lorentz equation will be used to derive the force between current-carrying wires.

The definition of the unit current in the electromagnetic system is based on the existence of magnetic charges. Since naturally occurring magnetic charges have not been found, this warrants a change to the definition of the unit current in electromagnetic units, the abampere. The force equation between current-carrying wires provides a path to the revised definition, as discussed later in the chapter.

55.1 Force Between Current-Carrying Wires

The Lorentz equation derived earlier from Faraday's law, states that a charge q moving in a magnetic field $\vec{B}$ with velocity $\vec{v}$, experiences a force $\vec{F}$,

$$\vec{F} = q\vec{v} \times \vec{B}. \tag{55.1}$$

When two current-carrying wires are placed adjacent to each other, each of the wires generates a magnetic field, and exerts this influence on the other wire. Current is the flow of charges, charges moving with some velocity. From Lorentz's equation, it comes as no surprise that there is a force exerted on the moving charges in each of the wires, by the magnetic field generated by the current in the other wire.

The convention followed is that the direction of current flow is in the direction of the flow of positive charges. Two current carrying wires adjacent to each other are shown in Figure 55.1. In Figure 55.1(a), the current flow in the wires are in the same direction, but opposite in Figure 55.1(b).

The direction of the cross product in the Lorentz force is given by the right-hand rule. The reader is referred to other algebra textbooks for more details on the direction of a cross product. Applying this rule, the direction of the resultant forces on the wires are shown by the block arrows.

Currents flowing in parallel wires in the same direction attract each other, while currents flowing in opposite directions repel. Qualitatively, the directions of the forces observed in the experiments are in agreement with the formulation of the Lorentz equation. This force will be quantified in the next section.

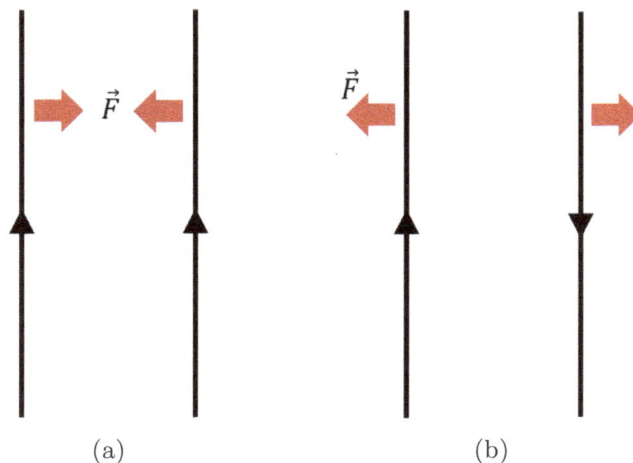

(a)   (b)

Figure 55.1: The force between current-carrying wires. (a) Currents flowing in the same direction attract, while (b) currents flowing in opposite directions repel.

55.2 Quantifying the Force Between Current-Carrying Wires

The electrical quantities q and $\vec{v}$ in the Lorentz equation will be rewritten with a different set of measurable variables. A wire is divided into small segments, one of which is shown in Figure 55.2(a) of length $\Delta \ell_p$. The wire may be of any shape, but drawn as a vertical line for illustration. The current i_p in the pth segment at some time t, flowing in the $+\hat{\ell}$ direction, is shown by the arrow. $\Delta \ell_p$ can be written as a vector, $\vec{\Delta \ell_p}$, whose direction is along the direction of the current flow.

The midpoint of the pth segment is marked A. The current in the wire segment, by definition of current, can be defined as the slope of the cumulative charge Q crossing the midpoint of the segment A. An example of the cumulative charge Q crossing Point A in Figure 55.2(a), as a function of time, is plotted in Figure 55.2(b). Points X and Y denote the values of Q at time instant t and $t + \Delta t$. After time Δt, charge Δq would have flown across Point A. By definition of current,

$$i_p = \frac{\Delta q}{\Delta t}.$$
(55.2)

Using this equation, the product $i_p \Delta \ell_p$ can be written as

$$i_p \vec{\Delta \ell_p} = \frac{\Delta q}{\Delta t} \vec{\Delta \ell_p}.$$
(55.3)

382

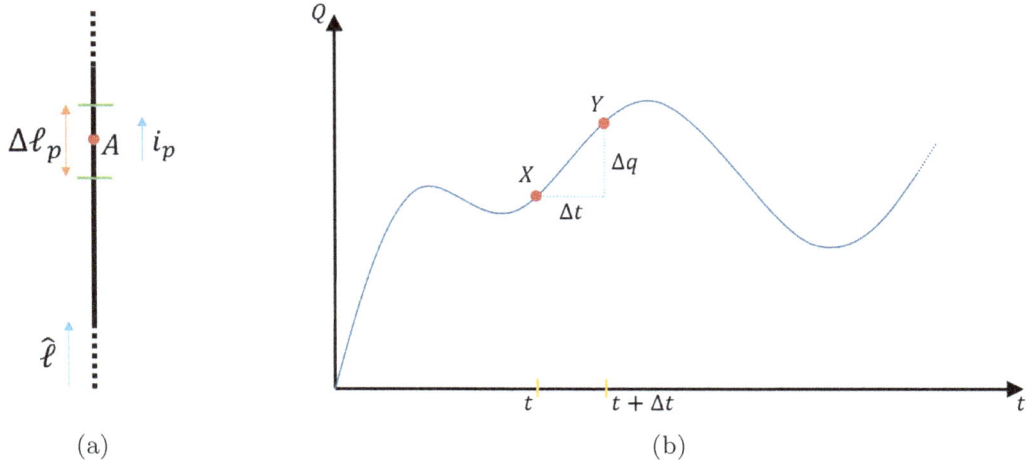

Figure 55.2: (a) A segment of a current-carrying wire. (b) An example of cumulative charge flow across Point A as a function of time.

The right-hand side of the equation is the same as

$$i_p \vec{\Delta\ell}_p = \Delta q \frac{\vec{\Delta\ell}_p}{\Delta t} \tag{55.4}$$

$$= \Delta q \, \vec{v}_p, \tag{55.5}$$

where

$$\vec{v}_p = \frac{\vec{\Delta\ell}_p}{\Delta t} \tag{55.6}$$

is a term that represents velocity. The above equations can be viewed as $i_p \Delta\ell_p$, representing the charge Δq moving with velocity $\vec{v}_p$ in the pth segment. If $\vec{B}_p$ is the magnetic-flux density in the pth segment, the force $\vec{F}_p$ experienced by the charges in the segment, using the Lorentz equation, is

$$\vec{F}_p = \Delta q \, \vec{v}_p \times \vec{B}_p \tag{55.7}$$

$$= i_p \vec{\Delta\ell}_p \times \vec{B}_p. \tag{55.8}$$

The force $\vec{F}$, at time t, experienced by a wire segment of length L, is the sum of all the forces experienced by the N segments making up L,

$$\vec{F} = \sum_{p=1}^{N} i_p \vec{\Delta\ell}_p \times \vec{B}_p. \tag{55.9}$$

In the limit, as $N \to \infty$, the above Riemann sum is the definite integral

$$\vec{F} = \int_L i \, \vec{d\ell} \times \vec{B}, \tag{55.10}$$

where i is the current flowing in the direction of the differential segment $\vec{dl}$, and is present in the magnetic-flux density $\vec{B}$.

In the remainder of this chapter, the above force equation will be applied to current-carrying wires in the steady state, where there is no time variation. In the special case where L is a straight line, not necessarily vertical or horizontal, which is present in a uniform constant field $\vec{B}$, and carrying a steady current i, the above equation simplifies to

$$\vec{F} = i\vec{L} \times \vec{B}, \tag{55.11}$$

where the magnitude of $\vec{L}$ is the length of the wire, and the direction of the vector is along the direction of the current flow.

55.3   Force Between Two Parallel Wires Carrying Steady Currents

Equation 55.11 will be used to formulate the force between two parallel segments of length L, shown in Figure 55.3. The wires marked 1 and 2 carry currents i_1 and i_2, which are not time varying, and are in the steady state. The right-hand coordinate system used is shown in the figure. The force $\vec{F}$ exerted on Wire 1 by Wire 2 is

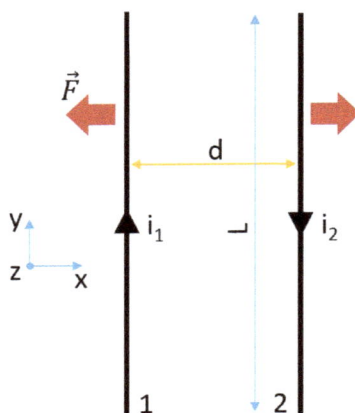

Figure 55.3: The calculation of the force between two parallel current-carrying wires.

$$\vec{F} = i_1\vec{L} \times \vec{B}, \tag{55.12}$$

where $\vec{B}$ is the magnetic-flux density generated by Wire 2 at the location of Wire 1, and

$$\vec{L} = L\hat{y}, \tag{55.13}$$

is the length vector of Wire 1 that is in the direction of the current flow. If the wires are in a non-magnetic medium, as discussed before in Chapter 41,

$$\vec{B} = \vec{H}. \tag{55.14, EMU}$$

From Biot-Savart's law, using Equation 23.19, the magnetic field generated by i_2 at Wire 1 is

$$\vec{H} = -\frac{2i_2}{d}\hat{z},$$ (55.15)

where d is the distance between the wires. From the directions of $\vec{B}$ and $\vec{L}$, Equation 55.12 simplifies to

$$\frac{F}{L} = i_1 B.$$ (55.16)

This equation will be used later in Chapter 58. From the above equations,

$$\frac{\vec{F}}{L} = -\frac{2i_1 i_2}{d}\hat{x}.$$ (55.17, EMU)

It is left as an exercise for the reader to verify that the magnitude of the force of attraction or repulsion in each of the cases shown in Figure 55.1 is

$$\frac{F}{L} = \frac{2i_1 i_2}{d}.$$ (55.18, EMU)

55.4   Revised Definition of Current in Electromagnetic Units

In the EMU definitions flowchart shown in Figure 26.1, magnetic charge was defined first, which was then used to define magnetic field using Equation 26.2. Biot-Savart's law, which relates magnetic field and current, was used to define electric current. The use of magnetic charges to define electrical quantities has been discontinued, and Equation 55.18 is a way by which current in EMU can be defined without the use of magnetic charges.

Using Equation 55.18, the unit current is defined as the current flowing in equal magnitudes in two thin parallel wires, separated by the unit distance $1\,cm$, exerting a force per unit length of $2\,dyne/cm$. The experiments to measure the force between current-carrying wires, using a Kelvin-Ampere current balance, will be demonstrated in the following chapter.

56

Measurement of the Force
Between Current-Carrying Wires

The Kelvin-Ampere current balance is used to measure the force between current-carrying wires. This apparatus is named after Lord Kelvin, who invented this device, and Ampere, who had first observed such a force. Lord Rayleigh used a different type of balance that serves a similar purpose, and will be described qualitatively.

The definition of current has been revised, as explained earlier in Section 55.4. Their measured values, therefore, has to be updated. Fortunately, the measurement techniques presented in the previous chapters, except for terrestrial magnetism, are still valid. This will become evident after the revised measurement techniques are presented in this chapter.

56.1   Kelvin-Ampere Current-Balance

The force between current-carrying wires, which also serves as the updated definition of current, repeating Equation 55.18, is

$$\frac{F}{L} = \frac{2i_1i_2}{d}.$$

(56.1, EMU)

The current balance similar to the ones in [92][93], which is used to measure the force in the above equation will be described in this section.

A sketch of the current balance is shown in Figure 56.1. The batteries connected to the wires are not shown for simplicity. The wires marked 1 and 2 making up the apparatus are more like thin metal rods, and are sturdy in the structure. A part of Wire 1 and Wire 2 of length L, drawn thicker than the rest of the wire segments, are routed parallel to each other in close proximity. Wire 2 is fixed, but Wire 1 can rotate about the axis marked R, when a force of repulsion exists between the wires along the length L.

The currents flowing in the wires along the length L are in opposite directions, creating a force of repulsion between the wire segments. It will be assumed that the force between the wires along L is dominant. The force on Wire 1 from any other segment of Wire 2 besides L, will be assumed to be negligible because of the larger distance between them.

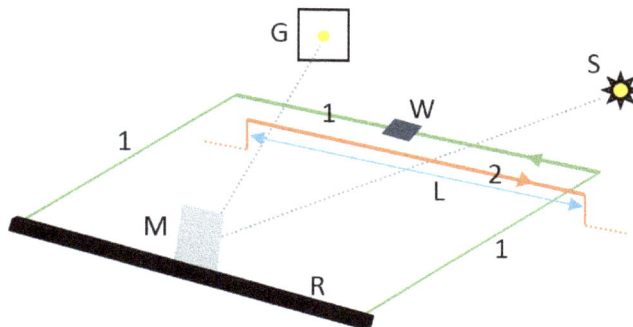

Figure 56.1: A Kelvin-Ampere current balance to measure the force between two parallel current-carrying wires.

A ray of light from the light source S, reflects off the mirror M, and its image falls on G. The light source may be a candle, for example, placed behind a cardboard with a small circular slot, to create a ray of light. The reference position of the image on G when no current flows in the wire is noted. The force of repulsion between the wires, moves Wire 1, causing the image on G to deviate from its reference position. Weights are placed on the tray W on Wire 1, to bring back Wire 1 to its reference position. The force of repulsion between the wires is equal to the total weight placed on the tray.

If $i_1 = i_2$ in Equation 56.1, solving for i,

$$i = \sqrt{\frac{d}{2}\frac{F}{L}}.$$
$$(56.2)$$

Current can be measured in absolute measure using the above equation, without the need of magnetic charges. The current in Wires 1 and 2 can be verified to be equal using tangent galvanometers that indicate the same angle of the magnetic needle.

56.2   Lord Rayleigh's Current Weigher

A variation of the current balance described earlier was created by Lord Rayleigh in 1884 [77][94], illustrated in Figure 56.2(a). The US National Bureau of Standards used such an apparatus, shown in Figure 56.2(b), to make an absolute measurement of current [94]. The balance will be described qualitatively in this section. The reader is referred to the papers cited for more details on the mathematical derivations. Coil C_2 is sandwiched between Coils C_1 and C_3, and connected to one of the pans of the balance. The currents flow in C_1 and C_3 such that it exerts a force on the current-carrying C_2, pulling it down. This force can be measured by adding weights W to the other pan, such that the balance returns to its reference position when no force is exerted on C_2.

The force on Coil C_2 is a few grams in the experiment documented in Reference [94]. The

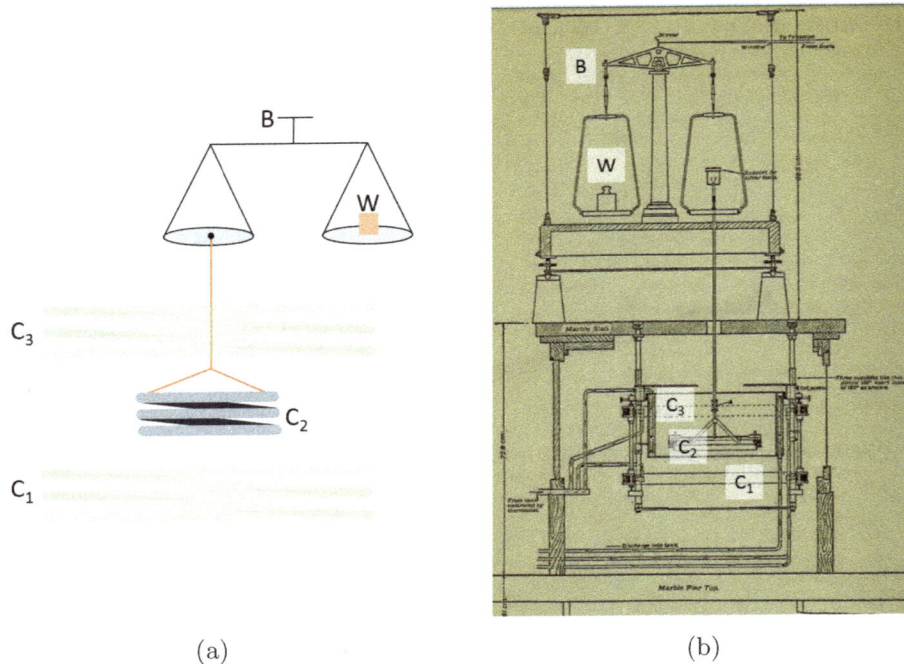

Figure 56.2: (a) Lord Rayleigh's current weigher. (b) The current weigher used by the US National Bureau of Standards to make an absolute measurement of current [94].

force can be quantified using the mutual inductance between the coils, and the reader is referred to Reference [77] for more details. The apparatus of Reference [94] has been marked with the same labels as Figure 56.2(a) for comparison.

In the force between current-carrying wires, one may ask if there also exists a Coulomb force of repulsion or attraction between the charges flowing in the wires? The wire is uncharged to begin with. A wire with current flowing in the direction marked by the arrows is shown in Figure 56.3. The wire is divided into small segments, three of which have been marked a, b, and c. If a steady stream of current is flowing in the wire, the current in the segments making up the wire are all equal. The charges flowing from b to c in time Δt, is compensated by the charges flowing from a to b. Therefore, Segment b does not accumulate a net charge. This is true for any segment in the wire, and the wire stays neutral. Therefore, the only force acting between the wires is the force due to the magnetic field acting on the charges in motion.

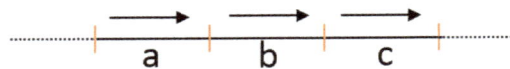

Figure 56.3: Three segments in a current-carrying wire.

388

The magnetic needle was modeled using magnetic charges at its end points, repeating Figure 14.1 in Figure 56.4. Such a representation of the magnetic needle with magnetic charges at its end

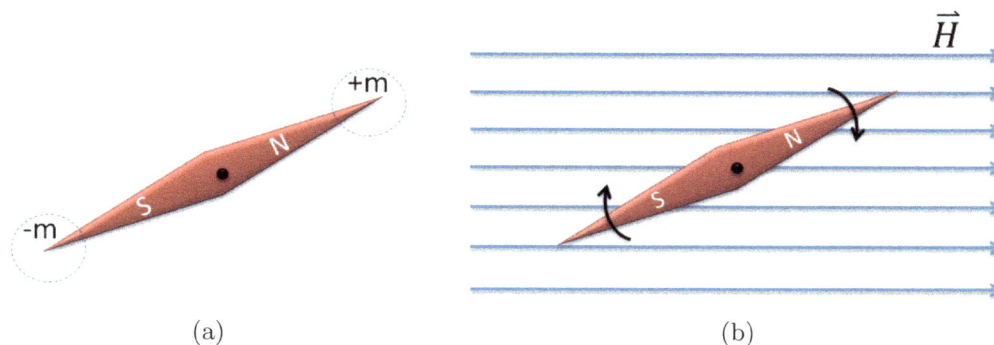

(a)   (b)

Figure 56.4: (a) The model of a compass with magnetic charges. (b) The torque on a magnetic compass causing it to align to the magnetic field of Earth.

points, shown in Figure 56.4(a), will be continued to be used for modeling. Any use of magnetic charges will be acceptable, as long as no electrical quantity is *defined* using magnetic charges.

The torque on the magnetic needle in the presence of a magnetic field, shown in Figure 56.4(b), is captured well using magnetic charges. The magnetic charge m is defined as Equation 14.1,

$$\vec{F} = m\vec{H},\tag{56.3}$$

where m experiences a force $\vec{F}$ in a magnetic field $\vec{H}$.

In Gauss's magnetometer, it was assumed that magnetic charges attracted or repelled each other in an inverse-square law behavior, similar to electric charges. With the revised definitions, there is no need for such a law to quantify the strength of terrestrial magnetism, or to define m. Earlier, the magnetic charge m was defined from the inverse square law of magnetic charges in Equation 13.1, and the magnetic field was defined using Equation 56.3. In the new definition flow, the magnetic field $\vec{H}$ is defined from current i using Biot-Savart's law, and m is defined using Equation 56.3, with the use of the inverse square law of magnetic charges discontinued. The revised flowchart of definitions in ESU and EMU will be presented in later chapters.

The measurement setup of terrestrial magnetism is shown in Figure 56.5 [95]. A battery B is connected to a Kelvin-Ampere current balance K, in series with a tangent galvanometer G. The current balance can be used to measure the current flowing in the circuit using Equation 56.1. The current balance can be setup such that the currents i_1 and i_2 are equal in Equation 56.1, making the setup simpler to solve for the current i.

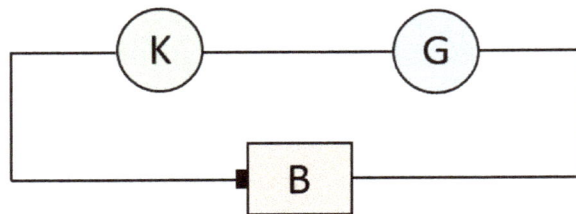

Figure 56.5: The revised setup of the measurement of the terrestrial magnetic-field strength.

The magnetic needle of the tangent galvanometer swings to an angle θ when the current flows through the galvanometer coil. Rearranging Equation 27.12 to solve for the tangential component of the terrestrial magnetic field H_E,

$$H_E = \frac{2N\pi i}{a \tan \theta},$$

(56.4)

where a is the radius of the galvanometer coil, and N is the number of turns in the galvanometer coil.

Biot-Savart's law relates a steady current i to the magnetic field $\vec{H}$ that it generates, and this has been used in the above expression. The values of the variables in the right-hand side of Equation 56.4 can be obtained from the measured data, and can be used to quantify terrestrial magnetism.

If the strength of the terrestrial magnetic field is known, one could solve for m using Equation 16.10, from the period of oscillation of the magnetic needle. Solving for m has no physical significance, but when used in conjunction with Equation 56.3, is useful for modeling a magnetic needle.

Once m has been determined, the magnetic-field strength at any point can be quantified from the period of oscillation of a magnetic needle, using Equation 16.10. This technique was used in Biot-Savart's two experiments in Chapter 24, and their results are valid even with the revised definitions. They used Equation 16.10 to make a relative comparison of magnetic-field strength, and did not require m to be known, or required the use of the inverse-square law of magnetic charges.

56.4 Revised Measurement of Current Using a Tangent Galvanometer

With the revised value of the terrestrial magnetic field in Section 56.3, current is measured using a tangent galvanometer in the same way it was done before. The updated value of the terrestrial magnetic field in Equation 56.4 is used in Equation 27.13 to measure current.

56.5 Revised Measurement of Charge in a Current Burst Using a Ballistic Galvanometer

The charge in a burst of current is measured using a ballistic galvanometer, in the same manner as before, except that the terrestrial magnetic field H in Equation 31.30 or Equation 31.77, is replaced by the revised value in Section 56.3.

Using Equation 31.30 or Equation 31.77 to calculate the charge implies that the magnetic needle is modeled as shown in Figure 56.4(a), with magnetic charges at its end points, and they follow the relation in Equation 56.3. Modeling the magnetic needle in such a manner allows a way by which calculations such as torque on the needle can be defined. None of the electrical quantities, however, have been defined using magnetic charges.

56.6 Revised Measurement of Resistance

The absolute measurement of resistance in Equation 46.12 is independent of the terrestrial magnetic field. Therefore, the measured resistance, prior to the revised measurement of terrestrial magnetism, does not affect the result.

56.7 Verifying the Formulation of Lorentz Equation and Faraday's Law

Equation 56.1 was derived from the Lorentz equation. The Lorentz equation in turn was derived from Faraday's law. By showing that the measurement results using a Kelvin-Ampere current balance satisfies Equation 56.1, one can show that the formulation of the Lorentz equation and Faraday's law are correct.

The current, the length of the parallel wires L, and their separation d can be varied in the Kelvin-Ampere current balance, showing that Equation 56.1 is satisfied, thereby proving the formulation of the Lorentz equation, and Faraday's law.

57

Inconsistency Between the Electrostatic and the Electromagnetic Systems of Units

The difference between the electrostatic and the electromagnetic systems of units was summarized in the flowchart in Figure 25.1 and Figure 26.1. The two different systems of units arise from how electric charge and electric current are defined. In Figure 25.1, in ESU, the unit electric charge is defined first, and the unit electric current is defined from the definition of the unit electric charge. However, in EMU, this order is reversed. In Figure 26.1, the unit electric current is defined before the unit electric charge.

The focus of this chapter is to show that the two definitions are not equal. For example, the unit electric charge (current) in ESU is not the same as the unit electric charge (current) in EMU, etc. The two systems of units are not consistent with each other.

With the use of magnetic charges discontinued, the definitions of the electrical quantities discussed until now, will be explained by a detailed flowchart in ESU and EMU in the following chapters. In this chapter, however, as an introduction, a snippet of this flowchart will be discussed.

57.1   Unit Electric Charge and Unit Electric Current Definitions in Electrostatic Units

The flowchart in Figure 57.1 shows how charge and current have been defined in the electrostatic system of units. The unit charge is defined from Coulomb's law,

$$F = \frac{q_1 q_2}{r^2},$$
(57.1, ESU)

as that quantity of charge, when separated by $1\,cm$ as equal point charges, repel each other with a force of $1\,dyne$, in CGS units.
The unit of charge in ESU is the statcoulomb, repeating Equation 4.2,

$$statcoulomb = \frac{gm^{\frac{1}{2}}cm^{\frac{3}{2}}}{s}.$$
(57.2, ESU)

Since current is the flow of charges, the definition of unit current follows from the definition of unit charge: unit current is unit charge flowing in unit time of $1\,second$. The unit of current in

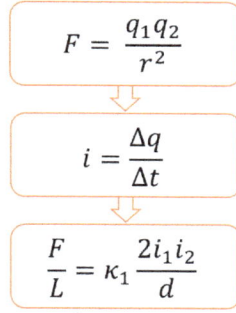

Figure 57.1: The flowchart of definitions in the electrostatic system of units.

ESU, *statampere*, is therefore,

$$statampere = \frac{gm^{\frac{1}{2}}cm^{\frac{3}{2}}}{s^2}.$$ (57.3, ESU)

Repeating Equation 55.18, the force between two parallel current-carrying wires,

$$\frac{F}{L} = \frac{2i_1i_2}{d},$$ (57.4, EMU)

has been derived from Faraday's law, which has been derived independent of the Coulomb's relation in Equation 57.1. Therefore, there is no reason to expect that current in ESU must satisfy the above equation. In ESU, the above equation is written as

$$\frac{F}{L} = \kappa_1\frac{2i_1i_2}{d},$$ (57.5, ESU)

where κ_1 is a positive constant that is yet to be determined, and the current values are in electrostatic measure. It is a positive constant, so that it does not change the direction of the force to the incorrect opposite direction. The above equation is the last block in Figure 57.1.

By dimensional analysis,

$$[\kappa_1] = \frac{s^2}{cm^2}.$$ (57.6)

If the force per unit length is $\frac{F}{L}$ between two parallel wires, equating the right-hand side of Equation 57.4 and Equation 57.5, the relation between current in ESU and EMU can be derived as

$$i_{EMU} = \sqrt{\kappa_1}\, i_{ESU}.$$ (57.7)

Substituting the equation for current as the flow of charges,

$$i = \frac{\Delta q}{\Delta t},$$ (57.8)

in the above equation, it must also be true that

$$q_{EMU} = \sqrt{\kappa_1}\, q_{ESU}.$$ (57.9)

From Equation 57.7 and Equation 57.9, it can be seen that the unit current in ESU is not the same as the unit current in EMU, and likewise for the unit electric charge. The current (charge) value in EMU is $\sqrt{\kappa_1} \times$ greater than the unit current (charge) in ESU.

57.2 Unit Electric Charge and Unit Electric Current Definitions in Electromagnetic Units

In EMU, shown in the flowchart in Figure 57.2, current is defined using Equation 57.4, without the need of magnetic charges. The electric current is defined first in EMU, and the electric charge is defined as the charge contained in unit current flowing for unit time. The unit current in EMU

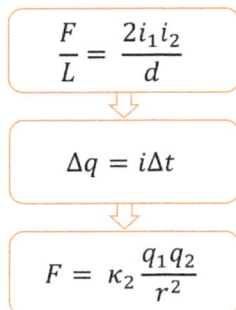

$$\frac{F}{L} = \frac{2i_1 i_2}{d}$$

$$\Delta q = i \Delta t$$

$$F = \kappa_2 \frac{q_1 q_2}{r^2}$$

Figure 57.2: The flowchart of definitions in the electromagnetic system of units.

is known as the abampere, as opposed to the statampere in ESU. It is left as an exercise for the reader to confirm using dimensional analysis applied to Equation 57.4 that the abampere has units, repeating Equation 27.7,

$$abampere = \frac{gm^{\frac{1}{2}} \, cm^{\frac{1}{2}}}{s}. \tag{57.10, EMU}$$

Note that Equation 57.4 is valid only in EMU. In ESU, this equation is written as Equation 57.5, where κ_1 is a constant that is yet to be determined.

The unit charge in EMU, the abcoulomb, is defined as the quantity of electric charge in unit current that flows for unit time,

$$\Delta q = i \Delta t. \tag{57.11}$$

Repeating Equation 27.14, the abcoulomb has units

$$abcoulomb = abampere \cdot s$$
$$= gm^{\frac{1}{2}} \, cm^{\frac{1}{2}}. \tag{57.12, EMU}$$

The definition of the unit charge in EMU has no dependency to the Coulomb's relation in Equation 57.1. Therefore, Coulomb's law is written in EMU as

$$F = \kappa_2 \frac{q_1 q_2}{r^2}, \tag{57.13, EMU}$$

where κ_2 is a positive constant that is still to be determined, and the charge values are in electromagnetic measure. It is restricted to positive values, so that it does not change the direction of the force to the incorrect opposite direction. κ_2 serves as a scaling factor to satisfy the above force relation between electric charges in electromagnetic units. By dimensional analysis,

$$[\kappa_2] = \frac{cm^2}{s^2}. \tag{57.14}$$

Given the same force between charges of equal magnitude, equating the right-hand side of Equation 57.13 and Equation 57.1, one arrives at the relation,

$$q_{ESU} = \sqrt{\kappa_2}\, q_{EMU}. \tag{57.15}$$

Since current is the flow of charges, from Equation 57.11 and the above relation, it must be true that

$$i_{ESU} = \sqrt{\kappa_2}\, i_{EMU}. \tag{57.16}$$

57.3   The Ratio of the Electrostatic Charge to the Electromagnetic Charge

From Equation 57.9 and Equation 57.15,

$$\frac{q_{ESU}}{q_{EMU}} = \frac{1}{\sqrt{\kappa_1}} \tag{57.17}$$

$$= \sqrt{\kappa_2}. \tag{57.18}$$

If the ratio of the electrostatic charge to the electromagnetic charge can be measured, the constants κ_1 and κ_2 can be determined. This will allow filling in the blanks in the set of equations in the ESU and the EMU systems of units. From the above equation,

$$\frac{1}{\sqrt{\kappa_1}} = \sqrt{\kappa_2}, \tag{57.19}$$

and simplifying,

$$\kappa_1\kappa_2 = 1. \tag{57.20}$$

The ratio of the electrostatic charge to the electromagnetic charge has units of velocity,

$$\left[\frac{q_{ESU}}{q_{EMU}}\right] = \frac{cm}{s}. \tag{57.21}$$

This is not a mere coincidence, but has a great significance. This will be explained in great detail in the next chapters. Determining this ratio is one of the most important milestones in electromagnetics. It was first measured by Weber and Kohlrausch in 1856 [96]-[99], whose experiment will be discussed in Chapter 63.

57.4   A General Form of the Electromagnetic Equations

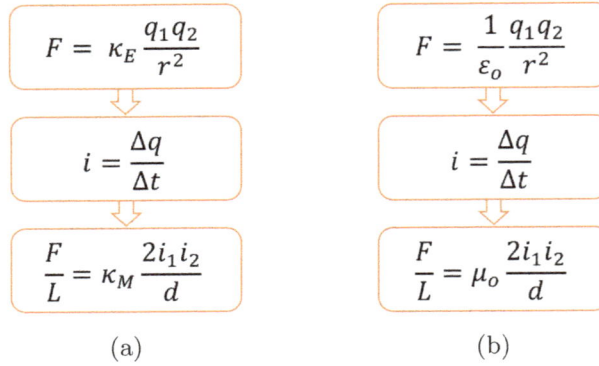

(a)   (b)

Figure 57.3: (a) The flowchart of definitions in ESU and EMU written as a common set of equations. (b) The equations in Figure 57.3(a) written using constants ϵ_o and μ_o.

The equations in ESU and EMU, shown in Figure 57.1 and Figure 57.2, can be written as a common set of equations, shown in Figure 57.3(a). The advantage of this is avoiding having to write two separate set of equations, one for ESU and one for EMU.

It is convenient to write κ_E and κ_M as

$$\kappa_E = \frac{1}{\epsilon_o} \tag{57.22}$$

$$\kappa_M = \mu_o, \tag{57.23}$$

the reasons for which will become clear in Chapter 58. The flowchart in Figure 57.3(a) is rewritten using the constants ϵ_o and μ_o in Figure 57.3(b).

In the case of ESU and EMU in Figure 57.1 and Figure 57.2, the values of κ_E and κ_M have been documented in Table 57.1. The units of κ_1 and κ_2 were derived in Equation 57.6 and Equation 57.14. $\epsilon_o = 1$ is a dimensionless number in ESU and $\mu_o = 1$ is dimensionless in EMU. In Figure 57.1 and Figure 57.2, the dimensions of κ_1 and κ_2 were derived in Equation 57.6 and Equation 57.14. Equating them to ϵ_o in EMU and μ_o in ESU, these have dimensions of inverse velocity squared.

It will be shown next that the ESU and EMU systems of equations are not the only way to formulate these equations. There are infinite ways to select the constants in the equations. It will be proven that Coulomb's law, and the force between current-carrying wires, can be written as

$$F = \frac{1}{\epsilon_o} \frac{q_1 q_2}{r^2} \tag{57.24}$$

$$\frac{F}{L} = \mu_o \frac{2 i_1 i_2}{d}, \tag{57.25}$$

where any values of ϵ_o and μ_o may be chosen, as long as it satisfies the relation

$$\frac{1}{\epsilon_o \mu_o} = \left(\frac{q_{ESU}}{q_{EMU}}\right)^2. \tag{57.26}$$

Coulomb's experiment, and other proofs that have been presented of Coulomb's law, would work

System of Units	ϵ_o	μ_o
ESU	1	κ_1
EMU	$1/\kappa_2$	1

Table 57.1: The values of ϵ_o and μ_o to tailor the common set of equations to ESU or EMU.

equally well for

$$F \propto \frac{q_1 q_2}{r^2}. \tag{57.27}$$

The proportionality constant has been chosen as 1 in ESU. However, any positive constant may be chosen.

Likewise, in the formulation of Faraday's law, it was explained that any positive constant may be chosen in the formulation. Since the force between current-carrying wires is derived from Faraday's law, any positive constant may be chosen in

$$\frac{F}{L} \propto \frac{2 i_1 i_2}{d}. \tag{57.28}$$

In EMU, the proportionality constant 1 is chosen in the above equation, although any positive constant may be chosen.

Lets assume that κ_1 in the system of equations in ESU, shown in the flowchart in Figure 57.4(a), has been determined. A new system of equations, which will be referred to as ESU′, is shown in Figure 57.4(b). In ESU′, κ_3 is chosen as the proportionality constant in Coulomb's relation, modifying the definition of the unit charge in ESU′ from ESU.
Equating the Coulomb force of the same magnitude between two charges in ESU and ESU′,

$$q_{ESU} = \sqrt{\kappa_3}\, q_{ESU'}. \tag{57.29}$$

The unit charge in ESU′ would be scaled by $\sqrt{\kappa_3}$ in ESU. As before, the above equation also means that

$$i_{ESU} = \sqrt{\kappa_3}\, i_{ESU'}. \tag{57.30}$$

The unit current in ESU′ would be scaled by $\sqrt{\kappa_3}$ in ESU. Substituting the above equation in Equation 57.5, the force per unit length between two wires in ESU′ is

$$\frac{F}{L} = \kappa_3 \kappa_1 \frac{2 i_1 i_2}{d}, \tag{57.31, ESU′}$$

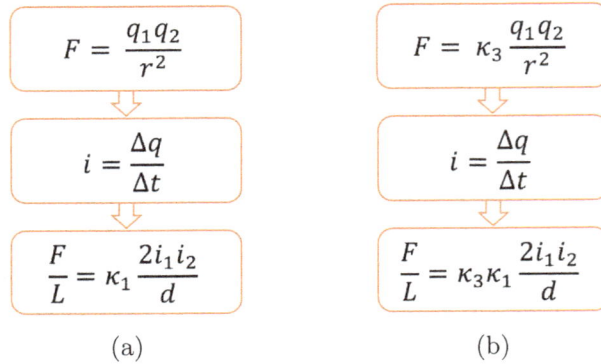

Figure 57.4: (a) The flowchart of definitions in ESU. (b) A new system of units ESU′ derived from ESU.

where the current measures are in ESU′ units.

The proportionality constant chosen in Coulomb's law in ESU′ is κ_3, which results in the proportionality constant of $\kappa_3\kappa_1$ in the force between current-carrying wires. Equating them to ϵ_o and μ_o in Figure 57.3(b),

$$\frac{1}{\epsilon_o\mu_o} = \frac{\kappa_3}{\kappa_3\kappa_1}$$
$$= \frac{1}{\kappa_1}. \tag{57.32}$$

Substituting Equation 57.17 in the above equation,

$$\frac{1}{\epsilon_o\mu_o} = \left(\frac{q_{ESU}}{q_{EMU}}\right)^2, \tag{57.33}$$

resulting in the same equation as Equation 57.26. This exercise shows that any value of ϵ_o and μ_o may be chosen, as long as it satisfies the above relation.

The same exercise can be repeated on the EMU system of equations in Figure 57.5(a). Lets assume that κ_2 has been determined. A new system of equations EMU′ can be created, shown in Figure 57.5(b), where

$$\frac{F}{L} = \kappa_4\frac{2i_1i_2}{d}. \tag{57.34}$$

Equating $\frac{F}{L}$ in wires carrying equal currents in EMU′ and Equation 57.4 in EMU,

$$i_{EMU} = \sqrt{\kappa_4}\, i_{EMU'}. \tag{57.35}$$

As argued earlier, the above equation implies

$$q_{EMU} = \sqrt{\kappa_4}\, q_{EMU'}. \tag{57.36}$$

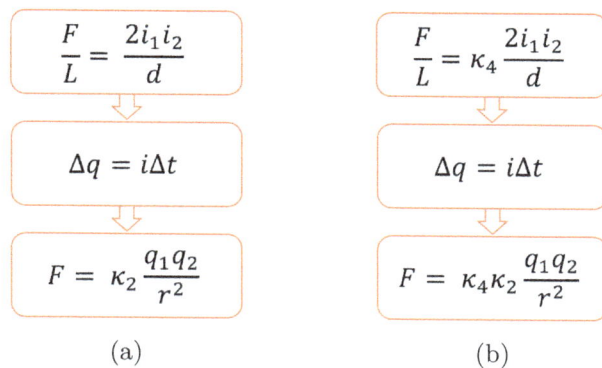

$$\frac{F}{L} = \frac{2i_1 i_2}{d}$$

$$\Delta q = i\Delta t$$

$$F = \kappa_2 \frac{q_1 q_2}{r^2}$$

(a)

$$\frac{F}{L} = \kappa_4 \frac{2i_1 i_2}{d}$$

$$\Delta q = i\Delta t$$

$$F = \kappa_4 \kappa_2 \frac{q_1 q_2}{r^2}$$

(b)

Figure 57.5: (a) The definitions flowchart in EMU. (b) A new system of units EMU′ derived from EMU.

Substituting the above equation in Coulomb's EMU relation in Equation 57.13,

$$F = \kappa_4 \kappa_2 \frac{q_1 q_2}{r^2}, \tag{57.37}$$

where the charge magnitudes are in EMU′ measures. Equating the EMU′ constants in Coulomb's relation and the force between wires equation, $\kappa_4 \kappa_2$ and κ_4, to $1/\epsilon_o$ and μ_o in Figure 57.3(b),

$$\frac{1}{\epsilon_o \mu_o} = \frac{\kappa_4 \kappa_2}{\kappa_4}$$

$$= \kappa_2. \tag{57.38}$$

Substituting Equation 57.18 in the above equation,

$$\frac{1}{\epsilon_o \mu_o} = \left(\frac{q_{ESU}}{q_{EMU}}\right)^2, \tag{57.39}$$

which is the same conclusion obtained earlier.

This shows that there is a lot of flexibility in picking the constants ϵ_o and μ_o. They only need to satisfy the above ratio. The set of equations shown in Figure 57.3(a) is written in a manner that is common to both ESU and EMU. The advantage of doing this is working with one set of equations, rather than multiple sets. Setting the values of the constants ϵ_o and μ_o, allows the equations to be customized for ESU, EMU, or any other system.

The values of ϵ_o and μ_o are different in ESU, EMU, MKS unrationalized, and SI systems of units, whose values will be presented in later chapters. In the present day form of these equations in the SI system, it will be shown in later chapters that ϵ_o and μ_o are set to values that are convenient for lab measurements, and cleverly chosen so that there are no "eruption of 4π's", in the words of Oliver Heaviside.

Definitions Flowchart in the Electrostatic System of Units

It was shown in Chapter 57 that the electrostatic and the electromagnetic systems of units are inconsistent with each other. The definitions of charge and current can be written in a manner shown in Figure 57.3(b), using constants ϵ_o and μ_o, which are common to both ESU and EMU. By setting their values to the ones shown in Table 57.1, the equations can be customized to ESU or EMU.

From the flowchart of the electrical definitions in ESU to be presented in this chapter, it will become clear that the definition of charge in ESU is the source of all the other definitions in ESU. In the next chapter, it will be shown that the definition of current is the source of all the other definitions in EMU.

In the previous chapters, as marked beside the equation number, some of the equations are valid only in ESU, some only in EMU, and some are common to both the systems. In this chapter, the electromagnetic equations will be cast in a form that is common to both the systems of units. It will then become clear, why some of the equations presented in the past chapters are valid only in ESU or EMU, or both. The reader is referred to the respective sections presented in the previous chapters for more details on the equations.

The ESU definitions flowchart is shown in Figure 58.1. Each of the electrical quantities discussed until now is shown in the rectangular box. The incoming arrow(s) to an electrical quantity represents all the variables used to define that electrical quantity. An outgoing arrow from a variable means that the variable is used to define the electrical quantity that the arrow is incoming to.

58.1 Definitions of Charge q and Current i

The definition of charge in ESU is the source of all the other definitions. The highlighted box marked 'Charge q' has only outgoing arrows, meaning that it is used to define all the other

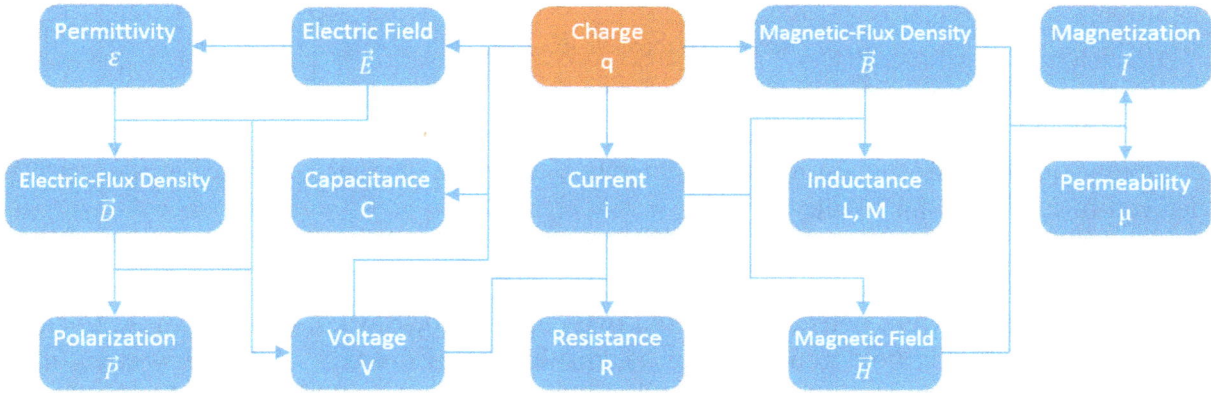

Figure 58.1: The definitions flowchart in the electrostatic system of units.

variables in the chart. The electric charge q is defined from Equation 57.24,

$$F = \frac{1}{\epsilon_o} \frac{q_1 q_2}{r^2}, \tag{58.1}$$

where $\epsilon_o = 1$ in ESU. Two point charges of magnitudes q_1 and q_2, separated by r, exert a force on each other, whose value is given by the above equation. Since the electric current i is the flow of electric charges, once charge has been defined, current can be defined as

$$i = \frac{\Delta q}{\Delta t}. \tag{58.2}$$

If charge Δq flows across a wire in time Δt, current is defined as written in the above equation.

58.2 Definitions of Magnetic Field $\vec{H}$ and Magnetic-Flux Density $\vec{B}$

From the definition of the electric current, the magnetic field $\vec{H}$ is defined from Biot-Savart's law as

$$\vec{H} = \int_L \frac{i}{r^2} \left(\vec{dl} \times \hat{r} \right). \tag{58.3}$$

The magnetic field at any point is calculated from the path integral along a wire carrying current i.

The Lorentz force was derived from Faraday's law as

$$\vec{F} = q\vec{v} \times \vec{B}. \tag{58.4}$$

Using this equation, the magnetic-flux density $\vec{B}$ can be defined as the field, whose value is such that a charge q, moving at velocity $\vec{v}$, experiences the force written in the above equation. It may seem that the definition of $\vec{B}$ is very abstract. However, the Lorentz force was used to derive the force between two current-carrying wires, which is a more practical form of the Lorentz equation that can be measured using the Kelvin-Ampere current balance.

401

The relation between $\vec{B}$ and $\vec{H}$ defines the permeability μ, which will be derived next. Using the Lorentz equation, the force F on a wire of length L carrying current i_1, by a parallel current i_2, was derived in Equation 55.16 as

$$\frac{F}{L} = i_1 B,\tag{58.5}$$

where B is the magnetic-flux density generated by i_2 at i_1. Biot-Savart's law was used to calculate the value of the magnetic field generated by i_2, resulting in Equation 55.18. However, in Chapter 57, Equation 55.18 was written in a form that is common to both ESU and EMU. Repeating Equation 57.25,

$$\frac{F}{L} = \mu_o \frac{2i_1 i_2}{d},\tag{58.6}$$

where μ_o is a constant, whose value will be determined in Chapter 63, and d is the distance between the parallel wires. Equating the value of B in the above equations,

$$B = \mu_o \frac{2i_2}{d}.\tag{58.7}$$

Using Biot-Savart's law,

$$H = \frac{2i_2}{d},\tag{58.8}$$

where H is the magnetic field generated by current i_2 at the wire carrying current i_1. From the above equations, the relation between B and H is

$$B = \mu_o H.\tag{58.9}$$

Since $\vec{B}$ and $\vec{H}$ are vectors, the above equation is the result of

$$\vec{B} = \mu_o \vec{H}.\tag{58.10}$$

Since Equation 58.6 assumes no magnetic material that modifies the magnetic field generated by i_2, the above equation captures the relation between $\vec{B}$ and $\vec{H}$ in free space or in a non-magnetic material. Note that in EMU, $\mu_o = 1$, resulting in $\vec{B} = \vec{H}$ in a non-magnetic material, as defined earlier. In ESU, however, they are related by the constant μ_o.

This is not the only way to formulate the equations. The above relation can be formulated such that $\vec{B} = \vec{H}$ in free space, in both ESU and EMU. However, the constant μ_o will show up in other equations, so that Equation 58.6 can be derived. For example, the force between two current-carrying wires was derived using Biot-Savart's law and the Lorentz force equation. It may have been very well formulated that $\vec{B} = \vec{H}$, and μ_o included in either Biot-Savart's law or in the Lorentz force equation. Equation 58.6 can also be derived in such a formulation. However, the advantage of Equation 58.10 is that it encapsulates μ_o in the $\vec{B} - \vec{H}$ relation, so that μ_o does not have to be carried around in other equations.

In the presence of a magnetic material, Equation 58.10 is modified as

$$\vec{B} = \mu_r \mu_o \vec{H} \tag{58.11}$$
$$= \mu \vec{H}, \tag{58.12}$$

where μ_r is the relative permeability of the material. A magnetic material modifies the $\vec{B} - \vec{H}$ relation by μ_r, as explained before.

μ_o in Figure 57.3(b) is chosen instead of κ_M in Figure 57.3(a), so that there is symmetry to μ_r, as written in the above equation. In free space, $\mu_r = 1$, and the relation between $\vec{B}$ and $\vec{H}$ is written in Equation 58.10. For this reason, μ_o is called the permeability of free space.

The permeability, or sometimes referred to as the absolute permeability of a material μ, is defined as

$$\mu = \mu_r \mu_o. \tag{58.13}$$

The measurement of the $\vec{B} - \vec{H}$ relation using the ballistic-galvanometer method was presented in Chapter 31. The technique is still valid, but the post-processing step of the measured data is slightly modified to reflect the change in Equation 58.11.

Equation 49.10 and Equation 49.11 are derived from Equation 49.9. In the case of a non-magnetic material, using the new formulation of the $\vec{B} - \vec{H}$ relation,

$$\Delta B = \mu_o \Delta H. \tag{58.14}$$

Equation 49.11 is modified as

$$\mu_o \Delta H \propto \theta. \tag{58.15}$$

As a result, Equation 49.12 is modified as

$$\Delta B = \left(\frac{\theta'}{\theta}\right) \mu_o \Delta H$$
$$= \mu_r \mu_o \Delta H. \tag{58.16}$$

This is the same result as what was obtained earlier in Equation 49.12, except for the factor μ_o. In the new formulation, the $B - H$ plot obtained earlier must be scaled by μ_o.

58.4 Magnetization Vector $\vec{I}$

The magnetization vector $\vec{I}$ is defined as

$$\vec{B} = \mu_o \vec{H} + 4\pi \mu_o \vec{I}. \tag{58.17}$$

Compared to Equation 41.6, there is a μ_o factor present. This is needed to derive Equation 58.11, as shown next. $\vec{I}$ is related to $\vec{H}$ using magnetic susceptibility χ_m,

$$\vec{I} = \chi_m \vec{H}. \tag{58.18}$$

From the above equations,

$$\vec{B} = \mu_o \vec{H} + 4\pi \mu_o \chi_m \vec{H}. \tag{58.19}$$

Factoring μ_o and $\vec{H}$,

$$\vec{B} = \mu_o \left(1 + 4\pi \chi_m\right) \vec{H}, \tag{58.20}$$

Equating the terms in the above equation with Equation 58.11, the relative permeability μ_r is

$$\mu_r = 1 + 4\pi \chi_m. \tag{58.21}$$

There is an additional μ_o factor, compared to the definition presented earlier in Chapter 41 in EMU. Note that when $\mu_o = 1$, tailoring the equations to EMU, the same equations as those in Chapter 41 are obtained.

58.5   Self Inductance L and Mutual Inductance M

The self inductance L is defined as before in Equation 44.2,

$$L = \frac{\Phi_{self}}{i}, \tag{58.22}$$

where Φ_{self} is the self-induced flux in a solenoid carrying current i. Mutual inductance is defined as before in Equation 45.2,

$$M = \frac{\Phi_S}{i_P}, \tag{58.23}$$

where Φ_S is the flux across the secondary coil, caused by the current i_P in the primary coil.

58.6   Electric Field $\vec{E}$ and Voltage V

The electric field $\vec{E}$ is defined as Equation 5.2,

$$\vec{E} = \frac{\vec{F}}{q}. \tag{58.24}$$

The electric field at a point is the force $\vec{F}$ experienced by a charge q at that point. In the case of the electrostatic (electromagnetic) system of units q is the charge in electrostatic (electromagnetic) measure.

The voltage V_{AB} between two points A and B, is the path integral of the electric field, as defined in Equation 10.6,

$$V_{AB} = \int_A^B \vec{E} \cdot \vec{dl}. \tag{58.25}$$

58.7 Electric-Flux Density $\vec{D}$, Permittivity ϵ, and the Permittivity of Free Space ϵ_o

In general, the Coulomb's relation is written as Equation 58.1, which is valid in both ESU and EMU. In this new formulation of Coulomb's law, it is left as an exercise for the reader to verify that Equation 32.20 needs to be modified as

$$\oint_S \epsilon_o \vec{E} \cdot d\vec{A} = 4\pi\, q_{enc},\tag{58.26}$$

where q_{enc} is the charge enclosed within the Gaussian surface S, and $\vec{E}$ is the value of the electric field at the Gaussian surface. In this formulation, it is assumed that the Gaussian surface is in free space, or in the absence of a dielectric material affecting the electric-field strength.

If an electric field is modified by a dielectric material, following the steps outlined in Chapter 36, and using the above equation, it is left as an exercise for the reader to show that Equation 36.6 needs to be modified as

$$\oint_S \epsilon_r \epsilon_o \vec{E} \cdot d\vec{A} = 4\pi\, q_{enc},\tag{58.27}$$

where ϵ_r is the permittivity of the dielectric material in which the Gaussian surface is present. If the electric-flux density $\vec{D}$ is defined as

$$\vec{D} = \epsilon_r \epsilon_o \vec{E}\tag{58.28}$$
$$= \epsilon \vec{E},\tag{58.29}$$

substituting this in Equation 58.27, Gauss's law stays the same as before,

$$\oint_S \vec{D} \cdot d\vec{A} = 4\pi\, q_{enc}.\tag{58.30}$$

The advantage of Equation 58.29 is that ϵ_o and ϵ_r are encapsulated within the definition of $\vec{D}$. If the definition of $\vec{D}$ is kept the same as before in Equation 36.7, the constant ϵ_o needs to be explicitly included in Gauss's law. The definition in Equation 58.29 avoids the need of explicitly including the constants.

$\frac{1}{\epsilon_o}$ in Figure 57.3(b) is chosen instead of κ_E in Figure 57.3(a), so that there is symmetry to ϵ_r, as written in Equation 58.29. In free space, $\epsilon_r = 1$, and the relation between $\vec{D}$ and $\vec{E}$ is

$$\vec{D} = \epsilon_o \vec{E}.\tag{58.31}$$

For this reason, ϵ_o is called the permittivity of free space.

From Equation 58.29, the permittivity ϵ, or absolute permittivity, as it is sometimes referred to, is defined as

$$\epsilon = \epsilon_r \epsilon_o.\tag{58.32}$$

58.8 Polarization $\vec{P}$

Comparison of Equation 58.29 and Equation 37.6, shows that the latter equation needs to be modified as

$$\vec{D} = \epsilon_o \left(1 + 4\pi\chi_e\right)\vec{E}, \tag{58.33}$$

where

$$\epsilon_r = 1 + 4\pi\chi_e, \tag{58.34}$$

and χ_e is the electric susceptibility. Multipying the terms within the parenthesis in the above equation,

$$\vec{D} = \epsilon_o\vec{E} + 4\pi\epsilon_o\chi_e\vec{E}. \tag{58.35}$$

The above equation can be written in a form similar to Equation 37.1,

$$\vec{D} = \epsilon_o\vec{E} + 4\pi\vec{P}, \tag{58.36}$$

where the polarization vector $\vec{P}$ is defined as

$$\vec{P} = \epsilon_o\chi_e\vec{E}. \tag{58.37}$$

58.9 Capacitance C

The definition of capacitance C is the same as Equation 35.8,

$$C = \frac{Q}{V}, \tag{58.38}$$

where $\pm Q$ is the charge stored in the two conductors making up the capacitor, whose potential difference is voltage V.

59

Definitions Flowchart in the Electromagnetic System of Units

The electromagnetic equations in Chapter 58, using constants ϵ_o and μ_o, are formulated in a way that they are common to both ESU and EMU. Setting the constants ϵ_o and μ_o to the values shown in Table 57.1, would tailor the equations to ESU or EMU.

Although the equations are common to both ESU and EMU, the order in which the electrical quantities are defined, are different. A consequence of this is that the units of the electrical quantities are different between ESU and EMU. The flowchart of definitions in EMU, and a comparison of the units will be presented in this chapter.

59.1 Definitions Flowchart in the Electromagnetic System of Units

In ESU, as seen in Chapter 58, the definition of charge is the source of all the other definitions. In EMU, however, the definition of current is the source of all the other definitions. The flowchart of definitions in EMU is shown in Figure 59.1. Using the same convention as Figure 58.1, the incoming arrows to an electrical quantity are all the variables used to define that electrical quantity. An outgoing arrow from a variable means that the variable is used to define the electrical quantity that the arrow is incoming to.

The variable 'Current i' has only outgoing arrows. This means that the current is the source of all the other definitions in EMU. Repeating Equation 58.6, current in EMU is defined as the flow of charges exerting a force between two parallel wires with current i_1 and i_2,

$$\frac{F}{L} = \mu_o \frac{2i_1 i_2}{d},$$

(59.1)

and L is the length of the parallel wires separated by d.
From this definition of current, the electric charge is defined using Equation 58.2,

$$\Delta q = i \Delta t.$$

(59.2)

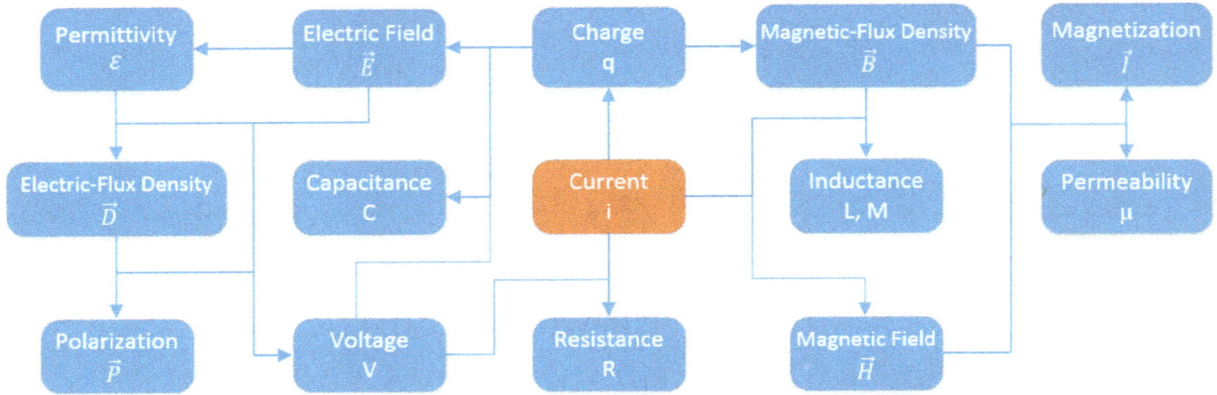

Figure 59.1: The definitions flowchart in the electromagnetic system of units.

Using the above equation, the electric charge is defined as the quantity contained in current i flowing for time Δt.

From the above definitions of charge and current, all the other definitions follow in the same way as stated in Chapter 58. To avoid repetition, this will not be stated again.

59.2   Comparison of Units in the Electrostatic and the Electromagnetic Systems

In ESU, the charge is defined first from Coulomb's law. Using dimensional analysis, repeating Equation 57.2, the unit of charge in ESU is

$$[q]_{ESU} = \frac{gm^{\frac{1}{2}}cm^{\frac{3}{2}}}{s}. \tag{59.3, ESU}$$

The unit of current in ESU, repeating Equation 57.3, is

$$[i]_{ESU} = \frac{gm^{\frac{1}{2}}cm^{\frac{3}{2}}}{s^2}. \tag{59.4, ESU}$$

In EMU, since current is first defined from the force between current carrying wires, the unit of current in EMU is

$$[i]_{EMU} = \frac{gm^{\frac{1}{2}} \, cm^{\frac{1}{2}}}{s}, \tag{59.5, EMU}$$

and the unit of charge in EMU, repeating Equation 57.12, is

$$\begin{aligned}
[q]_{EMU} &= abampere \cdot s \\
&= gm^{\frac{1}{2}} \, cm^{\frac{1}{2}}. \tag{59.6, EMU}
\end{aligned}$$

From the above units of charge and current, it can be seen that ESU and EMU don't have the same units for the electrical quantities. A summary of the units of some of the variables in ESU and EMU are shown in Table 59.1 and Table 59.2. It is left as an exercise for the reader to derive them from the flowchart of definitions, and using dimensional analysis. For a complete list of the units, the reader is referred to Reference [6].

Quantity	Symbol	Units in ESU
Charge	q	$\frac{\text{gm}^{1/2}\text{cm}^{3/2}}{\text{s}}$
Current	i	$\frac{\text{gm}^{1/2}\text{cm}^{3/2}}{\text{s}^2}$
Magnetic-Flux Density	$\vec{B}$	$\frac{\text{gm}^{1/2}}{\text{cm}^{3/2}}$
Magnetic Field	$\vec{H}$	$\frac{\text{gm}^{1/2}\text{cm}^{1/2}}{\text{s}^2}$
Electric-Flux Density	$\vec{D}$	$\frac{\text{gm}^{1/2}}{\text{cm}^{1/2}\text{s}}$
Electric Field	$\vec{E}$	$\frac{\text{gm}^{1/2}}{\text{cm}^{1/2}\text{s}}$

Table 59.1: A summary of the units of some of the commonly used electrical quantities in ESU.

Quantity	Symbol	Units in EMU
Charge	q	$\text{gm}^{1/2}\text{cm}^{1/2}$
Current	i	$\frac{\text{gm}^{1/2}\text{cm}^{1/2}}{\text{s}}$
Magnetic-Flux Density	$\vec{B}$	$\frac{\text{gm}^{1/2}}{\text{cm}^{1/2}\text{s}}$
Magnetic Field	$\vec{H}$	$\frac{\text{gm}^{1/2}}{\text{cm}^{1/2}\text{s}}$
Electric-Flux Density	$\vec{D}$	$\frac{\text{gm}^{1/2}}{\text{cm}^{3/2}}$
Electric Field	$\vec{E}$	$\frac{\text{gm}^{1/2}\text{cm}^{1/2}}{\text{s}^2}$

Table 59.2: A summary of the units of some of the commonly used electrical quantities in EMU.

60

Solving Electrostatic Problems Using Lord Kelvin's Method of Images

One of the greatest problems that confronted scientists in the nineteenth century was determining the ratio of the electrostatic to the electromagnetic unit charge. Weber and Kohlrausch were the first researchers to determine this ratio in 1856 [96]–[99]. Their experiment will be described in considerable detail in Chapter 63. Starting with the method of images to solving electrostatic problems in this chapter, the prerequisites to understanding their experiment will be developed in the following chapters, and finally leading up to their experiment.

The method of images was developed by Lord Kelvin in 1848 [100]. The importance of this technique in solving electrostatic problems cannot be overemphasized. Its powerful capability, yet simplicity, will become evident in the following chapters, and will be empirically verified in Chapter 61.

60.1 Electric Field of a Uniformly Charged Sphere

The method of images is a technique for solving voltage and electric field in a region, by replacing conductors with distributed charges on its surface, with point charges within the conductors. The values and locations of these point charges are chosen such that the voltage and the electric field outside of the conductors, are equal in the distributed and the point-charge cases. The electric field and the potential from point charges outside of the conductors can then be easily calculated using Equation 5.5 and Equation 10.15.

As explained in Section 36.4, Coulomb's law can be used to calculate the electric field generated by point charges in any uniform dielectric medium. However, for simplicity, air will be assumed. It will be assumed that Coulomb's law is applicable, including when conductors are present in a uniform dielectric medium, as done before in Section 32.1.

The application is not limited to solving for electric fields. This technique will be used in Chapter 61 to solve for the charge division between two spheres of different radii in contact. This result will be used in the Weber-Kohlrausch experiment.

The simplest example of method of images is calculating the electric field of a uniformly charged sphere. The 2D cross section of a charged spherical conductor A of radius r_A, is shown in Figure 60.1(a). For example, the sphere is positively charged with a total charge q_A, distributed uniformly over its surface. By spherical symmetry, the electric-field lines are radially outward, as shown. Applying Gauss's law to calculate the electric field,

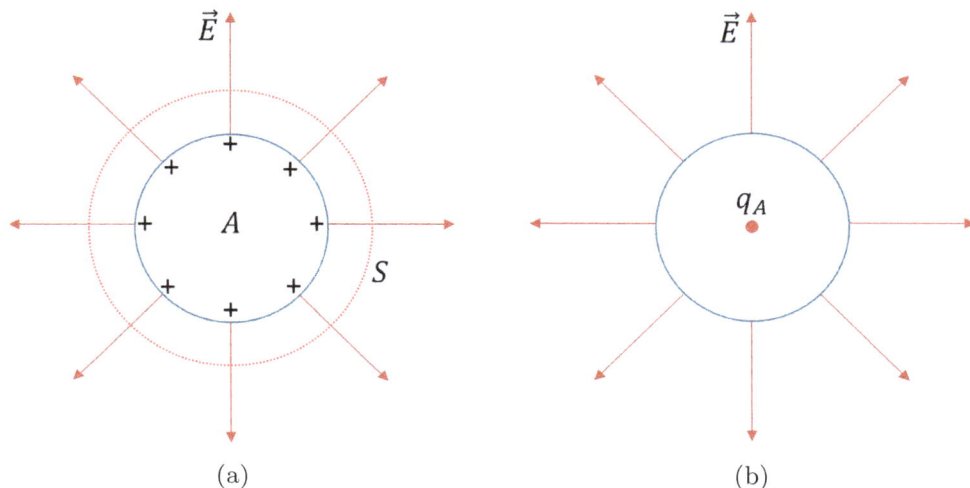

Figure 60.1: (a) The electric field surrounding a uniformly charged metal sphere. (b) The same electric-field pattern as Figure 60.1(a) can be obtained by replacing the total charge on the surface, by a point charge at the center of the sphere.

$$\oint_S \vec{D} \cdot d\vec{A} = 4\pi \, q_A, \tag{60.1}$$

where S is the dotted spherical Gaussian surface of radius r, marked in the figure.

Assuming that the dielectric material is air everywhere,

$$\vec{D} = \vec{E} \tag{60.2, ESU}$$

in ESU.

From the above equations, the electric field $\vec{E}$ outside the sphere is

$$\vec{E} = \frac{q_A}{r^2} \, \hat{r}, \ r \geq r_A. \tag{60.3, ESU}$$

Note that this result can also be obtained from Coulomb's law for $r \geq r_A$, if it is assumed that q_A is a point charge at the center of the sphere. This is shown in Figure 60.1(b).

Applying Equation 10.15, the voltage at the surface of the sphere $V(r_A)$, assuming the reference potential at ∞ is 0,

$$V(r_A) = \frac{q_A}{r_A}. \tag{60.4, ESU}$$

All points on the surface of the sphere are at the same potential. As this example demonstrates, a point charge can be placed within the conductor, such that the electric field outside the conductor is the same as the distributed-charge case.

60.2   Method of Images

The replacement of distributed charges by point charges, makes it easier to analytically solve for the electric field and potential. N conductors $\{C_1, ..., C_N\}$ are shown in Figure 60.2. The distributed charges on the conductor surfaces in Figure 60.2(a) are replaced by point charges in Figure 60.2(b). The region outside of the conductors in the distributed-charges case will be referred to as Region I, and Region II in the case of point charges, as marked. The two conditions that the point charges must satisfy, so that the electric field and the potential are equal in Region I and Region II, will be derived in this section.

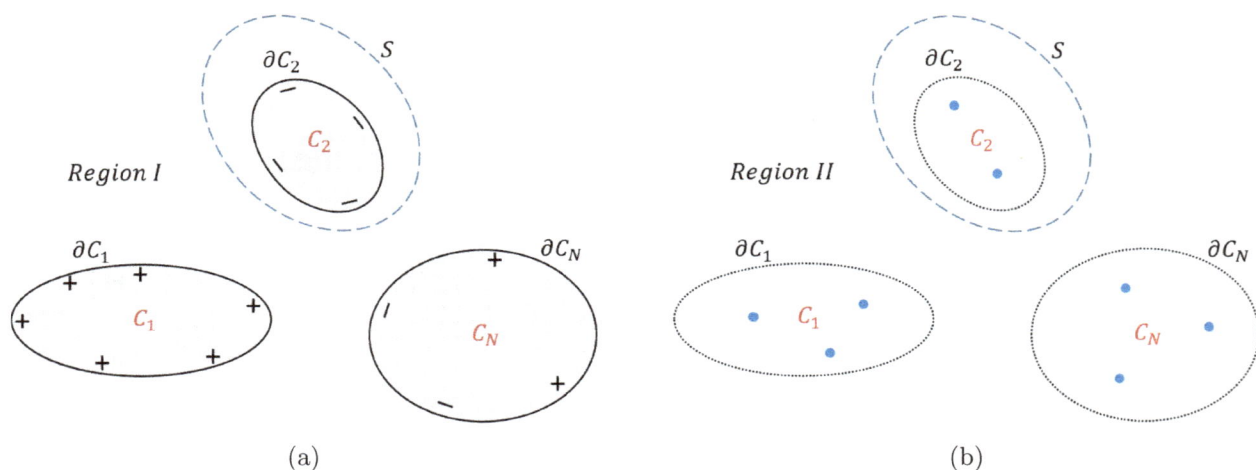

Figure 60.2: The distributed charges on the surfaces of conductors in (a), replaced by point charges in (b), such that the electric field in Region I and Region II are equal.

The first condition can be derived by noting that a conductor is an equipotential volume. The surface of a conductor and the volume enclosing the conductor are at the same potential. Therefore, the point charge(s) must be placed within a conductor, such that the conductor surface is equipotential. In the example shown in Figure 60.1, the point charge placed at the center of the sphere, results in an equipotential surface. From this example, also note that it is not necessary for points within the surface to be equipotential. An equipotential *surface* will suffice.

The second condition can be derived from Gauss's law. A Gaussian surface S enclosing conductor C_2 in Figure 60.2 is shown by the dashed line. If the fields in Region I and Region II are equal, by Gauss's law, the total charge enclosed by S must also be equal. Therefore, the total sum of the point charges placed within a conductor must equal the total distributed charge in that conductor. This is the second condition that the point charges need to satisfy. Clearly, this condition is met in the example in Figure 60.1, where the point charge placed within the conductor is equal to the total distributed charge.

Although not rigorously proven, it will be assumed that if the two conditions described are satisfied, then the electric field outside the conductors are the same in both the distributed and the point-charge cases. The method of images will be empirically verified in Chapter 61.

60.3   Point Charge Near a Spherical Conductor

A classic problem in method of images is calculating the electric field resulting from a point charge near a conducting sphere, as shown in Figure 60.3 [101]. The results from this exercise will be used to calculate charge sharing between two spheres in contact in Chapter 61.

Poisson developed a complex mathematical technique using solid harmonics that can be used to solve the problems described in this section and in Chapter 61 [56]. Poisson's method is beyond the scope of this book, but the same results arrived at using the method of images, validates both the techniques.

The point charge near the sphere will attract charges of the opposite polarity on the surface that is closer to the charge, creating an induced charged distribution. This is illustrated in Figure 60.3 for a positive point charge marked '+' near the sphere.

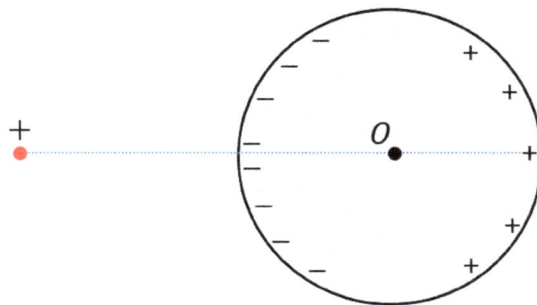

Figure 60.3: A point charge near a metal sphere.

It will be assumed that the reference potential at ∞ is 0. The potential at all other points will be calculated with respect to the 0 potential at ∞. With this assumption, Equation 10.15 can be used to calculate the potential generated by the point charges.

A point charge q is at distance d from the center of a sphere O of radius s, as shown in Figure 60.4. Using method of images, point charges are placed within the sphere, so that the surface of the sphere is an equipotential surface. The conducting sphere is uncharged, but a charged sphere can be easily accounted for using method of images, as explained towards the end of this section.

It is possible to place a point charge $\bar{q}$ at distance a from the center of the sphere, as shown in the figure, such that the surface of the sphere is at 0 potential. $\bar{q}$ lies along the line joining the point charge q to the center of the sphere O. One may ask, why is $\bar{q}$ located along this line, and cannot be placed anywhere else within the sphere? Different possibilities for the location of $\bar{q}$ can be analyzed, and by trial and error, the location shown in Figure 60.4 can be verified to set *all* the points on the spherical surface at 0 potential. It will be left as an exercise to show that it is not possible to set the surface to a potential of any other non-zero value V. The values of a and $\bar{q}$ will be solved next.

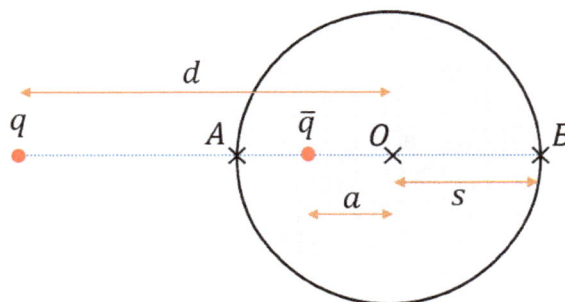

Figure 60.4: The placement of $\bar{q}$ such that the surface of the sphere is at 0 potential.

A and B are two points on the surface of the sphere, along the line through q and center O. Using Equation 10.15, V_A and V_B, the potentials at A and B caused by point charges q and $\bar{q}$, are

$$V_A = \frac{q}{d-s} + \frac{\bar{q}}{s-a} \qquad\qquad \text{(60.5, ESU)}$$

$$V_B = \frac{q}{d+s} + \frac{\bar{q}}{s+a}. \qquad\qquad \text{(60.6, ESU)}$$

Setting V_A and V_B to be 0,

$$0 = \frac{q}{d-s} + \frac{\bar{q}}{s-a} \qquad\qquad \text{(60.7)}$$

$$0 = \frac{q}{d+s} + \frac{\bar{q}}{s+a}, \qquad\qquad \text{(60.8)}$$

the values of $\bar{q}$ and a can be solved to set the surface of the sphere at 0 potential. It is left as an

414

exercise for the reader to verify that the solution to the above set of linear equations is

$$\bar{q} = -\frac{qs}{d} \tag{60.9}$$

$$a = \frac{s^2}{d}. \tag{60.10}$$

Although the 0 potential condition has been enforced only on A and B, it is left as an exercise for the reader to verify that all the points on the surface of the sphere are at 0 potential. If $V_A = V_B = V$ is set to a non-zero value V in Equation 60.5–60.6, it is left as an exercise for the reader to verify that the solution does not satisfy the requirement that the potential is V at all the points on the spherical surface. This shows that there is not much flexibility in how the point charges can be placed. Although not proven, it can be surmised that this must be a unique way.

The first condition in image theory is to place the image charges so that the surface of the sphere is at an equipotential surface. This is satisfied by placing charge $\bar{q}$ to set the sphere at 0 potential. The second condition is that the sum of the image charges in a conductor must equal the net distributed charge in the conductor. If the sphere is uncharged, then charge $-\bar{q}$ must be placed in the sphere to cancel $\bar{q}$, so that the net charge in the sphere is 0. $-\bar{q}$ must be placed at the center of the sphere, so that all points on the surface of the sphere are at the same potential. This is illustrated in Figure 60.5.

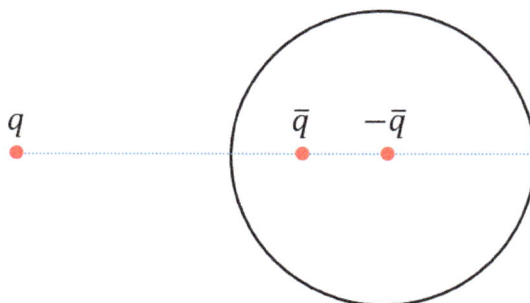

Figure 60.5: The placement of $\bar{q}$ and $-\bar{q}$ so that the sphere has 0 total charge, and the surface of the sphere is at a constant potential.

In this problem, it was assumed that the sphere is uncharged. If the sphere is charged, and contains Q statcoulombs or abcoulombs of charge, the results for $\bar{q}$ and a stay the same. But the point charge $Q - \bar{q}$ is now placed at the center of the sphere, so that the net charge in the sphere is $Q - \bar{q} + \bar{q} = Q$, which is the same as the total distributed charge on the sphere.

In this example, note that $\bar{q}$ and q set the potential of the sphere to 0. The final potential of the sphere is determined by the charge at the center of the sphere. Using Equation 60.4, the potential of the sphere is

$$V = \frac{-\bar{q}}{s}. \tag{60.11, ESU}$$

415

The electric field outside the conductors can be calculated using Equation 5.5, and superposition to add the field contributions of all the point charges. Once the electric field is calculated at the surface of the sphere, the charge distribution on the surface can be calculated using Equation 38.2.

61

Charge Division Between Two Spheres
of Different Radii in Contact

When a charged conducting sphere is put in contact with an uncharged conducting sphere, charges flow to the uncharged conductor, and charging it. The Weber-Kohlrausch experiment requires knowing in what ratio, charges divide between two spheres of different radii.

Coulomb measured the ratio in which charges divide between two spheres of different radii using a torsion balance. His experiment technique will be presented first in Section 61.1. Poisson theoretically calculated the ratio using a mathematical technique that he had developed [56]. Coulomb's empirical results match well with Poisson's calculations. This is a confirmation that both the data sets must be correct. The same exact result as Poisson's calculations can be obtained using Lord Kelvin's method of images. To appreciate the results of image theory, a highly inaccurate approximation will be presented first, followed by the solution using method of images [102].

61.1   Coulomb's Investigation of the Charge Distribution in Conducting Bodies

Coulomb documented his investigation of the charge distribution in conducting bodies, in his fourth–sixth memoirs, out of the seven that he read to the Paris Academy of Sciences [9]. His experiment to determine the ratio of the charge division between two spheres in contact will be presented in this section [103]. Coulomb constructed a slightly modified version of the torsion balance, shown in Figure 61.1, compared to the one in Figure 4.2, to accommodate for the larger spheres used in the experiment. The fixed and the movable charged spheres that repel each other are marked a and b, and the micrometer to control the twist in the wire is marked m.

A charged sphere $\bar{A}_1$ of radius r_{A1}, containing charge $\bar{q}_{A1}$, is shown in Figure 61.2(a). The naming convention followed is that the spheres and their related measurements, before they are put in contact, will be denoted using an overbar. $\bar{A}_1$ and $\bar{A}_2$ will be referred to as A_1 and A_2, after they are put in contact, as shown in Figure 61.2(b). The charge in A_1 is q_{A1}, and the charge in A_2, q_{A2}.

The top view of the torsion balance is shown in Figure 61.3(a). $\bar{A}_1$ is the fixed sphere in the

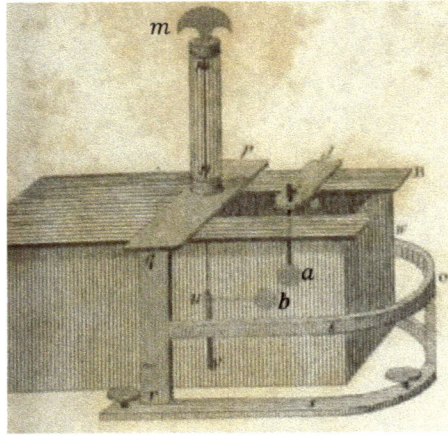

Figure 61.1: The torsion balance, Coulomb had used to calculate the ratio of charge division between two spheres in contact.

torsion balance, positioned a certain distance r from the movable sphere B_1. B_1 is at S, shown by the dotted outline, when there is no force of repulsion, and no twist in the wire. B_1 is charged independently, and is of the same polarity as $\bar{A}_1$, so that they repel each other. The charge value of B_1, q_{B1}, stays the same throughout the experiment.

The force of repulsion between B_1 and $\bar{A}_1$, swings the arm of the torsion bar of length ℓ to some angle $\bar{\alpha}$. The micrometer is turned by angle $\bar{\alpha}_m$ to increase the twist in the wire, and increasing the restoring torque, as seen before in Chapter 4, until B_1 is at its starting point S, as shown in Figure 61.3(b). The restoring torque τ_R is balanced by the torque caused by the force of repulsion τ_C between the spheres. The total twist in the wire is $\bar{\alpha} + \bar{\alpha}_m$. Since $\bar{\alpha} = 0$ when B_1 is at S, the twist in the wire is only created by the turn of the micrometer $\bar{\alpha}_m$.

Using Equation 3.21, the restoring torque created by the twist α_m in the wire is

$$\tau_R = -\kappa\bar{\alpha}_m, \tag{61.1}$$

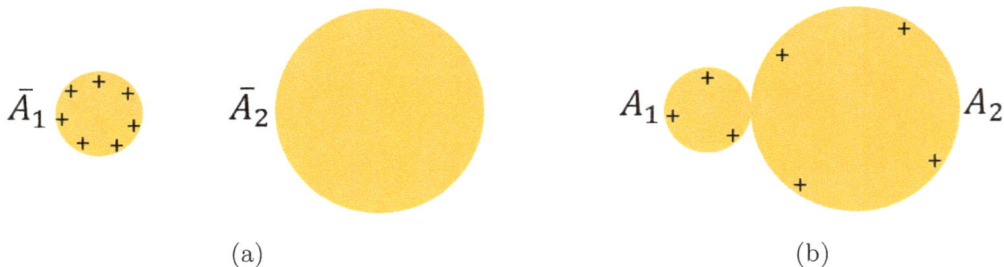

Figure 61.2: (a) A charged sphere $\bar{A}_1$ and an uncharged sphere $\bar{A}_2$. (b) The redistribution of charges after they are put in contact.

418

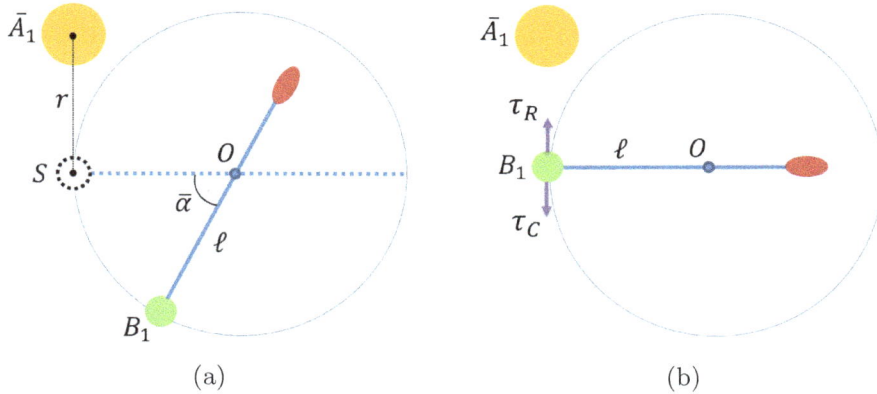

Figure 61.3: (a) The top view of the torsion balance. The force of repulsion between B_1 and $\bar{A}_1$, swings the arm of the torsion bar of length ℓ to some angle $\bar{\alpha}$. (b) The micrometer is turned to increase the twist in the wire, to bring back B_1 to its starting point S.

and the torque caused by the force of repulsion between the charges $\bar{q}_{A1}$ and q_{B1} is

$$\tau_C = \ell \frac{\bar{q}_{A1}\, q_{B1}}{r^2}. \tag{61.2, ESU}$$

Applying Newton's second law of rotational motion on an object in equilibrium, the sum of the torques acting on the object must be 0, resulting in

$$\kappa \bar{\alpha}_m = \ell \frac{\bar{q}_{A1}\, q_{B1}}{r^2}. \tag{61.3, ESU}$$

$\bar{A}_1$ is put in contact with the uncharged sphere $\bar{A}_2$, and the experiment is repeated on A_1, as shown in Figure 61.4. The charge q_{B1} stays the same as what was before. Part of the charge on $\bar{A}_1$ is transferred to $\bar{A}_2$, and therefore, $q_{A1} < \bar{q}_{A1}$. As a consequence, the torsion bar is repelled to an angle α, smaller than $\bar{\alpha}$, illustrated in Figure 61.4(a).

In Figure 61.4(b), the micrometer is turned by angle α_m to bring B_1 back to its starting point. Writing a torque equation similar to Equation 61.3,

$$\kappa \alpha_m = \ell \frac{q_{A1}\, q_{B1}}{r^2}. \tag{61.4, ESU}$$

Dividing Equation 61.4 by Equation 61.3,

$$q_{A1} = \bar{q}_{A1} \left(\frac{\alpha_m}{\bar{\alpha}_m} \right). \tag{61.5}$$

By the law of conservation of charge, the total charge before and after transfer must remain the same,

$$\bar{q}_{A1} = q_{A1} + q_{A2}. \tag{61.6}$$

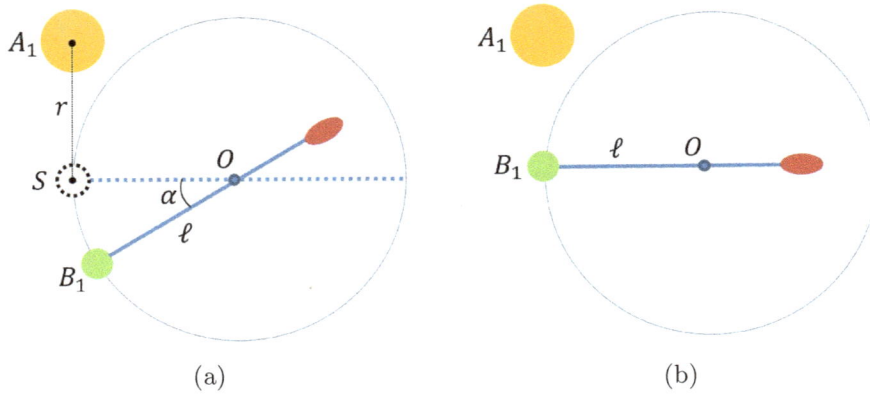

Figure 61.4: The experiment in Figure 61.3 repeated on the sphere A_1.

From the above equations,

$$q_{A2} = \bar{q}_{A1} \left(1 - \frac{\alpha_m}{\bar{\alpha}_m} \right). \tag{61.7}$$

Dividing Equation 61.5 by Equation 61.7, the ratio of the charges in A_1 and A_2 can now be calculated as

$$\frac{q_{A1}}{q_{A2}} = \frac{\alpha_m}{\bar{\alpha}_m - \alpha_m}. \tag{61.8}$$

If r_{A1} and r_{A2} are the radii of the spheres, the *average* surface charge densities in A_1 and A_2 are

$$\sigma_{A1} = \frac{q_{A1}}{4\pi r_{A1}^2}, \tag{61.9}$$

$$\sigma_{A2} = \frac{q_{A2}}{4\pi r_{A2}^2}. \tag{61.10}$$

The ratio of the average charge densities is

$$\frac{\sigma_{A1}}{\sigma_{A2}} = \frac{4\pi r_{A2}^2}{4\pi r_{A1}^2} \left(\frac{\alpha_m}{\bar{\alpha}_m - \alpha_m} \right). \tag{61.11}$$

An example of the experiment results from [103] is presented here. The radius r_{A1} of A_1 is 3.125″ and the radius r_{A2} of A_2 is 12″, for the trials presented in Table 61.1. The ratio of the average charge densities is calculated using Equation 61.11, and are almost the same value in each of the trials.

Using the experiment technique described, Coulomb's experiment results are summarized in Table 61.2. He used spheres with radii between $\frac{1}{12}″$ and 12″ in his experiment. The limiting radii ratio of ∞ in the last row of the table is obtained from spheres with $\frac{1}{12}″$ and 4″ radii [9]. Poisson's calculations match closely with the experiment observations. This was also verified by Plana [56]. The solution using the method of images are the same as Poisson's calculations, and will be solved in this chapter, following the solution presented in Reference [102]. Maxwell used the method of

Trial	$\bar{\alpha}_m$	α_m	Measured $\frac{\sigma_{A1}}{\sigma_{A2}}$
1	145°	12°	1.33
2	145°	12°	1.33
3	259°	21°	1.30
4	255°	21°	1.32
5	231°	19°	1.32

Table 61.1: The results from Coulomb's experiment of the average charge densities in the different trials.

Ratio of the Radii $\frac{r_{A2}}{r_{A1}}$	Measured $\frac{\sigma_{A1}}{\sigma_{A2}}$	Poisson's Calculation	Method of Images
1	1.00	1.00	1.00
2	1.08	1.16	1.16
4	1.30	1.32	1.32
8	1.65	1.44	1.44
∞	< 2.00	$\pi^2/6 \approx 1.65$	$\pi^2/6 \approx 1.65$

Table 61.2: The average charge densities obtained from Coulomb's experiment, Poisson's calculation, and the method of images are in excellent agreement.

inversion, which also results in the same solution as the method of images, but this technique will not be covered in this book.

61.2 A 0th Order Approximation to Calculate the Charge-Division Ratio

To appreciate the solution derived using the method of images, a highly inaccurate approximation is presented first in this section. Spheres A_1 and A_2 in Figure 61.2(b), containing charge q_{A1} and q_{A2}, have been redrawn in Figure 61.5. Region 1 to the left of the hemisphere A_1, and Region 2 to the right of A_2, are shown shaded.

It will be assumed that the charges in A_2 do not influence the electric field in Region 1. Although this is clearly not true, this assumption is made to simplify the calculations. Likewise, it will be assumed that the electric field generated by the charges in A_1, in Region 2, is negligible.

The second assumption that will be made is that the charges are uniformly distributed on the surface of the spheres. Clearly, this is also an incorrect assumption, since the charge distributions on the spheres will influence each other, creating a non-uniform distribution. However, this assumption is made to simplify the calculations.

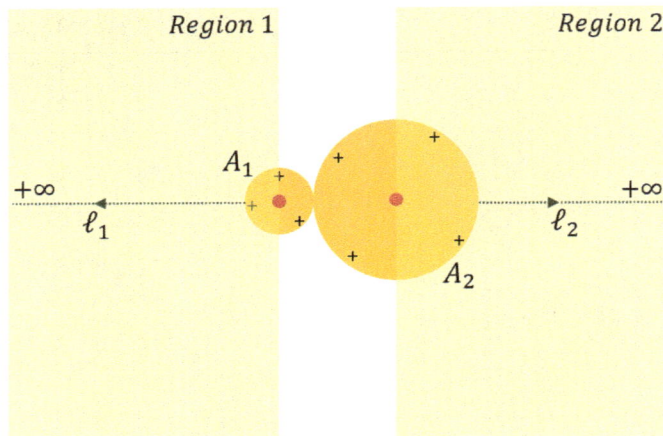

Figure 61.5: A highly inaccurate calculation of the charge division between two spheres.

In a sphere with a uniform charge distribution, the total charge on the surface of the sphere can be placed as a point charge at its center, as explained in Section 60.3. The electric field in Region 1 will be referred to as $\vec{E}_1$, and the electric field in Region 2, $\vec{E}_2$. Since q_{A2} (q_{A1}) does not change the field in Region 1 (Region 2), the electric field in the shaded regions can be calculated as

$$\vec{E}_1 = \frac{q_{A1}}{r^2} \, \hat{\ell}_1, \qquad\qquad\qquad \text{(61.12, ESU)}$$

and

$$\vec{E}_2 = \frac{q_{A2}}{r^2} \, \hat{\ell}_2. \qquad\qquad\qquad \text{(61.13, ESU)}$$

Assigning the reference voltage at ∞ to be 0, the voltage on the surface of sphere A_1, can be found by calculating the path integral in the radial direction shown by the arrow, along the dotted line ℓ_1. From the above equations,

$$\begin{aligned} V(A_1) &= \int_{r_{A1}}^{\infty} \vec{E}_1 \cdot \vec{dl} \\ &= \frac{q_{A1}}{r_{A1}}. \end{aligned} \qquad\qquad\qquad \text{(61.14, ESU)}$$

Likewise, the voltage on the surface of the sphere A_2, is the value of the path integral from r_{A2} to ∞ along ℓ_2, in the direction of the arrow,

$$\begin{aligned} V(A_2) &= \int_{r_{A2}}^{\infty} \vec{E}_2 \cdot \vec{dl} \\ &= \frac{q_{A2}}{r_{A2}}. \end{aligned} \qquad\qquad\qquad \text{(61.15, ESU)}$$

The above results are the same as Equation 10.15. Since the spheres in contact form an equipotential volume, the voltages calculated on the surfaces must be equal,

$$\frac{q_{A1}}{r_{A1}} = \frac{q_{A2}}{r_{A2}}. \qquad\qquad\qquad \text{(61.16)}$$

Rearranging the above equation, the ratio in which the charges divide between the spheres is the same as the ratio of their radii,

$$\frac{q_{A1}}{q_{A2}} = \frac{r_{A1}}{r_{A2}}. \tag{61.17}$$

The ratio of the average charge densities can be calculated using Equation 61.9–Equation 61.10,

$$\frac{\sigma_{A1}}{\sigma_{A2}} = \frac{r_{A2}}{r_{A1}}. \tag{61.18}$$

If $r_{A2} = 10\, r_{A1}$, for example, then

$$\sigma_{A1} = 10\, \sigma_{A2}. \tag{61.19}$$

The above equation states that if the radius of A_2 is 10× the radius of A_1, then A_1 will have 10× the average charge density as A_2. The results don't match with the experiment observations of Coulomb and Poisson's calculations in Table 61.2. As the ratio of the radii of A_2 to A_1 becomes greater and greater, the ratio of the average charge densities converges to $\pi^2/6 \approx 1.65$. The solution using the method of images, however, presented next, shows an excellent correlation to Coulomb's experiment results.

61.3 Method of Images to Calculate the Charge Division Between Two Spheres in Contact

The two spheres A_1 and A_2 in Figure 61.2(b), containing charges q_{A1} and q_{A2}, are redrawn in Figure 61.6. The ratio of their average charge densities will be calculated using the method of images.

The quantities related to A_1, such as charge, length, etc., will be denoted by lower-case alphabets, and the quantities related to A_2 will use upper-case alphabets.

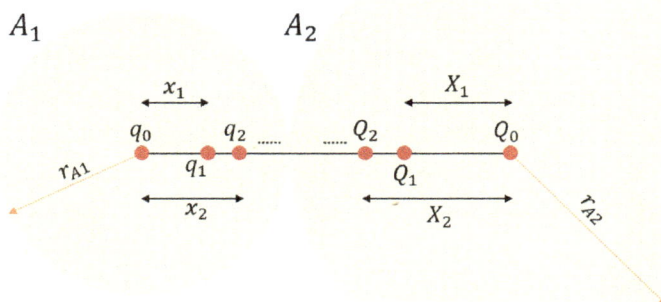

Figure 61.6: An accurate calculation of the charge division between two spheres using image theory.

The spheres in contact is an equipotential volume, and let its voltage with respect to the 0 potential at ∞ be any value V. In image theory, point charges are placed so that all the points on the surface of the spheres are at the same potential. Lets begin by placing q_0 (Q_0) at the center of A_1 (A_2), which sets the potential of the surface of A_1 (A_2) to V.

If the potential of the spheres is V, using Equation 60.4, the charge q_0 placed at the center of A_1 is

$$q_0 = V r_{A1}, \qquad \text{(61.20, ESU)}$$

where r_{A1} is the radius of A_1, and

$$Q_0 = V r_{A2}, \qquad \text{(61.21, ESU)}$$

where r_{A2} is the radius of A_2.

Following the example in Section 60.3, the presence of Q_0 (q_0), gives rise to the image charge q_1 (Q_1) within the sphere A_1 (A_2), so that the potential at the surface of the sphere is V. q_1 (Q_1) is placed at distance x_1 (X_1) from the center of A_1 (A_2). Applying Equation 60.9 and Equation 60.10 on A_1,

$$q_1 = -\frac{Q_0 \, r_{A1}}{r_{A1} + r_{A2}} \qquad \text{(61.22)}$$

$$x_1 = \frac{r_{A1}^2}{r_{A1} + r_{A2}}. \qquad \text{(61.23)}$$

Applying the equations simultaneously on A_2,

$$Q_1 = -\frac{q_0 \, r_{A2}}{r_{A1} + r_{A2}} \qquad \text{(61.24)}$$

$$X_1 = \frac{r_{A2}^2}{r_{A1} + r_{A2}}. \qquad \text{(61.25)}$$

$\{q_0, Q_0\}$ gives rise to the image charges $\{q_1, Q_1\}$, located at distance $\{x_1, X_1\}$ from the center of the spheres. From the exercise in Section 60.3, Q_0 and q_1 set the surface of A_1 at 0 potential. Likewise, q_0 and Q_1 set the surface of A_2 at 0 potential. $\{q_0, q_1, Q_0\}$ $(\{Q_0, Q_1, q_0\})$ set the potential of the surface of A_1 (A_2) to voltage V. The new image charge Q_1, however, makes the potential at the surface of A_1 non-uniform again. Similarly, the new image charge q_1 makes the potential at the surface of A_2 non-uniform.

$\{q_1, Q_1\}$, in turn, generate image charges $\{q_2, Q_2\}$, located at $\{x_2, X_2\}$, to set the surface potential of the spheres at voltage V. For the same reasons mentioned before, $\{q_2, Q_2\}$ spawn $\{q_3, Q_3\}$, and so on. This process continues indefinitely.

The total charge in sphere A_1 is

$$q_{A1} = q_0 + q_1 + q_2 + \dots. \qquad \text{(61.26)}$$

and the total charge in A_2,

$$q_{A2} = Q_0 + Q_1 + Q_2 + \dots. \qquad \text{(61.27)}$$

424

From the above equations, the ratios of the charges and the average charge densities can be calculated, similar to how it was done earlier.

The point charges $\{q_1, q_2, ..., Q_0, Q_1, ...\}$, except for q_0, set the potential of the surface of the sphere A_1 to 0. q_0 sets the surface of A_1 to voltage V. Likewise, $\{q_0, q_1, ..., Q_1, Q_2, ...\}$, except for Q_0, set the potential of the sphere A_2 to 0. Q_0 sets the surface of A_2 to voltage V.

The above calculations can be numerically solved, and the result is plotted in Figure 61.7. The x-axis is the ratio of the radii of A_2 to A_1, and the y-axis is the ratio of the average charge densities σ_{A1} to σ_{A2}. The plot marked (a), shown by the dotted line, is the approximation in Equation 61.18, and deviates significantly from the image-theory solution, marked (b). Note that the smaller sphere has a higher charge density than the larger sphere. The value of the asymptotic limit of $\pi^2/6$, shown by the dashed line, can be calculated analytically, and derived in Maxwell's treatise [56]. From the numerical solution, it is left as an exercise for the reader to verify that the point charges $\{q_0, q_1, q_2, ...\}$ lie within the boundary of A_1, and $\{Q_0, Q_1, Q_2, ...\}$ lie within the boundary of A_2. The image theory results and Poisson's results in Table 61.2, match well with Coulomb's measurement results.

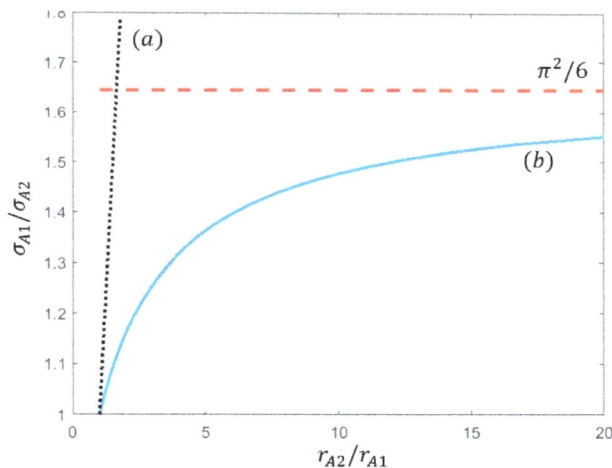

Figure 61.7: The ratio of the average charge densities as a function of the ratio of the radii. (a) The inaccurate calculation in Equation 61.18, and (b) the accurate solution using image theory.

The analysis presented in this chapter shows how powerful Lord Kelvin's method of images is in solving some of the electrostatic problems. This result will be used in the Weber-Kohlrausch experiment in Chapter 63.

62

Coulomb Electrometer

The Weber-Kohlrausch experiment requires knowing the quantity of charge in electrostatic units, present in a charged spherical conductor. The Coulomb torsion balance described in Chapter 4, which Coulomb used to prove the inverse-square law of electric charges, together with the results from Chapter 61, can be used to measure the charge in a spherical conductor, and its voltage. The torsion balance used for such a functionality will be referred to as the Coulomb electrometer.

The focus of this chapter is to present the experiment technique, and the calculations to determine the charge and the potential of a spherical conductor in electrostatic units.

62.1   Measurement of Charge in a Sphere Using a Coulomb Torsion Balance

In Chapter 4, the inverse-square behavior of Coulomb's law was treated as an unknown, and the Coulomb torsion balance was used to show that the inverse-square law must be the correct behavior of the force between charges. In Chapter 9, Coulomb's law was derived analytically. Now that Coulomb's law has been reasoned to be correct, the torsion balance can be used to measure the charge in a sphere.

A torsion balance has a fixed sphere and a movable sphere. In the torsion balance that Weber and Kohlrausch constructed, both these spheres are of equal radii r_B, and made of a metal. Any object attached to the spheres such as Rod Q, or the torsion bar, in Figure 4.2, are made of insulator materials, and do not affect the charge in the spheres.

A charged sphere A of radius r_A, with initial charge q_{Ai}, whose value is to be determined in electrostatic measure, is shown in Figure 62.1(a). A is put in contact with the initially uncharged fixed sphere B of the torsion balance, shown in Figure 62.1(b). Charge is divided between A and B, resulting in the new charge value q_{Af}, denoted using the subscript f that stands for "final". The initial charge acquired by B is q_{Bi}.

By the law of conservation of charge,

$$q_{Ai} = q_{Af} + q_{Bi}.$$ (62.1)

Using the result from the method of images in Equation 61.26 and Equation 61.27, the ratio of the charge in q_{Af} to q_{Bi} can be calculated. Let this value be γ,

$$q_{Af} = \gamma \, q_{Bi}. \tag{62.2}$$

Upon contact between the fixed and the movable spheres of the torsion balance, shown in Figure 62.1(c), by symmetry, half the charge on the fixed sphere is transferred to the movable sphere, since they have the same radii, resulting in

$$q_{Bi} = 2 \, q_{Bf}. \tag{62.3}$$

The above equation is in agreement with the calculations using the method of images.

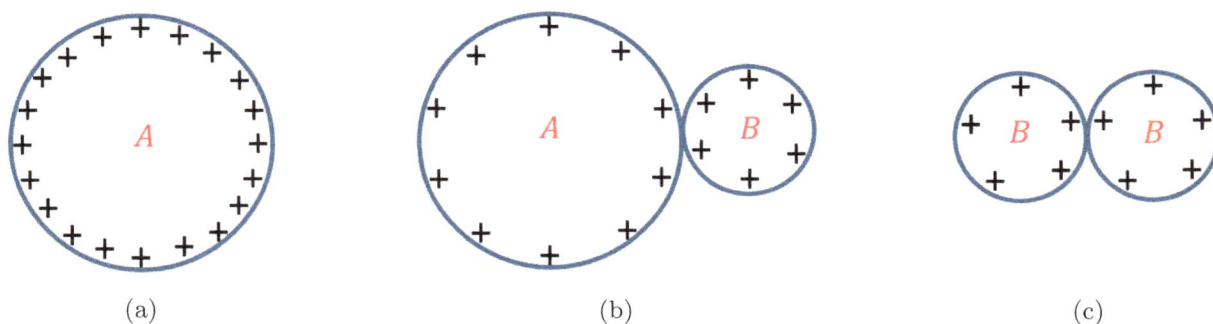

(a) (b) (c)

Figure 62.1: (a) A uniformly charged metal sphere A. (b) A is put in contact with the fixed sphere B of the torsion balance. (c) The charge division between the fixed and the movable, identical spheres, of the torsion balance.

If q_{Bf} can be measured, then q_{Bi} can be calculated using the above equation. This value can be used to calculate q_{Af} using Equation 62.2. Following this result, the values of q_{Bi} and q_{Af} can be used to calculate q_{Ai} using Equation 62.1, the charge in sphere A that is to be determined. The measurement of q_{Bf} will be presented next.

62.2 Calculation of the Charge in the Fixed and the Movable Spheres

The charge in the fixed and the movable spheres of the torsion balance q_{Bf}, can be determined using the analysis in Chapter 4. From the equations presented in the previous section, the charge in Sphere A to begin with, can be calculated.

Repeating Figure 4.3, the top view of the torsion balance is shown in Figure 62.2. The fixed and the movable spheres, drawn as unfilled and filled circles, being equal in their charge polarity, repel each other. The micrometer can be turned α_m to increase the twist in the wire, moving the movable sphere closer to the fixed sphere, to an angle α, as explained in Chapter 4.
In Equation 4.13, setting $n = 2$ to capture the inverse-square behavior, equating q_A and q_B, which

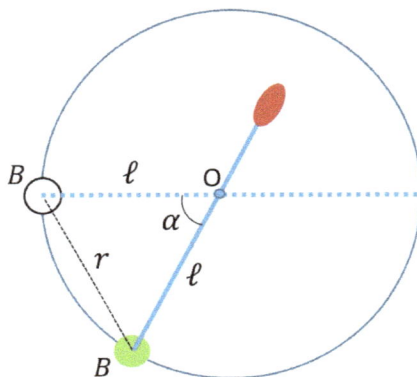

Figure 62.2: The top view of the torsion balance.

are the charges in the fixed and the movable spheres of the torsion balance, to q_{Bf}, and setting $k_e = 1$ in the electrostatic system of units, it is left as an exercise for the reader to verify that

$$q_{Bf} = 2\sin\left(\frac{\alpha}{2}\right)\sqrt{\frac{\kappa\,(\alpha + \alpha_m)\,\ell}{\cos\left(\frac{\alpha}{2}\right)}}. \tag{62.4, ESU}$$

In the above equation, κ is the torsion constant, whose value can be measured, as described in Chapter 4, ℓ is the length of the torsion bar to calculate the torque caused by the force of repulsion between the spheres.

To be specific, ℓ is the length from the center of the movable sphere to the pivot point about which the torsion bar turns. Assuming a uniform charge distribution on the movable sphere, the net charge can be placed as a point charge at its center. This is the point where the force of repulsion is applied on the torsion bar, creating a torque, and resulting in the above definition of ℓ.

All of the values needed to calculate q_{Bf} can be measured, and the charge in the fixed and movable spheres can thus be determined in electrostatic measure. From this result, q_{Ai} can be calculated as explained earlier.

A simple correction factor for the torsion balance, using the method of images, has been derived in Reference [104][105], but is not presented in this book. In the case where the dimensions of the fixed and the movable spheres are comparable to the separation between them, the spheres can no longer be treated as point charges that are placed at their centers. In this case, the charge distributions on the spheres will influence each other, creating a non-uniform distribution, and affecting the force of repulsion between the spheres. The correction factor adjusts the value of q_{Bf} to take this effect into account.

62.3   Voltage of a Charged Sphere

If the charge q_{Ai} is known, assuming uniform charge distribution, the voltage V_A of Sphere A, with respect to the reference potential of 0 at ∞, can be calculated using Equation 60.4,

$$V_A = \frac{q_{Ai}}{r_A}, \tag{62.5, ESU}$$

where r_A is the radius of the sphere.

63

The Weber-Kohlrausch Experiment to Determine the Ratio of q_{ESU} to q_{EMU}

It was noted in Chapter 57 that the electrostatic and the electromagnetic units are not consistent with each other: the unit charge in electrostatic units is not the same as the unit charge in electromagnetic units. Moreover, comparison of Table 59.1-Table 59.2 shows that their units are not equal. It was determined in Chapter 58-Chapter 59, if the ratio of the unit charge in electrostatic units to the unit charge in electromagnetic units is known, then all the electromagnetic equations in both these systems can be determined.

From these set of equations, and the ratio of the unit electrostatic to the unit electromagnetic charge, the electrical quantities in ESU and EMU can be converted from one to the other. The reader is referred to Reference [6] for more details on converting between units.

Weber and Kohlrausch are the first to determine this ratio in 1856 [96]–[98]. The chapters presented until now, come together in their experiment. Their experiment will be presented in this chapter, and requires knowing the capacitance of a Leyden jar in electrostatic measure, which will be calculated first.

63.1   Capacitance of a Leyden Jar in Electrostatic Measure

The Leyden jar, a capacitor, was described in Section 17.2, and shown in Figure 63.1(a). The shaded region is the overlapping metal foil on the inner and the outer glass surfaces, forming a cylindrical capacitor, illustrated in Figure 63.1(b). The capacitance of this structure will be derived in the remainder of this section.

The radius from the center of the cylinder to the inner metal foil is a, and the outer is b. The length of the foil is L. A cylindrical Gaussian surface S at radius $a < r < b$ is also shown. S is divided into 3 parts: the circular surfaces on the two sides of the Gaussian surface have been labeled S_1 and S_3, and the cylindrical part S_2, labeled in Figure 63.2(a). The side view of the cylindrical capacitor is drawn in Figure 63.2(b).

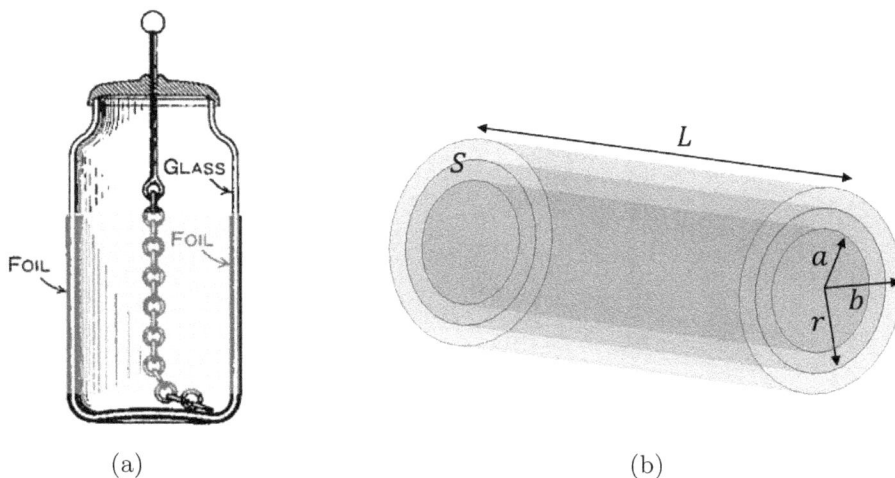

Figure 63.1: (a) A Leyden jar. (b) A cylindrical capacitor.

Without loss of generality, let the inner metal foil be charged positive, and the outer negative. The electric field, by cylindrical symmetry, will be assumed to be radially outward, perpendicular to the Gaussian surface, as shown by the dotted arrows in Figure 63.2(b). Therefore, the electric-field flux across S_1 and S_3 is 0. By cylindrical symmetry, for any radius r, the magnitude of the electric field E, and the electric-flux density magnitude D, have the same value at all points on S_2. This value can be determined from Gauss's law.

Applying Gauss's law on S,

$$\oint_S \vec{D} \cdot d\vec{A} = 4\pi \, q_{enc}. \tag{63.1}$$

Since the electric field is perpendicular to S_2, the surface integral can be simplified as

$$D \, 2\pi r \, L = 4\pi \, Q, \tag{63.2}$$

where Q is the enclosed charge on the inner metal foil, by the Gaussian surface, and D is the magnitude of the electric-flux density on S_2. The dielectric material filling the space between the metal foils in the Leyden jar is uniform, whose relative permittivity can be measured from Faraday's experiment in Chapter 35. Applying the relation between D and E in Equation 36.7, in the above equation,

$$E = \frac{1}{\epsilon_r} \frac{4\pi \, Q}{2\pi r \, L}. \tag{63.3, ESU}$$

The voltage difference between the inner foil and the outer foil is the path integral of the electric field from a to b. Since the voltage difference between two points is path independent, a radial path from a to b is chosen for simplicity of the calculation. The voltage difference between the

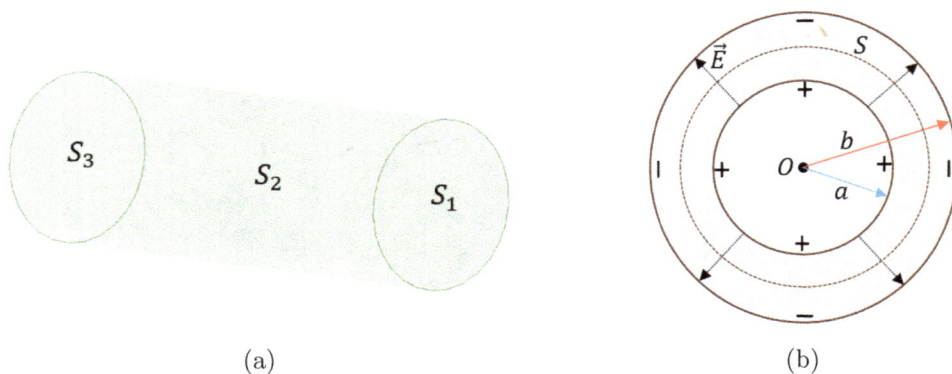

(a)   (b)

Figure 63.2: (a) The areas of the Gaussian surface. (b) The side view of the cylindrical capacitor.

foils V is

$$V = \int_a^b \vec{E} \cdot \vec{dl}$$

$$= \frac{2Q}{\epsilon_r L} \int_a^b \frac{1}{r}\, dr. \tag{63.4, ESU}$$

Evaluating the integral,

$$V = \frac{2Q}{\epsilon_r L} \ln\left(\frac{b}{a}\right). \tag{63.5, ESU}$$

By the definition of capacitance, from the above relation,

$$C = \frac{\epsilon_r L}{2\ln\left(\frac{b}{a}\right)}. \tag{63.6, ESU}$$

From the dimensions of the Leyden jar, its capacitance in electrostatic measure can be determined.

63.2   Weber-Kohlrausch Experiment

The setup of the experiment is shown in Figure 63.3. A sphere C of diameter $13''$ is coated with tin foil, and attached to one of the terminals of the Leyden jar L. The sphere and the Leyden jar are charged *together* using a Wimshurst machine. Several Leyden jars may be connected in parallel to increase the charge storage.

63.2.1   Measurement of Charge in Electrostatic Units

The voltage of the sphere V_C was measured in electrostatic units with a torsion balance, using the methodology presented in detail in Chapter 62. The fixed and the movable metal spheres of the torsion balance are identical, and $1''$ in diameter. The sphere and the terminal of the Leyden jar to which it is connected, form an equipotential volume, and are at the same potential. The terminal of the Leyden jar to which the sphere is connected is also therefore, at voltage V_C.

432

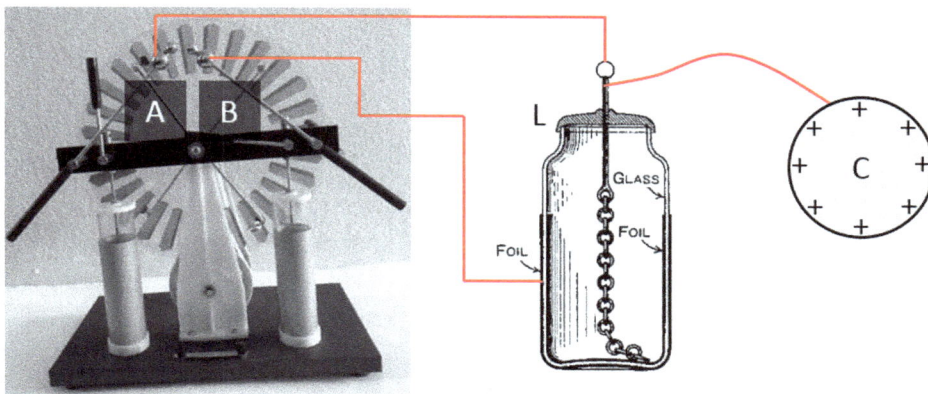

Figure 63.3: The setup of the Weber-Kohlrausch experiment.

By symmetry of the Wimshurst machine, the metal foils of the Leyden jar are charged equal and opposite. By symmetry of the Leyden jar, if one of the terminals of the Leyden jar is V_C, the other terminal must be at $-V_C$. The voltage difference between the terminals is therefore, $2V_C$.

The dimensions of the Leyden jar can be measured, and whose capacitance C in electrostatic measure can be determined using Equation 63.6. Using the capacitance relation in Equation 35.7, and the measured voltage across the capacitor terminals $2V_C$, the charge stored in the Leyden jar in electrostatic measure is

$$q_{ESU} = C\,(2V_C).$$
(63.7, ESU)

63.2.2 Measurement of Charge in Electromagnetic Units

The charge stored in the Leyden jar in electromagnetic units, can be measured using a ballistic galvanometer, as discussed in detail in Chapter 31. The Leyden jar was discharged through a ballistic galvanometer. A ballistic galvanometer was then called a *multiplier*, since it multiplies the strength of the magnetic field at the center of the loop, by the number of turns in the loop. They used a multiplier with 5635 windings. To increase the resistance of the wire, to avoid sparks during the discharge, a column of water, possibly mixed with salt to increase its conductivity, was added in series to the wire windings.

Repeating Equation 31.77,

$$q_{EMU} = 2\sin\left(\frac{\theta_1}{2}\right)\frac{T_{1/2}}{\pi}\frac{H}{G},$$
(63.8)

where θ_1 is the throw of the angle of the magnetic needle, $T_{1/2}$ is the half period of the oscillation of the magnetic needle, H is the terrestrial magnetic-field strength of the horizontal component that is parallel to the Earth's surface, and G is the galvanometer constant. The right-hand side are values that can be measured or calculated, and q_{EMU} determined.

Weber-Kohlrausch determined the ratio to be

$$\frac{q_{ESU}}{q_{EMU}} = 3.1074 \times 10^{10} \text{ cm/s}, \tag{63.9}$$

or 3.1074×10^8 m/s in SI units. Note that the ratio has units of velocity, which was derived in Equation 57.21. Weber and Kohlrausch did not know at the time that their result matches closely to the speed of light, although the speed of light was known at the time.

Many years prior to the experiment of Weber and Kohlrausch, in 1849, Fizeau measured the speed of light using a rotating toothed wheel, purely from mechanics, to be 3.15×10^8 m/s [108]. Foucault, who is noted for his beautiful pendulum experiment, improved upon Fizeau's method using a rotating mirror, and refined the value in 1862 to 2.98×10^8 m/s.

The result of the Weber-Kohlrausch experiment was later verified by many other experiments. Maxwell observed that a capacitor periodically charged and discharged behaves like a resistor. Today, circuits using such a component is called a *switched capacitor circuit* [106]. Maxwell proposed a switched capacitor circuit experiment to determine the ratio of the electrostatic to the electromagnetic charge. This experiment was carried out by Sir J. J. Thomson and Searle. Although this experiment is not presented in this book, the reader is referred to Reference [107] for the details of the experiment.

Kirchoff showed that voltages and currents propagate at the speed of light in a lossless wire in 1857 [99]. However, Kirchoff's work focused on electrical signals propagating in wires. Maxwell put forth a comprehensive theory of the electromagnetic field in 1864 that is valid anywhere electric and magnetic fields are present, and showed that light is an electromagnetic wave. This clearly explains the reason why the ratio measured by Weber and Kohlrausch is equal to the speed of light. Maxwell's equations in SI units will be presented in Chapter 67, and the reader is referred to other resources for the derivation of the wave equation from them.

If $q_{EMU} = 1\,abcoulomb$, substituting this value into Equation 63.9, $q_{ESU} = 3.0 \times 10^{10}\,statcoulomb$. The relation between abcoulomb and statcoulomb is

$$1\,abcoulomb = 3.0 \times 10^{10}\,statcoulomb. \tag{63.10}$$

Dividing the above equation by $1\,statcoulomb$,

$$\frac{1\,abcoulomb}{1\,statcoulomb} = 3.0 \times 10^{10}. \tag{63.11}$$

In some publications, the experiment result of Weber and Kohlrausch is referred to as the ratio of the unit electromagnetic charge to the unit electrostatic charge, as written in the above equation.

Substituting the base units of abcoulomb and statcoulomb in the above equation, and rearranging the units, it is left as an exercise for the reader to verify the resulting meaningless expression,

$$1\,s = 3.0 \times 10^{10}\,cm. \tag{63.12}$$

The dimension of the left-hand side of the above equation is time, while the right-hand side is length. Equating them has no physical meaning. In this case, the base units making up abcoulomb or statcoulomb must be kept together as one unit, without cancelling or rearranging the base units. To avoid this problem, Equation 63.10 will be treated similar to the equation relating meters m and centimeters cm,

$$1\,m = 100\,cm, \tag{63.13}$$

where abcoulomb and statcoulomb, similar to meters and centimeters, will be viewed as dimensions of charge that cannot be written in terms of other units.

64

Maxwell's Equations in the CGS System

James Clerk Maxwell was the first person to present a complete theory that describes the behavior of electric and magnetic fields. He wrote 3 famous papers [109]-[112] in electromagnetics. In the third paper of 1864 titled "On the Dynamical Theory of the Electromagnetic Field" [111], he derived the famous "wave equation".

The wave equation showed that light is an electromagnetic wave that propagates in free space at the speed of light. He predicted the generation of electromagnetic waves that can propagate in a medium, such as in free space, or a dielectric media. Maxwell's prediction was confirmed by Heinrich Hertz in the years 1885-1889 [113].

Maxwell translated Faraday's experiment results into a mathematical equation, known as Faraday's law. He used mechanical analogies in his earlier papers. Although his reasoning of the derivation of Faraday's law in Reference [111], and the way in which it is written, are different from what has been stated in Chapter 42, Faraday's law in Equation 42.25 can be derived from them. In his prior papers, Maxwell introduced the displacement current term in Ampere's law. Without these equations, the wave equation cannot be derived.

To appreciate his prediction of electromagnetic waves, one must go back in time, when it was thought that fields can only propagate in a wire. Today wireless communication is taken for granted. However, in Maxwell's time, his prediction was a quantum leap.

The behavior of electric and magnetic fields can be written as four equations, and these equations have all been derived in the previous chapters. These equations are called Maxwell's equations, in honor of James Clerk Maxwell. Maxwell's formulation is slightly different from how the electromagnetic equations are written today. He wrote them as eight equations, and have been reduced to four since then, but contains the same information.

The four equations are Gauss's law, Faraday's law, the divergence-free condition of magnetic-flux density $\vec{B}$, and Ampere's law,

$$\nabla \cdot \vec{D} = 4\pi\rho \tag{64.1}$$

$$\nabla \times \vec{E} = -\frac{\partial \vec{B}}{\partial t} \tag{64.2}$$

$$\nabla \cdot \vec{B} = 0 \tag{64.3}$$

$$\nabla \times \vec{H} = 4\pi\vec{J} + \frac{\partial \vec{D}}{\partial t}. \tag{64.4}$$

The remaining equations describe the relation between $\vec{D}$ and $\vec{E}$, as well as $\vec{B}$ and $\vec{H}$. But these are considered to be part of the above four equations. Accounting for a material medium,

$$\vec{D} = \epsilon_r\epsilon_o\vec{E} \tag{64.5}$$

$$\vec{B} = \mu_r\mu_o\vec{H}. \tag{64.6}$$

It was shown in Chapter 57 that any constants for ϵ_o and μ_o may be chosen, as long as they satisfy the relation

$$\frac{1}{\epsilon_o\mu_o} = \left(\frac{q_{ESU}}{q_{EMU}}\right)^2. \tag{64.7}$$

Weber and Kohlrausch first showed that the ratio of the electrostatic charge to the electromagnetic charge is c, the speed of light. The above relation becomes

$$\frac{1}{\epsilon_o\mu_o} = c^2. \tag{64.8}$$

Table 64.1 summarizes the values of ϵ_o and μ_o in ESU and EMU. MKS unrationalized and SI units have different values for ϵ_o and μ_o, which will be discussed in the next chapters.

System of Units	ϵ_o	μ_o
ESU	1	$(3.0 \times 10^{10})^{-2}\ s^2/cm^2$
EMU	$(3.0 \times 10^{10})^{-2}\ s^2/cm^2$	1

Table 64.1: The values of ϵ_o and μ_o to tailor the common set of equations in Chapter 58 to ESU or EMU.

Maxwell's equations have the same form in both electrostatic and electromagnetic units, although their units are different. These equations and units were presented in Chapter 58 and Chapter 59. Maxwell's equations can also be written in the integral form. Summarizing the differential equations in integral form,

$$\oint \vec{D} \cdot d\vec{A} = 4\pi q_{enc} \tag{64.9}$$

$$\oint \vec{E} \cdot d\vec{l} = -\frac{\partial \Phi}{\partial t} \tag{64.10}$$

$$\oint \vec{B} \cdot d\vec{A} = 0 \tag{64.11}$$

$$\oint \vec{H} \cdot d\vec{l} = 4\pi \int \vec{J} \cdot d\vec{A} + \frac{\partial}{\partial t} \int \vec{D} \cdot d\vec{A}. \tag{64.12}$$

The reader is referred to the previous chapters for more details on how the above equations are derived.

The differential equations describe the behavior of fields at a point, while the integral form describes the fields in a volume or an area. These equations can be used to numerically solve for the fields using techniques such as the finite-element method, finite-difference time domain, etc. Numerical methods, however, will not be covered in this book.

Ampere's law was derived from Biot-Savart's law, showing that the two equations are equivalent. Coulomb's law has been implicitly included in Gauss's law. Therefore, these equations are accounted for in Maxwell's equations. Maxwell's equations are cast in the above forms to solve for the fields using analytical and numerical mathematical techniques.

65

MKS Unrationalized System

The MKS unrationalized system of electromagnetic equations is the precursor to the SI system, as shown in Figure 2.2. The MKS system is named after the initials of the three base units of length, mass, and time: meter (m), kilograms (kg), and seconds (s).

The electromagnetic equations in the MKS unrationalized system have the same form as ESU and EMU in Chapter 58, but with different units, and different values of the constants μ_o and ϵ_o. The MKS unrationalized and rationalized systems, though differ slightly in how the equations are formulated, have the same units for the different electrical quantities. The rationalized system of equations and units, also called the SI system, is the formulation used today.

In EMU, resistance has the same unit as velocity, as noted in Equation 46.13, with no similarity between the two. For this reason, a new base unit for current, the ampere, is introduced in the MKS system, and have the same definitions in rationalized and unrationalized forms. The MKS system is also sometimes referred to as the MKSA system, to include the initial of ampere.

The flowchart of definitions in the MKS unrationalized system is identical to the electromagnetic system in Figure 59.1, repeated in Figure 65.1. The current is defined first. All other definitions follow from the definition of current. The force between two current-carrying wires was used as the definition of the unit current in electromagnetic units in Section 59.1. The force between two parallel current-carrying wires was derived earlier,

$$F = 2i_1 i_2 \frac{L}{d},$$

(65.1)

where i_1 and i_2 are the currents in the two parallel wires of negligible thickness of length L, separated by distance d. In this equation, i_1 and i_2 are in abamperes, and the force is in dynes. Since the ratios of the lengths cancel their units, it would not matter which unit is used for L and d.

In the CGS system, the unit of force is the dyne, repeating Equation 2.8,

$$1\,dyne = 1\,\frac{gm \cdot cm}{s^2}.$$

(65.2)

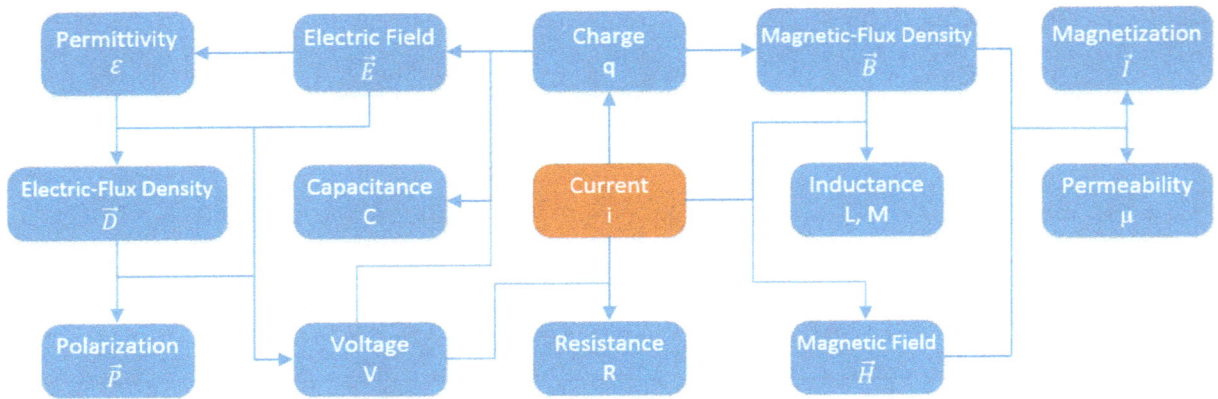

Figure 65.1: The flowchart of definitions in the MKS unrationalized system of units.

Converting the right-hand side into MKS base units,

$$1\,dyne = 1.0 \times 10^{-5} \frac{kg \cdot m}{s^2} \tag{65.3}$$

$$= 1.0 \times 10^{-5}\,N, \tag{65.4}$$

where N is the force in Newtons.

The absolute measurement of current was discussed before. The unit current in electromagnetic units, 1 abampere, turns out to be a large value that is inconvenient for lab measurements. By definition,

$$1\,ampere = 0.1\,abampere, \tag{65.5}$$

an ampere is one tenth of an abampere. The above equation also means that

$$i_{abA} = 0.1\,i_A, \tag{65.6}$$

where i_{abA} is the current in abamperes, and i_A is the current in amperes.

Likewise, variables related to length in cm, can be converted to MKS length in m. For example, length L_{cm} in cm can be converted into the variable L_m in m, from the relation

$$1\,m = 100\,cm. \tag{65.7}$$

This equation means that

$$L_{cm} = 100\,L_m. \tag{65.8}$$

Equation 65.1 can be converted into the MKS system using the above relations, by modifying the current variables from abamperes to amperes, and length variables from cm to m. This equation is modified as

$$F = 2\,(0.1i_1)\,(0.1i_2)\,\frac{100\,L}{100\,d}. \tag{65.9}$$

440

The above equation is no different from Equation 65.1, except that the currents are now in amperes, and the lengths are in meters. The results, however, are still in dynes. Using the conversion factor in Equation 65.4, and simplifying the equation,

$$F = 1.0 \times 10^{-7} \left(2i_1 i_2 \frac{L}{d} \right),$$

(65.10)

and the force is now in newtons, currents in amperes, and lengths in meters. The above equation is of the same form as Equation 58.6,

$$F = \mu_o \frac{2i_1 i_2}{d} L.$$

(65.11)

By dimensional analysis, the unit of μ_o is

$$[\mu_o] = \frac{N}{A^2}.$$

(65.12)

Equating the value of μ_o in Equation 65.10 to Equation 65.11, μ_o in the MKS unrationalized system is

$$\mu_o = 1.0 \times 10^{-7} \frac{N}{A^2}.$$

(65.13)

Using Equation 65.10, an ampere can be defined in the MKS unrationalized system, similar to the CGS system: if two parallel wires of negligible thickness, separated by $1\,m$, carrying equal currents, repel each other with a force per unit length of $2.0 \times 10^{-7}\,N/m$, each of the wires carries an ampere of current.

The unit charge can be defined from the unit current using Equation 58.2, as the quantity of charge contained in unit current flowing for unit time. In the MKS system, the unit charge is the coulomb C,

$$1\,C = 1A \cdot s.$$

(65.14)

The remaining definitions and equations follow the same flow as before in Chapter 58. To avoid repetition, the reader is referred to Chapter 58 for the list of electromagnetic equations.

An abcoulomb (abC) is the charge in an abampere (abA) flowing for 1 second,

$$1abC = 1abA \cdot s.$$

(65.15)

Substituting the relation between abampere and ampere in Equation 65.5 in the above equation,

$$1abC = 10A \cdot s.$$

(65.16)

From the above equations, a similar relation between abC and C, as abA and A, can be derived,

$$1\,C = 0.1\,abC,$$

(65.17)

and
$$q_{abC} = 0.1\,q_C, \tag{65.18}$$
where q_{abC} is the charge in abcoulombs, and q_C is the charge in coulombs.

Substituting the value of
$$\epsilon_o = \frac{1}{(3.0 \times 10^{10})^2}\frac{s^2}{cm^2} \tag{65.19}$$
in Table 64.1, in Coulomb's law in electromagnetic units,
$$F = (3.0 \times 10^{10})^2\,\frac{q_1\,q_2}{r^2}, \tag{65.20}$$
where q_1 and q_2 are in abcoulombs, r is in cm, and the force is in dynes. Converting the charge and length variables from CGS to MKS units, similar to how it was done in Equation 65.9,
$$F = (3.0 \times 10^{10})^2\,\frac{(0.1q_1)(0.1q_2)}{100^2 r^2}, \tag{65.21}$$
where the variables q_1, q_2, and r are in coulombs and meters. The above equation is no different from Equation 65.20, with the exception that the variables for charges and length are in MKS units. The resulting force, however, is still in dynes. Converting the force to newtons from dynes, and simplifying the result,
$$F = (3.0 \times 10^8)^2\,(1.0 \times 10^{-7})\,\frac{q_1\,q_2}{r^2}, \tag{65.22}$$
is the MKS version of Coulomb's law. Equating the value of ϵ_o in Equation 58.1 to the above equation,
$$\begin{aligned}
\epsilon_o &= \frac{1}{(3.0 \times 10^8)^2\,(1.0 \times 10^{-7})} \\
&\approx 1.11 \times 10^{-10},
\end{aligned} \tag{65.23}$$
in MKS units. Applying dimensional analysis to Equation 65.22, the unit of ϵ_o is
$$[\epsilon_o] = \frac{C^2}{N \cdot m^2}. \tag{65.24}$$
From the MKS values of μ_o and ϵ_o, note that
$$\frac{1}{\epsilon_o \mu_o} = (3.0 \times 10^8)^2\,\frac{m^2}{s^2}, \tag{65.25}$$
which is the same result in CGS units as well. Table 65.2 summarizes the differences between the values of ϵ_o and μ_o in ESU, EMU, and MKS systems.

A summary of the units of the electrical quantities in the MKS system is shown in Table 65.2. It is left as an exercise for the reader to derive them using dimensional analysis, using the flowchart of definitions in Figure 65.1. These set of units are common to both the unrationalized and the rationalized MKS (SI) systems, which will be covered in Chapter 66. For the complete list of MKS units, the reader is referred to Reference [6].

System of Units	ϵ_o	μ_o
ESU	1	$(3.0 \times 10^{10})^{-2}\ s^2/cm^2$
EMU	$(3.0 \times 10^{10})^{-2}\ s^2/cm^2$	1
MKS	$1.11 \times 10^{-10}\ C^2/(N \cdot m^2)$	$1.0 \times 10^{-7}\ N/A^2$

Table 65.1: The values of ϵ_o and μ_o in different systems of units.

Electrical Quantity (Symbol)	Unit (Symbol)	Units
Current (i)	ampere (A)	A
Current Density ($\vec{J}$)	–	$A\,m^{-2}$
Charge (q)	coulomb (C)	$A\,s$
Magnetic-flux density ($\vec{B}$)	tesla (T)	$N\,A^{-1}m^{-1}$
Magnetic-flux (Φ)	weber (Wb)	$N\,A^{-1}m$
Magnetic field ($\vec{H}$)	–	$A\,m^{-1}$
Inductance (H)	henry (H)	$Wb\,A^{-1}$
Magnetization ($\vec{I}$)	–	$A\,m^{-1}$
Permeability (μ)	–	$N\,A^{-2}$
Electric field ($\vec{E}$)	–	$N\,C^{-1}$
Voltage (V)	volt (V)	$N\,m\,C^{-1}$
Capacitance (C)	farad (F)	$C\,V^{-1}$
Permittivity (ϵ)	–	$C^2\,N^{-1}\,m^{-2}$
Electric-flux Density ($\vec{D}$)	–	$C\,m^{-2}$
Polarization ($\vec{P}$)	–	$C\,m^{-2}$

Table 65.2: A summary of the units in the MKS system.

66

SI (MKS Rationalized) System

Oliver Heaviside objected to the "spurious eruption of 4πs" in Maxwell's equations. There is a dangling 4π factor in Ampere's law, and Gauss's law. The origin of 4π can be traced to the surface area of a sphere. 4π is also present in the capacitance of a parallel-plate capacitor in Equation 38.9, which has no relation to the surface area of a sphere.

The electromagnetic equations can be formulated such that the 4π factors can be "removed" from some of the equations. This is known as rationalization. The MKS rationalized system of equations is the same as the SI system *(Système International)* that is presently used. It was developed from the work of Giorgi, Fessenden, Fleming, and others [114]. Some of the rationalized electromagnetic equations have a different form from the ones in Chapter 58. The electromagnetic equations will be derived in their rationalized form, from the MKS unrationalized system of equations.

The equations in the SI system will be derived following the definitions flowchart in Figure 58.1, repeated in Figure 66.1.

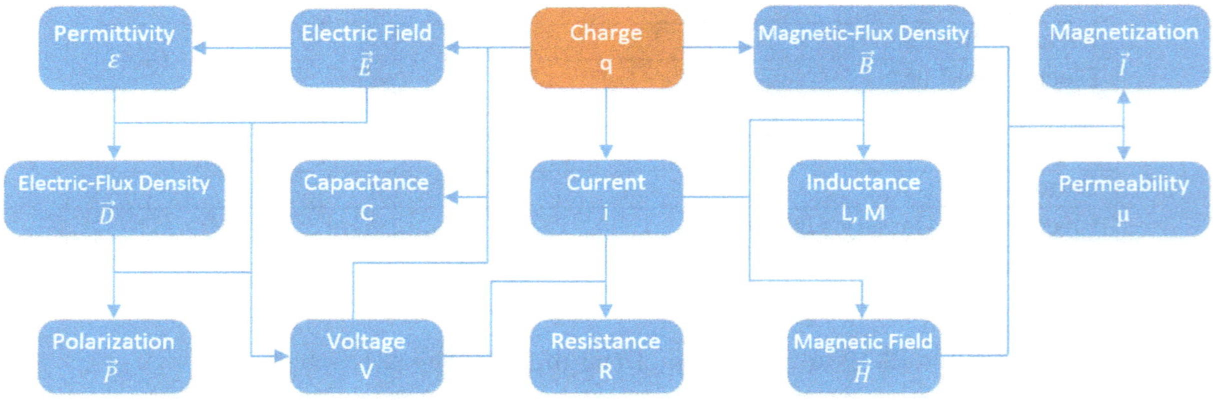

Figure 66.1: The flowchart of definitions in the SI system.

66.1 Definitions of Charge q and Current i

4π in Gauss's law can be removed by rewriting Coulomb's law in Equation 58.1 as

$$F = \frac{1}{4\pi\epsilon_o} \frac{q_1 q_2}{r^2}. \tag{66.1}$$

Following the MKS system, q_1, q_2 are in coulombs, and r is in m. ϵ_o in SI is $4\pi\times$ smaller than the MKS system, so that the multiplying factor in Coulomb's law stays the same. It will be shown later in the chapter, writing Coulomb's law in the manner shown above, with the factor 4π, removes the 4π factor from Gauss's law. The value of ϵ_o in SI is

$$\epsilon_o = \left(\frac{1}{4\pi}\right) 1.11 \times 10^{-10} \frac{C^2}{N\,m^2}$$
$$= 8.854 \times 10^{-12} \frac{C^2}{N\,m^2}. \tag{66.2}$$

The coulomb may also be defined as two equal charges placed $1\,m$ apart, which repel each other with a force of $\frac{1}{4\pi\epsilon_o}$ N. Since this multiplying factor is the same as the MKS unrationalized system, the unit coulomb is the same in both the SI and the MKS unrationalized systems,

$$q_{MKS} = q_{SI}, \tag{66.3}$$

where q_{MKS} is the charge in the MKS unrationalized system, and is equal to the charge q_{SI} in the SI system.

The definition of current is the same as before,

$$i = \frac{\Delta q}{\Delta t}. \tag{66.4}$$

As explained earlier, if charge Δq flows across a wire in time Δt, current is defined as written in the above equation.

66.2 Definitions of Magnetic Field $\vec{H}$ and Magnetic-Flux Density $\vec{B}$

Besides Gauss's law, Ampere's law also has the 4π term. Ampere's law was derived from Biot-Savart's law in Chapter 30. Biot-Savart's law is rewritten as

$$\vec{H} = \frac{1}{4\pi} \int_L \frac{i}{r^2} \left(\vec{dl} \times \hat{r}\right), \tag{66.5}$$

where i is the current in amperes, and the lengths are in meters. It is left as an exercise for the reader to follow the steps in Chapter 30, and show that Ampere's law becomes

$$\nabla \times \vec{H} = \vec{J}, \tag{66.6}$$

without the 4π factor multiplying the current density $\vec{J}$.

The Lorentz force was derived from Faraday's law as

$$\vec{F} = q\vec{v} \times \vec{B}. \tag{66.7}$$

Using this equation, the magnetic-flux density $\vec{B}$ can be defined as the field, whose value is such that a charge q, moving at velocity $\vec{v}$, experiences the force written in the above equation.

66.3 Definition of Permeability μ and Magnetization Vector $\vec{I}$

$\vec{B}$ and $\vec{H}$ are related by permeability μ, and this equation stays the same as before,

$$\vec{B} = \mu_r \mu_o \vec{H} \tag{66.8}$$
$$= \mu \vec{H}, \tag{66.9}$$

where μ_r is the relative permeability of the material, and μ_o is the permeability of free space. The value of μ_o in SI, however, is different from the MKS unrationalized system, and will be determined later in the chapter.

The magnetization vector $\vec{I}$ is defined as

$$\vec{B} = \mu_o \vec{H} + \mu_o \vec{I}, \tag{66.10}$$

without the 4π multiplying $\vec{I}$, where

$$\vec{I} = \chi_m \vec{H}. \tag{66.11}$$

χ_m is magnetic susceptibility. The 4π factor has been "absorbed" into the value of magnetic susceptibility χ_m.

From the above equations,

$$\vec{B} = \mu_o \vec{H} + \mu_o \chi_m \vec{H}. \tag{66.12}$$

Factoring μ_o and $\vec{H}$,

$$\vec{B} = \mu_o \left(1 + \chi_m\right) \vec{H}, \tag{66.13}$$

Equating the terms in the above equation with Equation 66.8, the relative permeability μ_r is

$$\mu_r = 1 + \chi_m. \tag{66.14}$$

66.4 Self Inductance L and Mutual Inductance M

Self inductance L is defined as before in Equation 58.22,

$$L = \frac{\Phi_{self}}{i}, \tag{66.15}$$

446

where Φ_{self} is the self-induced flux in a solenoid carrying current i. Mutual inductance is defined as before in Equation 58.23,

$$M = \frac{\Phi_S}{i_P},$$ (66.16)

where Φ_S is the flux across the secondary coil, caused by the current i_P in the primary coil.

66.5   Electric Field $\vec{E}$ and Voltage V

The electric field $\vec{E}$ is defined as Equation 5.2,

$$\vec{E} = \frac{\vec{F}}{q}.$$ (66.17)

The electric field at a point is the force $\vec{F}$ in newtons, experienced by a charge q in coulombs, at that point.

The voltage V_{AB} between two points A and B, is the path integral of the electric field, as defined in Equation 10.6,

$$V_{AB} = \int_A^B \vec{E} \cdot \vec{dl}.$$ (66.18)

66.6   Electric-Flux Density $\vec{D}$, Permittivity ϵ, and the Permittivity of Free Space ϵ_o

If Coulomb's law is written as Equation 66.1, it is left as an exercise for the reader to verify that Gauss's law in free space becomes

$$\oint_S \epsilon_o \vec{E} \cdot \vec{dA} = q_{enc},$$ (66.19)

where q_{enc} is the charge enclosed within the Gaussian surface S, and $\vec{E}$ is the value of the electric field at the Gaussian surface. The 4π factor multiplying q_{enc}, cancels with the 4π in Coulomb's law, in the above equation.

If an electric field is modified by a dielectric material, following the steps outlined in Chapter 36, and using the above equation, it is left as an exercise for the reader to show that Equation 36.6 needs to be modified as

$$\oint_S \epsilon_r \epsilon_o \vec{E} \cdot \vec{dA} = q_{enc},$$ (66.20)

where ϵ_r is the relative permittivity of the dielectric material in which the Gaussian surface is present. If the electric-flux density $\vec{D}$ is defined as

$$\vec{D} = \epsilon_r \epsilon_o \vec{E}$$ (66.21)

$$= \epsilon \vec{E},$$ (66.22)

substituting this in Equation 66.20, Gauss's law is written as,

$$\oint_S \vec{D} \cdot d\vec{A} = q_{enc}.$$ (66.23)

From Equation 66.22, the permittivity ϵ, or the absolute permittivity, as it is sometimes referred to, is defined as

$$\epsilon = \epsilon_r \epsilon_o.$$ (66.24)

66.7 Polarization $\vec{P}$

Equation 58.33 is written as

$$\vec{D} = \epsilon_o \left(1 + \chi_e \right) \vec{E},$$ (66.25)

where

$$\epsilon_r = 1 + \chi_e,$$ (66.26)

and χ_e is the electric susceptibility. As before, in the case of magnetic susceptibility, the 4π term multiplying χ_e is "absorbed" into the value of the electric susceptibility. Multipying the terms within the parenthesis in the above equation,

$$\vec{D} = \epsilon_o \vec{E} + \epsilon_o \chi_e \vec{E}.$$ (66.27)

The above equation can be written in a form similar to Equation 37.1,

$$\vec{D} = \epsilon_o \vec{E} + \vec{P},$$ (66.28)

where the polarization vector $\vec{P}$ is defined as

$$\vec{P} = \epsilon_o \chi_e \vec{E}.$$ (66.29)

66.8 Capacitance C

The definition of capacitance C is the same as Equation 35.8,

$$C = \frac{Q}{V},$$ (66.30)

where $\pm Q$ is the charge stored in two conductors, whose potential difference is voltage V.

66.9 SI Value of the Permeability of Free Space μ_o

The force between two parallel wires in the SI system of equations will be derived next. In the new formulation of Biot-Savart's law in Equation 66.5, the magnetic field around a long straight wire can be calculated as

$$\vec{H} = \frac{1}{4\pi} \frac{2i}{d},$$
$$= \frac{i}{2\pi d},$$ (66.31)

where d is the radial distance from the wire, and i is the current in the wire.

The force per unit length acting on a wire can be calculated using Equation 55.16,

$$\frac{F}{L} = i_1 B,$$

(66.32)

where B is the magnetic-flux density at Wire 1, carrying current i_1. In a non-magnetic material,

$$B = \mu_o H.$$

(66.33)

From the above equations,

$$\frac{F}{L} = \mu_o i_1 H.$$

(66.34)

If the magnetic field H is generated by a parallel wire, substituting Equation 66.31 in the above relation,

$$\frac{F}{L} = \frac{\mu_o}{2\pi} \frac{i_1 i_2}{d}.$$

(66.35)

Equation 66.3 also means that

$$i_{MKS} = i_{SI},$$

(66.36)

where i_{MKS} is the current in the MKS unrationalized system, and i_{SI} is the current in the SI system. Since the unit current is equal in both the systems, the force between two parallel current-carrying wires must also be equal. In the case of the MKS unrationalized system, the force between two parallel wires, repeating Equation 65.10,

$$\frac{F}{L} = 2.0 \times 10^{-7} \frac{i_1 i_2}{d}.$$

(66.37)

Equating the above equation in the MKS unrationalized system to Equation 66.35 in the SI system, the permeability of free space in the SI system is

$$\mu_o = 4\pi \times 10^{-7} \frac{N}{A^2}.$$

(66.38)

Note that the value of

$$\frac{1}{\mu_o \epsilon_o} = \left(3.0 \times 10^8\right)^2,$$

(66.39)

which is the same result in other systems of units, and is the speed of light squared. Table 66.2 summarizes the values of μ_o and ϵ_o in the different systems of units. The units of the different electrical quantities in the SI system are the same as the MKS system. This can be verified by applying dimensional analysis on the definitions. Repeating Table 65.2, the SI units are summarized in Table 66.2.

System of Units	ϵ_o	μ_o
ESU	1	$(3.0 \times 10^{10})^{-2}$ s^2/cm^2
EMU	$(3.0 \times 10^{10})^{-2}$ s^2/cm^2	1
MKS	1.11×10^{-10} $C^2/(N \cdot m^2)$	1.0×10^{-7} N/A^2
SI	8.85×10^{-12} $C^2/(N \cdot m^2)$	$4\pi \times 10^{-7}$ N/A^2

Table 66.1: A summary of the values of ϵ_o and μ_o in different systems of units.

Electrical Quantity (Symbol)	Unit (Symbol)	Units
Current (i)	ampere (A)	A
Current Density $(\vec{J})$	–	$A\,m^{-2}$
Charge (q)	coulomb (C)	$A\,s$
Magnetic-flux density $(\vec{B})$	tesla (T)	$N\,A^{-1}m^{-1}$
Magnetic-flux (Φ)	weber (Wb)	$N\,A^{-1}m$
Magnetic field $(\vec{H})$	–	$A\,m^{-1}$
Inductance (H)	henry (H)	$Wb\,A^{-1}$
Magnetization $(\vec{I})$	–	$A\,m^{-1}$
Permeability (μ)	–	$N\,A^{-2}$
Electric field $(\vec{E})$	–	$N\,C^{-1}$
Voltage (V)	volt (V)	$N\,m\,C^{-1}$
Capacitance (C)	farad (F)	$C\,V^{-1}$
Permittivity (ϵ)	–	$C^2\,N^{-1}\,m^{-2}$
Electric-flux Density $(\vec{D})$	–	$C\,m^{-2}$
Polarization $(\vec{P})$	–	$C\,m^{-2}$

Table 66.2: A summary of the units in the SI system of the commonly used electrical quantities.

67

Maxwell's Equations in the SI System

Maxwell's equations in the SI system are in the rationalized form, without the "spurious eruption of $4\pi s$". Maxwell's equations are the set of four equations: Gauss's law, Faraday's law, divergence-free condition of magnetic-flux density $\vec{B}$, and Ampere's law. In the differential form, these equations are

$$\nabla \cdot \vec{D} = \rho \tag{67.1}$$

$$\nabla \times \vec{E} = -\frac{\partial \vec{B}}{\partial t} \tag{67.2}$$

$$\nabla \cdot \vec{B} = 0 \tag{67.3}$$

$$\nabla \times \vec{H} = \vec{J} + \frac{\partial \vec{D}}{\partial t}. \tag{67.4}$$

The relation between $\vec{D}$, $\vec{E}$ and $\vec{B}$, $\vec{H}$ are

$$\vec{D} = \epsilon_r \epsilon_o \vec{E}, \ \epsilon_o = 8.85 \times 10^{-12} \ C^2/(N \cdot m^2) \tag{67.5}$$

$$\vec{B} = \mu_r \mu_o \vec{H}, \ \mu_o = 4\pi \times 10^{-7} \ N/A^2. \tag{67.6}$$

In the integral form, as derived in the previous chapters, the above equations are written as

$$\oint \vec{D} \cdot d\vec{A} = q_{enc} \tag{67.7}$$

$$\oint \vec{E} \cdot d\vec{l} = -\frac{\partial \Phi}{\partial t} \tag{67.8}$$

$$\oint \vec{B} \cdot d\vec{A} = 0 \tag{67.9}$$

$$\oint \vec{H} \cdot d\vec{l} = \int \vec{J} \cdot d\vec{A} + \frac{\partial}{\partial t} \int \vec{D} \cdot d\vec{A}. \tag{67.10}$$

Maxwell's equations are also valid for time-varying fields, and sources.

Bibliography

☐ James Clerk Maxwell, *A Treatise on Electricity and Magnetism*, Cambridge University Press, 2010.

☐ Sir J. J. Thomson, *Elements of the Mathematical Theory of Electricity and Magnetism*, London, UK: Cambridge University Press, 1909.

☐ Andrew Gray, *The Theory and Practice of Absolute Measurements in Electricity and Magnetism*, Vol. I and Vol. II, London: Macmillan and Co., 1888.

☐ Richard Fitzpatrick, *Classical Electromagnetism*, [Online], Available: http://farside.ph.ute xas.edu/teaching/jk1/Electromagnetism.pdf [Accessed: Feb. 23, 2018].

☐ John D. Jackson, *Classical Electrodynamics*, 3rd Edition, Wiley, Aug. 1998.

☐ David J. Griffiths, *Introduction to Electrodynamics*, 4th Edition, Cambridge University Press, July 2017.

☐ Douglas L. Cohen, *Demystifying Electromagnetic Equations: A Complete Explanation of EM Unit Systems and Equation Transformations*, Bellingham, Washington: The Society of Photo-Optical Instrumentation Engineers, 2001.

☐ Fawwaz T. Ulaby and Umberto Ravaioli, *Fundamentals of Applied Electromagnetics*, 7th Edition, Pearson, Oct. 2014.

☐ Matthew O. Sadiku, *Elements of Electromagnetics*, 6th Edition, Oxford University Press, Jan. 2014.

☐ Matthew O. Sadiku, *Numerical Techniques in Electromagnetics*, 2nd Edition, CRC Press, 2000.

☐ David K. Cheng, *Field and Wave Electromagnetics*, 2nd Edition, Addison-Wesley, Jan. 1989.

☐ Walter Lewin, 'Lec 16: Electromagnetic Induction 8.02 Electricity and Magnetism, Spring 2002', [Online], Available: https://www.youtube.com/watch?v=FUUMCT7FjaI [Accessed: May 26, 2016].

References

[1] 'Richard Feynman on Magnets', [Online] Available:https://www.youtube.com/watch?v=MO0r930Sn_8 [Accessed: Sep. 21, 2017].

[2] Marzieh, 'Electric Charges', [Online] Available: http://www.youtube.com/watch?v=QcBVa1VKUdc [Accessed: Apr. 18, 2013].

[3] RimstarOrg, 'Positive or Negative? How to Find Electric Charge Polarity', [Online] Available: https://www.youtube.com/watch?v=HupFY_24o-4 [Accessed: Oct. 26, 2017].

[4] 'Methods of Charging', [Online] Available: http://www.physicsclassroom.com/class/estatics/u8l2b.cfm [Accessed: Oct. 10, 2013].

[5] D. Tong, 'Interlude: Dimensional Analysis', [Online] Available: http://www.damtp.cam.ac.uk/user/tong/relativity/three.pdf, [Accessed: Dec. 30, 2014].

[6] Douglas L. Cohen, *Demystifying Electromagnetic Equations: A Complete Explanation of EM Unit Systems and Equation Transformations*, Bellingham, Washington: The Society of Photo-Optical Instrumentation Engineers, 2001.

[7] Richard Fitzpatrick, 'The Torsion Pendulum', [Online] Available: http://farside.ph.utexas.edu/teaching/301/lectures/node139.html, [Accessed: Apr. 12, 2014].

[8] David Joyce, 'The derivative of $\sin(x)$ equals $\cos(x)$ only when x is expressed in radian. Is that true and why?', [Online] Available: http://www.quora.com/The-derivative-of-sin-x-equals-cos-x-only-when-x-is-expressed-in-radian-Is-that-true-and-why [Accessed: Feb. 14, 2015].

[9] Stewart C. Gillmor, *Coulomb and the Evolution of Physics and Engineering in Eighteenth-Century France*, Princeton University Press, 2017.

[10] A. A. Martinez, "Replication of Coulomb's Torsion Balance Experiment", *Archive for History of Exact Sciences*, pp. 517-563, 2006.

[11] Johnnie T. Dennis, *The Complete Idiot's Guide to Physics*, Penguin, 2003.

[12] Keith Gibbs, "Gravitational Potential Gradient", [Online] 2012, http://www.schoolphysics.co.uk/age16-19/Mechanics/Gravitation/text/Gravitational_potential_gradient/index.html [Accessed: August 14, 2013].

[13] 'The Blue Marble', [Online] Available: https://commons.wikimedia.org/wiki/File:The_Earth_seen_from_Apollo_17.jpg [Accessed: Oct. 10, 2013].

[14] Richard Fitzpatrick, "Electric Potential Energy", [Online] 2007, http://farside.ph.utexas.edu/teaching/302l/lectures/node32.html [Accessed: April 18, 2015].

[15] Richard P. Feynman, *The Feynman Lectures on Physics: Mainly Electromagnetism and Matter*, Vol. 2, Addison-Wesley, Feb. 1, 1977.

[16] Thomas Kim, 'Friction Electrostatic Generator 2', [Online] Available: https://www.youtube.com/watch?v=uh-CUEfdiic [Accessed: Sep. 21, 2017].

[17] 'Electricity from Glass', [Online] Available: http://physics.kenyon.edu/EarlyApparatus/Static_Electricity/Electricity_from_Glass/Electricity_from_Glass.html [Accessed: Sep. 21, 2017].

[18] RimstarOrg, 'Kelvin Water Dropper and How it Works/Lord Kelvin's Thunderstorm', [Online] Available: https://www.youtube.com/watch?v=sArNxGnYhNU&t [Accessed: Sep. 23, 2017].

[19] Reinhard Schumacher, 'Kelvin Water Dropper With Electroscope: Thunderstorm Warning', [Online] Available: https://www.youtube.com/watch?v=8Jx1pvFiaoI&t [Accessed: Sep. 23, 2017].

[20] Thomas Kim, 'Lord Kelvin's Thunderstorm (Kelvin Water Dropper) Experiment', [Online] Available: https://www.youtube.com/watch?v=8OuonluJPw8 [Accessed: Sep. 23, 2017].

[21] Jus Woods, 'Electricity of Distilled Water vs Tap Water', [Online] Available: https://www.youtube.com/watch?v=7hQh8eg30IY [Accessed: Sep. 23, 2017].

[22] Sci-Supply: http://www.sci-supply.com/ [Accessed: Sep. 23, 2017].

[23] Antonio C. M. de Queiroz, 'Operation of the Wimshurst Machine', [Online] Available: http://www.coe.ufrj.br/~acmq/whyhow.html [Accessed: Sep. 26, 2017].

[24] RimstarOrg, 'Wimshurst Machine: How to Make Using CDs', [Online] Available: https://www.youtube.com/watch?v=puC6-UaT9Fk&t [Accessed: Sep. 30, 2017].

[25] A. P. French, *Special Relativity*, New York, W. W. Norton & Company Inc., 1968.

[26] MIT, 'Charge and Electric Field of a Hollow Conductor', [Online] Available: http://www.youtube.com/watch?v=LfJywoeIIUI [Accessed: Oct. 12, 2013].

[27] Fleeming Jenkin, *Electricity and Magnetism*, London, UK: Longmans, Green, and Co., 1914.

[28] 'Heating Effect of a Current', [Online] Available: https://www.youtube.com/watch?v=uSb yHcYb850 [Accessed: Apr. 20, 2018].

[29] Michael Melloch, 'Faraday Cage', [Online] Available: https://www.youtube.com/watch?v= gHaLdAYwQ-4 [Accessed: Sep. 15, 2017].

[30] Michael Melloch, 'Curvature and Charge Density', [Online] Available: https://www.youtu be.com/watch?v=VTkGMKAeNIA [Accessed: Sep. 15, 2017].

[31] 'Faraday's Ice Pail Experiment', [Online] Available: http://en.wikipedia.org/wiki/Faraday 's_ice_pail_experiment [Accessed: Nov. 2, 2013].

[32] "The Fundamental Theorem of Line Integrals", [Online], http://www.whitman.edu/mathe matics/calculus_online/section16.03.html [Accessed: Aug. 14, 2013].

[33] Amir D. Aczel, *The Riddle of the Compass: The Invention that Changed the World*, Orlando, USA: Harcourt, 2002.

[34] Children Learning Online, 'How to Make a Compass', [Online] Available: http://www.you tube.com/watch?v=VobcByagbPU [Accessed: Dec. 13, 2013].

[35] TheWajSity, 'Oersted Experiment', [Online] Available: http://www.youtube.com/watch?v =-w-1-4Xnjuw [Accessed: Dec. 13, 2013].

[36] 'Leyden Jar', [Online] Available: http://en.wikipedia.org/wiki/Leyden_jar [Accessed: Nov. 2, 2013].

[37] Herman Erlichson, "The experiments of Biot and Savart concerning the force exerted by a current on a magnetic needle", *American Association of Physics Teachers*, pp. 385-391, Vol. 66, No. 5, May 1998.

[38] 'Magnetic Field', [Online] Available: https://en.wikipedia.org/wiki/Magnetic_field [Accessed: Nov. 11, 2017].

[39] Thomas Kim, 'Magnetic field around electric wire', [Online] Available: https://www.youtu be.com/watch?v=opJYLFvI-RE [Accessed: Nov. 11, 2017].

[40] Grant Thompson-"The King of Random", 'How To Make a 3 Penny Battery', [Online] Available: http://www.youtube.com/watch?v=rIdPfDHeROI [Accessed: Dec. 13, 2013].

[41] ScienceOnline, 'Create a Lemon Battery', [Online] Available: http://www.youtube.com/w atch?v=AY9qcDCFeVI [Accessed: Dec. 13, 2013].

[42] Michael Melloch, 'Force between current carrying coils', [Online] Available: http://www.y outube.com/watch?v=FLWgm_j0XZc [Accessed: July 16, 2014].

[43] Jl Ang, 'Flat Square Coil Carrying a Current', [Online] Available: https://www.youtube.c om/watch?v=Ve7AH14xQ74 [Accessed: July 16, 2014].

[44] Michael Melloch, 'Magnetic Field of a Coil', [Online] Available: https://www.youtube.com/watch?v=bq6IhapfucE [Accessed: Nov. 11, 2017].

[45] A. Ganot, *Elementary Treatise on Physics, Experimental and Applied for the Use of Colleges and Schools*, W. Wood and Company, 1879.

[46] Christine Blondel Bertrand Wolff, Andrew Butrica, 'In Search of a Newtonian Law of Electrodynamics (1820-1826)', [Online] Available: http://www.ampere.cnrs.fr/parcourspedagogique/zoom/courant/formule/index-en.php [Accessed: May 31, 2014].

[47] A. Ganot, translated by E. Atkinson, *Elementary Treatise on Physics Experimental and Applied*, pp. 706, New York: William Wood and Co., Publishers, 1868.

[48] R. A. R. Tricker, *Early Electrodynamics: The First Law of Circulation*, Pergamon Press, 1965.

[49] J. B. Biot, Translated by O. M. Blunn, *Précis Élémentaire De Physique*, Vol. II, 3rd Ed., 1824.

[50] F. Jenkin, *Electricity and Magnetism*, p. 135, London: Longmans, Green and Co., 1914.

[51] Dept. of Mathematics, Oregon State University, 'Flux (Surface Integrals of Vector Fields)', [Online] Available: http://math.oregonstate.edu/home/programs/undergrad/CalculusQuestStudyGuides/vcalc/flux/flux.html [Accessed: Oct. 21, 2014].

[52] D. A. Fleisch, *A Student's Guide to Maxwell's Equations*, pp. 102-104, New York: Cambridge University Press, 2008

[53] D. V. Redžić, 'On the Laplacian of 1/r', [Online] Available: http://arxiv.org/pdf/1303.2567.pdf [Accessed: Oct. 26, 2014].

[54] J. D. Jackson, *Classical Electrodynamics*, 3rd Edition, New York: John Wiley & Sons, Inc., 1998.

[55] W3007-Electricity and Magnetism, 'Divergence and Curl of $\vec{B}$', [Online] Available: http://phys.columbia.edu/~nicolis/Div_and_Curl_of_B.pdf [Accessed: Oct. 26, 2014].

[56] J. C. Maxwell, *A Treatise on Electricity and Magnetism*, Vol. I-II, 3rd Edition, New York: Dover Publications Inc., 1954.

[57] D. Sober, 'Proof of Gauss's Law Using Solid Angle', [Online] Available: http://faculty.cua.edu/sober/536/Gauss_solid_angle.pdf, [Accessed: Apr. 27, 2015].

[58] Sir J. J. Thomson, *Elements of the Mathematical Theory of Electricity and Magnetism*, London, UK: Cambridge University Press, 1909.

[59] M. J. Hagmann, "Solution of Static Electric and Magnetic Problems Using the Integral Definitions of the Curl and Divergence Operators", *IEEE Trans. on Education*, Vol. 32, No. 1, pp. 25-28, Feb. 1989.

[60] J. R. Nagel, 'Solving the Generalized Poisson Equation Using the Finite-Difference Method (FDM)', [Online] Available: http://www.ece.utah.edu/ ece6340/LECTURES/Feb1/Nagel %202012%20-%20Solving%20the%20Generalized%20Poisson%20Equation%20using%20FD M.pdf[Accessed: May 13, 2017].

[61] J. C. Kolecki, 'An Introduction to Tensors for Students of Physics and Engineering', [Online] Available: https://www.grc.nasa.gov/www/k-12/Numbers/Math/documents/Tensors_TM 2002211716.pdf [Accessed: Dec. 31, 2015].

[62] C. R. Paul, *Analysis of Multiconductor Transmission Lines*, 2nd Edition, Wiley-IEEE Press, Oct. 2007.

[63] Morton L. Schagrin, "Resistance to Ohm's Law", American Journal of Physics, Vol. 31, pp. 536-547, 1963.

[64] J. C. Shedd and M. D. Hershey, "The History of Ohm's Law", *The Popular Science Monthly*, Vol. LXXXIII, pp. 599-614, The Science Press, New York, 1913.

[65] S. Ghosh, 'Laws of Thermocouple', [Online], Available: https://www.youtube.com/watch? v=ZwXtPW0gdD0 [Accessed: Oct 3, 2015].

[66] Michael Melloch, 'Electromagnetic Induction and Faraday's Law', [Online] Available: http s://www.youtube.com/watch?v=vwIdZjjd8fo [Accessed: Mar. 20, 2016].

[67] J. A. Ewing, *Magnetic Induction in Iron and Other Metals*, Third Edition Revised, "The Electrician" Printing and Publishing Company Limited, London, 1900.

[68] St. Mary's Physics Online, 'Generating Electricity - 3 Factors', [Online] Available: https:/ /www.youtube.com/watch?v=Q8t_12NQpZY [Accessed: Mar. 25, 2016].

[69] MIT Course Notes, 'Chapter 11 Inductance and Magnetic Energy', [Online] Available: http ://web.mit.edu/viz/EM/visualizations/coursenotes/modules/guide11.pdf [Accessed: Apr. 6, 2016].

[70] Kurt Nalty, 'Classical Calculation for Mutual Inductance of Two Coaxial Loops in MKS Units', [Online] Available: http://www.kurtnalty.com/Helmholtz.pdf [Accessed: Apr. 25, 2016].

[71] Kurt Nalty, 'Useful Elliptic Integral Formulas for Magnetics', [Online] Available: http://w ww.kurtnalty.com/UsefulEllipticIntegralFormulasSheet.pdf [Accessed: Apr. 25, 2016].

[72] Robert Weaver, 'Bob's Electron Bunker', [Online] Available: http://electronbunker.ca/eb/ Home.html [Accessed: Apr. 24, 2016].

[73] D. W. Knight, 'An Introduction to the Art of Solenoid Inductance Calculation', [Online] Available: http://www.g3ynh.info/zdocs/magnetics/Solenoids.pdf [Accessed: Apr. 24, 2016].

[74] L. Lorenz, *Über die Fortplanzung der Elektrizität, Annalen der Physik*, VII, pp. 161-193, 1879.

[75] H. Nagaoka, "The Inductance Coefficients of Solenoids", *Journal of the College of Science*, Vol. XXVII, Article 6, Imperial University, Tokyo, Japan, 1909. [Online] Available: http://repository.dl.itc.u-tokyo.ac.jp/index_e.html [Accessed: Apr. 24, 2016].

[76] L. Cohen, "An Exact Formula for the Mutual Inductance of Coaxial Solenoids", *Bulletin of the Bureau of Standards*, Vol. 3, pp. 295-303, 1907.

[77] A. Gray, *The Theory and Practice of Absolute Measurements in Electricity and Magnetism*, Vol. I and Vol. II, London: Macmillan and Co., 1888.

[78] A. Campbell, "On the Determination of Resistance in Terms of Mutual Inductance", *Proceedings of the Royal Society of London*, Vol. 107, No. 742, pp. 310-312, 1925.

[79] A. Campbell, "On the Determination of the Absolute Unit of Resistance by Alternating Current Methods", *Proceedings of the Royal Society of London*, Vol. 87, No. 597, pp. 391-414, 1912.

[80] J. L. Thomas, C. Peterson, I. L. Cooter, and F. Ralph Kotter, "An Absolute Measurement of Resistance by the Wenner Method", *Journal of Research of the National Bureau of Standards*, Vol. 43, pp. 291-353, Oct. 1949.

[81] Latimer Clark, "On a Voltaic Standard of Electromotive Force", *Proceedings of the Royal Society of London*, Vol. 20, pp. 444-448, 1871.

[82] Thanu Padmanabhan, *Sleeping Beauties in Theoretical Physics: 26 Surprising Insights*, Springer, 2015.

[83] Ruth W. Chabay and Bruce A. Sherwood, *Matter and Interactions, Volume II: Electric and Magnetic Interactions*, 3rd Edition, Wiley, Jan. 5, 2010.

[84] R. Fitzpatrick, 'Retarded Fields', [Online] Available: http://farside.ph.utexas.edu/teaching/em/lectures/node52.html [Accessed: Jan 2, 2018].

[85] Keith W. Whites, 'Lecture 5: Displacement Current and Ampere's Law', [Online] Available: http://whites.sdsmt.edu/classes/ee382/notes/382Lecture5.pdf [Accessed: July 31, 2016].

[86] Khan Academy, 'Green's, Stokes', and Divergence Theorems', [Online] Available: https://www.khanacademy.org/math/multivariable-calculus/greens-theorem-and-stokes-theorem/ [Accessed: August 2, 2016].

[87] Paul Dawkins, 'Paul's Online Math Notes', [Online] Available: http://tutorial.math.lamar.edu/Classes/CalcIII/StokesTheorem.aspx [Accessed: August 2, 2016].

[88] HooplaKidzLab, 'How to Make Electromagnet Experiment', [Online] Available: https://www.youtube.com/watch?v=wX9QBwJBI_Y [Accessed: Dec. 31, 2015].

[89] M. V. K. Chari, S. J. Salon, *Numerical Methods in Electromagnetism*, First Edition, Elsevier, Nov. 3, 1999.

[90] William P. Houser, "Deriving the Lorentz Force Equation from Maxwell's Equations", *Proc. IEEE SoutheastCon*, 2002.

[91] Christine Blondel Bertrand Wolff, Andrew Butrica, 'Ampere's Force Law: An Obsolete Formula?', [Online] Available: http://www.ampere.cnrs.fr/parcourspedagogique/zoom/courant/force/index-en.php [Accessed: Feb. 6, 2017].

[92] Sargent Welch, 'CENCO® Coulomb and Current Apparatus', [Online] Available: https://www.sargentwelch.com/store/product/8870879/cenco-coulomb-and-current-apparatus [Accessed: Jan. 27, 2017].

[93] MSTCouncilGradStudent's channel, 'Current Balance Lab', [Online] Available: https://www.youtube.com/watch?v=PqcHu1Gm5wA&t=30s [Accessed: Jan 27, 2017].

[94] E. B. Rosa, N. E. Dorsey, and J. M. Miller, "A Determination of the International Ampere in Absolute Measure", *Bulletin of the Bureau of Standards*, Vol. 8, pp. 269-393, June 1912.

[95] 'Physics 21L: Lab 7, Earth's Magnetic Field', [Online] Available: https://www.youtube.com/watch?v=SAco-U24_mY [Accessed: Mar. 21, 2017].

[96] W. Weber and R. Kohlrausch, "Ueber die Elektricitätsmenge, welche bei galvanischen Strömen durch den Querschnitt der Kette fliesst", *Annalen der Physik*, Vol. 99, pp. 10-25, 1856.

[97] W. Weber and R. Kohlrausch, "On the amount of electricity which flows through the cross-section of the circuit in galvanic currents", English translation by Susan P. Johnson, edited by L. Hecht, unpublished, [Online] Available: http://ppp.unipv.it/Collana/Pages/Libri/Saggi/Volta%20and%20the%20History%20of%20Electricity/V%26H%20Sect3/V%26H%20287-297.pdf [Accessed: Apr. 7, 2017].

[98] F. Kirchner, "Determination of the velocity of light from electromagnetic measurements according to W. Weber and R. Kohlrausch", American Journal of Physics, Vol. 25, pp. 623-629, 1957.

[99] A. K. T. Assis, 'On the First Electromagnetic Measurement of the Velocity of Light by Wilhelm Weber and Rudolf Kohlrausch', [Online] Available: http://www.ifi.unicamp.br/ assis/Weber-Kohlrausch(2003).pdf [Accessed: Apr. 7, 2017].

[100] Sir W. Thomson, *Reprint of Papers on Electrostatics and Magnetism*, second edition, London, Macmillan & Co., pp. 52-85, 1884.

[101] 'Finding Image Charges for a Grounded Conducting Sphere', [Online] Available: https://www.youtube.com/watch?v=KoQ3KP2oSMo [Accessed: July 22, 2017].

[102] G. Tong, "Charge Distributions of Two Conducting Spheres", *Eur. J. Phys.*, Vol. 13, pp. 186-188, 1992.

[103] D. Brewster, *The Edinburgh Encyclopaedia Conducted by David Brewster With the Assistance of Gentlemen Eminent in Science and Literature*, First American Edition, Vol. 8, pp. 235-375, 1832.

[104] C. O. Larson and E. W. Goss, "A Coulomb's Law Balance Suitable for Physics Majors and Nonscience Students", *American Journal of Physics*, Vol. 38, No. 11, pp. 1349-1352, Nov. 1970.

[105] J. Sliško, R. A. Brito-Orta, "On Approximate Formulas for the Electrostatic Force Between Two Conducting Spheres", Vol. 66, No. 4, pp. 352-355, Apr. 1998.

[106] R. Jacob Baker, *CMOS: Circuit Design, Layout, and Simulation*, Third Edition, Wiley-IEEE Press, 2010.

[107] J. J. Thomson and G. F. C. Searle, "A Determination of "v", the Ratio of the Electromagnetic Unit of Electricity to the Electrostatic Unit", *Philosophical Transactions of the Royal Society of London*, Vol. 181, pp. 583-621, 1890.

[108] D. Butler, 'Classroom Aid - Fizeau measures the speed of light', [Online] Available: https://www.youtube.com/watch?v=xqWnzaZHwf8 [Accessed: Sep. 2, 2017].

[109] J. C. Maxwell, "On Faraday's Lines of Force", *Trans. Cambridge Philosoph. Soc.*, Vol. X, Part I, 1855. Repr. in *The Scientific Papers of James Maxwell*, pp. 155-229, New York: Dover, 1890.

[110] J. C. Maxwell, "On Physical Lines of Force", *Philosoph. Mag*, Vol. XXI, Jan-Feb. 1862. Repr. in *The Scientific Papers of James Maxwell*, pp. 451-513, New York: Dover, 1890.

[111] J. C. Maxwell, "A Dynamical Theory of the Electromagnetic Field", in *Royal Soc. Trans.*, Vol. CLV, Dec. 8, 1864. Repr. in *The Scientific Papers of James Maxwell*, pp. 451-513, New York: Dover, 1890.

[112] James C. Rautio, "Maxwell's Legacy", *IEEE Microwave Magazine*, pp. 46-53, June 2005.

[113] Heinrich Hertz, *Electric Waves: Being Researches on the Propagation of Electric Action with Finite Velocity Through Space*, London, UK: Macmillan, 1893.

[114] F. W. Sears, C. C. Murdock, H. H. Nielsen, and H. C. Wolfe, "Rationalization of the Electromagnetic Equations", *American Journal of Physics*, Vol. 30, pp. 423-433, 1962.

[115] 'Antenna Basics', [Online] Available: highered.mheducation.com/sites/dl/free/0072321032 /62577/ch02_011_056.pdf [Accessed: March 13, 2015].

$\mathcal{A}$

Introduction to Solid Angles

Solid angles are angles in 3D. An introduction to solid angles will be presented in this appendix.

$\mathcal{A}$.1 Definition of the Solid Angle

The radian is defined using a circle shown in Figure $\mathcal{A}$.1(a). θ in radians, is defined as the ratio of the arc length ℓ to the radius r,

$$\theta = \frac{\ell}{r}. \tag{$\mathcal{A}$.1}$$

By this definition, the unit radian is the angle that an unit arc length subtends at the center of the circle with unit radius. From the above equation, the radian is the ratio of the lengths, and has no units.

The source of the above definition can be traced back to the ratio of the circumference of a circle P to its radius r, which is given by,

$$\begin{aligned} \frac{P}{r} &= \frac{2\pi r}{r} \\ &= 2\pi. \end{aligned} \tag{$\mathcal{A}$.2}$$

The 2D angle of a complete circle is therefore 2π radians. The ratio of half the perimeter to its radius is the famous constant π, and therefore, the semi circle subtends π radians at the center of the circle. Any arc length is a fraction of $2\pi r$. Therefore, the ratio of an arc length to the radius is always a numerical value, and independent of the radius of the circle. This property has been used to define the 2D angle in terms of the ratio of the arc length to the radius.

The definition of the 2D angle can be extended to 3D. Similar to how the ratio of the perimeter to the radius in the 2D case is a numerical value, the ratio of the surface area of a sphere to the square of the radius is always a numerical value in 3D, independent of the dimensions of the

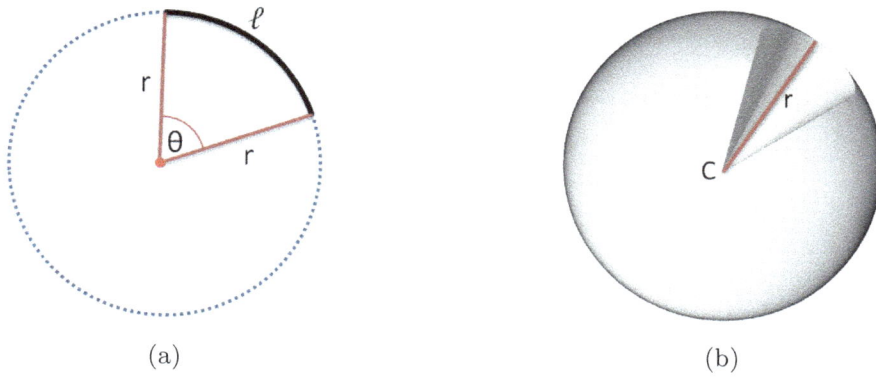

(a)   (b)

Figure $\mathcal{A}$.1: (a) A circle is used to define the 2D angle. (b) A sphere is used to define the 3D angle.

sphere. For a sphere with radius r, the ratio of the surface area A_{sphere} to the radius squared r^2 is

$$\frac{A_{sphere}}{r^2} = \frac{4\pi \, r^2}{r^2}$$
$$= 4\pi. \tag{$\mathcal{A}$.3}$$

From this result, in a hemisphere, the ratio of the surface area to r^2 is therefore, 2π. Since a portion of the surface area of a sphere is a fraction of the total surface area of the sphere, the ratio of a part of the surface area of a sphere to r^2 is always a numerical value. The angle that a part of the surface area of a sphere subtends at the center of the sphere, can be used as a measure of the 3D angle, called the solid angle, similar to how the 2D angle is defined in terms of the arc length.

For example, a cone with its vertex at the center of a sphere marked C, is shown in Figure $\mathcal{A}$.1(b). The cone is redrawn in Figure $\mathcal{A}$.2 and the solid angle shown by $d\Omega$. The solid angle is defined as

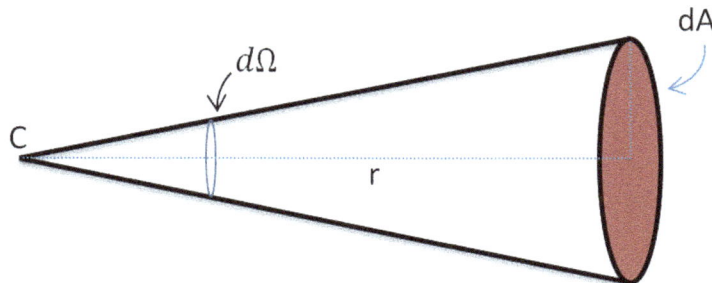

Figure $\mathcal{A}$.2: A cone with vertex C at the center of a sphere of radius r.

the ratio

$$d\Omega = \frac{dA}{r^2}, \tag{$\mathcal{A}$.4}$$

463

where dA is the area of the sphere intersecting the cap of the cone, shown shaded in the figure, and r is the radius of the sphere. For small values of the solid angle, dA can be approximated as planar, as shown. For larger values of the solid angle, however, the sphere area would be spherical, as shown in Figure $\mathcal{A}$.3(a).

The unit of solid angle is the *steradian* (sr), and similar to radian, is a dimensionless unit. By the above definition of the solid angle, the dimensions of the numerator and the denominator cancel, making *sr* unitless. If the spherical area intersecting the cap of the cone is equal to r^2, the solid angle is an unit steradian,

$$d\Omega = \frac{r^2}{r^2}$$

$$= 1 \ sr. \qquad (\mathcal{A}.5)$$

In the representation of a cone for a solid angle, it is not a requirement for the cone to have a circular cross-section. The "cone" is a generic term, whose cross-section can be of any shape, to represent the solid angle subtended by the cap area at its vertex. A patch area A of an arbitrary shape is drawn on the sphere with radius r in Figure $\mathcal{A}$.3(b). Using Equation $\mathcal{A}$.4, the solid angle subtended by A at the center of the sphere is

$$\Omega = \frac{A}{r^2}. \qquad (\mathcal{A}.6)$$

From this example, it can be noted that the "cone" of the solid angle can have any cross-section shape.

All points on the patch lie at a distance r from the center, making it easy to derive the solid angle subtended by A. A general definition of the solid angle will be derived in Section $\mathcal{A}$.3, to define the solid angle subtended at a point, by a surface area with non-uniform distance from the point.

$\mathcal{A}$.2   The Solid Angle Subtended by the Surface Area of a Sphere at its Center

The solid angle subtended by the entire surface area of a sphere at its center, can be calculated using Equation $\mathcal{A}$.4.

$$\Omega = \int_{sphere} \frac{dA}{r^2}, \qquad (\mathcal{A}.7)$$

where the integration is over the surface area of the sphere, and r is the radius of the sphere. Since r^2 does not change over the surface area of the sphere, it can be moved out of the integral,

$$\Omega = \frac{1}{r^2} \int_{sphere} dA. \qquad (\mathcal{A}.8)$$

The result of the integral is the surface area of the sphere, which is $4\pi r^2$, and therefore, the solid angle subtended by the area of the sphere at its center is

$$\Omega_{sphere} = 4\pi \ sr. \qquad (\mathcal{A}.9)$$

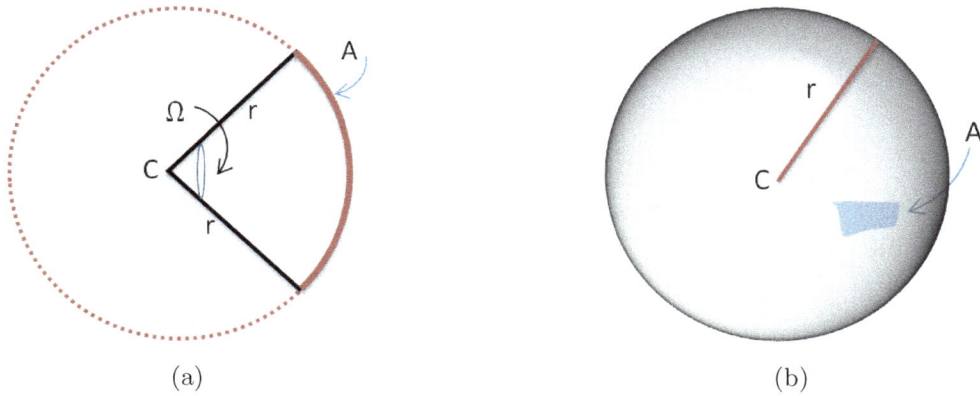

Figure $\mathcal{A}$.3: (a) A cone with a spherical cap area A. (b) A patch area A of an arbitrary shape on the surface of a sphere.

This result will be used to show in Section $\mathcal{A}$.4 that the solid angle subtended by *any* closed surface, not limited to a sphere, at any point within the surface is 4π. Following the above steps, the solid angle subtended by a hemisphere at its center is half of Ω_{sphere} or

$$\Omega_{hemisphere} = 2\pi \ sr. \tag{$\mathcal{A}$.10}$$

$\mathcal{A}$.3   General Definition of a Solid Angle

Although the general definition of a solid angle will be informally derived in this section, it will be indirectly proved in Chapter 32–Chapter 33, where Gauss's law will be derived in two different ways, one that uses the general definition of a solid angle, and one that does not. Both the derivations yield the same result, thereby verifying the general definition of a solid angle.

The solid angle definition in Equation $\mathcal{A}$.4 assumes that the surface area that subtends the solid angle at the center of the sphere, is equidistant from the center at radius r. This definition can be generalized to any surface area, not necessarily lying on a sphere, which subtends the solid angle at any point, not necessarily at the center of a sphere.

An example of such a case is shown in Figure $\mathcal{A}$.4(a), drawn as 2D for simplicity, which shows the 2D cross section of a closed volume, not necessarily a sphere, enclosing Point A that lies somewhere within. The solid angle subtended by an area on the surface, which intersects the cone at Point A, will be derived. Note that neither the enclosing surface is a sphere, nor A lies at the center of a sphere.

Let $d\Omega$ be a sufficiently small solid angle, such that the cap area of the cone that intersects with the surface, marked EF and redrawn in Figure $\mathcal{A}$.4(b), is planar. The length from Point A to the area EF has been marked r. Let $d\Omega$ be small enough that the variation of r over the area EF can be ignored. If an imaginary sphere is drawn centered at A, the intersection of the cap

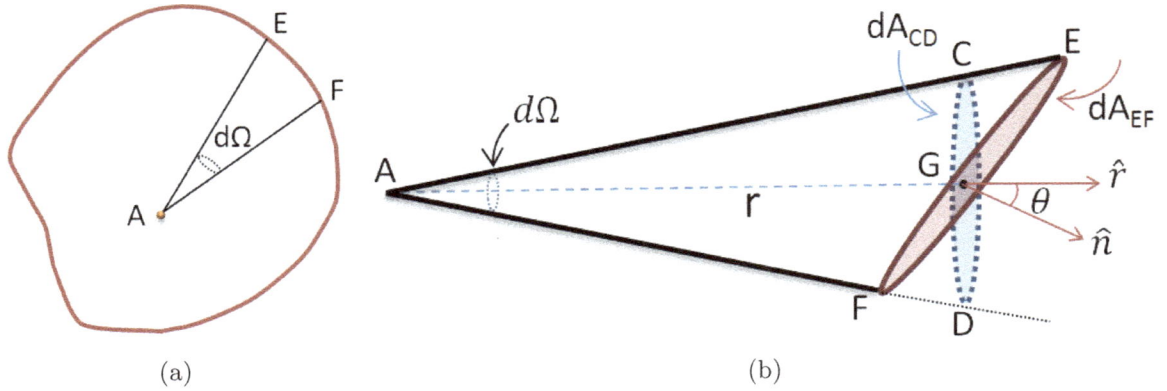

Figure A.4: (a) The solid angle subtended at Point A, somewhere within an arbitrary closed surface. (b) The cone in Figure A.4(a) redrawn.

area of the cone with the sphere is shown by the area CD.

The area CD is a circle, since the sphere slices the cone perpendicularly, as shown in the figure. The area EF is an ellipse, since it slices the cone at any angle. Let $\hat{r}$ be the unit vector normal to the circular area CD, and $\hat{n}$ be the unit normal vector to the elliptical area EF, as shown in the figure. The angle between $\hat{r}$ and $\hat{n}$ is θ. Point G is the center of the area CD and EF.

The front view of the areas are shown in Figure A.5(a). The diameters of the areas are redrawn in Figure A.5(b). From trigonometry,

$$\angle EGC = \angle FGD = \theta. \tag{A.11}$$

The relation between the diameter of the circle $\overline{CD}$ and the longer diameter of the ellipse $\overline{EF}$ is therefore,

$$\overline{CD} = \overline{EF}\cos\theta. \tag{A.12}$$

From Figure A.4(b), it can be noted that the smaller radius of the ellipse is approximately the same as the radius of the circle, marked ℓ in Figure A.5(a).

The area of a circle A_{circle} is

$$A_{circle} = \pi\,\ell^2, \tag{A.13}$$

where ℓ is the radius of the circle, and that of an ellipse is

$$A_{ellipse} = \pi\,\ell\,m, \tag{A.14}$$

where ℓ and m are the shorter and the longer radii of the ellipse. Since only the length of the longer radius of the ellipse scales by the factor $\frac{1}{\cos\theta}$, and the shorter radius of the area EF stays the same as the area CD, the area of the circle dA_{CD} relates to the area of the ellipse dA_{EF} by

$$dA_{CD} = dA_{EF}\cos\theta. \tag{A.15}$$

466

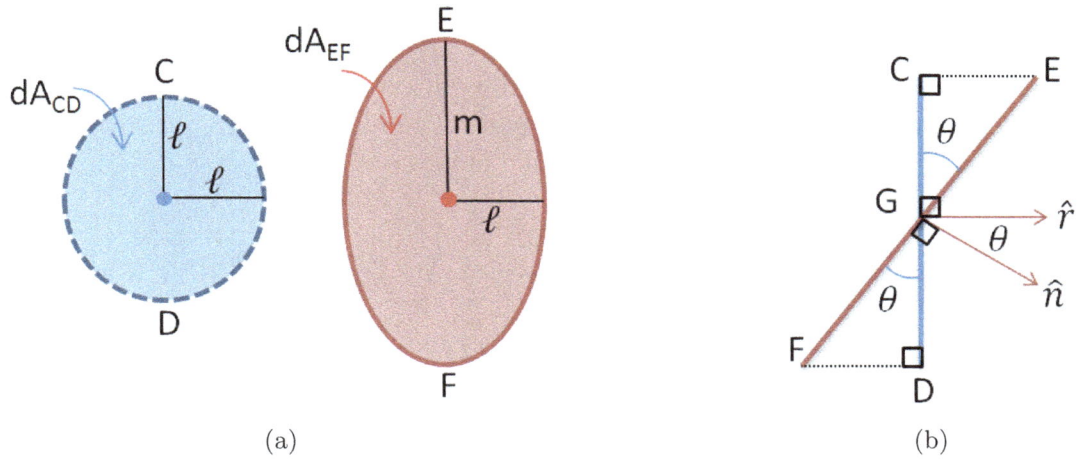

Figure $\mathcal{A}.5$: (a) The front view of areas dA_{CD} and dA_{EF} in Figure $\mathcal{A}.4$(b). (b) The side view showing the diameter of dA_{CD}, and the longer radius of the ellipse dA_{EF} in Figure $\mathcal{A}.4$(b).

dA_{CD} lies on an imaginary sphere centered at A, as mentioned earlier. dA_{CD} is the projection of the area dA_{EF} onto the surface of a sphere centered at Point A. By definition of the solid angle, the solid angle subtended by dA_{CD} at Point A is

$$d\Omega = \frac{dA_{CD}}{r^2}, \tag{$\mathcal{A}.16$}$$

and from the above equations,

$$d\Omega = \frac{dA_{EF}\cos\theta}{r^2}. \tag{$\mathcal{A}.17$}$$

Rewriting the above equation using the dot product of $\hat{r}\cdot\hat{n} = \cos\theta$,

$$d\Omega = \frac{dA_{EF}\,\hat{r}\cdot\hat{n}}{r^2}. \tag{$\mathcal{A}.18$}$$

The above relation becomes more accurate as the variation of r over dA_{EF} diminishes, or in other words, as $d\Omega$ and dA_{EF} become smaller and smaller.

From the above equation, the general definition of a solid angle is given by

$$d\Omega = \frac{dA\,\hat{r}\cdot\hat{n}}{r^2}, \tag{$\mathcal{A}.19$}$$

where dA is the elemental surface area that is at distance r from Point A, the point at which the solid angle that dA subtends is to be calculated. $\hat{r}$ is the unit vector from Point A to the mid-point of dA, and dA is small enough for the variation of r over dA to be ignored. $\hat{n}$ is the unit normal vector to dA and is *outward* to the solid angle.

Although the above derivation assumes that dA_{EF} is an ellipse, and dA_{CD} is a circle, the same derivation holds true if dA_{EF} is a rectangular area, for example. This conclusion holds true for any area dA, and $dA\,\hat{r}\cdot\hat{n}$ is the projection of dA onto the surface of a sphere centered at Point A. As mentioned in the beginning of the section, this derivation will be proved indirectly during the discussion of Gauss' law. Note that the above equation reduces to the special case of the solid angle definition in Equation $\mathcal{A}.4$, if $\hat{r}$ is equal to $\hat{n}$. This would be the case if dA is already lying on a sphere centered at Point A.

$\mathcal{A}.4$ Solid Angle Subtended by Any Closed Surface at Any Point Within

It will be shown in this section that the solid angle subtended by an arbitrary closed surface at any point within the surface is 4π. This result will be used in one of the proofs of Gauss' law in Chapter 33. An arbitrary closed 3D surface S_1 is shown in Figure $\mathcal{A}.6$(a), drawn as 2D for simplicity. The solid angle subtended by the elemental surface dA_1 on S_1 at Point P, has been marked $d\Omega$. A sphere with its center at Point P is drawn in Figure $\mathcal{A}.6$(b).

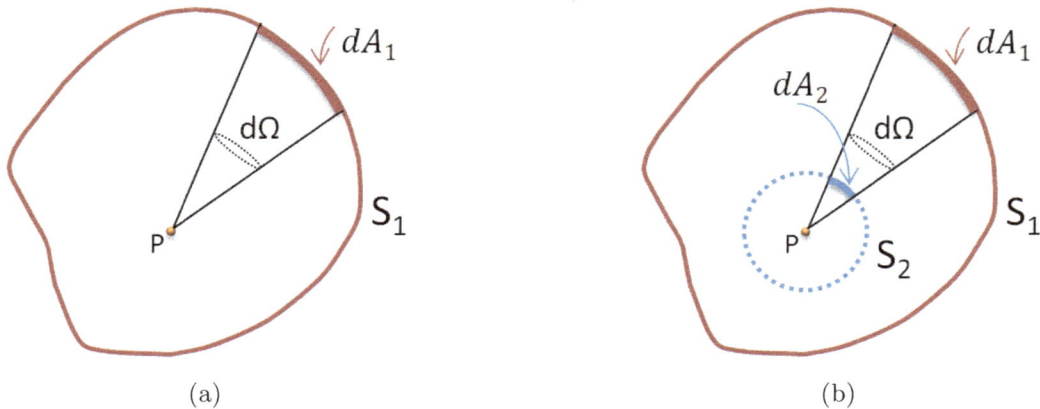

Figure $\mathcal{A}.6$: (a) The solid angle subtended by dA_1, at some Point P within the closed surface. (b) dA_1 subtends the same solid angle at Point P as dA_2.

The solid angle $d\Omega$ subtended by dA_1 is equal to the solid angle subtended by dA_2 at Point P, since the same cone that intersects S_1, also intersects S_2 at dA_2. Intuitively, it can be seen that the infinitesimal cap area dA_1, maps 1:1 to an infinitesimal area on S_2, marked dA_2. The infinitesimal cones covering the entire surface S_1 would cover the spherical surface S_2. Therefore, the solid angle subtended by the complete surface S_1 at Point P, is the same as the solid angle subtended by S_2 at Point P. By Equation $\mathcal{A}.4$, the solid angle subtended by the surface area of a sphere is 4π sr. This completes the proof that the solid angle subtended by any closed surface at any point within the surface is 4π.

$\mathcal{A}.5$ Solid Angle Subtended by an Infinite Plane

The result from this section will be used in the derivation of Ampere's law in Chapter 21. The title may sound like an abstract problem with no practical application. However, that is not the case. To a point located above and very close to a finite surface, the surface would appear to be an infinite plane. In this case, the solid angle subtended by the finite surface would be the same as the solid angle subtended by an infinite plane. The motivation for solving such a problem would be explained in considerable detail during the discussion of Ampere's law.

An infinite plane that subtends the solid angle at Point Q lies on the xy-plane, as shown in Figure $\mathcal{A}$.7(a). Let Point Q be a point anywhere on the z-axis as shown, whose coordinates are $(0, 0, z)$, as marked. The solid angle subtended by the infinite plane at Point Q will be calculated using Equation $\mathcal{A}$.19.

The result can be calculated intuitively. A hemisphere centered at Point Q is shown in Figure $\mathcal{A}$.7(b). A cone from Q to the area on the plane dA_1 is also shown. The area on the hemisphere that intersects with the cone is shown by the arrow labeled dA_2. The solid angle subtended by dA_1 is the same as the angle that dA_2 subtends, since its the same cone that intersects both the areas. The cap area from the infinitesimal cones covering the entire plane area, would map 1:1 to an area on the hemisphere, covering the entire surface area of the hemisphere. On each of these infinitesimal cones, the cap area on the plane subtends the same solid angle as the infinitesimal area on the hemisphere that intersects with the cone, since its the same cone that intersects both dA_1 and dA_2. The solid angle subtended by the infinite plane is the same as the solid angle subtended by the hemisphere, and is therefore, 2π sr. This conclusion will be verified next.

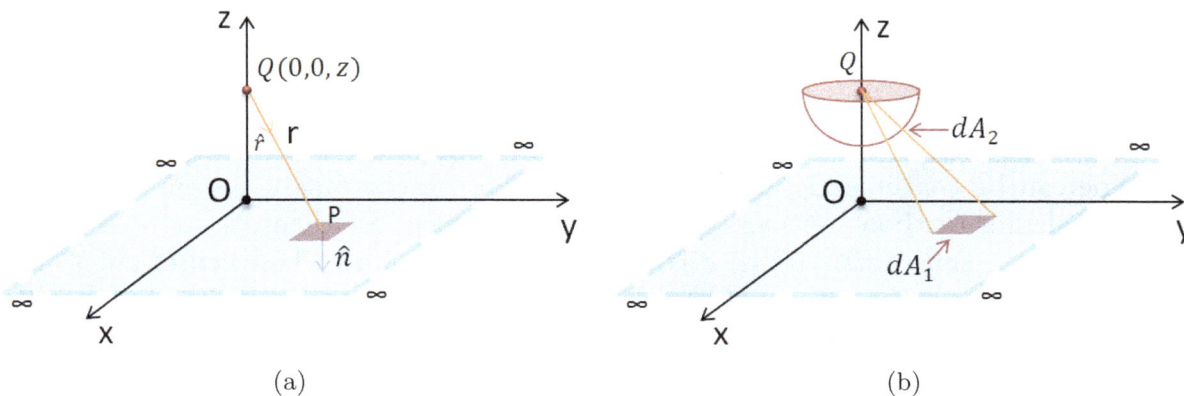

(a)   (b)

Figure $\mathcal{A}$.7: (a) The solid angle subtended by an infinite plane at Point Q. (b) A hemisphere centered at Point Q that subtends the same solid angle as the infinte plane.

Point P in Figure $\mathcal{A}$.7(a), lies on the infinite plane at the center of an infinitesimal patch, whose coordinates are $(x, y, 0)$. $\hat{n}$ is the unit area normal vector to the patch. Let r be the distance $\overline{QP}$, and $\hat{r}$ the unit vector from Q to P in the direction shown by the arrow. The solid angle that the patch with area dA subtends at Point Q is given by Equation $\mathcal{A}$.19.

Since the patch on the infinite plane lies on the xy-plane,

$$\hat{n} = -\hat{z}, \tag{$\mathcal{A}$.20}$$

where $\hat{z}$ is the unit vector on the z-axis. The unit vector $\hat{r}$ is

$$\begin{aligned} \hat{r} &= \frac{\vec{QP}}{|\vec{QP}|} \\ &= \frac{(x, y, -z)}{\sqrt{x^2 + y^2 + z^2}}. \end{aligned} \tag{$\mathcal{A}$.21}$$

The infinitesimal solid angle subtended by the patch is

$$d\Omega = \frac{\hat{r} \cdot \hat{n}}{r^2} \, dx \, dy. \tag{$\mathcal{A}$.22}$$

Integrating the above equation over the infinite plane,

$$\Omega = \int_{-\infty}^{+\infty} \int_{-\infty}^{+\infty} \frac{\hat{r} \cdot \hat{n}}{r^2} \, dx \, dy. \tag{$\mathcal{A}$.23}$$

It is left as an exercise for the reader to verify that the above integral evaluates to

$$\Omega = 2\pi, \tag{$\mathcal{A}$.24}$$

which is the same result as the intuitive derivation presented first.

$\mathcal{A}$.6   Units of solid angle: sr, rad^2, and $\square^\circ$ (square degree) [Optional]

A solid angle can be conveniently expressed in the spherical coordinate system. The reader is referred to a calculus textbook for more details on the spherical coordinate system. An infinitesimal patch lying on a sphere with radius r is shown in Figure $\mathcal{A}$.8(a), whose center point has been marked P with coordinates (r, θ, ϕ) in the spherical coordinates. Its dimensions are $r \sin \theta \, d\phi$ in the $\hat{\phi}$ direction and $r \, d\theta$ in the $\hat{\theta}$ direction. The infinitesimal area is therefore,

$$dA = r^2 \sin \theta \, d\theta \, d\phi. \tag{$\mathcal{A}$.25}$$

By definition of the solid angle, the infinitesimal solid angle subtended by the patch, at the center of the sphere is

$$\begin{aligned} d\Omega &= \frac{dA}{r^2} \\ &= \sin \theta \, d\theta \, d\phi. \end{aligned} \tag{$\mathcal{A}$.26}$$

Note that the surface area dA is

$$dA = r^2 \, d\Omega, \tag{$\mathcal{A}$.27}$$

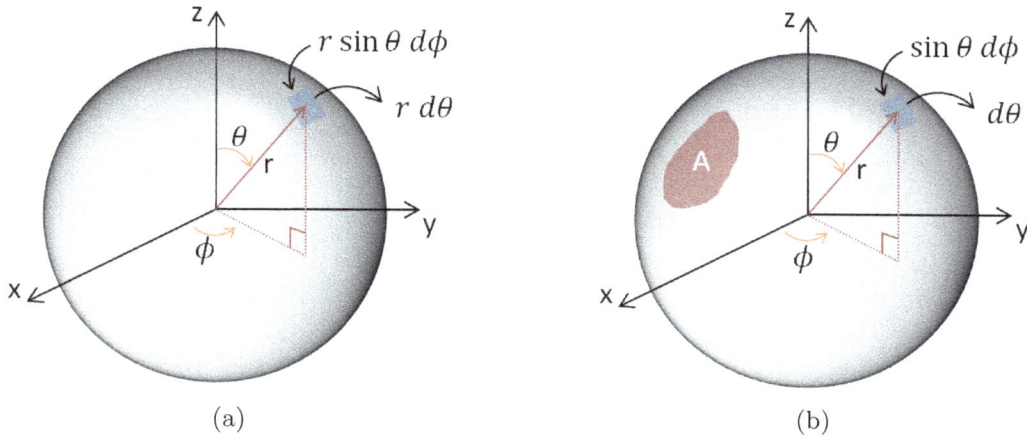

Figure $\mathcal{A}$.8: (a) An infinitesimal surface area in spherical coordinates. (b) An infinitesimal *angular* area in spherical coordinates.

and the surface area can be used to represent the solid angle in a figure, since the surface area is only the solid angle scaled by r^2. From dimensional analysis, $\sin\theta$ has no units, but the product of the infinitesimal angles $d\theta\,d\phi$, each with the dimensionless unit rad, has the dimensionless unit rad^2. Therefore, the unit of $[d\Omega] = rad^2$. Steradian (sr) is used as an alias for rad^2, and therefore,

$$1\,sr = 1\,rad^2. \tag{$\mathcal{A}$.28}$$

The steradian is a derived unit from the radian, and is not a fundamental unit. The steradian is an *angular area*. This can be easily seen in the spherical coordinate system.

Similar to how the degrees is more convenient to use than the radians, square degree deg^2, alternately written as $\square°$, may be used, rather than rad^2. If the patch dimensions in Figure $\mathcal{A}$.8(a) are written without the factor r, as shown in Figure $\mathcal{A}$.8(b), the width and the length of the patch $\sin\theta\,d\phi$ and $d\theta$ are each in radians and the area of the patch is an angular area in rad^2. The solid angle that the patch subtends

$$d\Omega = \sin\theta\,d\theta\,d\phi, \tag{$\mathcal{A}$.29}$$

is the angular area of the patch, which is the same as Equation $\mathcal{A}$.26. If the sphere is discretized into patches of equal angular area, given any area A on the surface of the sphere, as shown in Figure $\mathcal{A}$.8(b), the number of these patches lying on A, times the angular area per patch, is the solid angle subtended by A at the center of the sphere in rad^2 or sr.

2π radians is divided into $360°$. Using this conversion ratio, if the sphere is discretized into patches of area $1° \times 1°$, each of the patches would be one square degree $1\square°$. The patch dimensions in radians to represent $1\square°$ would be

$$1° \times 1° = \left(\frac{2\pi}{360}\right)rad \times \left(\frac{2\pi}{360}\right)rad. \tag{$\mathcal{A}$.30}$$

From the above equation, the conversion ratio between rad^2 and $\square°$ is

$$\left(\frac{2\pi}{360}\right)^2 rad^2 = 1\,\square°. \tag{$\mathcal{A}$.31}$$

If the area A in Figure $\mathcal{A}$.8(b) is discretized by patches, $1° \times 1°$ each, counting the number of such patches covering A would be the solid angle subtended by A at the center of the sphere in square degrees $\square°$.

This chapter is concluded with an example of $\square°$, modified from Reference [115]. A sphere is shown in Figure $\mathcal{A}$.9 with its center marked C. The θ and the ϕ axes are marked in the figure. An area is shown in Figure $\mathcal{A}$.9 that ranges from $\theta = 40° - 60°$ and $\phi = 30° - 70°$.

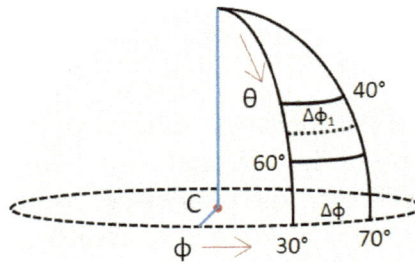

Figure $\mathcal{A}$.9: The solid angle subtended by the area $\{\theta = [40°, 60°], \phi = [30°, 70°]\}$ on a sphere, at its center.

An approximate value of the solid angle subtended by the area in $\square°$ will first be calculated, and later verified with a more accurate calculation. In the spherical coordinate system, the width of the dotted line at the center of the area is $r\sin\theta\,\Delta\phi$, but since the factor r is not included, as mentioned earlier,

$$\begin{aligned}
\Delta\phi_1 &= \Delta\phi\sin\theta \\
&= 40°\sin 50°,
\end{aligned} \tag{$\mathcal{A}$.32}$$

where $\theta = 50°$ is shown by the dotted line. The reader is referred to a calculus textbook for more details on the spherical coordinate system. The area can be approximated by a rectangle whose width is $\Delta\phi_1$, and length,

$$\begin{aligned}
\Delta\theta &= 60° - 40° \\
&= 20°.
\end{aligned} \tag{$\mathcal{A}$.33}$$

472

The solid angle subtended by the area in $\square°$ is the angular area of the rectangle, and is approximately,

$$\Omega = 40° \sin 50° \times 20°$$
$$\approx 613\,\square°, \tag{A.34}$$

This result will be verified using Equation A.29 to calculate the exact result in rad^2 first, and then converting it to $\square°$ using Equation A.31.

$$\Omega = \int_{30°}^{70°} d\phi \int_{40°}^{60°} \sin\theta\, d\theta$$
$$= \frac{2\pi}{360} \left(70° - 30°\right) \left[-\cos\theta\right]_{40°}^{60°}$$
$$= 0.1857\,rad^2 \;=\; 0.1857\,sr \;\approx\; 610\,\square°, \tag{A.35}$$

which is very close to the approximate result in the prior calculation.

Verifying the Transformation Steps from Biot-Savart's Law to Ampere's Law

As a checkpoint to verify the validity of the steps involved in the transformation in Chapter 30, such as having primed/unprimed coordinate systems and primed/unprimed operators, Equation 30.52

$$\vec{H}(\vec{r}) = - \int_{V'} \vec{J} \times \nabla \left(\frac{1}{R} \right) dV', \qquad (\mathcal{B}.1)$$

will be shown to be equivalent to Equation 30.55

$$\vec{H}(\vec{r}) = \nabla \times \int_{V'} \left(\frac{\vec{J}}{R} \right) dV', \qquad (\mathcal{B}.2)$$

for the specific case of an infinite line current.

$\mathcal{B}.1$ Representation of Line Currents Using Volume Current Density

A line current is current flowing in a line, as opposed to current flowing in an area. A line current can also be written as a volume current density. An example is presented for a steady line current i that extends in the infinite direction along the z-axis.

A planar surface S parallel to the xy-plane is shown in Figure $\mathcal{B}.1$, with the surface area marked ΔS. Let $\vec{J}$ be the volume current density to be determined that represents the line current i along the z-axis flowing in the $\hat{z}$ direction. $\vec{J} = J\,\hat{z}$ is perpendicular to the surface area ΔS, and therefore, using Equation 28.13, the current through ΔS is

$$i = J\,\Delta S. \qquad (\mathcal{B}.3)$$

For the line current i on the z-axis, regardless of how small ΔS might be, as long as it includes $x = 0$, $y = 0$, $J\,\Delta S$ must equal i. J is written as

$$J = i\,\delta(x, y), \qquad (\mathcal{B}.4)$$

474

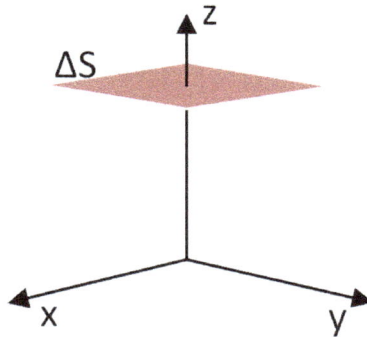

Figure B.1: A planar area parallel to the xy-plane at some location z, enclosing $x = 0$ and $y = 0$.

where $\delta(x, y)$ is the 2D Dirac-delta function that satisfies

$$\delta(x, y)\, \Delta S = 1, \quad if\ x = 0, y = 0 \in S$$
$$= 0, \quad otherwise. \tag{B.5}$$

The volume current density that represents the line current i on the z-axis is

$$\vec{J} = i\, \delta(x, y)\, \hat{z}. \tag{B.6}$$

B.2 Equivalence of Equation 30.52 and Equation 30.55

In Chapter 30, the source of the magnetic field in Biot-Savart's law in Equation 30.6 is volume current density. The source is viewed as being in the primed coordinate system and the point at which the field is calculated in the unprimed coordinate system, repeating Figure 30.4 shown below in Figure B.2. The source of the magnetic field, which is the volume current density, is

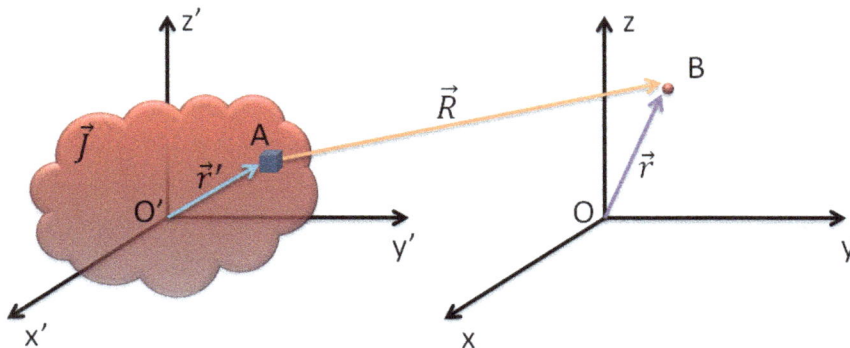

Figure B.2: The source, which is the volume current density, is present in the primed coordinate system. The point at which the magnetic field is calculated, using Biot-Savart's law, is present in the unprimed coordinate system.

illustrated by the cloud in the primed coordinate system, shown in the figure.

In Equation B.1 and Equation B.2, $R = |\vec{r} - \vec{r}'|$, where $\vec{r} = (x, y, z)$ is the point at which the magnetic field is calculated, and $\vec{r}' = (x', y', z')$ is the center of the differential volume element dV', as shown in the figure. The ∇ operator operates in the unprimed coordinate system, while ∇' in the primed coordinate system.

The variables representing the source, and the point at which the magnetic field is calculated, are common to both Equation B.1 and Equation B.2. The equivalence of Equation B.1 and Equation B.2 will be shown for the specific case of $\vec{J}$ being a line current along the z'-axis, repeating Equation B.4,

$$\vec{J} = i\,\delta(x', y')\,\hat{z}, \quad -\infty < z' < +\infty, \tag{B.7}$$

where the source lies in the primed coordinate system, as marked in Figure B.3. The point at

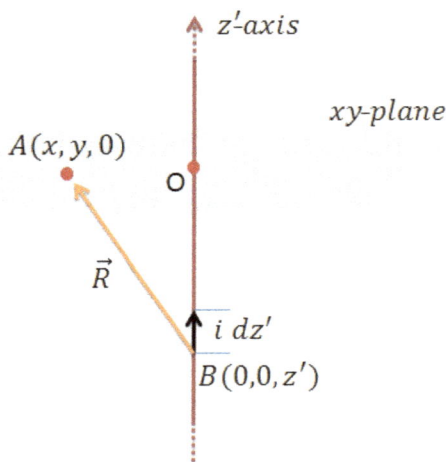

Figure B.3: A line current extending from $[-\infty, +\infty]$ on the z'-axis.

which the magnetic field is calculated is some point on the xy-plane of the unprimed coordinate system, marked $A(x, y, 0)$. For simplicity, the two separate coordinate systems in Figure B.2, have been merged into one in Figure B.3. $\vec{R}$ is the vector from the source on the z'-axis, some point $B(0, 0, z')$ to the point $A(x, y, 0)$. The value of R is therefore,

$$R = \sqrt{x^2 + y^2 + z'^2}. \tag{B.8}$$

Although a specific value of $\vec{J}$ has been assumed, it provides a good checkpoint to pause and reflect on the validity of the steps in the derivation. The evaluation of the integrals in Equation B.1 and Equation B.2 will be presented in the following two sections. It will be shown that the magnetic field calculated at Point A, using both the equations are equal to each other. This shows that the steps followed in Chapter 30, such as having primed/unprimed coordinate systems, etc. is correct.

B.3 Solving for $\vec{H}$ using Equation B.1

The magnetic field at Point A in Figure B.3 will be solved using Equation B.1. It is left as an exercise for the reader to show that

$$\nabla \left(\frac{1}{R} \right) = -\frac{x}{(x^2 + y^2 + z'^2)^{3/2}} \, \hat{x} - \frac{y}{(x^2 + y^2 + z'^2)^{3/2}} \, \hat{y}, \tag{B.9}$$

where ∇ is the gradient operator on the unprimed variables, as defined in Equation 30.8.

Using $\vec{J}$ defined in Equation B.7, the result of the cross-product operation is

$$\vec{J} \times \nabla \left(\frac{1}{R} \right) = \frac{i\delta(x', y') \, y}{(x^2 + y^2 + z'^2)^{3/2}} \, \hat{x} - \frac{i\delta(x', y') \, x}{(x^2 + y^2 + z'^2)^{3/2}} \, \hat{y}. \tag{B.10}$$

Integrating the above equation over all of the V' space,

$$\vec{H}(\vec{r}) = -\int_{-\infty}^{+\infty} \vec{J} \times \nabla \left(\frac{1}{R} \right) dx' \, dy' \, dz'. \tag{B.11}$$

The above equation is written as a single integral extending over $[-\infty, +\infty]$ for all the three variables x', y', z', rather than a triple integral for brevity.

$$\vec{H}(\vec{r}) = -\int_{-\infty}^{+\infty} \left(\frac{i\,y}{(x^2 + y^2 + z'^2)^{3/2}} \, \hat{x} - \frac{i\,x}{(x^2 + y^2 + z'^2)^{3/2}} \, \hat{y} \right) dz'$$
$$\int_{-\infty}^{+\infty} \int_{-\infty}^{+\infty} \delta(x', y') \, dx' \, dy'. \tag{B.12}$$

Note that in the integration on the primed coordinate space, the unprimed variables are treated as constants, or as being fixed. By definition of the Dirac delta function,

$$\int_{-\infty}^{+\infty} \int_{-\infty}^{+\infty} \delta(x', y') \, dx' \, dy' = 1, \tag{B.13}$$

and the integral can be simplified to

$$\vec{H}(\vec{r}) = -\int_{-\infty}^{+\infty} \left(\frac{i\,y}{(x^2 + y^2 + z'^2)^{3/2}} \, \hat{x} - \frac{i\,x}{(x^2 + y^2 + z'^2)^{3/2}} \, \hat{y} \right) dz'. \tag{B.14}$$

Using table of integrals, it is left as an exercise for the reader to show that the result of the above integral is

$$\vec{H}(\vec{r}) = -\frac{2i\,y}{x^2 + y^2} \, \hat{x} + \frac{2i\,x}{x^2 + y^2} \, \hat{y}. \tag{B.15}$$

B.4 Solving for $\vec{H}$ using Equation B.2

Equation B.1 was transformed to Equation B.2 by a series of operations in Chapter 30. Applying the line-current example to Equation B.2, it will be shown that the result is the same as Equation B.15, obtained using Equation B.1.

The integrand in Equation B.2 is

$$\frac{\vec{J}}{R} = \frac{i\,\delta(x', y')}{\sqrt{x^2 + y^2 + z'^2}}\,\hat{z}. \tag{B.16}$$

Integrating the above expression over all of V' space,

$$\int_{V'} \frac{\vec{J}}{R}\, dV' = \int_{-\infty}^{+\infty} \frac{i\,\hat{z}}{\sqrt{x^2 + y^2 + z'^2}}\, dz' \int_{-\infty}^{+\infty} \int_{-\infty}^{+\infty} \delta(x', y')\, dx'\, dy'. \tag{B.17}$$

Using Equation B.13, the above equation can be simplified to

$$\int_{V'} \frac{\vec{J}}{R}\, dV' = \hat{z} \int_{-\infty}^{+\infty} \frac{i}{\sqrt{x^2 + y^2 + z'^2}}\, dz'. \tag{B.18}$$

In the above integral, the integration is with respect to z', and the variables x and y are treated as constants. From table of integrals,

$$\frac{d}{dz'}\left(\operatorname{arcsinh}\left(\frac{z'}{a}\right)\right) = \frac{d}{dz'}\left(\ln\left(z' + \sqrt{z'^2 + a^2}\right) - \ln a\right) \tag{B.19}$$

$$= \frac{1}{\sqrt{z'^2 + a^2}}, \tag{B.20}$$

where $a = \sqrt{x^2 + y^2}$. In Equation B.19, the identity

$$\operatorname{arcsinh}(q) = \ln\left(q + \sqrt{q^2 + 1}\right), \tag{B.21}$$

where q is any real number, has been used to derive the relation shown. From Equation B.2, and using the above equations to evaluate and simplify the integral in Equation B.18,

$$\vec{H}(\vec{r}) = \lim_{z_o \to \infty} \nabla \times \left[2i \ln\left(\frac{z_o}{\sqrt{x^2 + y^2}} + \sqrt{\frac{z_o^2}{x^2 + y^2} + 1}\right)\hat{z}\right]. \tag{B.22}$$

In the above equation, the limit as $z_o \to \infty$ will be evaluated after the curl operator has been applied. It is left as an exercise for the reader to complete the above operation to show that

$$\vec{H}(\vec{r}) = -\frac{2i\,y}{x^2 + y^2}\,\hat{x} + \frac{2i\,x}{x^2 + y^2}\,\hat{y}, \tag{B.23}$$

which is the same result as that obtained using Equation B.1 in the previous section. This exercise has been a good checkpoint, midway in the transformation process in Chapter 30, to verify that the steps are correct.

www.ingramcontent.com/pod-product-compliance
Lightning Source LLC
Chambersburg PA
CBHW081239220326
41597CB00023BA/4072